NUREG-1576
EPA 402-B-04-001A
NTIS PB2004-105421

Multi-Agency Radiological Laboratory Analytical Protocols Manual

Volume I: Chapters 1 – 9 and Appendices A – E

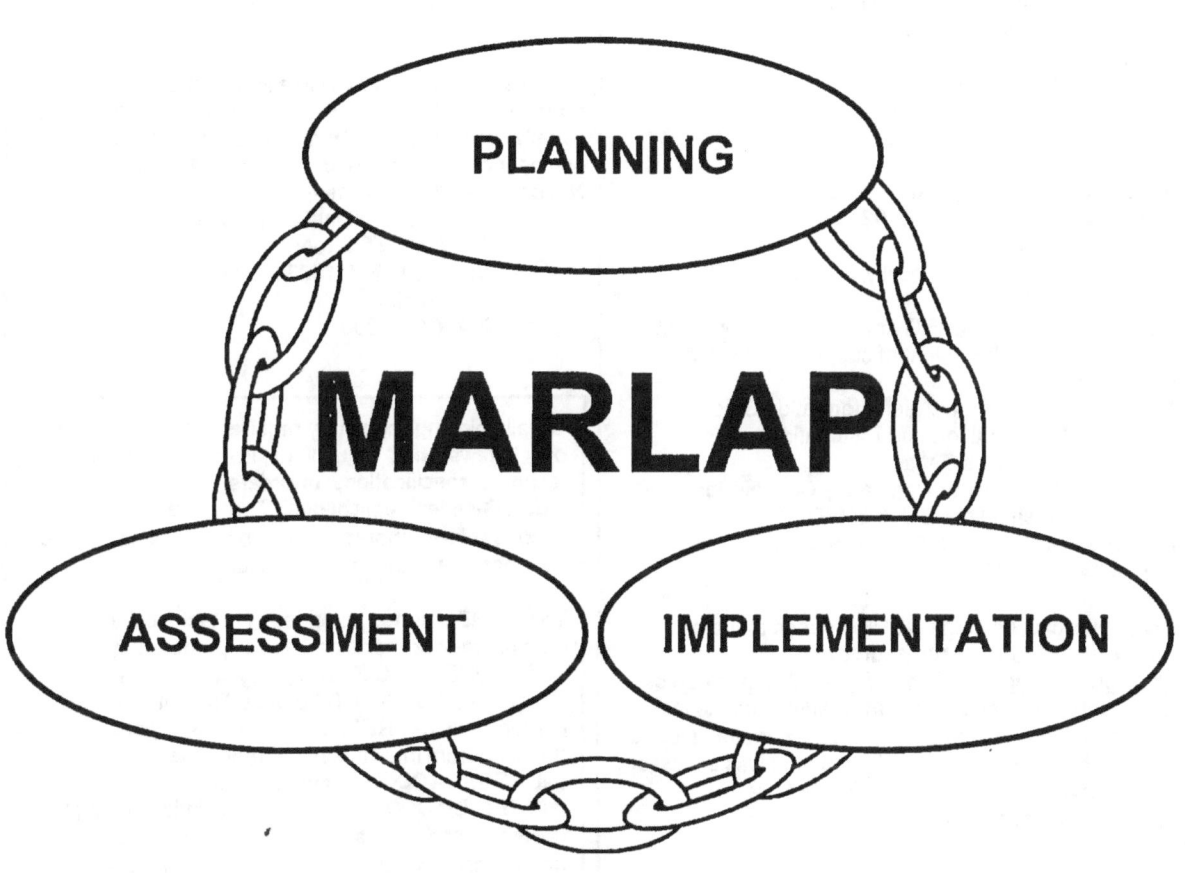

July 2004

NUREG-1576
EPA 402-B-04-001A
NTIS PB2004-105421

Multi-Agency Radiological Laboratory Analytical Protocols Manual (MARLAP)

Part I: Chapters 1 – 9
Appendices A – E
(Volume I)

United States Environmental Protection Agency
United States Department of Defense
United States Department of Energy
United States Department of Homeland Security
United States Nuclear Regulatory Commission
United States Food and Drug Administration
United States Geological Survey
National Institute of Standards and Technology

July 2004

ABSTRACT

The Multi-Agency Radiological Laboratory Analytical Protocols (MARLAP) manual provides guidance for the planning, implementation, and assessment of projects that require the laboratory analysis of radionuclides. MARLAP's basic goal is to provide guidance for project planners, managers, and laboratory personnel to ensure that radioanalytical laboratory data will meet a project's or program's data requirements. To attain this goal, the manual offers a framework for national consistency in the form of a performance-based approach for meeting data requirements that is scientifically rigorous and flexible enough to be applied to a diversity of projects and programs. The guidance in MARLAP is designed to help ensure the generation of radioanalytical data of known quality, appropriate for its intended use. Examples of data collection activities that MARLAP supports include site characterization, site cleanup and compliance demonstration, decommissioning of nuclear facilities, emergency response, remedial and removal actions, effluent monitoring of licensed facilities, environmental site monitoring, background studies, and waste management activities.

MARLAP is organized into two parts. Part I, intended primarily for project planners and managers, provides the basic framework of the directed planning process as it applies to projects requiring radioanalytical data for decision making. The nine chapters in Part I offer recommendations and guidance on project planning, key issues to be considered during the development of analytical protocol specifications, developing measurement quality objectives, project planning documents and their significance, obtaining laboratory services, selecting and applying analytical methods, evaluating methods and laboratories, verifying and validating radiochemical data, and assessing data quality. Part II is intended primarily for laboratory personnel. Its eleven chapters provide detailed guidance on field sampling issues that affect laboratory measurements, sample receipt and tracking, sample preparation in the laboratory, sample dissolution, chemical separation techniques, instrumentation for measuring radionuclides, data acquisition, reduction, and reporting, waste management, laboratory quality control, measurement uncertainty, and detection and quantification capability. Seven appendices provide complementary information and additional details on specific topics.

MARLAP was developed by a workgroup that included representatives from the U.S. Environmental Protection Agency (EPA), Department of Energy (DOE), Department of Defense (DOD), Department of Homeland Security (DHS), Nuclear Regulatory Commission (NRC), National Institute of Standards and Technology (NIST), U.S. Geological Survey (USGS), and Food and Drug Administration (FDA), and from the Commonwealth of Kentucky and the State of California.

FOREWORD

MARLAP is organized into two parts. Part I, consisting of Chapters 1 through 9, is intended primarily for project planners and managers. Part I introduces the directed planning process central to MARLAP and provides guidance on project planning with emphasis on radioanalytical planning issues and radioanalytical data requirements. Part II, consisting of Chapters 10 through 20, is intended primarily for laboratory personnel and provides guidance in the relevant areas of radioanalytical laboratory work. In addition, MARLAP contains seven appendices—labeled A through G—that provide complementary information, detail background information, or concepts pertinent to more than one chapter. Six chapters and one appendix are immediately followed by one or more attachments that the authors believe will provide additional or more detailed explanations of concepts discussed within the chapter. Attachments to chapters have letter designators (e.g, Attachment "6A" or "3B"), while attachments to appendices are numbered (e.g., "B1"). Thus, "Section B.1.1" refers to section 1.1 of appendix B, while "Section B1.1" refers to section 1 of attachment 1 to appendix B. Cross-references within the text are explicit in order to avoid confusion.

Because of its length, the printed version of MARLAP is bound in three volumes. Volume I (Chapters 1 through 9 and Appendices A through E) contains Part I. Because of its length, Part II is split between Volumes II and III. Volume II (Chapters 10 through 17 and Appendix F) covers most of the activities performed at radioanalytical laboratories, from field and sampling issues that affect laboratory measurements through waste management. Volume III (Chapters 18 through 20 and Appendix G) covers laboratory quality control, measurement uncertainty and detection and quantification capability. Each volume includes a table of contents, list of acronyms and abbreviations, and a complete glossary of terms.

MARLAP and its periodic revisions are available online at www.epa.gov/radiation/marlap and www.nrc.gov/reading-rm/doc-collections/nuregs/staff/sr1576/. The online version is updated periodically and may differ from the last printed version. Although references to material found on a web site bear the date the material was accessed, the material available on the date cited may subsequently be removed from the site. Printed and CD-ROM versions of MARLAP are available through the National Technical Information Service (NTIS). NTIS may be accessed online at www.ntis.gov. The NTIS Sales Desk can be reached between 8:30 a.m. and 6:00 p.m. Eastern Time, Monday through Friday at 1-800-553-6847; TDD (hearing impaired only) at 703-487-4639 between 8:30 a.m. and 5:00 p.m Eastern Time, Monday through Friday; or fax at 703-605-6900.

MARLAP is a living document, and future editions are already under consideration. Users are urged to provide feedback on how MARLAP can be improved. While suggestions may not always be acknowledged or adopted, commentors may be assured that they will be considered carefully. Comments may be submitted electronically through a link on EPA's MARLAP web site (www.epa.gov/radiation/marlap).

ACKNOWLEDGMENTS

The origin of the Multi-Agency Radiological Laboratory Analytical Protocols (MARLAP) manual can be traced to the recognition by a number of agencies for the need to have a nationally consistent approach to producing radioanalytical data that meet a program's or project's needs. A workgroup was formed with representatives from the U.S. Environmental Protection Agency (EPA), Department of Energy (DOE), Department of Homeland Security (DHS), Nuclear Regulatory Commission (NRC), Department of Defense (DOD), U.S. Geological Survey (USGS), National Institute of Standards and Technology (NIST), and Food and Drug Administration (FDA) to develop guidance for the planning, implementation, and assessment of projects that require the laboratory analysis of radionuclides. Representatives from the Commonwealth of Kentucky and the State of California also contributed to the development of the manual. Contractors and consultants of EPA, DOE, and NRC—and members of the public—have been present during open meetings of the MARLAP workgroup.

MARLAP would not have been possible without the workgroup members who contributed their time, talent, and efforts to develop this guidance document:

John Griggs*, EPA, Chair

EPA: H. Benjamin Hull
      Marianne Lynch*
      Keith McCroan*
      Eric Reynolds
      Jon Richards

FDA: Edmond Baratta

DOE: Emile Boulos*
      Stan Morton*
      Stephanie Woolf*

NRC: Rateb (Boby) Abu Eid
      Tin Mo
      George Powers

DOD: Andrew Scott (Army)
      Ronald Swatski* (Army)
      Jan Dunker (Army Corps of Engineers)
      William J. Adams (Navy)
      Troy Blanton (Navy)
      David Farrand (Navy)
      Dale Thomas (Air Force)

DHS[†]: Carl Gogolak*
      Pamela Greenlaw*
      Catherine Klusek*
      Colin Sanderson*

NIST: Kenneth G.W. Inn*

USGS: Ann Mullin*

* These workgroup members also served as chapter chairs.
† All with the Environmental Measurements Laboratory, which was part of DOE prior to the establishment of DHS on March 1, 2003.

Special recognition is given to John Volpe, Commonwealth of Kentucky, and Penny Leinwander, State of California, for their contributions to the development of the MARLAP manual. The following federal agency contractors provided assistance in developing the MARLAP manual:

EPA: N. Jay Bassin (Environmental Management Support, Inc.)
       U. Hans Behling (S. Cohen & Associates, Inc.)
       Richard Blanchard (S. Cohen & Associates, Inc.)
       Leca Buchan (Environmental Management Support, Inc.)
       Jessica Burns (Environmental Management Support, Inc.)
       Harry Chmelynski
       Diane Dopkin (Environmental Management Support, Inc.)
       Scott Hay (S. Cohen & Associates, Inc.)
       Patrick Kelly (S. Cohen & Associates, Inc.)
       Robert Litman
       David McCurdy
       Charles (Chick) Phillips (S. Cohen & Associates, Inc.)
       William Richardson III (S. Cohen & Associates, Inc.)
       Steven Schaffer (S. Cohen & Associates, Inc.)
       Michael Schultz
       Robert Shannon (Time Solutions Corp.)

DOE: Stan Blacker (MACTEC, Inc.)
       Pat Harrington (MACTEC, Inc.)
       David McCurdy
       John Maney (Environmental Measurements Assessments)
       Mike Miller (MACTEC, Inc.)
       Lisa Smith (Argonne National Laboratory)

NRC: Eric W. Abelquist (ORISE)
       Dale Condra (ORISE)

The MARLAP workgroup was greatly aided in the development of the manual by the contributions and support provided by the individuals listed below.

John Arnold (USGS)	David Friedman (EPA)	Jim Mitchell (EPA)
David Bottrell (DOE)	Lino Fragoso (Navy)	Colleen Petullo (EPA)
Lloyd Currie (NIST)	Richard Graham (EPA)	Steve Pia (EPA)
Mike Carter (EPA)	Patricia Gowland (EPA)	Phil Reed (NRC)
Mary Clark (EPA)	Larry Jensen (EPA)	Cheryl Trottier (NRC)
Ron Colle (NIST)	K. Jack Kooyoomjian (EPA)	Mary C. Verwolf (DOE)
Mark Doehnert (EPA)	Jim Kottan (NRC)	John Warren (EPA)
Steve Domotor (DOE)	Ed Messer (EPA)	Mary L. Winston (EPA)
Joan Fisk (EPA)	Kevin Miller (DOE)	Tony Wolbarst (EPA)

EPA's Science Advisory Board (SAB) Radiation Advisory Committee's Review Subcommittee that conducted an extensive peer review of the MARLAP includes:

Chair
    Dr. Janet A. Johnson, Shepherd Miller, Inc.

SAB Members
    Dr. Lynn R. Anspaugh, University of Utah
    Dr. Bruce B. Boecker, (Scientist Emeritus), Lovelace Respiratory Research Institute
    Dr. Gilles Y. Bussod, Science Network International
    Dr. Thomas F. Gesell, Idaho State University
    Dr. Helen Ann Grogan, Cascade Scientific, Inc.
    Dr. Richard W. Hornung, University of Cincinnati
    Dr. Jill Lipoti, New Jersey Department of Environmental Protection
    Dr. Genevieve S. Roessler, Radiation Consultant

SAB Consultants
    Dr. Vicki M. Bier, University of Wisconsin
    Dr. Stephen L. Brown, R2C2 (Risks of Radiation and Chemical Compounds)
    Dr. Michael E. Ginevan, M.E. Ginevan & Associates
    Dr. Shawki Ibrahim, Colorado State University
    Dr. Bernd Kahn, Georgia Institute of Technology
    Dr. June Fabryka-Martin, Los Alamos National Laboratory
    Dr. Bobby R. Scott, Lovelace Respiratory Research Institute

Science Advisory Board Staff
    Dr. K. Jack Kooyoomjian, Designated Federal Officer, EPA
    Ms. Mary L. Winston, Management Assistant, EPA

Dozens of individuals and organizations offered hundreds of valuable comments in response to the Call for Public Comments on the draft MARLAP between August 2001 and January 2002, and their suggestions contributed greatly to the quality and consistency of the final document. While they are too numerous to mention individually, Jay A. MacLellan, Pacific Northwest National Laboratory; Daniel J. Strom, Pacific Northwest National Laboratory; and James H. Stapleton, Department of Statistics and Probability of Michigan State University are especially acknowledged for their comments and suggestions.

CONTENTS

                                                                                          Page

Abstract III

Foreword V

Acknowledgments VII

Contents of Appendices XXXVI

List of Figures XLI

List of Tables XLV

Acronyms and Abbreviations XLIX

Unit Conversion Factors LVII

1 Introduction to MARLAP 1-1
  1.1 Overview 1-1
  1.2 Purpose of the Manual 1-2
  1.3 Use and Scope of the Manual 1-3
  1.4 Key MARLAP Concepts and Terminology 1-4
    1.4.1   Data Life Cycle 1-4
    1.4.2   Directed Planning Process 1-5
    1.4.3   Performance-Based Approach 1-5
    1.4.4   Analytical Process 1-6
    1.4.5   Analytical Protocol 1-7
    1.4.6   Analytical Method 1-7
    1.4.7   Uncertainty and Error 1-7
    1.4.8   Precision, Bias, and Accuracy 1-9
    1.4.9   Performance Objectives: Data Quality Objectives and Measurement Quality
             Objectives 1-10
    1.4.10 Analytical Protocol Specifications 1-11
    1.4.11 The Assessment Phase 1-11
  1.5 The MARLAP Process 1-12
  1.6 Structure of the Manual 1-13
    1.6.1   Overview of Part I 1-16
    1.6.2   Overview of Part II 1-17
    1.6.3   Overview of the Appendices 1-18
  1.7 References 1-19

Page

2 Project Planning Process 2-1
   2.1 Introduction 2-1
   2.2 The Importance of Directed Project Planning 2-2
   2.3 Directed Project Planning Processes 2-3
       2.3.1 A Graded Approach to Project Planning 2-4
       2.3.2 Guidance on Directed Planning Processes 2-4
       2.3.3 Elements of Directed Planning Processes 2-5
   2.4 The Project Planning Team 2-6
       2.4.1 Team Representation 2-7
       2.4.2 The Radioanalytical Specialists 2-7
   2.5 Directed Planning Process and Role of the Radioanalytical Specialists 2-8
       2.5.1 State the Problem 2-11
       2.5.2 Identify the Decision 2-12
            2.5.2.1 Define the Action Level 2-12
            2.5.2.2 Identify Inputs to the Decision 2-13
            2.5.2.3 Define the Decision Boundaries 2-13
            2.5.2.4 Define the Scale of the Decision 2-14
       2.5.3 Specify the Decision Rule and the Tolerable Decision Error Rates 2-14
       2.5.4 Optimize the Strategy for Obtaining Data 2-15
            2.5.4.1 Analytical Protocol Specifications 2-16
            2.5.4.2 Measurement Quality Objectives 2-16
   2.6 Results of the Directed Planning Process 2-17
       2.6.1 Output Required by the Radioanalytical Laboratory: The Analytical Protocol
             Specifications 2-18
       2.6.2 Chain of Custody 2-19
   2.7 Project Planning and Project Implementation and Assessment 2-19
       2.7.1 Documenting the Planning Process 2-19
       2.7.2 Obtaining Analytical Services 2-20
       2.7.3 Selecting Analytical Protocols 2-20
       2.7.4 Assessment Plans 2-21
            2.7.4.1 Data Verification 2-21
            2.7.4.2 Data Validation 2-22
            2.7.4.3 Data Quality Assessment 2-22
   2.8 Summary of Recommendations 2-22
   2.9 References 2-23

3 Key Analytical Planning Issues and Developing Analytical Protocol Specifications 3-1
   3.1 Introduction 3-1
   3.2 Overview of the Analytical Process 3-2
   3.3 General Analytical Planning Issues 3-2

3.3.1   Develop Analyte List 3-3
3.3.2   Identify Concentration Ranges 3-5
3.3.3   Identify and Characterize Matrices of Concern 3-5
3.3.4   Determine Relationships Among the Radionuclides of Concern 3-6
3.3.5   Determine Available Project Resources and Deadlines 3-7
3.3.6   Refine Analyte List and Matrix List 3-7
3.3.7   Method Performance Characteristics and Measurement Quality Objectives ... 3-7
          3.3.7.1 Develop MQOs for Select Method Performance Characteristics 3-9
          3.3.7.2 The Role of MQOs in the Protocol Selection and Evaluation Process ... 3-14
          3.3.7.3 The Role of MQOs in the Project's Data Evaluation Process 3-14
3.3.8   Determine Any Limitations on Analytical Options 3-15
          3.3.8.1 Gamma Spectrometry 3-16
          3.3.8.2 Gross Alpha and Beta Analyses 3-16
          3.3.8.3 Radiochemical Nuclide-Specific Analysis 3-17
3.3.9   Determine Method Availability 3-17
3.3.10 Determine the Type and Frequency of, and Evaluation Criteria for, Quality
          Control Samples 3-17
3.3.11 Determine Sample Tracking and Custody Requirements 3-18
3.3.12 Determine Data Reporting Requirements 3-19
3.4 Matrix-Specific Analytical Planning Issues 3-20
3.4.1   Solids 3-21
          3.4.1.1 Removal of Unwanted Materials 3-21
          3.4.1.2 Homogenization and Subsampling 3-21
          3.4.1.3 Sample Dissolution 3-22
3.4.2   Liquids 3-22
3.4.3   Filters and Wipes 3-23
3.5 Assembling the Analytical Protocol Specifications 3-23
3.6 Level of Protocol Performance Demonstration 3-24
3.7 Project Plan Documents 3-24
3.8 Summary of Recommendations 3-27
3.9 References 3-27
Attachment 3A: Measurement Uncertainty 3-29
3A.1   Introduction 3-29
3A.2   Analogy: Political Polling 3-29
3A.3   Measurement Uncertainty 3-30
3A.4   Sources of Measurement Uncertainty 3-31
3A.5   Uncertainty Propagation 3-32
3A.6   References 3-32
Attachment 3B: Analyte Detection 3-33
3B.1 Introduction 3-33

Contents

3B.2   The Critical Value 3-34
3B.3   The Minimum Detectable Value 3-35
3B.4   Sources of Confusion 3-36
3B.5   Implementation Difficulties 3-37

4 Project Plan Documents 4-1
4.1 Introduction 4-1
4.2 The Importance of Project Plan Documents 4-2
4.3 A Graded Approach to Project Plan Documents 4-3
4.4 Structure of Project Plan Documents 4-3
    4.4.1   Guidance on Project Plan Documents 4-4
    4.4.2   Approaches to Project Plan Documents 4-5
4.5 Elements of Project Plan Documents 4-6
    4.5.1   Content of Project Plan Documents 4-6
    4.5.2   Plan Documents Integration 4-9
    4.5.3   Plan Content for Small Projects 4-9
4.6 Linking the Project Plan Documents and the Project Planning Process 4-10
    4.6.1   Planning Process Report 4-14
    4.6.2   Data Assessment 4-15
        4.6.2.1 Data Verification 4-15
        4.6.2.2 Data Validation 4-15
        4.6.2.3 Data Quality Assessment 4-16
4.7 Summary of Recommendations 4-17
4.8 References 4-17

5 Obtaining Laboratory Services 5-1
5.1 Introduction 5-1
5.2 Importance of Writing a Technical and Contractual Specification Document 5-2
5.3 Statement of Work—Technical Requirements 5-2
    5.3.1   Analytes 5-3
    5.3.2   Matrix 5-3
    5.3.3   Measurement Quality Objectives 5-3
    5.3.4   Unique Analytical Process Requirements 5-4
    5.3.5   Quality Control Samples and Participation in External Performance Evaluation
           Programs 5-4
    5.3.6   Laboratory Radiological Holding and Turnaround Times 5-5
    5.3.7   Number of Samples and Schedule 5-5
    5.3.8   Quality System 5-6
    5.3.9   Laboratory's Proposed Methods 5-6
    5.4   Request for Proposal—Generic Contractual Requirements 5-7

5.4.1 Sample Management 5-7
5.4.2 Licenses, Permits and Environmental Regulations 5-8
    5.4.2.1 Licenses 5-8
    5.4.2.2 Environmental and Transportation Regulations 5-8
5.4.3 Data Reporting and Communications 5-9
    5.4.3.1 Data Deliverables 5-9
    5.4.3.2 Software Verification and Control 5-10
    5.4.3.3 Problem Notification and Communication 5-10
    5.4.3.4 Status Reports 5-11
5.4.4 Sample Re-Analysis Requirements 5-11
5.4.5 Subcontracted Analyses 5-11
5.5 Laboratory Selection and Qualification Criteria 5-11
5.5.1 Technical Proposal Evaluation 5-12
    5.5.1.1 Scoring and Evaluation Scheme 5-12
    5.5.1.2 Scoring Elements 5-13
5.5.2 Pre-Award Proficiency Evaluation 5-14
5.5.3 Pre-Award Assessments and Audits 5-15
5.6 Summary of Recommendations 5-15
5.7 References 5-16
5.7.1 Cited References 5-16
5.7.2 Other Sources 5-16

6 Selection and Application of an Analytical Method 6-1
6.1 Introduction 6-1
6.2 Method Definition 6-3
6.3 Life Cycle of Method Application 6-5
6.4 Generic Considerations for Method Development and Selection 6-9
6.5 Project-Specific Considerations for Method Selection 6-11
6.5.1 Matrix and Analyte Identification 6-11
    6.5.1.1 Matrices 6-11
    6.5.1.2. Analytes and Potential Interferences 6-14
6.5.2 Process Knowledge 6-14
6.5.3 Radiological Holding and Turnaround Times 6-15
6.5.4 Unique Process Specifications 6-16
6.5.5 Measurement Quality Objectives 6-17
    6.5.5.1 Method Uncertainty 6-17
    6.5.5.2 Quantification Capability 6-18
    6.5.5.3 Detection Capability 6-19
    6.5.5.4 Applicable Analyte Concentration Range 6-20
    6.5.5.5 Method Specificity 6-20

Contents

6.5.5.6 Method Ruggedness 6-21
6.5.5.7 Bias Considerations 6-21
6.6 Method Validation 6-22
  6.6.1   General Method Validation 6-24
  6.6.2   Project Method Validation Protocol 6-25
  6.6.3   Tiered Approach to Project Method Validation 6-26
    6.6.3.1 Existing Methods Requiring No Additional Validation 6-28
    6.6.3.2 Routine Methods Having No Project Method Validation 6-28
    6.6.3.3 Use of a Validated Method for Similar Matrices 6-28
    6.6.3.4 New Application of a Validated Method 6-29
    6.6.3.5 Newly Developed or Adapted Methods 6-30
  6.6.4   Testing for Bias 6-31
    6.6.4.1 Absolute Bias 6-31
    6.6.4.2 Relative Bias 6-32
  6.6.5   Project Method Validation Documentation 6-32
6.7 Analyst Qualifications and Demonstrated Proficiency 6-32
6.8 Method Control 6-33
6.9 Continued Performance Assessment 6-34
6.10 Documentation To Be Sent to the Project Manager 6-35
6.11 Summary of Recommendations 6-36
6.12 References 6-36
Attachment 6A: Bias-Testing Procedure 6-39
  6A.1 Introduction 6-39
  6A.2 The Test 6-39
  6A.3 Bias Tests at Multiple Concentrations 6-42

7 Evaluating Methods and Laboratories 7-1
7.1 Introduction 7-1
7.2 Evaluation of Proposed Analytical Methods 7-2
  7.2.1   Documentation of Required Method Performance 7-2
    7.2.1.1 Method Validation Documentation 7-3
    7.2.1.2 Internal Quality Control or External PE Program Reports 7-4
    7.2.1.3 Method Experience, Previous Projects, and Clients 7-5
    7.2.1.4 Internal and External Quality Assurance Assessments 7-5
  7.2.2   Performance Requirements of the SOW—Analytical Protocol Specifications . 7-5
    7.2.2.1 Matrix and Analyte Identification 7-6
    7.2.2.2 Radiological Holding and Turnaround Times 7-7
    7.2.2.3 Unique Processing Specifications 7-8
    7.2.2.4 Measurement Quality Objectives 7-8
    7.2.2.5 Bias Considerations 7-13

7.3 Initial Evaluation of a Laboratory 7-15
  7.3.1 Review of Quality System Documents 7-15
  7.3.2 Adequacy of Facilities, Instrumentation, and Staff Levels 7-17
  7.3.3 Review of Applicable Prior Work 7-17
  7.3.4 Review of General Laboratory Performance 7-18
      7.3.4.1 Review of Internal QC Results 7-18
      7.3.4.2 External PE Program Results 7-19
      7.3.4.3 Internal and External Quality Assessment Reports 7-20
  7.3.5 Initial Audit 7-20
7.4 Ongoing Evaluation of the Laboratory's Performance 7-20
  7.4.1 Quantitative Measures of Quality 7-21
      7.4.1.1 MQO Compliance 7-22
      7.4.1.2 Other Parameters 7-27
  7.4.2 Operational Aspects 7-28
      7.4.2.1 Desk Audits 7-28
      7.4.2.2 Onsite Audits 7-30
7.5 Summary of Recommendations 7-32
7.6 References 7-33

8 Radiochemical Data Verification and Validation 8-1
8.1 Introduction 8-1
8.2 Data Assessment Process 8-2
  8.2.1 Planning Phase of the Data Life Cycle 8-2
  8.2.2 Implementation Phase of the Data Life Cycle 8-3
      8.2.2.1 Project Objectives 8-3
      8.2.2.2 Documenting Project Activities 8-4
      8.2.2.3 Quality Assurance/Quality Control 8-4
  8.2.3 Assessment Phase of the Data Life Cycle 8-5
8.3 Validation Plan 8-7
  8.3.1 Technical and Quality Objectives of the Project 8-8
  8.3.2 Validation Tests 8-9
  8.3.3 Data Qualifiers 8-9
  8.3.4 Reporting and Documentation 8-10
8.4 Other Essential Elements for Data Validation 8-11
  8.4.1 Statement of Work 8-11
  8.4.2 Verified Data Deliverables 8-12
8.5 Data Verification and Validation Process 8-12
  8.5.1 The Sample Handling and Analysis System 8-13
      8.5.1.1 Sample Descriptors 8-14
      8.5.1.2 Aliquant Size 8-15

Contents

8.5.1.3 Dates of Sample Collection, Preparation, and Analysis 8-16
8.5.1.4 Preservation 8-16
8.5.1.5 Tracking 8-17
8.5.1.6 Traceability 8-17
8.5.1.7 QC Types and Linkages 8-18
8.5.1.8 Chemical Separation (Yield) 8-18
8.5.1.9 Self-Absorption 8-19
8.5.1.10 Efficiency, Calibration Curves, and Instrument Background 8-19
8.5.1.11 Spectrometry Resolution 8-20
8.5.1.12 Dilution and Correction Factors 8-20
8.5.1.13 Counts and Count Time (Duration) 8-21
8.5.1.14 Result of Measurement, Uncertainty, Minimum Detectable Concentration,
      and Units 8-21
  8.5.2   Quality Control Samples 8-22
8.5.2.1 Method Blank 8-23
8.5.2.2 Laboratory Control Samples 8-23
8.5.2.3 Laboratory Replicates 8-24
8.5.2.4 Matrix Spikes and Matrix Spike Duplicates 8-24
  8.5.3   Tests of Detection and Unusual Uncertainty 8-25
8.5.3.1 Detection 8-25
8.5.3.2 Detection Capability 8-26
8.5.3.3 Large or Unusual Uncertainty 8-27
  8.5.4   Final Qualification and Reporting 8-27
8.6 Validation Report 8-29
8.7 Summary of Recommendations 8-31
8.8 Bibliography 8-31

9 Data Quality Assessment 9-1
9.1 Introduction 9-1
9.2 Assessment Phase 9-2
9.3 Graded Approach to Assessment 9-3
9.4 The Data Quality Assessment Team 9-3
9.5 Data Quality Assessment Plan 9-4
9.6 Data Quality Assessment Process 9-5
  9.6.1   Review of Project Documents 9-7
9.6.1.1 The Project DQOs and MQOs 9-7
9.6.1.2 The DQA Plan 9-8
9.6.1.3 Summary of the DQA Review 9-8
  9.6.2   Sample Representativeness 9-9
9.6.2.1 Review of the Sampling Plan 9-9

Page

9.6.2.2 Sampling Plan Implementation 9-12
9.6.2.3 Data Considerations 9-13
9.6.3   Data Accuracy 9-14
9.6.3.1 Review of the Analytical Plan 9-18
9.6.3.2 Analytical Plan Implementation 9-19
9.6.4   Decisions and Tolerable Error Rates 9-21
9.6.4.1 Statistical Evaluation of Data 9-21
9.6.4.2 Evaluation of Decision Error Rates 9-24
9.7 Data Quality Assessment Report 9-25
9.8 Summary of Recommendations 9-26
9.9 References 9-27
9.9.1   Cited Sources 9-27
9.9.2   Other Sources 9-27

Volume II

10 Field and Sampling Issues That Affect Laboratory Measurements 10-1
Part A: Generic Issues 10-1
10.1   Introduction 10-1
10.2   Field Sampling Plan: Non-Matrix-Specific Issues 10-3
10.2.1 Determination of Analytical Sample Size 10-3
10.2.2 Field Equipment and Supply Needs 10-3
10.2.3 Selection of Sample Containers 10-4
10.2.3.1   Container Material 10-4
10.2.3.2   Container Opening and Closure 10-5
10.2.3.3   Sealing Containers 10-5
10.2.3.4   Precleaned and Extra Containers 10-5
10.2.4 Container Label and Sample Identification Code 10-6
10.2.5 Field Data Documentation 10-7
10.2.6 Field Tracking, Custody, and Shipment Forms 10-8
10.2.7 Chain of Custody 10-9
10.2.8 Field Quality Control 10-10
10.2.9 Decontamination of Field Equipment 10-10
10.2.10 Packing and Shipping 10-11
10.2.11 Worker Health and Safety Plan 10-12
10.2.11.1   Physical Hazards 10-13
10.2.11.2   Biohazards 10-15
Part B: Matrix-Specific Issues That Impact Field Sample Collection, Processing, and
    Preservation 10-16
10.3   Liquid Samples 10-17

10.3.1 Liquid Sampling Methods 10-18
10.3.2 Liquid Sample Preparation: Filtration 10-18
    10.3.2.1 Example of Guidance for Ground-Water Sample Filtration 10-19
    10.3.2.2 Filters 10-21
10.3.3 Field Preservation of Liquid Samples 10-22
    10.3.3.1 Sample Acidification 10-22
    10.3.3.2 Non-Acid Preservation Techniques 10-23
10.3.4 Liquid Samples: Special Cases 10-25
    10.3.4.1 Radon-222 in Water 10-25
    10.3.4.1 Milk 10-26
10.3.5 Nonaqueous Liquids and Mixtures 10-26
10.4 Solids 10-28
10.4.1 Soils 10-29
    10.4.1.1 Soil Sample Preparation 10-29
    10.4.1.2 Sample Ashing 10-30
10.4.2 Sediments 10-30
10.4.3 Other Solids 10-31
    10.4.3.1 Structural Materials 10-31
    10.4.3.2 Biota: Samples of Plant and Animal Products 10-31
10.5 Air Sampling 10-34
10.5.1 Sampler Components and Operation 10-34
10.5.2 Filter Selection Based on Destructive Versus Nondestructive Analysis 10-35
10.5.3 Sample Preservation and Storage 10-36
10.5.4 Special Cases: Collection of Gaseous and Volatile Air Contaminants 10-36
    10.5.4.1 Radioiodines 10-36
    10.5.4.2 Gases 10-37
    10.5.4.3 Tritium Air Sampling 10-38
    10.5.4.4 Radon Sampling in Air 10-39
10.6 Wipe Sampling for Assessing Surface Contamination 10-41
10.6.1 Sample Collection Methods 10-42
    10.6.1.1 Dry Wipes 10-42
    10.6.1.2 Wet Wipes 10-43
10.6.2 Sample Handling 10-44
10.6.3 Analytical Considerations for Wipe Material Selection 10-44
10.7 References 10-45

11 Sample Receipt, Inspection, and Tracking 11-1
11.1 Introduction 11-1
11.2 General Considerations 11-1
11.2.1 Communication Before Sample Receipt 11-1

11.2.2 Standard Operating Procedures 11-3
11.2.3 Laboratory License 11-4
11.2.4 Sample Chain-of-Custody 11-4
11.3   Sample Receipt 11-5
11.3.1 Package Receipt 11-5
11.3.2 Radiological Surveying 11-6
11.3.3 Corrective Action 11-8
11.4   Sample Inspection 11-8
11.4.1 Physical Integrity of Package and Sample Containers 11-8
11.4.2 Sample Identity Confirmation 11-9
11.4.3 Confirmation of Field Preservation 11-9
11.4.4 Presence of Hazardous Materials 11-9
11.4.5 Corrective Action 11-10
11.5   Laboratory Sample Tracking 11-11
11.5.1 Sample Log-In 11-11
11.5.2 Sample Tracking During Analyses 11-11
11.5.3 Storage of Samples 11-12
11.6   References 11-13

12 Laboratory Sample Preparation 12-1
12.1   Introduction 12-1
12.2   General Guidance for Sample Preparation 12-2
12.2.1 Potential Sample Losses During Preparation 12-2
        12.2.1.1 Losses as Dust or Particulates 12-2
        12.2.1.2 Losses Through Volatilization 12-3
        12.2.1.3 Losses Due to Reactions Between Sample and Container 12-5
12.2.2 Contamination from Sources in the Laboratory 12-6
        12.2.2.1 Airborne Contamination 12-7
        12.2.2.2 Contamination of Reagents 12-7
        12.2.2.3 Contamination of Glassware and Equipment 12-8
        12.2.2.4 Contamination of Facilities 12-8
12.2.3 Cleaning of Labware, Glassware, and Equipment 12-8
        12.2.3.1 Labware and Glassware 12-8
        12.2.3.2 Equipment 12-10
12.3   Solid Samples 12-12
12.3.1 General Procedures 12-12
        12.3.1.1 Exclusion of Material 12-14
        12.3.1.2 Principles of Heating Techniques for Sample Pretreatment 12-14
        12.3.1.3 Obtaining a Constant Weight 12-23
        12.3.1.4 Subsampling 12-24

12.3.2 Soil/Sediment Samples 12-27
   12.3.2.1 Soils 12-28
   12.3.2.2 Sediments 12-28
12.3.3 Biota Samples 12-28
   12.3.3.1 Food 12-29
   12.3.3.2 Vegetation 12-29
   12.3.3.3 Bone and Tissue 12-30
12.3.4 Other Samples 12-30
12.4   Filters 12-30
12.5   Wipe Samples 12-31
12.6   Liquid Samples 12-32
12.6.1 Conductivity 12-32
12.6.2 Turbidity 12-32
12.6.3 Filtration 12-33
12.6.4 Aqueous Liquids 12-33
12.6.5 Nonaqueous Liquids 12-34
12.6.6 Mixtures 12-35
   12.6.6.1 Liquid-Liquid Mixtures 12-35
   12.6.6.2 Liquid-Solid Mixtures 12-35
12.7   Gases 12-36
12.8   Bioassay 12-36
12.9   References 12-37
12.9.1 Cited Sources 12-37
12.9.2 Other Sources 12-43

13 Sample Dissolution 13-1
13.1   Introduction 13-1
13.2   The Chemistry of Dissolution 13-2
13.2.1 Solubility and the Solubility Product Constant, Ksp 13-2
13.2.2 Chemical Exchange, Decomposition, and Simple Rearrangement Reactions . 13-3
13.2.3 Oxidation-Reduction Processes 13-4
13.2.4 Complexation 13-5
13.2.5 Equilibrium: Carriers and Tracers 13-6
13.3   Fusion Techniques 13-6
13.3.1 Alkali-Metal Hydroxide Fusions 13-9
13.3.2 Boron Fusions 13-11
13.3.3 Fluoride Fusions 13-12
13.3.4 Sodium Hydroxide Fusion 13-12
13.4   Wet Ashing and Acid Dissolution Techniques 13-12
13.4.1 Acids and Oxidants 13-13

13.4.2 Acid Digestion Bombs 13-20
13.5 Microwave Digestion 13-21
13.5.1 Focused Open-Vessel Systems 13-21
13.5.2 Low-Pressure, Closed-Vessel Systems 13-22
13.5.3 High-Pressure, Closed-Vessel Systems 13-22
13.6 Verification of Total Dissolution 13-23
13.7 Special Matrix Considerations 13-23
13.7.1 Liquid Samples 13-23
13.7.2 Solid Samples 13-24
13.7.3 Filters 13-24
13.7.4 Wipe Samples 13-24
13.8 Comparison of Total Dissolution and Acid Leaching 13-25
13.9 References 13-27
13.9.1 Cited References 13-27
13.9.2 Other Sources 13-29

14 Separation Techniques 14-1
14.1 Introduction 14-1
14.2 Oxidation-Reduction Processes 14-2
14.2.1 Introduction 14-2
14.2.2 Oxidation-Reduction Reactions 14-3
14.2.3 Common Oxidation States 14-6
14.2.4 Oxidation State in Solution 14-10
14.2.5 Common Oxidizing and Reducing Agents 14-11
14.2.6 Oxidation State and Radiochemical Analysis 14-13
14.3 Complexation 14-18
14.3.1 Introduction 14-18
14.3.2 Chelates 14-20
14.3.3 The Formation (Stability) Constant 14-22
14.3.4 Complexation and Radiochemical Analysis 14-23
14.3.4.1 Extraction of Laboratory Samples and Ores 14-23
14.3.4.2 Separation by Solvent Extraction and Ion-Exchange Chromatography 14-23
14.3.4.3 Formation and Dissolution of Precipitates 14-24
14.3.4.4 Stabilization of Ions in Solution 14-24
14.3.4.5 Detection and Determination 14-25
14.4 Solvent Extraction 14-25
14.4.1 Extraction Principles 14-25
14.4.2 Distribution Coefficient 14-26
14.4.3 Extraction Technique 14-27
14.4.4 Solvent Extraction and Radiochemical Analysis 14-30

14.4.5 Solid-Phase Extraction 14-32
    14.4.5.1   Extraction Chromatography Columns 14-33
    14.4.5.2   Extraction Membranes 14-34
14.4.6 Advantages and Disadvantages of Solvent Extraction 14-35
    14.4.6.1   Advantages of Liquid-Liquid Solvent Extraction 14-35
    14.4.6.2   Disadvantages of Liquid-Liquid Solvent Extraction 14-35
    14.4.6.3   Advantages of Solid-Phase Extraction Media 14-35
    14.4.6.4   Disadvantages of Solid-Phase Extraction Media 14-36
14.5   Volatilization and Distillation 14-36
    14.5.1 Introduction 14-36
    14.5.2 Volatilization Principles 14-36
    14.5.3 Distillation Principles 14-38
    14.5.4 Separations in Radiochemical Analysis 14-39
    14.5.5 Advantages and Disadvantages of Volatilization 14-40
        14.5.5.1   Advantages 14-40
        14.5.5.2   Disadvantages 14-40
14.6   Electrodeposition 14-41
    14.6.1 Electrodeposition Principles 14-41
    14.6.2 Separation of Radionuclides 14-42
    14.6.3 Preparation of Counting Sources 14-43
    14.6.4 Advantages and Disadvantages of Electrodeposition 14-43
        14.6.4.1   Advantages 14-43
        14.6.4.2 Disadvantages 14-43
14.7   Chromatography 14-44
    14.7.1 Chromatographic Principles 14-44
    14.7.2 Gas-Liquid and Liquid-Liquid Phase Chromatography 14-45
    14.7.3 Adsorption Chromatography 14-45
    14.7.4 Ion-Exchange Chromatography 14-46
        14.7.4.1   Principles of Ion Exchange 14-46
        14.7.4.2   Resins 14-48
    14.7.5 Affinity Chromatography 14-54
    14.7.6 Gel-Filtration Chromatography 14-54
    14.7.7 Chromatographic Laboratory Methods 14-55
    14.7.8 Advantages and Disadvantages of Chromatographic Systems 14-56
        14.7.8.1   Advantages 14-56
        14.7.8.2   Disadvantages 14-56
14.8   Precipitation and Coprecipitation 14-56
    14.8.1 Introduction 14-56
    14.8.2 Solutions 14-57
    14.8.3 Precipitation 14-59

14.8.3.1   Solubility and the Solubility Product Constant, K_{sp} 14-59
14.8.3.2   Factors Affecting Precipitation 14-64
14.8.3.3   Optimum Precipitation Conditions 14-69
14.8.4 Coprecipitation 14-69
14.8.4.1   Coprecipitation Processes 14-70
14.8.4.2   Water as an Impurity 14-74
14.8.4.3   Postprecipitation 14-74
14.8.4.4   Coprecipitation Methods 14-75
14.8.5 Colloidal Precipitates 14-78
14.8.6 Separation of Precipitates 14-81
14.8.7 Advantages and Disadvantages of Precipitation and Coprecipitation 14-82
14.8.7.1   Advantages 14-82
14.8.7.2   Disadvantages 14-82
14.9   Carriers and Tracers 14-82
14.9.1 Introduction 14-82
14.9.2 Carriers 14-83
14.9.2.1   Isotopic Carriers 14-83
14.9.2.2   Nonisotopic Carriers 14-84
14.9.2.3   Common Carriers 14-85
14.9.2.4   Holdback Carriers 14-89
14.9.2.5   Yield of Isotopic Carriers 14-89
14.9.3 Tracers 14-90
14.9.3.1   Characteristics of Tracers 14-92
14.9.3.2   Coprecipitation 14-93
14.9.3.3   Deposition on Nonmetallic Solids 14-93
14.9.3.4   Radiocolloid Formation 14-94
14.9.3.5   Distribution (Partition) Behavior 14-95
14.9.3.6   Vaporization 14-95
14.9.3.7   Oxidation and Reduction 14-96
14.10   Analysis of Specific Radionuclides 14-97
14.10.1   Basic Principles of Chemical Equilibrium 14-97
14.10.2   Oxidation State 14-100
14.10.3   Hydrolysis 14-100
14.10.4   Polymerization 14-102
14.10.5   Complexation 14-103
14.10.6   Radiocolloid Interference 14-103
14.10.7   Isotope Dilution Analysis 14-104
14.10.8   Masking and Demasking 14-105
14.10.9   Review of Specific Radionuclides 14-109
14.10.9.1   Americium 14-109

Contents

Page

        14.10.9.2 Carbon 14-114
        14.10.9.3 Cesium 14-116
        14.10.9.4 Cobalt 14-119
        14.10.9.5 Iodine 14-125
        14.10.9.6 Neptunium 14-132
        14.10.9.7 Nickel 14-136
        14.10.9.8 Plutonium 14-139
        14.10.9.9 Radium 14-148
        14.10.9.10 Strontium 14-155
        14.10.9.11 Sulfur and Phosphorus 14-160
        14.10.9.12 Technetium 14-163
        14.10.9.13 Thorium 14-169
        14.10.9.14 Tritium 14-175
        14.10.9.15 Uranium 14-180
        14.10.9.16 Zirconium 14-191
        14.10.9.17 Progeny of Uranium and Thorium 14-198
   14.11   References 14-201
   14.12   Selected Bibliography 14-218
       14.12.1   Inorganic and Analytical Chemistry 14-218
       14.12.2   General Radiochemistry 14-219
       14.12.3   Radiochemical Methods of Separation 14-219
       14.12.4   Radionuclides 14-220
       14.12.5   Separation Methods 14-222
   Attachment 14A Radioactive Decay and Equilibrium 14-223
      14A.1 Radioactive Equilibrium 14-223
        14A.1.1   Secular Equilibrium 14-223
        14A.1.2   Transient Equilibrium 14-225
        14A.1.3   No Equilibrium 14-226
        14A.1.4   Summary of Radioactive Equilibria 14-227
      14A.1.5 Supported and Unsupported Radioactive Equilibria 14-228
      14A.2 Effects of Radioactive Equilibria on Measurement Uncertainty 14-229
        14A.2.1   Issue 14-229
        14A.2.2   Discussion 14-229
        14A.2.3   Examples of Isotopic Distribution: Natural, Enriched, and Depleted
                 Uranium 14-231
      14A.3 References 14-232

15 Quantification of Radionuclides 15-1
   15.1   Introduction 15-1
   15.2   Instrument Calibrations 15-2

15.2.1 Calibration Standards 15-3
15.2.2 Congruence of Calibration and Test-Source Geometry 15-3
15.2.3 Calibration and Test-Source Homogeneity 15-5
15.2.4 Self-Absorption, Attenuation, and Scattering Considerations for Source
       Preparations 15-5
15.2.5 Calibration Uncertainty 15-7
15.3   Methods of Source Preparation 15-8
15.3.1 Electrodeposition 15-8
15.3.2 Precipitation/Coprecipitation 15-11
15.3.3 Evaporation 15-12
15.3.4 Thermal Volatilization/Sublimation 15-15
15.3.5 Special Source Matrices 15-16
       15.3.5.1   Radioactive Gases 15-16
       15.3.5.2   Air Filters 15-17
       15.3.5.3   Swipes 15-18
15.4   Alpha Detection Methods 15-18
15.4.1 Introduction 15-18
15.4.2 Gas Proportional Counting 15-20
       15.4.2.1 Detector Requirements and Characteristics 15-20
       15.4.2.2   Calibration and Test Source Preparation 15-25
       15.4.2.3   Detector Calibration 15-25
       15.4.2.4   Troubleshooting 15-27
15.4.3 Solid-State Detectors 15-29
       15.4.3.1   Detector Requirements and Characteristics 15-30
       15.4.3.2   Calibration- and Test-Source Preparation 15-33
       15.4.3.3   Detector Calibration 15-33
       15.4.3.4   Troubleshooting 15-34
       15.4.3.5   Detector or Detector Chamber Contamination 15-35
       15.4.3.6   Degraded Spectrum 15-37
15.4.4 Fluorescent Detectors 15-38
       15.4.4.1   Zinc Sulfide 15-38
       15.4.4.2   Calibration- and Test-Source Preparation 15-40
       15.4.4.3   Detector Calibration 15-41
       15.4.4.4   Troubleshooting 15-41
15.4.5 Photon Electron Rejecting Alpha Liquid Scintillation (PERALS®) 15-42
       15.4.5.1   Detector Requirements and Characteristics 15-42
       15.4.5.2   Calibration- and Test-Source Preparation 15-44
       15.4.5.3   Detector Calibration 15-45
       15.4.5.4   Quench 15-45
       15.4.5.5   Available Cocktails 15-46

15.4.5.6   Troubleshooting 15-46
15.5   Beta Detection Methods 15-46
15.5.1 Introduction 15-46
15.5.2 Gas Proportional Counting/Geiger-Mueller Tube Counting 15-49
15.5.2.1   Detector Requirements and Characteristics 15-49
15.5.2.2   Calibration- and Test-Source Preparation 15-53
15.5.2.3   Detector Calibration 15-54
15.5.2.4.   Troubleshooting 15-57
15.5.3 Liquid Scintillation 15-57
15.5.3.1   Detector Requirements and Characteristics 15-58
15.5.3.2   Calibration- and Test-Source Preparation 15-61
15.5.3.3   Detector Calibration 15-62
15.5.3.4   Troubleshooting 15-68
15.6   Gamma Detection Methods 15-68
15.6.1 Sample Preparation Techniques 15-70
15.6.1.1   Containers 15-71
15.6.1.2   Gases 15-71
15.6.1.3   Liquids 15-72
15.6.1.4   Solids 15-72
15.6.2 Sodium Iodide Detector 15-73
15.6.2.1   Detector Requirements and Characteristics 15-73
15.6.2.2   Operating Voltage 15-76
15.6.2.3   Shielding 15-76
15.6.2.4   Background 15-76
15.6.2.5   Detector Calibration 15-77
15.6.2.6   Troubleshooting 15-77
15.6.3 High Purity Germanium 15-78
15.6.3.1   Detector Requirements and Characteristics 15-78
15.6.3.2   Gamma Spectrometer Calibration 15-82
15.6.3.3   Troubleshooting 15-84
15.6.4 Extended Range Germanium Detectors 15-88
15.6.4.1   Detector Requirements and Characteristics 15-89
15.6.4.2   Detector Calibration 15-89
15.6.4.3   Troubleshooting 15-90
15.6.5 Special Techniques for Radiation Detection 15-90
15.6.5.1   Other Gamma Detection Systems 15-90
15.6.5.2   Coincidence Counting 15-91
15.6.5.3   Anti-Coincidence Counting 15-93
15.7   Specialized Analytical Techniques 15-94
15.7.1 Kinetic Phosphorescence Analysis by Laser (KPA) 15-94

15.7.2 Mass Spectrometry 15-95
    15.7.2.1 Inductively Coupled Plasma-Mass Spectrometry 15-96
    15.7.2.2 Thermal Ionization Mass Spectrometry 15-99
    15.7.2.3 Accelerator Mass Spectrometry 15-100
15.8 References 15-101
  15.8.1 Cited References 15-101
  15.8.2 Other Sources 15-115

16 Data Acquisition, Reduction, and Reporting for Nuclear Counting Instrumentation 16-1
16.1 Introduction 16-1
16.2 Data Acquisition 16-2
  16.2.1 Generic Counting Parameter Selection 16-3
    16.2.1.1 Counting Duration 16-4
    16.2.1.2 Counting Geometry 16-5
    16.2.1.3 Software 16-5
  16.2.2 Basic Data Reduction Calculations 16-6
16.3 Data Reduction on Spectrometry Systems 16-8
  16.3.1 Gamma-Ray Spectrometry 16-9
    16.3.1.1 Peak Search or Identification 16-10
    16.3.1.2 Singlet/Multiplet Peaks 16-13
    16.3.1.3 Definition of Peak Centroid and Energy 16-14
    16.3.1.4 Peak Width Determination 16-15
    16.3.1.5 Peak Area Determination 16-17
    16.3.1.6 Calibration Reference File 16-19
    16.3.1.7 Activity and Concentration 16-20
    16.3.1.8 Summing Considerations 16-21
    16.3.1.9 Uncertainty Calculation 16-22
  16.3.2 Alpha Spectrometry 16-23
    16.3.2.1 Radiochemical Yield 16-27
    16.3.2.2 Uncertainty Calculation 16-28
  16.3.3 Liquid Scintillation Spectrometry 16-29
    16.3.3.1 Overview of Liquid Scintillation Counting 16-29
    16.3.3.2 Liquid Scintillation Spectra 16-29
    16.3.3.3 Pulse Characteristics 16-29
    16.3.3.4 Coincidence Circuitry 16-30
    16.3.3.5 Quenching 16-30
    16.3.3.6 Luminescence 16-31
    16.3.3.7 Test-Source Vials 16-31
    16.3.3.8 Data Reduction for Liquid Scintillation Counting 16-31
16.4 Data Reduction on Non-Spectrometry Systems 16-32

16.5   Internal Review of Data by Laboratory Personnel 16-36
    16.5.1 Primary Review 16-37
    16.5.2 Secondary Review 16-37
16.6   Reporting Results 16-38
    16.6.1 Sample and Analysis Method Identification 16-38
    16.6.2 Units and Radionuclide Identification 16-38
    16.6.3 Values, Uncertainty, and Significant Figures 16-39
16.7   Data Reporting Packages 16-39
16.8   Electronic Data Deliverables 16-41
16.9   References 16-41
    16.9.1 Cited References 16-41
    16.9.2 Other Sources 16-44

17 Waste Management in a Radioanalytical Laboratory 17-1
17.1   Introduction 17-1
17.2   Types of Laboratory Wastes 17-1
17.3   Waste Management Program 17-2
    17.3.1 Program Integration 17-3
    17.3.2 Staff Involvement 17-3
17.4   Waste Minimization 17-3
17.5   Waste Characterization 17-6
17.6   Specific Waste Management Requirements 17-6
    17.6.1 Sample/Waste Exemptions 17-9
    17.6.2 Storage 17-9
        17.6.2.1 Container Requirements 17-10
        17.6.2.2 Labeling Requirements 17-10
        17.6.2.3 Time Constraints 17-11
        17.6.2.4 Monitoring Requirements 17-11
    17.6.3 Treatment 17-12
    17.6.4 Disposal 17-12
17.7   Contents of a Laboratory Waste Management Plan/Certification Plan 17-13
    17.7.1 Laboratory Waste Management Plan 17-13
    17.7.2 Waste Certification Plan/Program 17-14
17.8   Useful Web Sites 17-15
17.9   References 17-17
    17.9.1 Cited References 17-17
    17.9.2 Other Sources 17-17

Volume III

18 Laboratory Quality Control 18-1
  18.1   Introduction 18-1
     18.1.1 Organization of Chapter 18-2
     18.1.2 Format 18-2
  18.2   Quality Control 18-3
  18.3   Evaluation of Performance Indicators 18-3
     18.3.1 Importance of Evaluating Performance Indicators 18-3
     18.3.2 Statistical Means of Evaluating Performance Indicators — Control Charts .. 18-5
     18.3.3 Tolerance Limits 18-7
     18.3.4 Measurement Uncertainty 18-8
  18.4   Radiochemistry Performance Indicators 18-9
     18.4.1 Method and Reagent Blank 18-9
     18.4.2 Laboratory Replicates 18-13
     18.4.3 Laboratory Control Samples, Matrix Spikes, and Matrix Spike Duplicates . 18-16
     18.4.4 Certified Reference Materials 18-18
     18.4.5 Chemical/Tracer Yield 18-21
  18.5   Instrumentation Performance Indicators 18-24
     18.5.1 Instrument Background Measurements 18-24
     18.5.2 Efficiency Calibrations 18-26
     18.5.3 Spectrometry Systems 18-29
          18.5.3.1   Energy Calibrations 18-29
          18.5.3.2   Peak Resolution and Tailing 18-32
     18.5.4 Gas Proportional Systems 18-36
          18.5.4.1   Voltage Plateaus 18-36
          18.5.4.2   Self-Absorption, Backscatter, and Crosstalk 18-37
     18.5.5 Liquid Scintillation 18-38
     18.5.6 Summary Guidance on Instrument Calibration, Background,
            and Quality Control 18-40
          18.5.6.1   Gas Proportional Counting Systems 18-42
          18.5.6.2   Gamma-Ray Detectors and Spectrometry Systems 18-45
          18.5.6.3   Alpha Detector and Spectrometry Systems 18-49
          18.5.6.4   Liquid Scintillation Systems 18-51
     18.5.7 Non-Nuclear Instrumentation 18-53
  18.6   Related Concerns 18-54
     18.6.1 Detection Capability 18-54
     18.6.2 Radioactive Equilibrium 18-54
     18.6.3 Half-Life 18-57
     18.6.4 Interferences 18-58

18.6.5 Negative Results 18-60
18.6.6 Blind Samples 18-61
18.6.7 Calibration of Apparatus Used for Mass and Volume Measurements 18-63
18.7 References 18-65
18.7.1 Cited Sources 18-65
18.7.2 Other Sources 18-67
Attachment 18A: Control Charts 18-69
18A.1 Introduction 18-69
18A.2 X Charts 18-69
18A.3 $\overline{X}$ Charts 18-72
18A.4 R Charts 18-74
18A.5 Control Charts for Instrument Response 18-75
18A.6 References 18-79
Attachment 18B: Statistical Tests for QC Results 18-81
18B.1 Introduction 18-81
18B.2 Tests for Excess Variance in the Instrument Response 18-81
18B.3 Instrument Background Measurements 18-88
18B.3.1 Detection of Background Variability 18-88
18B.3.2 Comparing a Single Observation to Preset Limits 18-90
18B.3.3 Comparing the Results of Consecutive Measurements 18-93
18B.4 Negative Activities 18-96
18B.5 References 18-96

19 Measurement Uncertainty 19-1
19.1 Overview 19-1
19.2 The Need for Uncertainty Evaluation 19-1
19.3 Evaluating and Expressing Measurement Uncertainty 19-3
19.3.1 Measurement, Error, and Uncertainty 19-3
19.3.2 The Measurement Process 19-4
19.3.3 Analysis of Measurement Uncertainty 19-6
19.3.4 Corrections for Systematic Effects 19-7
19.3.5 Counting Uncertainty 19-7
19.3.6 Expanded Uncertainty 19-7
19.3.7 Significant Figures 19-8
19.3.8 Reporting the Measurement Uncertainty 19-9
19.3.9 Recommendations 19-10
19.4 Procedures for Evaluating Uncertainty 19-11
19.4.1 Identifying Sources of Uncertainty 19-12
19.4.2 Evaluation of Standard Uncertainties 19-13

19.4.2.1 Type A Evaluations 19-13
19.4.2.2 Type B Evaluations 19-16
19.4.3 Combined Standard Uncertainty 19-20
19.4.3.1 Uncertainty Propagation Formula 19-20
19.4.3.2 Components of Uncertainty 19-24
19.4.3.3 Special Forms of the Uncertainty Propagation Formula 19-25
19.4.4 The Estimated Covariance of Two Output Estimates 19-26
19.4.5 Special Considerations for Nonlinear Models 19-29
19.4.5.1 Uncertainty Propagation for Nonlinear Models 19-29
19.4.5.2 Bias due to Nonlinearity 19-31
19.4.6 Monte Carlo Methods 19-34
19.5 Radiation Measurement Uncertainty 19-34
19.5.1 Radioactive Decay 19-34
19.5.2 Radiation Counting 19-35
19.5.2.1 Binomial Model 19-35
19.5.2.2 Poisson Approximation 19-36
19.5.3 Count Time and Count Rate 19-38
19.5.3.1 Dead Time 19-39
19.5.3.2 A Confidence Interval for the Count Rate 19-40
19.5.4 Instrument Background 19-41
19.5.5 Radiochemical Blanks 19-42
19.5.6 Counting Efficiency 19-43
19.5.7 Radionuclide Half-Life 19-47
19.5.8 Gamma-Ray Spectrometry 19-48
19.5.9 Balances 19-48
19.5.10 Pipets and Other Volumetric Apparatus 19-52
19.5.11 Digital Displays and Rounding 19-54
19.5.12 Subsampling 19-55
19.5.13 The Standard Uncertainty for a Hypothetical Measurement 19-56
19.6 References 19-58
19.6.1 Cited Sources 19-58
19.6.2 Other Sources 19-61
Attachment 19A: Statistical Concepts and Terms 19-63
19A.1 Basic Concepts 19-63
19A.2 Probability Distributions 19-66
19A.2.1 Normal Distributions 19-67
19A.2.2 Log-normal Distributions 19-68
19A.2.3 Chi-squared Distributions 19-69
19A.2.4 *T*-Distributions 19-70
19A.2.5 Rectangular Distributions 19-71

Contents

19A.2.6 Trapezoidal and Triangular Distributions 19-72
19A.2.7 Exponential Distributions 19-73
19A.2.8 Binomial Distributions 19-73
19A.2.9 Poisson Distributions 19-74
19A.3 References 19-76
Attachment 19B: Example Calculations 19-77
19B.1 Overview 19-77
19B.2 Sample Collection and Analysis 19-77
19B.3 The Measurement Model 19-78
19B.4 The Combined Standard Uncertainty 19-80
Attachment 19C: Multicomponent Measurement Models 19-83
19C.1 Introduction 19-83
19C.2 The Covariance Matrix 19-83
19C.3 Least-Squares Regression 19-83
19C.4 References 19-84
Attachment 19D: Estimation of Coverage Factors 19-85
19D.1 Introduction 19-85
19D.2 Procedure 19-85
19D.2.1 Basis of Procedure 19-85
19D.2.2 Assumptions 19-85
19D.2.3 Effective Degrees of Freedom 19-86
19D.2.4 Coverage Factor 19-87
19D.3 Poisson Counting Uncertainty 19-88
19D.4 References 19-91
Attachment 19E: Uncertainties of Mass and Volume Measurements 19-93
19E.1 Purpose 19-93
19E.2 Mass Measurements 19-93
19E.2.1 Considerations 19-93
19E.2.2 Repeatability 19-94
19E.2.3 Environmental Factors 19-95
19E.2.4 Calibration 19-97
19E.2.5 Linearity 19-98
19E.2.6 Gain or Loss of Mass 19-98
19E.2.7 Air-Buoyancy Corrections 19-99
19E.2.8 Combining the Components 19-103
19E.3 Volume Measurements 19-105
19E.3.1 First Approach 19-105
19E.3.2 Second Approach 19-108
19E.3.3 Third Approach 19-111
19E.4 References 19-111

20 Detection and Quantification Capabilities 20-1
20.1 Overview 20-1
  20.2   Concepts and Definitions 20-1
    20.2.1 Analyte Detection Decisions 20-1
    20.2.2 The Critical Value 20-3
    20.2.3 The Blank 20-5
    20.2.4 The Minimum Detectable Concentration 20-5
    20.2.6 Other Detection Terminologies 20-9
    20.2.7 The Minimum Quantifiable Concentration 20-10
  20.3   Recommendations 20-11
  20.4   Calculation of Detection and Quantification Limits 20-12
    20.4.1 Calculation of the Critical Value 20-12
      20.4.1.1 Normally Distributed Signals 20-13
      20.4.1.2 Poisson Counting 20-13
      20.4.1.3 Batch Blanks 20-17
    20.4.2 Calculation of the Minimum Detectable Concentration 20-18
      20.4.2.1 The Minimum Detectable Net Instrument Signal 20-19
      20.4.2.2 Normally Distributed Signals 20-20
      20.4.2.3 Poisson Counting 20-24
      20.4.2.4 More Conservative Approaches 20-28
      20.4.2.5 Experimental Verification of the MDC 20-28
    20.4.3 Calculation of the Minimum Quantifiable Concentration 20-29
  20.5   References 20-33
    20.5.1 Cited Sources 20-33
    20.5.2 Other Sources 20-35
Attachment 20A: Low-Background Detection Issues 20-37
  20A.1 Overview 20-37
  20A.2 Calculation of the Critical Value 20-37
    20A.2.1   Normally Distributed Signals 20-37
    20A.2.2   Poisson Counting 20-39
  20A.3 Calculation of the Minimum Detectable Concentration 20-53
    20A.3.1   Normally Distributed Signals 20-54
    20A.3.2   Poisson Counting 20-58
  20A.4 References 20-62

Glossary .... *End of each volume*

Appendices – Volume I

A Directed Planning Approaches A-1
   A.1   Introduction A-1
   A.2   Elements Common to Directed Planning Approaches A-2
   A.3   Data Quality Objectives Process A-2
   A.4   Observational Approach A-3
   A.5   Streamlined Approach for Environmental Restoration A-4
   A.6   Technical Project Planning A-4
   A.7   Expedited Site Characterization A-4
   A.8   Value Engineering A-5
   A.9   Systems Engineering A-6
   A.10 Total Quality Management A-6
   A.11 Partnering A-7
   A.12 References A-7
      A.12.1 Data Quality Objectives A-7
      A.12.2 Observational Approach A-9
      A.12.3 Streamlined Approach for Environmental Restoration (Safer) A-10
      A.12.4 Technical Project Planning A-11
      A.12.5 Expedited Site Characterization A-11
      A.12.6 Value Engineering A-12
      A.12.7 Systems Engineering A-13
      A.12.8 Total Quality Management A-15
      A.12.9 Partnering A-16

Appendix B The Data Quality Objectives Process B-1
  B.1 Introduction B-1
  B.2 Overview of the DQO Process B-2
  B.3 The Seven Steps of the DQO Process B-3
     B.3.1 DQO Process Step 1: State the Problem B-3
     B.3.2 DQO Process Step 2: Identify the Decision B-4
     B.3.3 DQO Process Step 3: Identify Inputs to the Decision B-5
     B.3.4 DQO Process Step 4: Define the Study Boundaries B-6
     B.3.5 Outputs of DQO Process Steps 1 through 4 Lead Into Steps 5 through 7 ..... B-7
     B.3.6 DQO Process Step 5: Develop a Decision Rule B-7
     B.3.7 DQO Process Step 6: Specify the Limits on Decision Errors B-9
     B.3.8 DQO Process Step 7: Optimize the Design for Obtaining Data B-22
  B.4 References B-24
  Attachment B1: Decision Error Rates and the Gray Region for Decisions About
  Mean Concentrations B-26

Page

B1.1   Introduction B-26
B1.2   The Region of Interest B-26
B1.3   Measurement Uncertainty at the Action Level B-26
B1.4   The Null Hypothesis B-29
       Case 1: Assume the True Concentration is Over 1.0 .... B-30
       Case 2: Assume the True Concentration is 0.9 .... B-32
B1.5   The Gray Region B-32
B1.6   Summary B-34
Attachment B2: Decision Error Rates and the Gray Region for Detection Decisions B-36
B2.1   Introduction B-36
B2.2   The DQO Process Applied to the Detection Limit Problem B-36
B2.3   Establish the Concentration Range of Interest B-37
B2.4   Estimate the Measurement Variability when Measuring a Blank B-41

Appendix C Measurement Quality Objectives for Method Uncertainty and Detection and
       Quantification Capability C-1
C.1 Introduction C-1
C.2 Hypothesis Testing C-1
C.3 Development of MQOs for Analytical Protocol Selection C-3
C.4 The Role of the MQO for Method Uncertainty in Data Evaluation C-9
    C.4.1 Uncertainty Requirements at Various Concentrations C-9
    C.4.2 Acceptance Criteria for Quality Control Samples C-11
C.5 References C-17

Appendix D Content of Project Plan Documents D-1
D.1 Introduction D-1
D.2 Group A: Project Management D-5
    D.2.1 Project Management (A1): Title and Approval Sheet D-5
    D.2.2 Project Management (A2): Table of Contents D-7
    D.2.3 Project Management (A3): Distribution List D-7
    D.2.4 Project Management (A4): Project/Task Organization D-7
    D.2.5 Project Management (A5): Problem Definition/Background D-8
    D.2.6 Project Management (A6): Project/Task Description D-9
    D.2.7 Project Management (A7): Quality Objectives and Criteria for Measurement
           Data D-11
           D.2.7.1   Project's Quality Objectives D-11
           D.2.7.2   Specifying Measurement Quality Objectives D-12
           D.2.7.3   Relation between the Project DQOs, MQOs, and QC Requirements D-13
    D.2.8 Project Management (A8): Special Training Requirements/Certification ... D-13
    D.2.9 Project Management (A9): Documentation and Record D-13

D.3 Group B: Measurement/Data Acquisition D-14
    D.3.1 Measurement/Data Acquisition (B1): Sampling Process Design D-15
    D.3.2 Measurement/Data Acquisition (B2): Sampling Methods Requirements ... D-16
    D.3.3 Measurement/Data Acquisition (B3): Sample Handling and Custody
    Requirements D-18
    D.3.4 Measurement/Data Acquisition (B4): Analytical Methods Requirements .. D-19
    D.3.5 Measurement/Data Acquisition (B5): Quality Control Requirements D-21
    D.3.6 Measurement/Data Acquisition (B6): Instrument/Equipment Testing, Inspection,
    and Maintenance Requirements D-22
    D.3.7 Measurement/Data Acquisition (B7): Instrument Calibration and Frequency D-23
    D.3.8 Measurement/Data Acquisition (B8): Inspection/Acceptance Requirements for
    Supplies and Consumables D-23
    D.3.9 Measurement/Data Acquisition (B9): Data Acquisition Requirements for Non-
    Direct Measurement Data D-24
    D.3.10 Measurement/Data Acquisition (B10): Data Management D-25
D.4 Group C: Assessment/Oversight D-26
    D.4.1 Assessment/Oversight (C1): Assessment and Response Actions D-26
    D.4.2 Assessment/Oversight (C2): Reports to Management D-27
D.5 Group D: Data Validation and Usability D-28
    D.5.1 Data Validation and Usability (D1): Verification and Validation
    Requirements D-28
    D.5.2 Data Validation and Usability (D2): Verification and Validation Methods . D-29
        D.5.2.1   Data Verification D-29
        D.5.2.2   Data Validation D-30
    D.5.3 Data Validation and Usability (D3): Reconciliation with Data Quality
    Objectives D-30
D.6 References D-31

Appendix E Contracting Laboratory Services E-1
E.1 Introduction E-1
E.2 Procurement of Services E-4
    E.2.1 Request for Approval of Proposed Procurement Action E-5
    E.2.2 Types of Procurement Mechanisms E-6
E.3 Request for Proposals—The Solicitation E-7
    E.3.1 Market Research E-8
    E.3.2 Period of Contract E-9
    E.3.3 Subcontracts E-9
E.4 Proposal Requirements E-10
    E.4.1 RFP and Contract Information E-11
    E.4.2 Personnel E-13

E.4.3   Instrumentation E-15
     E.4.3.1   Type, Number, and Age of Laboratory Instruments E-16
     E.4.3.2   Service Contract E-16
E.4.4   Narrative to Approach E-16
     E.4.4.1   Analytical Methods or Protocols E-16
     E.4.4.2   Meeting Contract Measurement Quality Objectives E-17
     E.4.4.3   Data Package E-17
     E.4.4.4   Schedule E-17
     E.4.4.5   Sample Storage and Disposal E-18
E.4.5   Quality Manual E-18
E.4.6   Licenses and Accreditations E-19
E.4.7   Experience E-20
     E.4.7.1   Previous or Current Contracts E-20
     E.4.7.2   Quality of Performance E-20
E.5 Proposal Evaluation and Scoring Procedures E-21
E.5.1   Evaluation Committee E-21
E.5.2   Ground Rules — Questions E-22
E.5.3   Scoring/Evaluating Scheme E-22
     E.5.3.1   Review of Technical Proposal and Quality Manual E-23
     E.5.3.2   Review of Laboratory Accreditation E-25
     E.5.3.3   Review of Experience E-25
E.5.4   Pre-Award Proficiency Samples E-25
E.5.5   Pre-Award Audit E-26
E.5.6   Comparison of Prices E-30
E.5.7   Debriefing of Unsuccessful Vendors E-30
E.6 The Award E-31
E.7 For the Duration of the Contract E-31
E.7.1   Managing a Contract E-32
E.7.2   Responsibility of the Contractor E-32
E.7.3   Responsibility of the Agency E-32
E.7.4   Anomalies and Nonconformance E-33
E.7.5   Laboratory Assessment E-33
     E.7.5.1   Performance Testing and Quality Control Samples E-33
     E.7.5.2   Laboratory Performance Evaluation Programs E-34
     E.7.5.3   Laboratory Evaluations Performed During the Contract Period E-35
E.8 Contract Completion E-36
E.9 References E-36

Appendices – Volume II

Appendix F Laboratory Subsampling F-1
  F.1 Introduction F-1
  F.2 Basic Concepts F-2
  F.3 Sources of Measurement Error F-3
      F.3.1 Sampling Bias F-4
      F.3.2 Fundamental Error F-5
      F.3.3 Grouping and Segregation Error F-6
  F.4 Implementation of the Particulate Sampling Theory F-9
      F.4.1 The Fundamental Variance F-10
      F.4.2 Scenario 1 – Natural Radioactive Minerals F-10
      F.4.3 Scenario 2 – Hot Particles F-11
      F.4.4 Scenario 3 – Particle Surface Contamination F-13
  F.5 Summary F-15
  F.6 References F-16

Appendices – Volume III

G   Statistical Tables G-1

List of Figures

Figure 1.1 The data life cycle 1-4
Figure 1.2 Typical components of an analytical process 1-6
Figure 1.3 The MARLAP process 1-14
Figure 1.4 Key MARLAP terms and processes 1-15

Figure 3.1 Typical components of an analytical process 3-2
Figure 3.2 Analytical protocol specifications 3-25
Figure 3.3 Example analytical protocol specifications 3-26
Figure 3B.1 The critical value of the net signal 3-35

Figure 6.1 Analytical process 6-2
Figure 6.2 Method application life cycle 6-6
Figure 6.3 Expanded Figure 6.2 addressing the laboratory's method evaluation process 6-7
Figure 6.4 Relationship between level of laboratory effort, method validation level, and degree
    of assurance of method performance under the tiered approach to method validation ... 6-27

Figure 7.1 Considerations for the initial evaluation of a laboratory 7-16

Figure 8.1 The assessment process 8-5

Figure 9.1 Using physical samples to measure a characteristic of the population
    representatively. 9-10
Figure 9.2 Types of sampling and analytical errors. 9-17

Volume II

Figure 10.1 Example of chain-of-custody record 10-9

Figure 11.1 Overview of sample receipt, inspection, and tracking 11-2

Figure 12.1 Degree of error in laboratory sample preparation relative to other activities ... 12-1
Figure 12.2 Laboratory sample preparation flowchart (for solid samples) 12-13

Figure 14.1 Ethylene diamine tetraacetic acid (EDTA) 14-20
Figure 14.2 Crown ethers 14-21
Figure 14.3 The behavior of elements in concentrated hydrochloric acid on cation-exchange
    resins 14-52

Figure 14.4 The behavior of elements in concentrated hydrochloric acid on anion-exchange resins 14-53
Figure 14.5 The electrical double layer. 14-79
Figure 14A.1 Decay chain for ^{238}U 14-224
Figure 14A.2 Secular equilibrium of ^{210}Pb/^{210}Bi 14-225
Figure 14A.3 Transient equilibrium of ^{95}Zr/^{95}Nb 14-226
Figure 14A.4 No equilibrium of ^{239}U/^{239}Np 14-227

Figure 15.1 Alpha plateau generated by a ^{210}Po source on a GP counter using P-10 gas ... 15-23
Figure 15.2 Gas proportional counter self-absorption curve for ^{230}Th 15-28
Figure 15.3 Beta plateau generated by a ^{90}Sr/Y source on a GP counter using P-10 gas ... 15-52
Figure 15.4 Gas proportional counter self-absorption curve for ^{90}Sr/Y 15-56
Figure 15.5 Representation of a beta emitter energy spectrum 15-65
Figure 15.6 Gamma-ray interactions with high-purity germanium 15-70
Figure 15.7 NaI(Tl) spectrum of ^{137}Cs 15-75
Figure 15.8 Energy spectrum of ^{22}Na 15-80
Figure 15.9 Different geometries for the same germanium detector and the same sample in different shapes or position 15-83
Figure 15.10 Extended range coaxial germanium detector 15-88
Figure 15.11 Typical detection efficiencies comparing extended range with a normal coaxial germanium detector 15-90
Figure 15.12 Beta-gamma coincidence efficiency curve for ^{131}I 15-93

Figure 16.1 Gamma-ray spectrum 16-9
Figure 16.2 Gamma-ray analysis flow chart and input parameters 16-11
Figure 16.3 Low-energy tailing 16-16
Figure 16.4 Photopeak baseline continuum 16-17
Figure 16.5 Photopeak baseline continuum-step function 16-18
Figure 16.6 Alpha spectrum (^{238}U, ^{235}U, ^{234}U, $^{239/240}$Pu, ^{241}Am) 16-23

Volume III

Figure 18.1 Problems leading to loss of analytical control 18-4
Figure 18.2 Control chart for daily counting of a standard reference source, with limits corrected for decay 18-7
Figure 18.3 Three general categories of blank changes 18-12
Figure 18.4 Failed performance indicator: replicates 18-15
Figure 18.5 Failed performance indicator: chemical yield 18-23

Figure 19.1 Addition of uncertainty components 19-25

Figure 19.2 Expected fraction of atoms remaining at time t 19-35
Figure 19.3 A symmetric distribution 19-64
Figure 19.4 An asymmetric distribution 19-65
Figure 19.5 A normal distribution 19-67
Figure 19.6 A log-normal distribution 19-68
Figure 19.7 Chi-squared distributions 19-69
Figure 19.8 The t-distribution with 3 degrees of freedom 19-70
Figure 19.9 A rectangular distribution 19-72
Figure 19.10 A trapezoidal distribution 19-72
Figure 19.11 An exponential distribution 19-73
Figure 19.12a Poisson distribution vs. normal distribution, $\mu = 3$ 19-75
Figure 19.12b Poisson distribution vs. normal distribution, $\mu = 100$ 19-76
Figure 19.13 Nonlinear balance response curve 19-98

Figure 20.1 The critical net signal, S_C, and minimum detectable net signal, S_D 20-6
Figure 20.2 Type I error rates for Table 20.1 20-41
Figure 20.3 Type I error rate for the Poisson-normal approximation ($t_B = t_S$) 20-42
Figure 20.4 Type I error rates for Formula A 20-44
Figure 20.5 Type I error rates for Formula B 20-45
Figure 20.6 Type I error rates for Formula C 20-46
Figure 20.7 Type I error rates for the Stapleton approximation 20-48
Figure 20.8 Type I error rates for the nonrandomized exact test 20-49

Figure B.1 Seven steps of the DQO process. B-2
Figure B.2(a) Decision performance goal diagram null hypothesis: the parameter exceeds the
    action level. B-11
Figure B.2(b) Decision performance goal diagram null hypothesis: the parameter is less than the
    action level. B-11
Figure B.3 Plot is made showing the range of the parameter of interest on the x-axis B-15
Figure B.4 A line showing the action level, the type of decision error possible at a given value of
    the true concentration, and a y-axis showing the acceptable limits on making a decision error
    have been added to Figure B.3. .... B-16
Figure B.5 The gray region is a specified range of values of the true concentration where the
    consequences of a decision error are considered to be relatively minor B-17
Figure B.6 Three possible ways of setting the gray region. B-17
Figure B.7 Example decision performance goal diagram B-19
Figure B.8 A power curve constructed from the decision performance goal diagram in Figure
    B.7. .... B-20
Figure B.9 Example power curve showing the key parameters used to determine the appropriate
    number of samples to take in the survey unit. B-21

Page

Figure B.10 How proximity to the action level determines what is an acceptable level of uncertainty B-24
Figure B1.1 The action level is 1.0 . . . . B-26
Figure B1.2 The true mean concentration is 1.0. . . . . B-27
Figure B1.3 The true mean concentration is 0.9. . . . . B-28
Figure B1.4 If 0.95 is measured, is the true mean concentration 1.0 (right) or 0.9 (left)? . . . B-28
Figure B1.5 When the true mean concentration is 1.0, and the standard uncertainty of the distribution of measured concentrations is 0.1, a measured concentration of 0.84 or less will be observed only about 5 percent of the time B-30
Figure B1.6 When the true mean concentration is 0.84, and the standard uncertainty of the distribution of measured concentrations is 0.1, a measured concentration of 0.84 or less will be observed only about half the time B-31
Figure B1.7 When the true mean concentration is 0.68 and the standard uncertainty of the distribution of measured concentrations is 0.1, a measured concentration over 0.84 will be observed only about 5 percent of the time B-31
Figure B1.8 The true mean concentration is 0.9 (left) and 1.22 (right). B-32
Figure B1.9 The true mean concentration is 0.84 (left) and 1.0 (right) B-34
Figure B2.1 Region of interest for the concentration around the action level of 1.0 . . . . B-38
Figure B2.2 (a) The distribution of blank (background) readings. (b) The true concentration is 0.0. The standard deviation of the distribution of measured concentrations is 0.2. . . . . . B-38
Figure B2.3 The true concentration is 0.0, and the standard deviation of the distribution of measured concentrations is 0.2 . . . . B-39
Figure B2.4 The true concentration is 0.2 and the standard deviation of the distribution of measured concentrations is 0.2 . . . . B-39
Figure B2.5 The true value of the concentration is 0.66 and the standard deviation of the distribution of measured concentrations is 0.2 . . . . B-41
Figure B2.6 The true value of the measured concentration is 0.0 and the standard deviation of the measured concentrations is 0.2. . . . . B-41
Figure B2.7 The standard deviation of the normally distributed measured concentrations is 0.2. . . . . B-42

Figure C.1 Required analytical standard deviation (u_{Req}) C-10

Figure E.1 General sequence initiating and later conducting work with a contract laboratory E-4

List of Tables

Table 2.1 — Summary of the directed planning process and radioanalytical specialists participation 2-9

Table 3.1 Common matrix-specific analytical planning issues 3-20

Table 4.1 Elements of project plan documents 4-7
Table 4.2 Crosswalk between project plan document elements and directed planning process 4-10

Table 6.1 Tiered project method validation approach 6-26

Table 7.1 Cross reference of information available for method evaluation 7-3

Table 9.1 Summary of the DQA process 9-6

Volume II

Table 10.1 Summary of sample preservation techniques 10-25

Table 11.1 Typical topics addressed in standard operating procedures related to sample receipt, inspection, and tracking 11-3

Table 12.1 Examples of volatile radionuclides 12-4
Table 12.2 Properties of sample container materials 12-5
Table 12.3 Examples of dry-ashing temperatures (platinum container) 12-23
Table 12.4 Preliminary ashing temperature for food samples 12-29

Table 13.1 Common fusion fluxes 13-7
Table 13.2 Examples of acids used for wet ashing 13-13
Table 13.3 Standard reduction potentials of selected half-reactions at 25 °C 13-14

Table 14.1 Oxidation states of elements 14-8
Table 14.2 Oxidation states of selected elements 14-10
Table 14.3 Redox reagents for radionuclides 14-13
Table 14.4 Common ligands 14-19
Table 14.5 Radioanalytical methods employing solvent extraction 14-32

Contents

Table 14.6 Radioanalytical methods employing extraction chromatography 14-33
Table 14.7 Elements separable by volatilization as certain species 14-37
Table 14.8 Typical functional groups of ion-exchange resins 14-49
Table 14.9 Common ion-exchange resins 14-50
Table 14.10 General solubility behavior of some cations of interest 14-58
Table 14.11 Summary of methods for utilizing precipitation from homogeneous solution . 14-68
Table 14.12 Influence of precipitation conditions on the purity of precipitates 14-69
Table 14.13 Common coprecipitating agents for radionuclides 14-76
Table 14.14 Coprecipitation behavior of plutonium and neptunium 14-78
Table 14.15 Atoms and mass of select radionuclides equivalent to 500 dpm 14-83
Table 14.16 Masking agents for ions of various metals 14-106
Table 14.17 Masking agents for anions and neutral molecules 14-108
Table 14.18 Common radiochemical oxidizing and reducing agents for iodine 14-129
Table 14.19 Redox agents in plutonium chemistry 14-142
Table 14A.1 Relationships of radioactive equilibria 14-228

Table 15.1 Radionuclides prepared by coprecipitation or precipitation 15-12
Table 15.2 Nuclides for alpha calibration 15-20
Table 15.3 Typical gas operational parameters for gas proportional alpha counting 15-22
Table 15.4 Nuclides for beta calibration 15-48
Table 15.5 Typical operational parameters for gas proportional beta counting 15-50
Table 15.6 Typical FWHM values as a function of energy 15-79
Table 15.7 Typical percent gamma-ray efficiencies for a 55 percent HPGe detector with various
counting geometries 15-83
Table 15.8 AMS detection limits for selected radionuclides 15-100

Table 16.1 Units for data reporting 16-39
Table 16.2 Example elements of a radiochemistry data package 16-40

Table 17.1 Examples of laboratory-generated wastes 17-2

Volume III

Table 18.1a Certified Massic activities for natural radionuclides with a normal distribution of
measurement results 18-20
Table 18.1b Certified Massic activities for anthropogenic radionuclides with a Weibull
distribution of measurement results 18-20
Table 18.1c Uncertified Massic activities 18-20
Table 18.2 Instrument background evaluation 18-26
Table 18.3 Root-cause analysis of performance check results for spectrometry systems . . . 18-35

Table 18.4 Some causes of excursions in liquid scintillation analysis 18-40
Table 18.5 Example gas proportional instrument calibration, background frequency, and
performance criteria 18-44
Table 18.6 Example gamma spectrometry instrument calibration, background frequency, and
performance criteria 18-48
Table 18.7 Example alpha spectrometry instrument calibration, background frequency, and
performance criteria 18-50
Table 18.8 Example liquid scintillation counting systems calibration, background frequency, and
performance criteria 18-53
Table 18A.1 Bias-correction factor for the experimental standard deviation 18-70

Table 19.1 Differentiation rules 19-21
Table 19.2 Applications of the first-order uncertainty propagation formula 19-21
Table 19.3 95 % confidence interval for a Poisson mean 19-75
Table 19.4 Input estimates and standard uncertainties 19-81
Table 19.5 Coefficients of cubical expansion 19-107
Table 19.6 Density of air-free water 19-109

Table 20.1 Critical gross count (well-known blank) 20-40
Table 20.2 Bias factor for the experimental standard deviation 20-55
Table 20.3 Estimated and true values of S_D $(t_B = t_S)$ 20-62

Table B.1 Possible decision errors B-12
Table B.2 Example of possible decision errors with null hypothesis that the average
concentration in a survey unit is above the action level B-15
Table B.3 Example decision error limits table B-19

Table D.1 QAPP groups and elements D-2
Table D.2 Comparison of project plan contents D-3
Table D.3 Content of the three elements that constitute the project description D-8

Table E.1 Examples of procurement options to obtain materials or services E-6
Table E.2 SOW checklists for the agency and proposer E-12
Table E.3 Laboratory technical supervisory personnel listed by position title and examples for
suggested minimum qualifications E-14
Table E.4 Laboratory technical personnel listed by position title and examples for suggested
minimum qualifications and examples of optional staff members E-14
Table E.5 Laboratory technical personnel listed by position title and examples for suggested
minimum qualifications E-15
Table E.6 Example of a proposal evaluation plan E-23

Contents

		Page
Table G.1	Quantiles of the standard normal distribution	G-1
Table G.2	Quantiles of Student's t distribution	G-3
Table G.3	Quantiles of chi-square	G-5
Table G.4	Critical values for the nonrandomized exact test	G-7
Table G.5	Summary of probability distributions	G-11

ACRONYMS AND ABBREVIATIONS

AC .... alternating current
ADC .... analog to digital convertor
AEA .... Atomic Energy Act
AL .... action level
AMS .... accelerator mass spectrometry
ANSI .... American National Standards Institute
AOAC ...... Association of Official Analytical Chemists
APHA ...... American Public Health Association
APS .... analytical protocol specification
ARAR ...... applicable or relevant and appropriate requirement (CERCLA/Superfund)
ASL .... analytical support laboratory
ASQC .... American Society for Quality Control
ASTM ...... American Society for Testing and Materials
ATD .... alpha track detector

BGO .... bismuth germanate [detector]
BNL .... Brookhaven National Laboratory (DOE)
BOA .... basic ordering agreement

CAA .... Clean Air Act
CC .... charcoal canisters
CEDE .... committed effective dose equivalent
CERCLA .... Comprehensive Environmental Response, Compensation, and Liability Act of 1980 ("Superfund")
c.f. .... carrier free [tracer]
cfm .... cubic feet per minute
CFR .... *Code of Federal Regulations*
CL .... central line (of a control chart)
CMPO ...... [octyl(phenyl)]-N,N-diisobutylcarbonylmethylphosphine oxide
CMST .... Characterization, Monitoring, and Sensor Technology Program (DOE)
CO .... contracting officer
COC .... chain of custody
COR .... contracting officer's representative
cpm .... counts per minute
cps .... counts per second
CRM .... (1) continuous radon monitor; (2) certified reference material
CSU .... combined standard uncertainty
CV .... coefficient of variation
CWA .... Clean Water Act
CWLM ...... continuous working level monitor

d . . . . day[s]
D . . . . homogeneous distribution coefficient
DAAP . . . . diamylamylphosphonate
DC . . . . direct current
DCGL . . . . derived concentration guideline level
DHS . . . . U.S. Department of Homeland Security
DIN . . . . di-isopropylnaphthalene
DL . . . . discrimination limit
DoD . . . . U.S. Department of Defense
DOE . . . . U.S. Department of Energy
DOELAP . . . . DOE Laboratory Accreditation Program
DOT . . . . U.S. Department of Transportation
DOP . . . . dispersed oil particulate
dpm . . . . disintegrations per minute
DPPP . . . . dipentylpentylphosphonate
DQA . . . . data quality assessment
DQI . . . . data quality indicator
DQO . . . . data quality objective
DTPA . . . . diethylene triamine pentaacetic acid
DVB . . . . divinylbenzene

E_e . . . . emission probability per decay event
$E_{\beta max}$ . . . . maximum beta-particle energy
EDD . . . . electronic data deliverable
EDTA . . . . ethylene diamine tetraacetic acid
EGTA . . . . ethyleneglycol bis(2-aminoethylether)-tetraacetate
EMEDD . . . . . environmental management electronic data deliverable (DOE)
EPA . . . . U.S. Environmental Protection Agency
ERPRIMS . . . Environmental Resources Program Management System (U.S. Air Force)
ESC . . . . expedited site characterization; expedited site conversion
eV . . . . electron volts

FAR . . . . *Federal Acquisition Regulations*, CFR Title 48
FBO . . . . *Federal Business Opportunities* [formerly *Commerce Business Daily*]
FDA . . . . U.S. Food and Drug Administration
FEP . . . . full energy peak
fg . . . . femtogram
FOM . . . . figure of merit
FWHM . . . . . . full width of a peak at half maximum
FWTM . . . . . . full width of a peak at tenth maximum

GC .... gas chromatography
GLPC .... gas-liquid phase chromatography
GM .... Geiger-Mueller [detector]
GP .... gas proportional [counter]
GUM .... *Guide to the Expression of Uncertainty in Measurement* (ISO)
Gy .... gray[s]

h .... hour[s]
H_0 .... null hypothesis
H_A, H_1 .... alternative hypothesis
HDBP .... dibutylphosphoric acid
HDEHP ..... bis(2-ethylhexyl) phosphoric acid
HDPE .... high-density polyethylene
HLW .... high-level [radioactive] waste
HPGe .... high-purity germanium
HPLC .... high-pressure liquid chromatography; high-performance liquid chromatography
HTRW ...... hazardous, toxic, and radioactive waste

IAEA .... International Atomic Energy Agency
ICRU .... International Commission on Radiation Units and Measurements
ICP-MS ..... inductively coupled plasma-mass spectroscopy
IPPD .... integrated product and process development
ISO .... International Organization for Standardization
IUPAC ...... International Union of Pure and Applied Chemistry

k .... coverage factor
keV .... kilo electron volts
KPA .... kinetic phosphorimeter analysis

LAN .... local area network
LANL .... Los Alamos National Laboratory (DOE)
LBGR .... lower bound of the gray region
LCL .... lower control limit
LCS .... laboratory control samples
LDPE .... low-density polyethylene
LEGe .... low-energy germanium
LIMS .... laboratory information management system
LLD .... lower limit of detection
LLNL .... Lawrence Livermore National Laboratory (DOE)
LLRW ...... low-level radioactive waste
LLRWPA .... Low Level Radioactive Waste Policy Act

LOMI . . . . low oxidation-state transition-metal ion
LPC . . . . liquid-partition chromatography; liquid-phase chromatography
LS . . . . liquid scintillation
LSC . . . . liquid scintillation counter
LWL . . . . lower warning limit

MAPEP . . . . . Mixed Analyte Performance Evaluation Program (DOE)
MARSSIM . . . *Multi-Agency Radiation Survey and Site Investigation Manual*
MCA . . . . multichannel analyzer
MCL . . . . maximum contaminant limit
MDA . . . . minimum detectable amount; minimum detectable activity
MDC . . . . minimum detectable concentration
MDL . . . . method detection limit
MeV . . . . mega electron volts
MIBK . . . . methyl isobutyl ketone
min . . . . minute[s]
MPa . . . . megapascals
MQC . . . . minimum quantifiable concentration
MQO . . . . measurement quality objective
MS . . . . matrix spike; mass spectrometer
MSD . . . . matrix spike duplicate
MVRM . . . . . . method validation reference material

NAA . . . . neutron activation analysis
NaI(Tl) . . . . . . thallium-activated sodium iodide [detector]
NCP . . . . National Oil and Hazardous Substances Pollution Contingency Plan
NCRP . . . . National Council on Radiation Protection and Measurement
NELAC . . . . . National Environmental Laboratory Accreditation Conference
NESHAP . . . . National Emission Standards for Hazardous Air Pollutants (EPA)
NIM . . . . nuclear instrumentation module
NIST . . . . National Institute of Standards and Technology
NPL . . . . National Physics Laboratory (United Kingdom); National Priorities List (United
             States)
NRC . . . . U.S. Nuclear Regulatory Commission
NRIP . . . . NIST Radiochemistry Intercomparison Program
NTA (NTTA) . nitrilotriacetate
NTU . . . . nephelometric turbidity units
NVLAP . . . . . National Voluntary Laboratory Accreditation Program (NIST)

OA . . . . observational approach
OFHC . . . . oxygen-free high-conductivity

OFPP . . . . Office of Federal Procurement Policy

φ_{MR} . . . . required relative method uncertainty
Pa . . . . pascals
PARCC . . . . . precision, accuracy, representativeness, completeness, and comparability
PBBO . . . . 2-(4'-biphenylyl) 6-phenylbenzoxazole
PCB . . . . polychlorinated biphenyl
pCi . . . . picocurie
pdf . . . . probability density function
PE . . . . performance evaluation
PERALS . . . . . Photon Electron Rejecting Alpha Liquid Scintillation®
PFA . . . . perfluoroalcoholoxil™
PIC . . . . pressurized ionization chamber
PIPS . . . . planar implanted passivated silicon [detector]
PM . . . . project manager
PMT . . . . photomultiplier tube
PT . . . . performance testing
PTB . . . . Physikalisch-Technische bundesanstalt (Germany)
PTFE . . . . polytetrafluoroethylene
PUREX . . . . . plutonium uranium reduction extraction
PVC . . . . polyvinyl chloride

QA . . . . quality assurance
QAP . . . . Quality Assessment Program (DOE)
QAPP . . . . quality assurance project plan
QC . . . . quality control

rad . . . . radiation absorbed dose
RCRA . . . . Resource Conservation and Recovery Act
REE . . . . rare earth elements
REGe . . . . reverse-electrode germanium
rem . . . . roentgen equivalent: man
RFP . . . . request for proposals
RFQ . . . . request for quotations
RI/FS . . . . remedial investigation/feasibility study
RMDC . . . . . . required minimum detectable concentration
ROI . . . . region of interest
RPD . . . . relative percent difference
RPM . . . . remedial project manager
RSD . . . . relative standard deviation
RSO . . . . radiation safety officer

s . . . . second[s]
SA . . . . spike activity
S_c . . . . critical value
SAFER . . . . . . Streamlined Approach for Environmental Restoration Program (DOE)
SAM . . . . site assessment manager
SAP . . . . sampling and analysis plan
SEDD . . . . staged electronic data deliverable
SI . . . . international system of units
SMO . . . . sample management office[r]
SOP . . . . standard operating procedure
SOW . . . . statement of work
SQC . . . . statistical quality control
SPE . . . . solid-phase extraction
SR . . . . unspiked sample result
SRM . . . . standard reference material
SSB . . . . silicon surface barrier [alpha detector]
SSR . . . . spiked sample result
Sv . . . . sievert[s]

$t_{\frac{1}{2}}$ . . . . half-life
TAT . . . . turnaround time
TBP . . . . tributylphosphate
TC . . . . to contain
TCLP . . . . toxicity characteristic leaching procedure
TD . . . . to deliver
TEC . . . . technical evaluation committee
TEDE . . . . total effective dose equivalent
TEC . . . . technical evaluation committee (USGS)
TES . . . . technical evaluation sheet (USGS)
TFM . . . . tetrafluorometoxil™
TIMS . . . . thermal ionization mass spectrometry
TIOA . . . . triisooctylamine
TLD . . . . thermoluminescent dosimeter
TnOA . . . . tri-n-octylamine
TOPO . . . . trioctylphosphinic oxide
TPO . . . . technical project officer
TPP . . . . technical project planning
TPU . . . . total propagated uncertainty
TQM . . . . Total Quality Management
TRUEX . . . . . trans-uranium extraction
TSCA . . . . Toxic Substances Control Act

TSDF .... treatment, storage, or disposal facility
tSIE .... transfomed spectral index of the external standard
TTA .... thenoyltrifluoroacetone

U .... expanded uncertainty
u_{MR} .... required absolute method uncertainty
$u_c(y)$ .... combined standard uncertainty
UBGR ...... upper bound of the gray region
UCL .... upper control limit
USACE ..... United States Army Corps of Engineers
USGS .... United States Geological Survey
UV .... ultraviolet
UWL .... upper warning limit

V .... volt[s]

WCP .... waste certification plan

XML .... extensible mark-up language
XtGe® .... extended-range germanium

y .... year[s]
Y .... response variable

ZnS(Ag) ..... silver-activated zinc sulfide [detector]

UNIT CONVERSION FACTORS

To Convert	To	Multiply by	To Convert	To	Multiply by
Years (y)	Seconds (s) Minutes (min) Hours (h)	3.16×10^7 5.26×10^5 8.77×10^3	s min h	y	3.17×10^{-8} 1.90×10^{-6} 1.14×10^{-4}
Disintegrations per second (dps)	Becquerels (Bq)	1.0	Bq	dps	1.0
Bq Bq/kg Bq/m^3 Bq/m^3	Picocuries (pCi) pCi/g pCi/L Bq/L	27.03 2.7×10^{-2} 2.7×10^{-2} 10^3	pCi pCi/g pCi/L Bq/L	Bq Bq/kg Bq/m^3 Bq/m^3	3.7×10^{-2} 37 37 10^{-3}
Microcuries per milliliter (µCi/mL)	pCi/L	10^9	pCi/L	µCi/mL	10^{-9}
Disintegrations per minute (dpm)	µCi pCi	4.5×10^{-7} 4.5×10^{-1}	pCi	dpm	2.22
Gallons (gal)	Liters (L)	3.78	Liters	Gallons	0.265
Gray (Gy)	rad	100	rad	Gy	10^{-2}
Roentgen Equivalent Man (rem)	Sievert (Sv)	10^{-2}	Sv	rem	10^2

1 INTRODUCTION TO MARLAP

1.1   Overview

Each year, hundreds of millions of dollars are spent on projects and programs that rely, to varying degrees, on radioanalytical data for decisionmaking. These decisions often have a significant impact on human health and the environment. Of critical importance to informed decisionmaking are data of known quality, appropriate for their intended use. Making incorrect decisions due to data inadequacies, such as failing to remediate a radioactively contaminated site properly, necessitates the expenditure of additional resources, causes delays in project completions and, depending on the nature of the project, can result in the loss of public trust and confidence. The Multi-Agency Radiological Laboratory Analytical Protocols (MARLAP) Manual addresses the need for a nationally consistent approach to producing radioanalytical laboratory data that meet a project's or program's data requirements. MARLAP provides guidance for the planning, implementation, and assessment phases of those projects that require the laboratory analysis of radionuclides. The guidance provided by MARLAP is both scientifically rigorous and flexible enough to be applied to a diversity of projects and programs. This guidance is intended for project planners, managers, and laboratory personnel.

MARLAP is divided into two main parts. Part I is primarily for project planners and managers and provides guidance on project planning with emphasis on analytical planning issues and analytical data requirements. Part I also provides guidance on preparing project plan documents and radioanalytical statements of work (SOWs), obtaining and evaluating radioanalytical laboratory services, data validation, and data quality assessment. Part I of MARLAP covers the entire life of a project that requires the laboratory analysis of radionuclides from the initial project planning phase to the assessment phase.

Part II of MARLAP is primarily for laboratory personnel and provides guidance in the relevant areas of radioanalytical laboratory work. Part II offers information on the laboratory analysis of radionuclides. The chapters in Part II cover the range of activities performed at radioanalytical laboratories, including sample preservation, shipping and handling, sample preparation, sample dissolution, separation techniques, instrument measurements, data reduction, quality control, statistics, and waste management. Part II is not a compilation of analytical procedures but rather is intended to provide information on many of the radioanalytical options available to laboratories and discuss the advantages and disadvantages of each.

MARLAP was developed collaboratively by the following federal agencies: the Environmental Protection Agency (EPA), the Department of Energy (DOE), the Department of Homeland

Contents	
1.1 Overview 1-1	
1.2 Purpose of the Manual 1-2	
1.3 Use and Scope of the Manual 1-3	
1.4 Key MARLAP Concepts and Terminology .. 1-4	
1.5 The MARLAP Process 1-12	
1.6 Structure of the Manual 1-13	
1.7 References 1-19	

Security (DHS), the Nuclear Regulatory Commission (NRC), the Department of Defense (DOD), the National Institute of Standards and Technology (NIST), the United States Geological Survey (USGS), and the Food and Drug Administration (FDA). State participation in the development of MARLAP involved contributions from representatives from the Commonwealth of Kentucky and the State of California.

1.2 Purpose of the Manual

MARLAP's basic goal is to provide guidance for project planners, managers, and laboratory personnel to ensure that radioanalytical laboratory data will meet a project's or program's data requirements and needs. To attain this goal, MARLAP provides the necessary framework for national consistency in radioanalytical work in the form of a performance-based approach for meeting a project's data requirements. In general terms, a performance-based approach to laboratory analytical work involves clearly defining the analytical data needs and requirements of a project in terms of measurable goals during the planning phase of a project. These project-specific analytical data needs and requirements then serve as measurement performance criteria for decisions as to exactly how the laboratory analysis will be conducted during the implementation phase of a project. They are used subsequently as criteria for evaluating analytical data during the assessment phase. The manual focuses on activities performed at radioanalytical laboratories as well as on activities and issues that direct, affect, or can be used to evaluate activities performed at radioanalytical laboratories.

Specific objectives of MARLAP include:

- Promoting a directed planning process for projects involving individuals from relevant disciplines including radiochemistry;

- Highlighting common radioanalytical planning issues;

- Providing a framework and information resource for using a performance-based approach for planning and conducting radioanalytical work;

- Providing guidance on linking project planning, implementation, and assessment;

- Providing guidance on obtaining and evaluating radioanalytical laboratory services;

- Providing guidance for evaluating radioanalytical laboratory data, i.e., data verification, data validation, and data quality assessment;

- Promoting high quality radioanalytical laboratory work; and

- Making collective knowledge and experience in radioanalytical work widely available.

1.3   Use and Scope of the Manual

The guidance contained in MARLAP is for both governmental and private sectors. Users of MARLAP include project planners, project managers, laboratory personnel, regulators, auditors, inspectors, data evaluators, decisionmakers, and other end users of radioanalytical laboratory data.

Because MARLAP uses a performance-based approach to laboratory measurements, the guidance contained in the manual is applicable to a wide range of projects and activities that require radioanalytical laboratory measurements. Examples of data collection activities that MARLAP supports include:

- Site characterization activities;
- Site cleanup and compliance demonstration activities;
- License termination activities;
- Decommissioning of nuclear facilities;
- Remedial and removal actions;
- Effluent monitoring of licensed facilities;
- Emergency response activities;
- Environmental site monitoring;
- Background studies;
- Routine ambient monitoring; and
- Waste management activities.

MARLAP and the *Multi-Agency Radiation Survey and Site Investigation Manual* (MARSSIM, 2000) are complementary guidance documents in support of cleanup and decommissioning activities. MARSSIM provides guidance on how to plan and carry out a study to demonstrate that a site meets appropriate release criteria. It describes a methodology for planning, conducting, evaluating, and documenting environmental radiation surveys conducted to demonstrate compliance with cleanup criteria. MARLAP provides guidance and a framework for both project planners and laboratory personnel to ensure that radioanalytical data will meet the needs and requirements of cleanup and decommissioning activities.

While MARLAP supports a wide range of projects, some topics are not specifically discussed in the manual. These include high-level waste, mixed waste, and medical applications involving radionuclides. While they are not specifically addressed, much of MARLAP's guidance may be applicable in these areas. Although the focus of the manual is to provide guidance for those projects that require the laboratory analysis of radionuclides, much of the guidance on the planning and assessment phases can be applied wherever the measurement process is conducted, for example, in the field. In addition, MARLAP does not provide specific guidance on sampling design issues, sample collection, field measurements, or laboratory health and safety practices. However, a brief discussion of some aspects of these activities has been included in the manual because of the effect these activities often have on the laboratory analytical process.

1.4   Key MARLAP Concepts and Terminology

Some of the terms used in MARLAP were developed for the purpose of this manual, while others are commonly used terms that have been adopted by MARLAP. Where possible, every effort has been made to use terms and definitions from consensus-based organizations (e.g., International Organization for Standardization [ISO], American National Standards Institute [ANSI], American Society for Testing and Materials [ASTM], International Union of Pure and Applied Chemistry [IUPAC]).

The following sections are intended to familiarize the reader with the key terms and concepts used in MARLAP. In general, each term or concept is discussed individually in each section without emphasizing how these terms and concepts are linked. Section 1.5 ties these terms and concepts together to provide an overview of the MARLAP process.

1.4.1   Data Life Cycle

The data life cycle (EPA, 2000) approach provides a structured means of considering the major phases of projects that involve data collection activities (Figure 1.1). The three phases of the data life cycle are *planning, implementation*, and *assessment*.

DATA LIFE CYCLE		
	PROCESS	PROCESS OUTPUTS
Planning	Directed Planning Process	Development of Data Quality Objectives and Measurement Quality Objectives (Including Optimized Sampling and Analytical Design)
Planning	Plan Documents	Project Plan Documents Including Quality Assurance Project Plan (QAPP); Work Plan or Sampling and Analysis Plan (SAP); Data Validation Plan; Data Quality Assessment Plan
Planning	Contracting Services	Statement of Work (SOW) and Other Contractual Documents
Implementation	Sampling	Laboratory Samples
Implementation	Analysis	Laboratory Analysis (Including Quality Control [QC] Samples) Complete Data Package
Assessment	Verification	Verified Data Data Verification Report
Assessment	Validation	Validated Data Data Validation Report
Assessment	Data Quality Assessment	Assessment Report
Data of Known Quality Appropriate for the Intended Use		

FIGURE 1.1 — The data life cycle

Although the diagram represents the data life cycle in a linear fashion, it is important to note that the actual process is an iterative one, with feedback loops. MARLAP provides information on all three phases for two major types of activities: those performed at radioanalytical laboratories and those that direct, affect, or evaluate activities performed at radioanalytical laboratories (such as project planning, development of plan documents, data verification and data validation).

One of MARLAP's specific objectives is to emphasize the importance of establishing the proper linkages among the three phases of the data life cycle. This results in an integrated and iterative process that translates the expectations and requirements of data users into measurement performance criteria for data suppliers. The integration of the three phases of the data life cycle is critical to ensuring that the analytical data requirements (defined during the planning phase) can serve as measurement performance criteria during the implementation phase and subsequently as data evaluation criteria during the assessment phase.

Without the proper linkages and integration of the three phases, there is a significant likelihood that the analytical data will not meet a project's data requirements. The data may be evaluated using criteria that have little relation to their intended use. Therefore, failure to integrate and adequately link the three phases of the data life cycle increases the likelihood of project cost escalation or project failure.

1.4.2 Directed Planning Process

MARLAP recommends the use of a directed or systematic planning process. A directed planning process is an approach for setting well-defined, achievable objectives and developing a cost-effective, technically sound sampling and analysis design that balances the data user's tolerance for uncertainty in the decision process with the resources available for obtaining data to support a decision. While MARLAP recommends and promotes the use of a directed planning process, it does not recommend or endorse any particular directed planning process. However, MARLAP employs many of the terms and concepts associated with the data quality objective (DQO) process (ASTM D5792; EPA, 2000). This was done to ensure consistent terminology throughout the manual, and also because many of the terms and concepts of this process are familiar to those engaged in environmental data collection activities.

1.4.3 Performance-Based Approach

MARLAP provides the necessary guidance for using a performance-based approach to meet a project's analytical data requirements. In a performance-based approach, the project-specific analytical data requirements that are determined during directed planning serve as measurement performance criteria for analytical selections and decisions. The project-specific analytical data requirements also are used for the initial, ongoing, and final evaluation of the laboratory's performance and the laboratory's data. MARLAP provides guidance for using a performance-based approach for all three phases of the data life cycle for those projects that require radioanalytical laboratory data. This involves not only using a performance-based approach for selecting an analytical protocol, but also using a performance-based approach for other project activities, such as developing acceptance criteria for laboratory quality control samples, laboratory evaluations, data verification, data validation, and data quality assessment.

There are three major steps associated with a performance-based approach. The first is clearly and accurately defining the analytical data requirements for the project. This process is discussed in more detail in Section 1.4.9 of this chapter. The second step uses an organized, interactive process to select or develop analytical protocols to meet the specified analytical data requirements and to demonstrate the protocols' abilities to meet the analytical data requirements (Section 1.4.10). The third major step uses the analytical data requirements as measurement performance criteria for the ongoing and final evaluation of the laboratory data, including data verification, data validation, and data quality assessment (Section 1.4.11). Within the constraints of other factors, such as cost, a performance-based approach allows for the use of any analytical protocol that meets the project's analytical data requirements. For all relevant project activities, the common theme of a performance-based approach is the use of project-specific analytical data requirements that are developed during project planning and serve as measurement performance criteria for selections, evaluations, and decisionmaking.

1.4.4   Analytical Process

Most environmental data collection efforts center around two major processes: the sampling process and the analytical process. MARLAP does not provide guidance on the sampling process, except for brief discussions of certain activities that often affect the analytical process (field processing, preservation, etc.). The analytical (or measurement) process is a general term used by MARLAP to refer to a compilation of activities starting from the time a sample is collected and ending with the reporting of data. Figure 1.2 illustrates the major components of an analytical process. A particular analytical process for a project may not include all of the activities listed. For example, if a project involves the analysis of

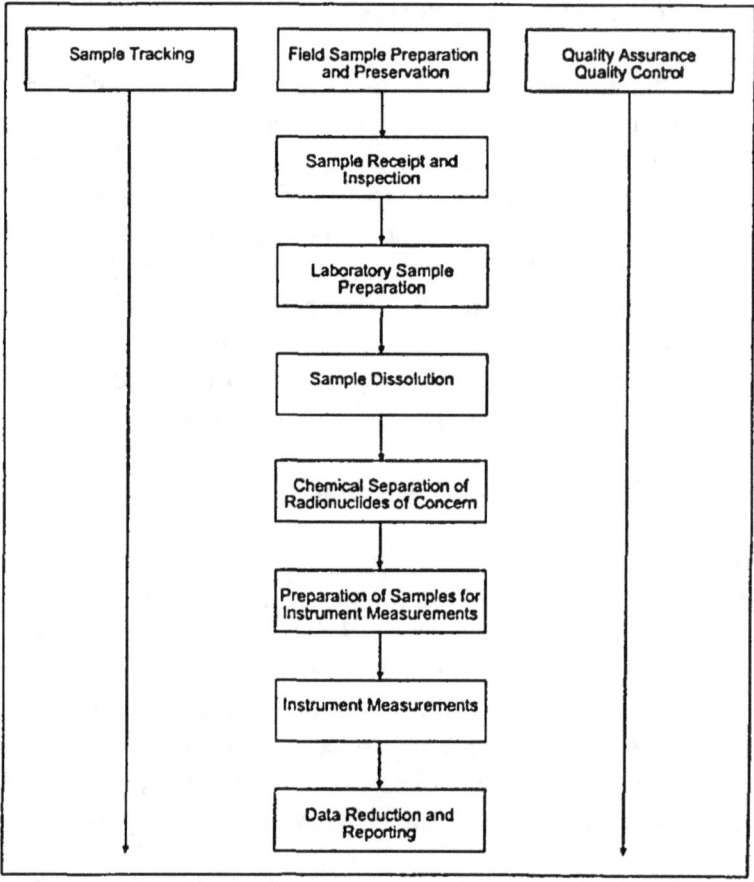

FIGURE 1.2 — Typical components of an analytical process

tritium in drinking water, then the analytical process for the project will not include sample dissolution and the chemical separation of the radionuclide of concern. It is important to identify the relevant activities of the analytical process for a particular project early in the planning phase. Once the activities have been identified, the analytical requirements of the activities can be established, which will ultimately lead to defining how the activities will be accomplished through the selection or development of written procedures.

1.4.5 Analytical Protocol

MARLAP uses the term "analytical protocol" to refer to a compilation of specific procedures and methods that are performed in succession for a particular analytical process. For example, a protocol for the analysis of drinking water samples for tritium would be comprised of the set of procedures that describe the relevant activities, such as sample tracking, quality control, field sample preparation and preservation, sample receipt and inspection, laboratory sample preparation (if necessary), preparing the samples for counting, counting the samples, and data reduction and reporting. A written procedure may cover one or more of the activities, but it is unlikely that a single procedure will cover all of the activities of a given analytical process. With a performance-based approach, there may be a number of alternative protocols that might be appropriate for a particular analytical process. Selecting or developing an analytical protocol requires knowledge of the particular analytical process, as well as an understanding of the analytical data requirements developed during the project planning phase.

1.4.6 Analytical Method

A major component of an analytical protocol is the *analytical method*, which normally includes written instructions for sample digestion, chemical separation (if required), and counting. It is recognized that in many instances the analytical method may cover many of the activities of a particular analytical process. Therefore attention is naturally focused on the selection or development of an analytical method. However, many analytical methods do not address activities such as field preparation and preservation, certain aspects of laboratory preparation, laboratory subsampling, etc., which are often important activities within an analytical process. The analytical protocol is generally more inclusive of the activities that make up the analytical process than the analytical method.

1.4.7 Uncertainty and Error

An important aspect of sampling and measurement is uncertainty. The term "uncertainty" has different shades of meaning in different contexts, but generally the word refers to a lack of complete knowledge about something of interest. In the context of metrology (the science of measurement), the more specific term "measurement uncertainty" often will be used. "Uncertainty (of measurement)" is defined in the *Guide to the Expression of Uncertainty in Measurement* (ISO 1995—"GUM") as a "parameter, associated with the result of a measurement, that characterizes the dispersion of values that could reasonably be attributed to the measurand." The "measurand" is the quantity being measured. MARLAP recommends the terminology and methods of GUM for describing, evaluating, and reporting measurement uncertainty. The uncertainty of a measured value is typically expressed as an estimated standard deviation, called a "standard uncertainty" (or "one-sigma uncertainty"). The standard uncertainty of a calculated result usually is obtained by propagating the standard uncertainties of a number of other measured values, and in this case, the standard uncertainty is called a "combined standard uncertainty." The combined standard uncertainty may be multiplied by a specified factor called a "coverage factor" (e.g., 2 or 3) to obtain an "expanded uncertainty" (a "two-sigma" or "three-sigma" uncertainty), which describes an interval about the result that can be expected to contain the true value with a specified high probability. MARLAP recommends that either the combined standard uncertainty or an expanded uncertainty be reported with every result. Chapter 19 discusses the terminology, notation, and methods of GUM in more detail and provides guidance for applying the concepts to radioanalytical measurements.

While measurement uncertainty is a parameter associated with an individual result and is calculated after a measurement is performed, MARLAP uses the term "method uncertainty" to refer to the predicted uncertainty of a measured value that likely would result from the analysis of a sample at a specified analyte concentration. Method uncertainty is a method performance characteristic much like the detection capability of a method. Reasonable values for both characteristics can be predicted for a particular method based on typical values for certain parameters and on information and assumptions about the samples to be analyzed. These predicted values can be used in the method selection process to identify the most appropriate method based on a project's data requirements. Chapter 3 provides MARLAP's recommendations for deriving analytical protocol selection criteria based on the required method uncertainty and other analytical requirements.

When a decisionmaker bases a decision on the results of measurements, the measurement uncertainties affect the probability of making a wrong decision. When sampling is involved, sampling statistics also contribute to the probability of a wrong decision. Because decision errors are possible, there is uncertainty in the decisionmaking process. MARLAP uses the terms "decision uncertainty" or "uncertainty of the decision" to refer to this type of uncertainty. Decision uncertainty is usually expressed as the estimated probability of a decision error under specified assumptions. Appendix B discusses decision uncertainty further in the context of the DQO process.

A concept that should not be confused with uncertainty is error. In general, error refers to something that deviates from what is correct, right or true. In terms of measurements such as laboratory analyses, the difference between the measured result and the actual value of the measurand is the error of the measurement. Because the actual value of the measurand is generally not known, the measurement error cannot be determined. Therefore, the error of a measurement is primarily a theoretical concept with little practical use. However, the measurement uncertainty, which provides an estimated bound for the likely size of the measurement error, is very useful and plays a key role in MARLAP's performance-based approach.

1.4.8 Precision, Bias, and Accuracy

Analytical data requirements often have been described in terms of precision and bias. Precision is usually expressed as a standard deviation, which measures the dispersion of measured values about their mean. It is sometimes more natural to speak of "imprecision," because larger values of the standard deviation indicate less precision. MARLAP considers bias to be a persistent difference between the measured result and the true value of the quantity being measured, which does not vary if the measurement is repeated. If the measurement process is in statistical control, then precision may be improved by averaging the results of many independent measurements of the same quantity. Bias is unaffected by averaging (see Section 6.5.5.7).

A bias in a data set may be caused by measurement errors that occur in steps of the measurement process that are not repeated, such as the determination of a half-life. Imprecision may be caused by measurement errors in steps that are repeated many times, such as weighing, pipetting, and radiation counting. However, distinguishing between bias and precision is complicated by the fact that some steps in the process, such as instrument calibration or tracer preparation, are repeated at frequencies less than those of other steps, and the measurement errors in seldom repeated steps may affect large blocks of data. Consequently, measurement errors that produce apparent biases in small data sets might adversely affect precision in larger data sets.

Because the same type of measurement error may produce either bias or precision, depending on one's point of view, the concept of measurement uncertainty, described in Section 1.4.7, treats all types of measurement error alike and combines estimates of their magnitudes into a single numerical parameter (i.e., combined standard uncertainty). The concepts of precision and bias are useful in context when a measurement process or a data set consisting of many measurement results is considered. When one considers only a single measurement result, the concept of measurement uncertainty tends to be more useful than the concepts of precision and bias. Therefore, it is probably best to consider precision and bias to be characteristics of the measurement process or of the data set, and to consider measurement uncertainty to be an aspect of each individual result.

Quality control samples are analyzed for the purpose of assessing precision and bias. Spiked samples and method blanks are typically used to assess bias, and duplicates are used to assess precision. Because a single measurement of a spike or blank cannot in principle distinguish between precision and bias, a reliable estimate of bias requires a data set that includes many such measurements.

Different authors have given the word *accuracy* different technical definitions, expressed in terms of bias and precision. MARLAP avoids all of these technical definitions and uses the term "accuracy" in its common, ordinary sense, which is consistent with its definition in the *International Vocabulary of Basic and General Terms in Metrology* (ISO, 1993). In MARLAP's terminology, the result of a measurement is "accurate" if it is close to the true value of the quantity being measured. Inaccurate results may be caused either by bias or precision in the measurement process.

While it is recognized that the terms bias, precision, and accuracy are commonly used in data collection activities, these terms are used somewhat sparingly in this manual. MARLAP emphasizes and provides guidance in the use of measurement uncertainty as a means of establishing analytical data requirements and in the evaluation of single measurement results.

1.4.9   Performance Objectives: Data Quality Objectives and Measurement Quality Objectives

One of the outputs of a directed planning process is DQOs for a project or program. DQOs are qualitative and quantitative statements that clarify the study objectives, define the most appropriate type of data to collect, determine the most appropriate conditions from which to collect the data, and specify tolerable limits on decision error rates (ASTM D5792; EPA, 2000). DQOs apply to all data collection activities associated with a project or program, including sampling and analysis. In particular, DQOs should encompass the "total uncertainty" resulting from all data collection activities, including analytical and sampling activities.

From an analytical perspective, a process of developing the analytical data requirements from the DQOs of a project is essential. These analytical data requirements serve as measurement performance criteria or objectives of the analytical process. MARLAP refers to these performance objectives as "measurement quality objectives" (MQOs). The MARLAP Manual provides guidance on developing the MQOs from the overall project DQOs (Chapter 3). MQOs can be viewed as the analytical portion of the DQOs and are therefore project-specific. MARLAP provides guidance on developing MQOs during project planning for select method performance characteristics, such as method uncertainty at a specified concentration; detection capability; quantification capability; specificity, or the capability of the method to measure the analyte of concern in the presence of interferences; range; ruggedness, etc. An MQO is a statement of a performance objective or requirement for a particular method performance characteristic. Like DQOs, MQOs can be quantitative and qualitative statements. An example of a quantitative MQO would be a statement of a required method uncertainty at a specified radionuclide concentration, such as the action level—i.e., "a method uncertainty of 3.7 Bq/kg (0.10 pCi/g) or less is required at the action level of 37 Bq/kg (1.0 pCi/g)." An example of a qualitative MQO would be a statement of the required specificity of the analytical protocol—the ability to analyze for the radionuclide of concern given the presence of interferences—i.e., "the protocol must be able to quantify the amount of ^{226}Ra present given high levels of ^{235}U in the samples."

The MQOs serve as measurement performance criteria for the selection or development of analytical protocols and for the initial evaluation of the analytical protocols. Once the analytical protocols have been selected and evaluated, the MQOs serve as criteria for the ongoing and final evaluation of the laboratory data, including data verification, data validation, and data quality assessment. In a performance-based approach, analytical protocols are either selected or rejected for a particular project, to a large measure, based on their ability or inability to achieve the stated MQOs. Once selected, the performance of the analytical protocols is evaluated using the project-specific MQOs.

1.4.10 Analytical Protocol Specifications

MARLAP uses the term "analytical protocol specifications" (APSs) to refer to the output of a directed planning process that contains the project's analytical data requirements in an organized, concise form. In general, there will be an APS developed for each analysis type. These specifications serve as the basis for the evaluation and selection of the analytical protocols that will be used for a particular project. In accordance with a performance-based approach, the APSs contain only the minimum level of specificity required to meet the project's analytical data requirements without dictating exactly how the requirements are to be met. At a minimum, the APSs should indicate the analyte of interest, the matrix of concern, the type and frequency of quality control (QC) samples, and provide the required MQOs and any specific analytical process requirements, such as chain-of-custody for sample tracking. In most instances, a particular APS document would be a one-page form (see Chapter 3, Figure 3.2). Depending on the particular project, a number of specific analytical process requirements may be included. For example, if project or process knowledge indicates that the radionuclide of interest exists in a refractory form, then the APSs may require a fusion step for sample digestion.

Within the constraints of other factors, such as cost, MARLAP's performance-based approach allows the use of any analytical protocol that meets the requirements in the APSs. The APSs—in particular the MQOs—are used to select and evaluate the analytical protocols. Once the analytical protocols have been selected and evaluated, the APSs then serve as criteria for the ongoing and final evaluation of the laboratory data, including data verification, data validation, and data quality assessment.

1.4.11 The Assessment Phase

The MARLAP Manual provides guidance for the assessment phases for those projects that require the laboratory analysis of radionuclides. The guidance on the assessment phase of projects focuses on three major activities: data verification, data validation, and data quality assessment.

Data verification assures that laboratory conditions and operations were compliant with the statement of work and any appropriate project plan documents (e.g., Quality Assurance Project

Plan), which may reference laboratory documents such as laboratory standard operating procedures. Verification compares the material delivered by the laboratory to these requirements (compliance) and checks for consistency and comparability of the data throughout the data package, correctness of calculations, and completeness of the results to ensure that all necessary documentation is available. The verification process usually produces a report identifying which requirements are not met. The verification report may be used to determine payment for laboratory services and to identify problems that should be investigated during data validation. Verification works iteratively and interactively with the generator (i.e., laboratory) to assure receipt of all available, necessary data. Although the verification process identifies specific problems, the primary function should be to apply appropriate feedback resulting in corrective action improving the analytical services before the work is completed.

Validation addresses the reliability of the data. The validation process begins with a review of the verification report and laboratory data package to screen the areas of strength and weakness of the data set. The validator evaluates the data to determine the presence or absence of an analyte and the uncertainty of the measurement process for contaminants of concern. During validation, the technical reliability and the degree of confidence in reported analytical data are considered. Validation "flags" (i.e., qualifiers) are applied to data that do not meet the acceptance criteria established to assure data meet the needs of the project. The product of the validation process is a validation report noting all data sufficiently inconsistent with the validation acceptance criteria in the expert opinion of the validator. The appropriate data validation tests should be established during the project planning phase.

Data quality assessment (DQA), the third and final step of the assessment phase, is defined as the "scientific and statistical evaluation of data to determine if data are of the right type, quality, and quantity to support their intended use." DQA is more global in its purview than the previous verification and validation steps. DQA, in addition to reviewing the issues raised during verification and validation, may be the first opportunity to review other issues, such as field activities and their impact on data quality and usability. DQA should consider the combined impact of all project activities in making a data usability determination, which is documented in a DQA report.

1.5   The MARLAP Process

An overarching objective of the MARLAP Manual is to provide a framework and information for the selection, development, and evaluation of analytical protocols and the resulting laboratory data. The MARLAP process is a performance-based approach that develops APSs and uses these requirements as criteria for the analytical protocol selection, development and evaluation processes, and for the evaluation of the resulting laboratory data. This process, which spans the three phases of the data life cycle for a project—planning, implementation and assessment—is the basis for achieving MARLAP's basic goal of ensuring that radioanalytical data will meet a project's data requirements. A brief overview of this process, which is referred to as the MARLAP process and is the focus of Part I of the manual, is provided below.

The MARLAP process starts with a directed planning process. Within a directed planning process, key analytical issues based on the project's particular analytical processes are discussed and resolved. The resolution of these key analytical issues produces the APSs, which include the MQOs. The APSs are documented in project plan documents (e.g., Quality Assurance Project Plans, Sampling and Analysis Plans). A SOW is then developed that contains the APSs. The laboratories receiving the SOW respond with proposed analytical protocols based on the requirements of the APSs and provide evidence that the proposed protocols meet the performance criteria in the APSs. The proposed analytical protocols are initially evaluated by the project manager or designee to determine if they will meet the requirements in the APSs. If the proposed analytical protocols are accepted, the project plan documents are updated by the inclusion or referencing of the actual analytical protocols to be used. During analyses, resulting sample and QC data will be evaluated primarily using MQOs from the respective APSs. Once the analyses are completed, an evaluation of the data will be conducted, including data verification, data validation, and data quality assessment with the respective MQOs serving as criteria for evaluation. The role of the APSs (particularly the MQOs, which make up an essential part of the APSs) in the selection, development, and evaluation of the analytical protocols and the laboratory data is to provide a critical link between the three phases of the data life cycle of a project. This linkage helps to ensure that radioanalytical laboratory data will meet a project's data requirements, and that the data are of known quality appropriate for their intended use. The MARLAP process is illustrated in Figure 1.3. Although the diagram represents the MARLAP process in a linear fashion, it is important to note that the process is an iterative one, and there can be many variations on this stylized diagram. Also, the phases shown at the right of Figure 1.3 only illustrate the relationship of the MARLAP process to the data life cycle.

1.6   Structure of the Manual

MARLAP is divided into two main parts. Part I provides guidance on implementing the MARLAP process as described in Section 1.5. This part of the manual focuses on the sequence of steps involved when using a performance-based approach for projects requiring radioanalytical laboratory work starting with a directed planning process and ending with DQA. Part I provides the overall guidance for using a performance-based approach for all three phases of a project. A more detailed overview of Part I is provided in Section 1.6.1. While the primary users for most of the Part I chapters are project managers and planners, other groups can benefit from the guidance in Part I.

Part II of the manual provides information on the laboratory analysis of radionuclides to support a performance-based approach. Part II provides guidance and information on the various activities performed at radioanalytical laboratories, such as sample preparation, sample

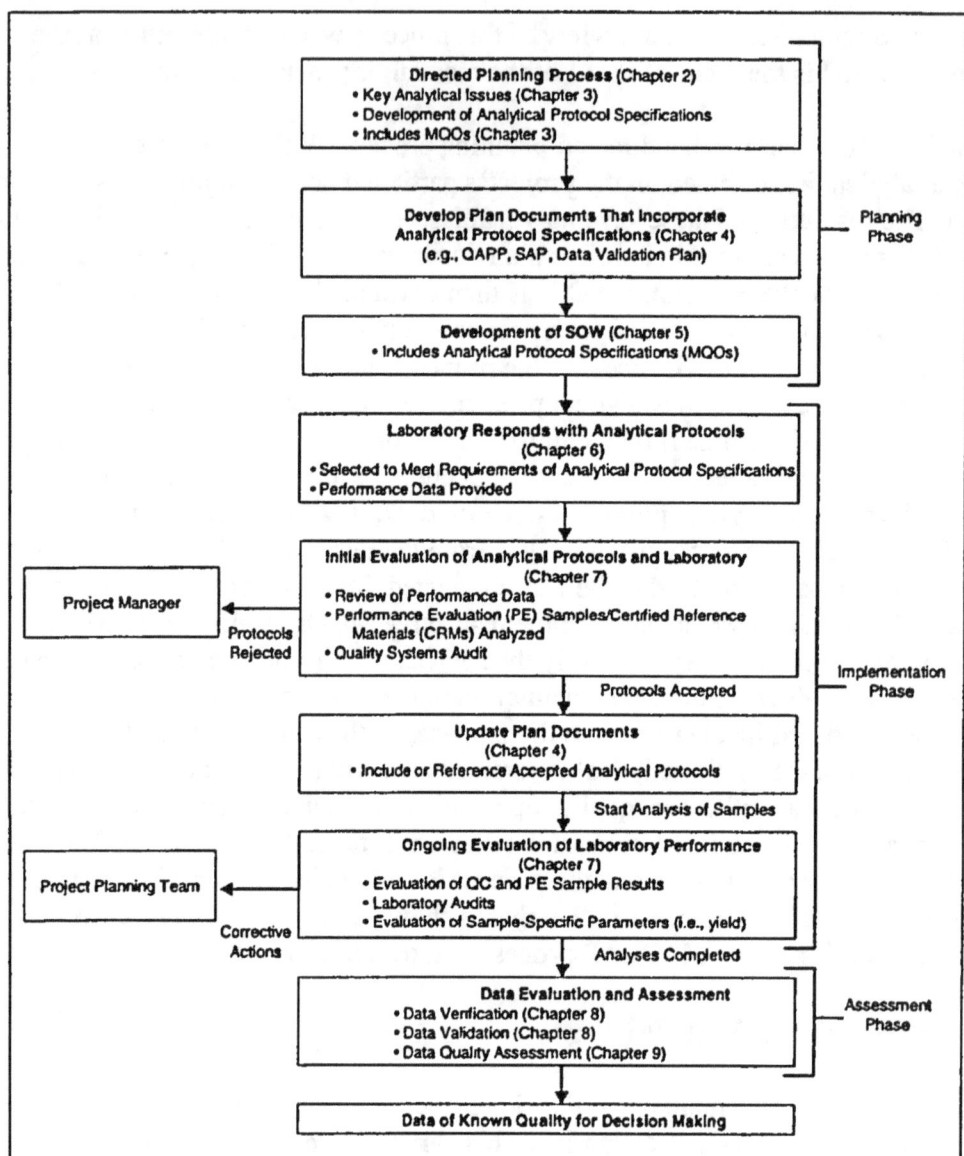

FIGURE **1.3** — The MARLAP process

dissolution, chemical separations, preparing sources for counting, nuclear counting, etc. The primary users for Part II are laboratory personnel. Using the overall framework provided in Part I, the material in Part II can be used to assist project planners, managers, and laboratory personnel in the selection, development, evaluation, and implementation of analytical protocols for a particular project or program. Figure 1.4 illustrates the interaction of the project manager and the laboratory using key MARLAP terms and processes. A more detailed overview of Part II is provided in Section 1.6.2. In addition to Part I and Part II, MARLAP has several appendices that support both Part I and Part II of the manual. An overview of the appendices is provided in Section 1.6.3 of this chapter.

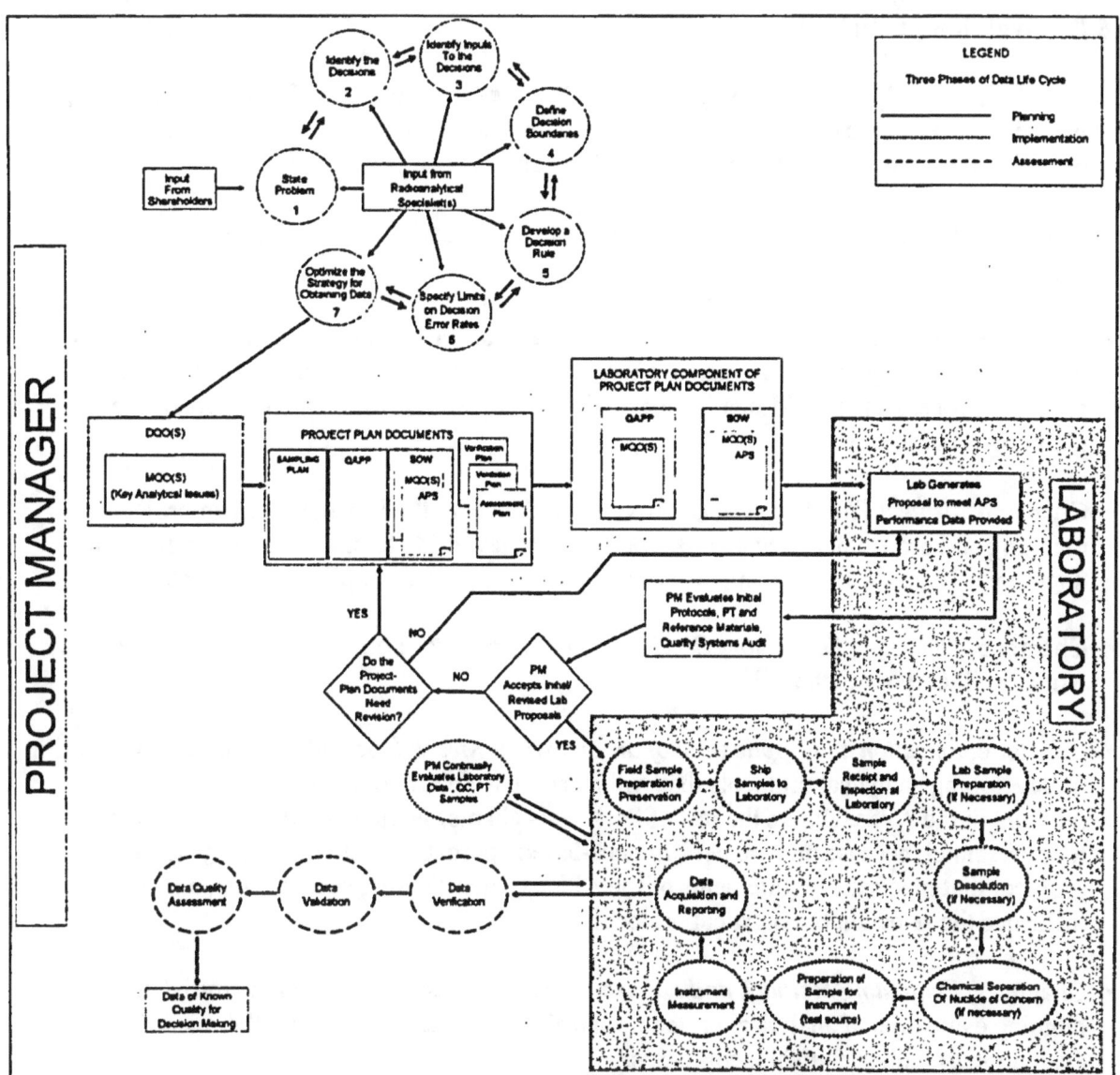

FIGURE 1.4 — Key MARLAP terms and processes

Because of the structure and size of the manual, most individuals will naturally focus on those chapters that provide guidance in areas directly related to their work. Therefore, to help ensure that key concepts are conveyed to the readers, there is some material is repeated, often in very similar or even the same language, throughout the manual.

1.6.1 Overview of Part I

Figure 1.3, the MARLAP Process on page 1-14, illustrates the sequence of steps that make up a performance-based approach for the planning, implementation, and assessment phases of radioanalytical projects. The remainder of Part I closely tracks this sequence:

- Chapter 2, *Project Planning Process,* provides an overview of the directed planning process and its outputs.

- Chapter 3, *Key Analytical Planning Issues and Developing Analytical Protocol Specifications,* describes key analytical planning issues that need to be addressed during a directed planning process and provides guidance on developing APSs, which are outputs of the planning process.

- Chapter 4, *Project Plan Documents,* provides guidance on the linkage between project planning and project plan documents, with an overview of different types of project plan documents (e.g., work plans, quality assurance project plans, sampling and analysis plans).

- Chapter 5, *Obtaining Laboratory Services,* provides guidance on developing a statement of work that incorporates the APSs.

- Chapter 6, *Selection and Application of an Analytical Method,* provides guidance on selecting or developing analytical protocols that will meet the MQOs and other requirements as outlined in the APSs. Unlike the rest of Part I, this chapter is intended primarily for laboratory personnel, because under a performance-based approach, a laboratory may use any protocol that meets the requirements of the APSs. (Other factors, such as cost, also will influence the selection of analytical protocols.)

- Chapter 7, *Evaluating Methods and Laboratories,* provides guidance on the initial and ongoing evaluation of analytical protocols and also provides guidance on the overall evaluation of radioanalytical laboratories.

- Chapter 8, *Radiochemical Data Verification and Validation,* provides an overview of the data evaluation process, provides general guidelines for data verification and validation, and provides "tools" for data validation.

- The last chapter of Part I, Chapter 9, *Data Quality Assessment,* discusses data quality assessment and provides guidance on linking data quality assessment to the planning process.

1.6.2 Overview of Part II

The chapters in Part II are intended to provide information on the laboratory analysis of radionuclides. The chapters provide information on many of the options available for analytical protocols, and discuss common advantages and disadvantages of each. The chapters highlight common analytical problems and ways to identify and correct them. The chapters also serve to educate the reader by providing a detailed explanation of the typical activities performed at a radioanalytical laboratory. Consistent with a performance-based approach, the chapters in Part II do not contain detailed step-by-step instructions on how to perform certain laboratory tasks, such as the digestion of a soil sample. The chapters do contain information and guidance intended to assist primarily laboratory personnel in deciding on the best approach for a particular laboratory task. For example, while the chapter on sample dissolution does not contain step-by-step instructions on how to dissolve a soil sample, it does provide information on acid digestion, fusion techniques, and microwave digestion, which is intended to help the reader select the most appropriate technique or approach for a particular project.

The primary audience for Part II is laboratory personnel and the chapters generally contain a significant amount of technical information. While the primary target audience is laboratory personnel, other groups, such as project planners and managers, can benefit from the guidance in Part II. Listed below are the chapters that make up Part II of the manual. It should be noted that Part II of the manual does not provide specific guidance for some laboratory activities that are common to all laboratories, such as laboratory quality assurance, and laboratory health and safety practices. This is primarily due to the fact that these activities are not unique to radioanalytical laboratories and considerable guidance in these areas already exists.

Chapter 10	Field and Sampling Issues That Affect Laboratory Measurements
Chapter 11	Sample Receipt, Inspection, and Tracking
Chapter 12	Laboratory Sample Preparation
Chapter 13	Sample Dissolution
Chapter 14	Separation Techniques
Chapter 15	Quantification of Radionuclides
Chapter 16	Data Acquisition, Reduction, and Reporting for Nuclear Counting Instrumentation
Chapter 17	Waste Management in a Radioanalytical Laboratory
Chapter 18	Laboratory Quality Control
Chapter 19	Measurement Uncertainty
Chapter 20	Detection and Quantification Capabilities

Chapters 10 through 16 provide information on the typical components of an analytical process in the order in which activities that make up an analytical process are normally performed. While not providing step-by-step procedures for activities such as sample preservation, sample digestion, nuclear counting, etc., the chapters do provide an overview of options available for the various activities and importantly, provide information on the appropriateness of the assorted options under a variety of conditions.

Chapter 17, *Waste Management in a Radioanalytical Laboratory*, provides an overview of many of the regulations for waste disposal and provides guidance for managing wastes in a radioanalytical laboratory. Chapter 18, *Laboratory Quality Control,* provides guidance on monitoring key laboratory performance indicators as a means of determining if a laboratory's measurement processes are in control. The chapter also provides information on likely causes of excursions for selected laboratory performance indicators, such as chemical yield, instrument background, quality control samples, etc.

Chapters 19, *Measurement Uncertainty*, and 20, *Detection and Quantification Capabilities*, provide information on statistical principles and methods applicable to radioanalytical measurements, calibrations, data interpretation, and quality control. Topics covered in the chapter include detection and quantification, measurement uncertainty, and procedures for estimating uncertainty.

1.6.3   Overview of the Appendices

Seven appendices provide additional details on specific topics discussed in Part I and Part II chapters. Appendices A through E primarily support Part I chapters (project planning issues) and Appendices F and G primarily support the chapters in Part II (laboratory implementation issues).

* Appendix A, *Directed Planning Approaches,* provides an overview of a number of directed planning processes and discusses some common elements of the different approaches.

* Appendix B, *The Data Quality Objective Process,* provides an expanded discussion of the Data Quality Objectives Process including detailed guidance on setting up a "gray region" and establishing tolerable decision error rates.

* Appendix C, *Measurement Quality Objectives for Method Uncertainty and Detection and Quantification Capability,* provides the rationale and guidance for developing MQOs for select method performance characteristics.

* Appendix D, *Content of Project Plan Documents*, provides guidance on the appropriate content of plan documents.

* Appendix E, *Contracting Laboratory Services,* contains detailed guidance on contracting laboratory services.

* Appendix F, *Laboratory Subsampling,* provides information on improving and evaluating laboratory subsampling techniques.

• Appendix G, *Statistical Tables,* provides a compilation of statistical tables.

1.7   References

American Society for Testing and Materials (ASTM) D5792. *Standard Practice for Generation of Environmental Data Related to Waste Management Activities: Development of Data Quality Objectives*, 1995.

International Organization for Standardization (ISO). 1993. *International Vocabulary of Basic and General Terms in Metrology.* ISO, Geneva, Switzerland.

International Organization for Standardization (ISO). 1995. *Guide to the Expression of Uncertainty in Measurement.* ISO, Geneva, Switzerland.

MARSSIM. 2000. *Multi-Agency Radiation Survey and Site Investigation Manual, Revision 1.* NUREG-1575 Rev 1, EPA 402-R-97-016 Rev1, DOE/EH-0624 Rev1. August. Available at www.epa.gov/radiation/marssim/.

U.S. Environmental Protection Agency (EPA). 2000. *Guidance for the Data Quality Objective Process* (EPA QA/G-4). EPA/600/R-96/055, Washington, DC. Available at www.epa.gov/quality1/qa_docs.html.

2 PROJECT PLANNING PROCESS

2.1 Introduction

Efficient environmental data collection activities depend on successfully identifying the type, quantity, and quality of data needed, as well as how the data will be used to support the decision-making process. MARLAP recommends the use of a directed or systematic planning process. These planning processes provide a logical framework for establishing well-defined, achievable objectives within a cost-effective, technically sound, and defensible sampling and analysis design. They also balance the data user's tolerance for uncertainty in the decision process with the available resources for obtaining data to support a decision. *MARLAP uses the term "directed planning" to emphasize that the planning process, in addition to having a framework or structure (i.e., it is systematic), is focused on defining the data needed to support an informed decision for a specific project.*

This chapter provides an overview of the directed planning process. It promotes:

1. Directed planning as a tool for project management to identify and document the data quality objectives (DQOs)—qualitative and quantitative statements that define the project objectives and the tolerable rate of making decision errors, which in turn will be used to establish the quality and quantity of data needed to support the decision—and the measurement quality objectives (MQOs) that define the analytical data requirements appropriate for decision-making;

2. The involvement of technical experts—radioanalytical specialists, in particular—in the planning process; and

3. Integration of the outputs from the directed planning process into the implementation and assessment phases of the project through documentation in project plan documents, the analytical statement of work (SOW), and the data assessment plans (e.g., for data verification, data validation, and data quality assessment).

MARLAP uses the terms "DQOs" and "MQOs," as defined above and in Section 1.4.9, because of their widespread use in environmental data collection activities. These concepts may be expressed by other terms, such as "decision performance criteria" or "project quality objectives" for DQOs and "measurement performance criteria" or "data quality requirements" for MQOs.

Contents	
2.1 Introduction	2-1
2.2 The Importance of Directed Project Planning	2-2
2.3 Directed Project Planning Processes	2-3
2.4 The Project Planning Team	2-6
2.5 Directed Planning Process and Role of the Radioanalytical Specialists	2-8
2.6 Results of the Directed Planning Process	2-17
2.7 Project Planning and Project Implementation and Assessment	2-19
2.8 Summary of Recommendations	2-22
2.9 References	2-23

Section 2.2 discusses the importance of directed project planning. The approach, guidance, and common elements of directed planning are discussed in Section 2.3. The project planning team is addressed in Section 2.4, Section 2.5 describes the elements of project planning from the perspective of the radioanalytical specialists. The results of the planning process are discussed in Section 2.6. Section 2.7 presents the next steps of the planning phase of the project, which document the results of the planning process and link the results of the planning process to the implementation and assessment phases of data collection activities. Additional discussion on the planning process in Chapter 3, *Key Analytical Planning Issues and Developing Analytical Protocol Specifications*, focuses on project planning from the perspective of the analytical process and the development of Analytical Protocol Specifications (APSs).

The environmental data collection process consists of a series of elements: planning, developing, and updating project plan documents; contracting for services; sampling; analysis; data verification; data validation; and data quality assessment (see Section 1.4.1, "Data Life Cycle"). These elements are interrelated because sampling and analysis cannot be performed efficiently or resources allocated effectively without first identifying data needs during planning. Linkage and integration of the data collection process elements are essential to the success of the environmental data collection activity.

2.2   The Importance of Directed Project Planning

A directed planning process has several notable strengths. It brings together the stakeholders (see box), decisionmakers, and technical experts at the beginning of the project to obtain commitment for the project and a consensus on the nature of the problem and the desired decision. MARLAP recognizes the need for a directed planning process that involves radioanalytical and other technical experts as principals to ensure the decisionmakers' data requirements and the results from the field and radioanalytical laboratory are linked effectively. Directed planning enables each participant to play a constructive role in clearly defining:

- The problem that requires resolution;
- What type, quantity, and quality of data the decisionmaker needs to resolve that problem;
- Why the decisionmaker needs that type and quality of data;
- What are the tolerable decision error rates; and
- How the decisionmaker will use the data to make a defensible decision.

A directed planning process encourages efficient planning by framing and organizing complex issues. The process promotes timely, open, and effective communication among the stakeholders, resulting in well-conceived and documented plans. Because of the emphasis on documentation, directed planning also provides project management with a more efficient and consistent transfer of knowledge to new project members.

Example of Stakeholders for a Cleanup Project

A stakeholder is anyone with an interest in the outcome of an activity. For a cleanup project, some of the stakeholders could be:

- Federal, regional, state, and tribal environmental agencies with regulatory interests (e.g., NRC and EPA).

- States with direct interest in transportation, storage and disposition of wastes, and other related issues.

- City and county governments concerned with the operations and safety at sites as well as economic development and site transition.

- Site Advisory Boards, citizens groups, licensees, special interest groups, responsible parties, and other members of the public with interest in cleanup activities at the site.

A directed planning process focuses on collection of only those data needed to address the appropriate questions and support defensible decisions. Directed planning helps to eliminate poor or inadequate sampling and analysis designs that require analysis of (1) too few or too many samples, (2) samples that will not meet the needs of the project, or (3) inappropriate quality control (QC) samples. During directed planning, which is an iterative process, the sufficiency of existing data is evaluated, and the need for additional data to fill the gaps, as well as the desired quality of the additional data, are determined. By defining the MQOs, directed planning provides input for obtaining appropriate radioanalytical services, which balance constraints and the required data quality.

The time invested in preliminary planning can greatly reduce resource expenditure in the more resource-intensive execution phase of the project. Less overall time (and money) is expended when early efforts are focused on defining (and documenting) the project's objectives (DQOs), technically based, project-specific analytical data needs (MQOs and any specific analytical process requirements), and measures of performance for the assessment phase of the data collection activity.

2.3 Directed Project Planning Processes

The recognition of the importance of project planning has resulted in the development of a variety of directed planning approaches. MARLAP does not endorse any one planning approach. Users of this manual are encouraged to consider the available approaches and choose a directed planning process that is appropriate to their project and agency. Appendix A, *Directed Planning Approaches*, provides brief descriptions of several directed planning processes.

Section 2.3.1 discusses a graded approach to project planning, and existing standards and guidance are briefly summarized in Section 2.3.2. An overview of common elements of project planning is discussed in Section 2.3.3. The elements of project planning are discussed in detail in Section 2.5.

2.3.1 A Graded Approach to Project Planning

The sophistication, the level of QC and oversight, and the resources invested should be appropriate to the project (i.e., a "graded approach"). Directed planning for small or less complex projects follows the logic of the process but will proceed faster and involve fewer people. The goal still is to (1) plan properly to collect only the data needed to meet the objectives of the project and (2) establish the measures of performance for the implementation and assessment phases of the data life cycle of the project.

2.3.2 Guidance on Directed Planning Processes

The following national standards related to directed project planning for environmental data collection are available:

- *Standard Practice for Generation of Environmental Data Related to Waste Management Activities: Development of Data Quality Objectives* (ASTM D5792), which addresses the process of development of data quality objectives for the acquisition of environmental data. This standard describes the DQO process in detail.

- *Standard Provisional Guide for Expedited Site Characterization of Hazardous Waste Contaminated Sites* (ASTM PS85), which describes the Expedited Site Characterization (ESC) process used to identify all relevant contaminant migration pathways and determine the distribution, concentration and fate of the contaminants for the purpose of evaluating risk, determining regulatory compliance, and designing remediation systems.

- *Standard Guide for Site Characterization for Environmental Purposes with Emphasis on Soil, Rock, the Vadose Zone and Ground Water* (ASTM D5730), which covers a general approach to planning field investigations using the process of defining one or more conceptual site models that is useful for any type of environmental reconnaissance or investigation plan with a primary focus on the surface and subsurface environment.

- *Standard Guide for Quality Planning and Field Implementation of a Water Quality Measurements Program* (ASTM D5612), which defines criteria and identifies activities that may be required based on the DQOs.

- *Standard Guide for Planning and Implementing a Water Monitoring Program* (ASTM D5851), which provides a procedural flowchart for planning the monitoring of point and nonpoint sources of pollution of water resources (surface or ground water, rivers, lakes or estuaries).

Several directed planning approaches have been implemented by the federal sector for environmental data collection activities. MARLAP does not endorse a single planning approach and project planners should be cognizant of their agency's requirements for planning. The following guidance is available:

- EPA developed the DQO process (EPA, 2000a) and has tailored DQO process guidance for specific programmatic needs of project planning under the Comprehensive Environmental Response, Compensation, and Liability Act of 1980 (CERCLA/Superfund) (EPA, 1993) and for site-specific remedial investigation feasibility study activities (EPA, 2000b).

- The U. S. Army Corps of Engineers Technical Project Planning (TPP) Process (USACE, 1998) was developed for technical projects planning for hazardous, toxic and radioactive waste sites.

- DOE has developed the Streamlined Approach for Environmental Restoration (SAFER) (DOE, 1993) for its environmental restoration activities.

- Planning guidance, including decision frameworks, for projects demonstrating compliance with a dose- or risk-based regulation is available for final status radiological surveys (MARSSIM, 2000) and radiological criteria for license termination (NRC, 1998a; NRC, 1998b).

Additional information on the DQO process (ASTM D5792; EPA, 2000a) is presented in Appendix B, *The Data Quality Objectives Process.*

2.3.3 Elements of Directed Planning Processes

Environmental data collection activities require planning for the use of data in decisionmaking. The various directed planning approaches, when applied to environmental data collection activities, address common planning considerations. Some common elements of the planning processes are:

1. *State the problem*: Describe clearly the problem(s) facing the stakeholder or customer.

2. *Identify the Decision*: Define the decision(s) or the alternative actions that will address the problem(s) or concern and satisfy the stakeholder/customer, and define the inputs and boundaries to the decision.

3. *Specify the Decision Rule and the Tolerable Decision Error Rates*: Develop a decision rule to get from the problem or concern to the desired decision and define the limits on the decision error rates that are acceptable to the stakeholder/customer. The decision rule can take the form of "if ...then..." statements for choosing among decisions or alternative actions.

4. *Optimize the Strategy for Obtaining Data*: Determine the optimum, cost-effective way to reach the decision while satisfying the desired quality of the decision. Define the quality of the data that are required for the decision by establishing specific, quantitative and qualitative analytical performance measures (e.g, MQOs). Define the process and criteria to evaluate the suitability of the data to support their intended use (data quality assessment).

The objective of the directed project planning process for environmental data collection activities is to reach consensus among the stakeholders on defining the problem, the full range of possible solutions, the desired decision, the optimal data collection strategy, and performance measures for implementation and assessment phases of the project. If only a cursory job is done defining the problem or the desired results, the consequence will be the development of a design that may be technically sound but answers the wrong question, may answer the question only after the collection of significant quantities of unnecessary data, or may collect insufficient data to answer the question.

The key outputs of the directed planning process are DQOs: qualitative and quantitative statements that define the project objectives and the tolerable decision error rates that will be used as the basis for establishing the quality and quantity of data needed to support the decision. *The MQOs and the decisions on key analytical planning issues will provide the framework for Analytical Protocol Specifications*. The MQOs and the tolerable decision error rates will provide the basis for the data assessment phase (data validation and data quality assessment). Important analytical planning issues and APSs are discussed in Chapter 3, *Key Analytical Planning Issues and Developing Analytical Protocol Specifications*.

2.4   The Project Planning Team

The number of participants in the project planning process, and their respective disciplines, will vary depending on the nature and scope of the project, but in most cases a multidisciplinary team will be required. The project planning team should consist of all the parties who have a vested interest or can influence the outcome (stakeholders). A key to successful directed planning of environmental projects is getting the data users and data suppliers to work together early in the process to understand each other's needs and requirements, to agree on the desired end product, and to establish lines of communication. Equally important is having integrated teams of operational and technical experts. These experts will determine whether the problem has been sufficiently defined and if the desired outcomes are achievable. With the input of technical experts early in the planning process, efforts are focused on feasible solutions, and resources are not wasted pursuing unworkable solutions.

2.4.1 Team Representation

Members of the project planning team may include program and project managers, regulators, public representatives, project engineers, health and safety advisors, and specialists in statistics, health physics, chemical analysis, radiochemical analysis, field sampling, quality assurance/ quality control (QA/QC), data assessment, contract and data management, field operation, and other technical specialists. The program or project manager(s) may be a remedial project manager (RPM), a site assessment manager (SAM), or a technical project officer (TPO). Some systematic planning processes, such as Expedited Site Characterization, utilize a core technical team supported as needed by members of larger technical and operational teams. Throughout this document, the combined group of decisionmakers and technical experts is referred to as the "project planning team."

The duration of service for the project planning team members can vary, as can the level of participation required of each member during the various planning phases. While the project planning team may not meet as frequently once the project objectives and the sampling and analysis design have been established, a key point to recognize is that the project planning team should not disband. Rather, the team or a "core group" of the team (including the project manager and other key members) should continue to meet at agreed upon intervals to review the project's progress and to deal with actual project conditions that require changes to the original plan. The availability of a core team also provides the mechanism for the radioanalytical laboratory to receive needed information to clarify questions as they arise.

A key concept built into directed planning approaches is the ability to revisit previous decisions after the initial planning is completed (i.e., during the implementation phases of the environmental data collection process). Even when objectives are clearly established by the project planning team and contingency planning was included in the plan development, the next phases of the project may uncover new information or situations, which require alterations to the data collection strategy. For example, finding significantly different levels of analytes or different analytes than were anticipated based on existing information may require changes in the process. To respond to unexpected events, the project planning team (or the core group) should remain accessible during other phases of the data collection process to respond to questions raised, revisit and revise project requirements as necessary, and communicate the basis for previous assumptions.

2.4.2 The Radioanalytical Specialists

Depending on the size and complexity of the project, MARLAP recognizes that a number of key technical experts should participate on the project planning team and be involved throughout the project as needed. When the problem or concern involves radioactive analytes, it is important that the radioanalytical specialist(s) are part of the project planning team, in addition to radiation health and safety specialists. MARLAP recommends that the radioanalytical specialists be a part of the integrated effort of the project planning team. Throughout this manual, the term "radio-analytical specialists" is used to refer to the radioanalytical expertise needed.

Radioanalytical specialists may provide expertise in (1) radiochemistry and radiation/nuclide measurement systems and (2) the knowledge of the chemical characteristics of the analyte of concern. In particular, the radioanalytical specialist plays a key role in the development of MQOs. The radioanalytical specialists may also provide knowledge about sample transportation issues, preparation, preservation, sample size, subsampling, available analytical protocols and achievable analytical data quality. If more than one person is needed, the specialists members need not be from the same organization. The radioanalytical specialists need not be from the contractual radioanalytical laboratory. *The participation of the radioanalytical specialists is critical to the success of the planning process and the effective use of resources available to the project.*

2.5   Directed Planning Process and Role of the Radioanalytical Specialists

The importance of technical input in a directed planning process becomes apparent when one examines the common difficulties facing the radioanalytical laboratory. Without sufficient input, there is often a disconnect in translating the project planning team's analytical data requirements into laboratory requirements and products. Radioanalytical advice and input during planning, however, help to assure that the analytical protocols selected will satisfy the data requirements, including consideration of time, cost and relevance to the data requirements and budget. The role of the radioanalytical specialists during the early stage of the directed planning process is to focus on whether the desired radionuclides can be measured and the practicality of obtaining the desired analytical data. During the latter part of the process, the radioanalytical specialists can provide specific direction and fine tuning for defining the analytical performance requirements (MQOs) and other items of the APSs.

Planning with input from radioanalytical specialists can help ensure that the data received by the data users will meet the project's DQOs. Common areas that are improved with radioanalytical specialists' participation in project planning include:

- The correct radionuclide is measured;

- MQOs are adequately established and achievable;

- Consideration is given to the impact of half-life and parent/progeny factors;

- The data analysis is not compromised by interferences;

- Unnecessary or overly sophisticated analytical techniques are avoided in favor of analytical techniques appropriate to the required level of measurement uncertainty;

- Optimum radioanalytical variables, such as count time and sample volume, are considered;

- Environmental background levels are considered;

- Chemical speciation is addressed; and

- Consideration is given to laboratory operations (e.g., turnaround time, resources).

These improvements result in an appropriate data collection design, with specified MQOs and any specific analytical process requirements to be documented in the project plan documents and SOWs.

The following sections, using the common planning elements outlined in Section 2.3.3, will discuss the process and results of directed planning in more detail and emphasize the input of radioanalytical specialists. Table 2.1 provides a summary of (1) the information needed by the project planning team, (2) how the radioanalytical specialists participate, and (3) the output or product for each element of the directed planning process. It must be emphasized that a directed planning process is an *iterative*, rather than linear, process. Although the process is presented in discrete sections, the project planning may not progress in such an orderly fashion. The planning team will more precisely define decisions and data needs as the planning progresses and use new information to modify or change earlier decisions until the planning team has determined the most resource-effective approach to the problem. The common planning elements are used for ease of presentation and to delineate what should be covered in planning, not the order of discussion.

TABLE 2.1 — Summary of the directed planning process and radioanalytical specialists participation

Element	Information Needed by The Project Planning Team	Radioanalytical Specialists Participation/Input	Output/Product
1. State the problem	• Key stakeholders and their concerns. • Facts relevant to current situation (e.g., site history, ongoing studies). • Analytes of concern or analytes driving risk. • Matrix of concern. • Regulatory requirements and related issues. • Existing data and its reliability. • Known sampling constraints. • Resources and relevant deadlines.	• Evaluate existing radiological data for use in defining the issues (e.g., analytes of concern). • Assure that the perceived problem is really a concern by reviewing the underlying data that are the basis for the problem definition. • Consider how resource limitations and deadlines will impact measurement choices. • Use existing data to begin to define the analyte of concern and the potential range of concentrations.	• Problem defined with specificity. • Identification of the primary decisionmaker, the available resources, and constraints.

Element	Information Needed by The Project Planning Team	Radioanalytical Specialists Participation/Input	Output/Product
2a. Identify the decision(s)	• Analytical aspects related to the decision. • Possible alternative actions. • Sequence and priority for addressing the problem.	• Provide focus on what analytes need to be measured, considering analyte relationships and background. • Begin to address the feasibility of different analytical protocols. • Begin to identify the items of the APSs. • Begin to determine how sample collection and handling will affect MQOs.	• Statements that link the defined problem to the associated decision(s) and alternative actions.
2b. Identify inputs to the decisions	• All useful existing data. • The general basis for establishing an action level. • Acquisition strategy options (if new data is needed).	• Review the quality and sufficiency of the existing radiological data. • Identify alternate analytes.	• Defined list of needed new data. • Define the characteristic or parameter of interest (analyte/matrix). • Define the action level. • Identify estimated concentration range for analyte(s) of interest.
2c. Define the decision boundaries	• Sampling or measurement timeframe. • Sampling areas and boundaries. • Subpopulations. • Practical constraints on data collection (season, equipment, turnaround time, *etc.*). • Available protocols.	• Identify temporal trends and spatial heterogeneity using existing data. • With the sampling specialists, identify practical constraints that impact sampling and analysis. • Determine feasibility of obtaining new data with current methodology. • Identify limitations of available protocols.	• Temporal and spatial boundaries. • The scale of decision.
3a. Develop a decision rule	• Statistical parameter to describe the parameter of interest and to be compared to the action level. • The action level (quantitative). • The scale of decision-making.	• Identify potentially useful methods. • Estimate measurement uncertainty and detection limits of available analytical protocols.	• A logical, sequential series of steps ("if...then") to resolve the problem.
3b. Specify limits on decision error rates	• Potential consequences of making wrong decisions. • Possible range of the parameter of interest. • Allowable differences between the action level and the actual value. • Acceptable level of decision errors or confidence.	• Assess variability in existing data for decisions on hypothesis testing or statistical decision theory. • Evaluate whether the tolerable decision error rates can be met with available laboratory protocols, or if the error tolerance needs to be relaxed or new methods developed.	• Defined baseline condition (null hypothesis) and quantitative estimates of acceptable decision error rates. • Defined range of possible parameter values where the consequence of a Type II decision error is relatively minor (gray region).

Element	Information Needed by The Project Planning Team	Radioanalytical Specialists Participation/Input	Output/Product
4. Optimize the strategy for obtaining data	• All outputs from all previous elements including parameters (analytes and matrix) of concern, action levels, anticipated range of concentration, tolerable decision error rates, boundaries, resources and practical constraints. • Available protocols for sampling and analysis.	With sampling specialists, consider the potential combinations of sampling and analytical methods, in relation to: • Sample preparation, compositing, subsampling. • Available protocols. • Methods required by regulations (if any). • Detection and quantitation capability. • MQOs achievable by method, matrix and analyte. • QC sample types, frequencies, and evaluation criteria. • Sample volume, field processing, preservatives, and container requirements. • Assure that the MQOs for sample analysis are realistic. • Assure that the parameters for the APSs are complete. • Resources and time frame to develop and validate new method(s), if required.	• The most resource-effective sampling and analysis design that meets the established constraints (i.e., number of samples needed to satisfy the DQOs and the tolerable decision error rates). • A method for testing the hypothesis. • The MQOs and the statement(s) of the APSs. • The process and criteria for data assessment.

2.5.1 State the Problem

The first and most important step of the project planning process is a clear statement of the fundamental issue to be addressed by the project. Correctly implemented, directed planning ensures that a clear definition of the problem is developed before any additional resources are committed. The project planning team should understand clearly the conditions or circumstances that are causing the problem and the reason for making a decision (e.g., threat to human health or environment).

Many projects present a complex interaction of technical, economic and political factors. The problem definition should include a summary of the study objectives, regulatory context, funding and other resources available, relevant deadlines, previous study results, and any obvious data collection design constraints. By participating in the initial stages of the project planning, the radioanalytical specialists will understand the context of the facts and logic used to state the problem and begin to formulate information on applicable protocols based on the project's resources (time and budget).

Existing data (e.g., monitoring data, radioactive materials license, emergency actions, site permit files, operating records) may provide specific details about the identity, concentrations, and geographic, spatial, or temporal distribution of analytes. However, these data should be examined carefully. Conditions may have changed since the data were collected. For example, additional waste disposal may have occurred, the contaminant may have been released or migrated, or decontamination may have been performed. In some cases, a careful review of the historical data by the project planning team will show that a concern is not a problem or the problem can be adequately addressed using the available data.

2.5.2   Identify the Decision

The project planning team will define the decision(s) to be made (or the question the project will attempt to resolve) and the inputs and boundaries to the decision. There may also be multiple decision criteria that have to be met, and each should be clearly defined. For example, the decision may be for an individual survey area rather than the site as a whole, or a phase of the site closure project (scoping, characterization, cleanup operation, or final status survey) rather than the project as a whole because of the different objectives and data requirements.

The decision should be clear and unambiguous. It may be useful to state specifically what conclusions may and may not be drawn from the data. If the study is to be designed, for example, to investigate whether or not a site may be released for use by the general public, then the project planning team may want to specifically exclude other possible uses for the data.

The project planning team also should determine possible alternative actions that may be taken. Consideration should be given to the option of taking no action, as this option is frequently overlooked but still may be the optimal course of action (e.g., no technology available, too costly, relocation will create problems). After examining the alternative actions, the project planning team should develop a decision statement that expresses a choice among alternative actions.

During these discussions of the directed planning process, the role of the radioanalytical specialists is to ensure that the analytical aspects of the project have been clearly defined and incorporated into the decision(s). The radioanalytical specialists focus on defining: (1) the parameter (analyte/matrix) of interest; (2) what analytical information could resolve the problem; and (3) the practicality of obtaining the desired field and laboratory data. Sections 3.3.1 through 3.3.7 in Chapter 3 discuss in more detail the analytical aspects of the decision (or question) and determining the characteristic or parameter of concern. This information is incorporated into the APS.

2.5.2.1 Define the Action Level

The term "action level" is used in this document to mean the numerical value that will cause the decisionmaker to choose one of the alternative actions. The action level may be a derived concentration guideline level (see below), background level, release criterion, regulatory decision limit, etc. The action level is often associated with the type of medium, analyte and concentration limit.

Some action levels, such as the release criteria for license termination, are expressed in terms of dose or risk. The release criterion is typically based on the total effective dose equivalent (TEDE), the committed effective dose equivalent (CEDE), risk of cancer incidence (morbidity) or risk of cancer death (mortality) and generally cannot be measured directly. For example, in site cleanup, a radionuclide-specific predicted concentration or surface area concentration of specific nuclides that can result in a dose (TEDE or CEDE) or specific risk equal to the release criterion is called the "derived concentration guideline level" (DCGL). A direct comparison can be made between the project's analytical measurements and the DCGL (MARSSIM, 2000). For drinking water analysis, an example of an action level would be a radionuclide-specific concentration based on the Maximum Contaminant Level (MCL) under the Safe Drinking Water Act (42 U.S.C. §300f-300j-26).

2.5.2.2 Identify Inputs to the Decision

The project planning team should determine the specific information required for decisionmaking and this should include a list of the specific data requirements (e.g., number, type, quality). The statistical parameter (e.g., mean concentration) that will be used in the comparison to the action level should be established. An estimate of the expected variability of the data will be needed in order to specify controls on decision-error rates. Existing data, experience and scientific judgment can be used to establish the estimate. Information on environmental background levels and variability may be needed.

The project planning team establishes whether the existing data are sufficient or whether new data are needed to resolve the problem. The radioanalytical specialist can play a key role in this effort by evaluating the quality of the existing radiological data.

2.5.2.3 Define the Decision Boundaries

The project planning team should clearly define the spatial boundaries for the project as well as the time frame for collecting data and making the decision. The spatial boundaries define the physical area to be studied and generally where samples will be collected. Temporal boundaries describe the time frame the study data will represent and when samples should be taken. Any practical constraints that could interfere with sampling should also be identified since these constraints may limit the spatial and/or temporal boundaries of the study.

During these discussions, the radioanalytical specialist can:

- Review existing data for spatial and temporal trends;
- Identify practical constraints that can impact sampling and analysis; and
- Determine feasibility of obtaining new data with current analytical methodologies.

2.5.2.4 Define the Scale of the Decision

The project planning team should clearly define the scale of the decision. The scale of the decision should be the smallest, most appropriate subset of the population for which decisions will be made, based on spatial or temporal boundaries. For example, at a remediation site, a survey unit is generally formed by grouping contiguous site areas with a similar use history and the same classification of potential concentration of the analyte of interest. The survey unit will be defined with a specified size and shape for which a separate decision will be made as to whether the unit attains the site-specific reference-based cleanup standard for the designated analyte of interest (MARSSIM, 2000; NRC, 1998c).

The survey unit is established to delineate areas or volumes of similar composition and history for which a single decision can be made based on the statistical analysis of the data. The variability in the measurement data for a survey unit is a combination of the imprecision of the measurement process and the real spatial and temporal variability of the analyte concentration. If the measurement data include a background contribution, the spatial variability of the background adds to the overall measurement variability.

2.5.3 Specify the Decision Rule and the Tolerable Decision Error Rates

A decision statement or rule is developed by combining the decisions and the alternative actions (see Appendix B, *The Data Quality Objectives Process*). The decision rule presents the strategy or logical basis for choosing among the alternative decisions, generally by use of a series of "if...then" statements. For a complex problem, it may be helpful to develop a logic flow diagram (also called a "decision tree" or "decision framework"), arraying each element of the issue in its proper sequence along with the possible actions. The decision rule identifies (1) the action level that will be a basis for decision, (2) the statistical parameter that is to be compared to the action level, and (3) the decision that would be made and the action that would be taken.

Example of a Decision Rule
General form: "If the value of parameter A over the area B, is greater than C, then take action D, otherwise take action D*."
Example: "If the mean concentration of x in the upper y cm of surface soil of the site is greater than z Bq/g, then action will be taken to remove the soil from the site; otherwise, the soil will be left in place."

The radioanalytical specialists play a key role in the development of technical alternatives that are realistic and satisfy the programmatic and regulatory needs. (See Chapter 3, *Key Analytical Planning Issues and Developing Analytical Protocol Specifications*, for additional discussion on background.)

For each proposed alternative technical action, the radioanalytical specialists can:

- Focus the project planning team on what radionuclides will need to be measured and what types of analytical techniques are available;

- Address whether it is feasible to obtain the necessary analytical results;

- Present the technical limitations (i.e., the minimum detectable concentrations—MDCs) of available measurement systems; and

- Address how sample collection and handling will affect what measurement techniques can be used.

The project planning team also assesses the potential consequences of making a wrong decision. While the possibility of a decision error can never be totally eliminated, it can be controlled. The potential consequences of a decision error are used to establish tolerable limits on the probability that the data will mislead the decisionmaker into making an incorrect decision. (See Appendix B, *The Data Quality Objectives Process*, for a discussion of hypothesis testing, action levels, and decision errors). The decision rule and decisionmaker's limits on the decision error rates are used to establish performance criteria for a data collection design.

In choosing the tolerable decision error rates, the team needs to look at alternative measurement approaches, the sources of error in field and laboratory handling of samples and analysis, factors that would influence the likelihood of a decision error, estimates of the cost of analysis, and judicious use of resources. Realistic decision error rates should be determined during the planning process in order to develop and optimize the sampling and analysis design process.

2.5.4 Optimize the Strategy for Obtaining Data

During the process of developing and optimizing the sampling and analysis plans, the technical team members should determine the project-specific sampling and analytical requirements and associated quality control that will meet all the requirements (desired outputs) established by the project planning team. Optimizing the data collection design generally requires extensive coordination between the radioanalytical and sampling specialists on the planning team. The technical team may not know the most effective analytical protocols at this stage.

Typical considerations during the development of the analysis portion of the data collection design include the number of samples required and the APSs, which include the MQOs (e.g., a statement of the required method uncertainty) required of the analytical procedures (see Sections 2.5.4.1 and 2.5.4.2). In general, the more certainty required in the DQOs, the greater the number of samples or the more precise and unbiased the measurements need to be. During planning, the costs and time for field and analytical procedures must be balanced against the level of certainty that is needed to arrive at an acceptable decision.

The radioanalytical specialists are involved in evaluating the technical options and their effect on the sources of decision error, their resource requirements and the ability to meet the project's objectives. The radioanalytical specialists can identify an array of potential analytical methods, which can be combined in analytical protocols to meet the defined data needs and MQOs. Working with the sampling specialists, potential sampling methods are identified based on the sample requirements of the potential analytical protocols and other sampling constraints. The planning team specialists need to consider sources of bias and imprecision that will impact the representativeness of the samples and the accuracy of the data collected. Appropriate combinations of sampling methods, analytical protocols and sampling constraints can then be assessed with regard to resource effectiveness.

It may be useful at this point for the project planning team to perform a sensitivity analysis on the input parameters that contribute to the final analytical result. The final analytical result directly impacts the decision, so this sensitivity analysis will allow the project planning team to identify the portions of the analytical protocols that potentially have the most impact on the decision. Once identified, these portions of the analytical protocols can be targeted to receive a proportionally larger share of the resources available for developing the protocols.

2.5.4.1 Analytical Protocol Specifications

Requirements of the desired analytical protocols should be based on the intended use of the data. That is, project-specific critical parameters should be considered, including the type of radioactivity and the nuclides of concern, the anticipated range of concentrations, the matrix type and complexity, regulatory required methods, the measurement uncertainty required at some activity concentration, detection limits required, necessary chemical separation, qualification or quantification requirements, QC requirements, and turnaround time needed. MQOs are a key component of the APSs and are discussed on the next page. Chapter 3, *Key Analytical Planning Issues and Developing Analytical Protocol Specifications*, contains more detailed discussion on some of the key decisions and needed input to successfully optimize the sampling and analysis design and develop APSs. Chapter 6, *Selection and Application of an Analytical Method*, discusses the selection of an analytical protocol from the laboratory's perspective.

The project planning team should ensure that there are analytical methods available to provide acceptable measurements. If analytical methods do not exist, the project planning team will need to consider the resources needed to develop a new method, reconsider the approach for providing input data, or perhaps reformulate the decision statement.

2.5.4.2 Measurement Quality Objectives

When additional data are to be obtained, the project planning process should establish measures of performance for the analysis (MQOs) and evaluation of the data. Without these measures of performance, data assessment is difficult and arbitrary.

A MQO is a statement of a performance objective or requirement for a particular method performance characteristic, such as the required method uncertainty at some concentration. MQOs can be both quantitative and qualitative performance objectives. Quantitative and qualitative MQOs are used for real-time compliance monitoring by field and laboratory staff and during subsequent assessments and data usability determinations. Quantitative MQOs provide numerical criteria for field and laboratory QC samples or procedure performance (e.g., specifications for measurement uncertainty, detection limit, yield, spikes, blanks and duplicates). Precision, bias, completeness, and sensitivity are common data quality indicators for which quantitative MQOs could be developed during the planning process (ANSI/ASQC E-4). Thus, quantitative MQOs are statements that contain specific units of measure, such as: x percent recovery, x percent relative standard uncertainty, a standard deviation of x Bq/L, or a MDC of x Bq/g. The specificity of the MQOs allows specific comparisons of the data to an MQO. Chapter 3, *Key Analytical Planning Issues and Developing Analytical Protocol Specifications*, provides detailed guidance on developing MQOs for several method performance characteristics.

The MQOs for the analytical data should be documented in the project plan documents (e.g., the QA Project Plan). MQOs are also the basis for the data verification and validation criteria (see Appendix D, Section D2.7, for a discussion of MQOs and QA project plans).

2.6   Results of the Directed Planning Process

By the end of the directed planning process, the project planning team has established its priority of concerns, the definition of the problem, the decision(s) or outcome to address the posed problem, the inputs and boundaries to the decision(s), and the tolerable decision error rates. It has also agreed on decision rules that incorporate all this information into a logic statement about what must be done to obtain the desired answer. The key output of the planning process is the DQOs: qualitative and quantitative statements that clarify study objectives, define the appropriate type of data, and specify the tolerable rate of making decision errors that will be used as the basis for establishing the quantity and quality of data needed to support the decisions and the criteria for data assessment.

If new data are required, then the project planning team has defined the desired analytical quality of the data (MQOs). That is, the project planning team has determined the type, quantity, and quality of data needed to support a decision. The directed planning process has clearly linked sampling and analysis efforts to a decision and an action level. This linkage allows the project planning team to determine when enough data have been collected.

If new data are to be obtained, the project planning team has developed the most resource-effective sampling and analysis design that will provide adequate data for decisionmaking. Based on the DQOs, the project planning team specifies the sampling collection design and APSs, including:

- The type and quantity of samples to be collected;
- Where, when, and under what conditions they should be collected;
- What radionuclides are to be measured; and
- The MQOs to ensure that the analytical errors are controlled sufficiently to meet the tolerable decision error rates specified in the DQOs.

2.6.1 Output Required by the Radioanalytical Laboratory: The Analytical Protocol Specifications

As a result of directed planning, the description of the DQOs for the project and the APSs (which contain the MQOs and any specific analytical process requirements for additional data) will provide the radioanalytical laboratory with a clear and definitive description of the desired data, as well as the purpose and use of the data. This information will be provided to the project implementation team through the SOW and the project plan documents. Precise statements of analytical needs may prevent the radioanalytical laboratory from:

- Having to make a "best guess" as to what data are really required;
- Using the least costly or most routine protocol, which may not meet the needed data quality;
- Independently developing solutions for unresolved issues without direction from the project planning team; and
- Having "moving targets" and "scope creep" that stem from ambiguous statements of work.

The output of the planning process, from the perspective of the radioanalytical laboratory, is the APSs. The APSs should contain the minimum level of specificity required to meet the project data requirements. In accordance with a performance based measurement approach the laboratory will use this information to select or develop (specific) analytical protocols that will meet the MQOs. The APSs should present the resolution of the project planning team on both general issues and matrix-specific issues. Chapter 3, *Key Analytical Planning Issues and Developing Analytical Protocol Specifications*, addresses some of the common radioanalytical planning issues.

The APSs should include, but not be limited to:

- The radionuclide(s) of concern;
- The matrix of concern, with information on chemical, explosive and other hazardous components;
- The anticipated concentration range (estimate, maximum or detection capability);
- The MQOs desired for the radionuclides of concern;
- The sample preparation and preservation requirements (laboratory and field);
- The type and frequency of QC samples required of each radionuclide of concern;
- The sample transport, tracking, and custody requirements;

- The required analytical turnaround time for the project and the anticipated budget for the analysis; and
- The data reporting requirements.

2.6.2 Chain of Custody

Requirements for formal chain of custody (COC) should be specified in the APSs if required. COC procedures provide the means to trace possession and handling of the sample from collection to data reporting. COC will impact how the field and laboratory personnel handle the sample. COC is discussed in Chapter 10 (*Field and Sampling Issues that Affect Laboratory Measurements*) and Chapter 11 (*Sample Receipt, Inspection, and Tracking*).

2.7 Project Planning and Project Implementation and Assessment

A directed planning process generally is considered complete with the approval of an optimal data collection design approach or when historical data are deemed sufficient to support the desired decision. However to complete the process, the project planning team clearly should document the results of the planning process and link DQOs and MQOs to the implementation and assessment processes. The directed planning process is the first activity in the project's planning phase (see Figure 1.1, "The Data Life Cycle"). The planning process outputs are key inputs to the implementation and assessment processes of the data collection activities. That is, the outputs of the directed planning process are the starting point for developing plan documents, obtaining analytical services, selecting specific analytical protocols and assessing the data collected. This section will provide an overview of the next steps of the planning phase and the linkage to the implementation and assessment phases and to other Part I chapters in MARLAP.

2.7.1 Documenting the Planning Process

A concept inherent in directed planning approaches is the establishment of a formal process to document both the decisions and supporting logic established by the team during the project planning process. Establishing this documentation process is not only good management practice, but also tends to prevent situations where new team members recreate the past logic for activities being performed upon the departure of their predecessors. As actual field conditions or other situations force changes to the original plans, the documentation can then be updated through a change control process to continue to maintain the technically defensible basis for the actions being taken.

When properly documented, the directed planning process:

- Provides a background narrative of the project;
- Defines the necessary input needed (nuclides, matrices, estimate of concentration range, etc.);

- Defines the constraints and boundaries within which the project would have to operate;
- Defines the decision rule, which states the action level that will be the basis for the decision and the statistical parameter that is to be compared to the action level;
- Identifies the tolerable decision error rates;
- Identifies MQOs for new analytical data; and
- Identifies processes and criteria for evaluating the usability of the data.

The results of the project planning process are also needed for the development of project plan documents required for implementing the sampling and analysis activities. These project plan documents may include a quality assurance project plan (QAPP), work plan, or sampling and analysis plan (SAP). The format and title of plan documents are usually a function of the authoring organization's experience, the controlling federal or state regulations, or the controlling agency. Project plan documents are discussed in Chapter 4, *Project Plan Documents*, and in Appendix D, *Content of Project Plan Documents*. The project plan documents will rely on the planning process outputs, including the MQOs, to describe in comprehensive detail the necessary QA, QC, and other technical activities that must be implemented to ensure that the results of the work performed will satisfy the stated DQOs. The project plan documents should also document the processes and criteria developed for data assessment. MARLAP recommends that the planning process rationale is documented and the documentation integrated with the project plan documents. Documentation of the planning process can be incorporated directly in the project plan documents or through citation to a separate report on the planning process.

2.7.2   Obtaining Analytical Services

If laboratory services are required, a SOW should be developed based on the planning process statements of required data and data quality. The SOW is the contractual agreement that describes the project scope and requirements (i.e., what work is to be accomplished). MARLAP recommends that a SOW be developed even if a contract is not involved, for example when an agency employs one of its own laboratories. Contracting laboratory services is discussed in Chapter 5, *Obtaining Laboratory Services*, and Chapter 7, *Evaluating Methods and Laboratories*. Developing a SOW is discussed in Chapter 5.

2.7.3   Selecting Analytical Protocols

From an analytical perspective, one of the most important functions of a directed planning process is the identification and resolution of key analytical planning issues for a project. A key analytical planning issue may be defined as one that has the potential to be a significant contributor of uncertainty to the analytical process and ultimately the resulting data. Identifying key analytical issues for a particular process requires a clear understanding of the analytical process. It is the role of the radioanalytical specialist on the project planning team to ensure that key analytical planning issues have been clearly defined and articulated and incorporated into the principal decision or principal study question. Chapter 3 discusses the key analytical planning issues.

The selection of radioanalytical protocols by the laboratory is made in response to the APSs (for each analyte/matrix) developed by the project planning team as documented in the SOW. Unless required by regulatory policy, rarely will a radioanalytical method be specifically stated. A number of radioanalytical methods are available but no one method provides a general solution; all have advantages and disadvantages. The selection of a method involves a broad range of considerations, including analyte and matrix characteristics; technical complexity and practicality; quality requirements; availability of equipment, facility, and staff resources; regulatory and economic considerations; and previous use of the method. Chapter 6 discusses the selection of an analytical method as well as the modification of an existing analytical method to meet project requirements.

2.7.4 Assessment Plans

Concurrent with the development of MQOs and other specifications of the optimized analytical design, is the development of the data assessment plans. *Data assessment is difficult and arbitrary when attempted at the end of the project without planning and well defined, project specific criteria.* The development of these plans during the project planning process should ensure that the appropriate documentation will be available for assessment and that those implementing and assessing data will be aware of how the data will be assessed. Assessment of environmental data consists of three separate and identifiable phases: data verification, data validation, and data quality assessment (DQA). Verification and validation pertain to evaluation of analytical data generated by the laboratory. DQA considers all sampling, analytical, and data handling details, and other historical project data when determining the usability of data in the context of the decisions to be made. *The focus of verification and validation is on the analytical process and a data point by data point review, while DQA considers the data set as a whole, including the sampling and analytical protocols used to produce them.* Verification, validation, and DQA assure the technical strengths and weaknesses of the overall project data are known, and therefore, establishes the technical defensibility of the data.

2.7.4.1 Data Verification

The data verification process should be defined during the project planning process and documented in a data verification plan or the project plan documents (e.g., the QAPP). The verification plan should specify the types of documentation needed for verification. Analytical data verification assures that laboratory conditions and operations were compliant with the SOW and project plan (i.e., SAP or QAPP). The contract for analytical services and the project plan determine the procedures the laboratory must use to produce data of acceptable quality (as prescribed by the MQOs) and the content of the analytical data package. Verification compares the material delivered by the laboratory to these requirements and checks for consistency of the data throughout the data package, correctness of calculations, and completeness of the results to ensure all documentation is available. Compliance, exceptions, missing documentation and the resulting inability to verify compliance must be recorded in the data verification report. Data verification is discussed in more detail in Chapter 8, *Radiological Data Verification and Validation.*

2.7.4.2 Data Validation

Performance objectives and criteria for data validation should be developed during the project planning process and documented in a separate plan or included in the project plan documents (e.g., QAPP). Guidance on Data Validation Plans is provided in Chapter 8. After the data are collected, data validation activities will rely on the MQOs and other requirements of the APSs to confirm whether the obtained data meet the requirements of the project.

2.7.4.3 Data Quality Assessment

The DQA process evaluates whether the quality and quantity of data will support their intended use. The DQA process determines whether the data meet the assumptions under which the DQOs and the data collection design were developed and whether the analytical uncertainty in the data will allow the decisionmaker to use the data to support the decision within the tolerable decision error rates established during the directed planning process. Guidance on the DQA process and plan development is provided in Chapter 9, *Data Quality Assessment.* The process and criteria to be used for the DQA process should be developed by the project planning team and documented in the project plan documents or in a stand alone plan that is cited or appended to the project plan documents.

2.8   Summary of Recommendations

- MARLAP recommends the use of a directed project planning process.

- MARLAP recommends that the radioanalytical specialists be a part of the integrated effort of the project planning team.

- MARLAP recommends that the planning process rationale be documented and the documentation integrated with the project plan documents.

- MARLAP recommends using a graded approach in which the sophistication, level of QC and oversight, and resources applied are appropriate to the project.

2.9   References

American Society for Quality Control (ANSI/ASQC) E-4. *Specifications and Guidelines for Quality Systems for Environmental Data Collection and Environmental Technology Programs*. 1995. American Society for Quality Control, Milwaukee, Wisconsin.

American Society for Testing and Materials (ASTM) D5612. *Standard Guide for Quality Planning and Field Implementation of a Water Quality Measurements Program*, 1994. West Conshohocken, PA.

American Society for Testing and Materials (ASTM) D5730. *Standard Guide for Site Characterization for Environmental Purposes with Emphasis on Soil, Rock, the Vadose Zone and Ground Water*, 1996. West Conshohocken, PA.

American Society for Testing and Materials (ASTM) D5792. *Standard Practice for Generation of Environmental Data Related to Waste Management Activities: Development of Data Quality Objectives*, 1995. West Conshohocken, PA.

American Society for Testing and Materials (ASTM) D5851. *Standard Guide for Planning and Implementing a Water Monitoring Program*, 1995. West Conshohocken, PA.

American Society for Testing and Materials (ASTM) PS85. *Standard Provisional Guidance for Expedited Site Characterization of Hazardous Waste Contaminated Sites*, 1996. West Conshohocken, PA.

U.S. Army Corps of Engineers (USACE). 1998. *Technical Project Planning (TPP) Process*. Engineer Manual EM-200-1-2.

U.S. Department of Energy (DOE). December 1993. *Remedial Investigation/Feasibility Study (RI/FS) Process, Elements and Techniques Guidance, Module 7 Streamlined Approach for Environmental Restoration*, Office of Environmental Guidance, RCRA/CERCLA Division and Office of Program Support, Regulatory Compliance Division Report DOE/EH-94007658.

U.S. Environmental Protection Agency (EPA). September 1993. *Data Quality Objective Process for Superfund: Interim Final Guidance*. EPA/540/G-93/071, Washington, DC.

U.S. Environmental Protection Agency (EPA). 2000a. *Guidance for the Data Quality Objective Process (EPA QA/G-4)*. EPA/600/R-96/055, Washington, DC. Available at www.epa.gov/quality1/qa_docs.html.

U.S. Environmental Protection Agency (EPA). 2000b. *Data Quality Objectives Process for Hazardous Waste Site Investigations* (Quality Assurance/G-4HW), EPA 600/R-00/007, Washington, DC. Available at: http://www.epa.gov/quality1/qa_docs.html.

MARSSIM. 2000. *Multi-Agency Radiation Survey and Site Investigation Manual, Revision 1.* NUREG-1575 Rev 1, EPA 402-R-97-016 Rev1, DOE/EH-0624 Rev1. August. Available at www.epa.gov/radiation/marssim/.

U.S. Nuclear Regulatory Commission (NRC). 1998a. *Decision Methods for Dose Assessment to Comply with Radiological Criteria for License Termination.* NUREG-1549 (Draft).

U.S. Nuclear Regulatory Commission (NRC). 1998b. *Demonstrating Compliance with the Radiological Criteria for License Termination.* Regulatory Guide DG-4006.

U.S. Nuclear Regulatory Commission (NRC). 1998c. *A Nonparametric Statistical Methodology for the Design and Analysis of Final Status Decommissioning Surveys.* NUREG-1505, Rev.1.

3 KEY ANALYTICAL PLANNING ISSUES AND DEVELOPING ANALYTICAL PROTOCOL SPECIFICATIONS

3.1 Introduction

This chapter provides an overview of key analytical planning issues that should be addressed and resolved during a directed planning process (see Chapter 2). *A key analytical planning issue is one that has a significant effect on the selection and development of analytical protocols, or one that has the potential to be a significant contributor of uncertainty to the analytical process and, ultimately, the resulting data.* It should be noted that a key analytical planning issue for one project may not be a key issue for another project. One of the most important functions of a directed planning process is the identification and resolution of these key issues for a project. The resolution of these issues results in the development of analytical protocol specifications (APSs).

In accordance with a performance-based approach, APSs should contain only the minimum level of specificity required to meet the project or program data requirements and resolve the key analytical planning issues. While Chapter 2 provides an oversight of the project planning process, this chapter provides a focused examination of analytical planning issues and the development of APSs.

In order to assist the project planning team in identifying issues, this chapter provides a list of potential key analytical planning issues. Neither the list nor discussion of these potential issues is an exhaustive examination of all possible issues for a project. However, this chapter does provide a framework and a broad base of information that can assist in the identification of key analytical planning issues for a particular project during a directed planning process.

Analytical planning issues can be divided into two broad categories—those that tend to be matrix-specific and those that are more general in nature. While there is certainly some overlap between these two broad categories, MARLAP divides analytical planning issues along these lines because of the structure and logic it provides in developing APSs. This approach involves identifying key analytical planning issues from the general (non-matrix-specific) issues first and then proceeding on to the matrix-specific issues. Examples of non-matrix-specific analytical planning issues include sample tracking and

Contents

3.1	Introduction	3-1
3.2	Overview of the Analytical Process	3-2
3.3	General Analytical Planning Issues	3-2
3.4	Matrix-Specific Analytical Planning Issues	3-20
3.5	Assembling the Analytical Protocol Specifications	3-23
3.6	Level of Protocol Performance Demonstration	3-24
3.7	Project Plan Documents	3-24
3.8	Summary of Recommendations	3-27
3.9	References	3-27
Attachment 3A: Measurement Uncertainty		3-29
Attachment 3B: Analyte Detection		3-33

custody issues. These general issues are discussed in detail in Section 3.3. Examples of matrix-specific issues include filtration and preservation of water samples. Matrix-specific analytical planning issues are discussed in detail in Section 3.4. Section 3.5 provides guidance on assembling the APSs from the resolution of these issues. Section 3.6 discusses defining the level of protocol performance that must be demonstrated for a particular project, and Section 3.7 discusses incorporating the APSs into the project plan documents.

3.2 Overview of the Analytical Process

Identifying key analytical issues for a particular project requires a clear understanding of the analytical process. The analytical process (see Section 1.4.4 and Figure 3.1) starts with field-sample preparation and preservation and continues with sample receipt and inspection, laboratory sample preparation, sample dissolution, chemical separations, instrument measurements, data reduction and reporting, and sample tracking and quality control. It should be noted that a particular project's analytical process may not include all of the activities mentioned. For example, if the project's analytical process involves performing gamma spectrometry on soil samples, sample dissolution and chemical separation activities normally are not required. Each

step of a particular analytical process contains potential planning issues that may be key analytical planning issues, depending on the nature and data requirements of the project. Therefore, it is important to identify the relevant activities of the analytical process for a particular project early in the directed planning process. Once the analytical process for a particular project has been established, key analytical planning issues, including both general and matrix-specific ones, can be identified.

3.3 General Analytical Planning Issues

There are a number of general analytical planning issues that are common to many types of projects. They may often become key planning issues, depending on the

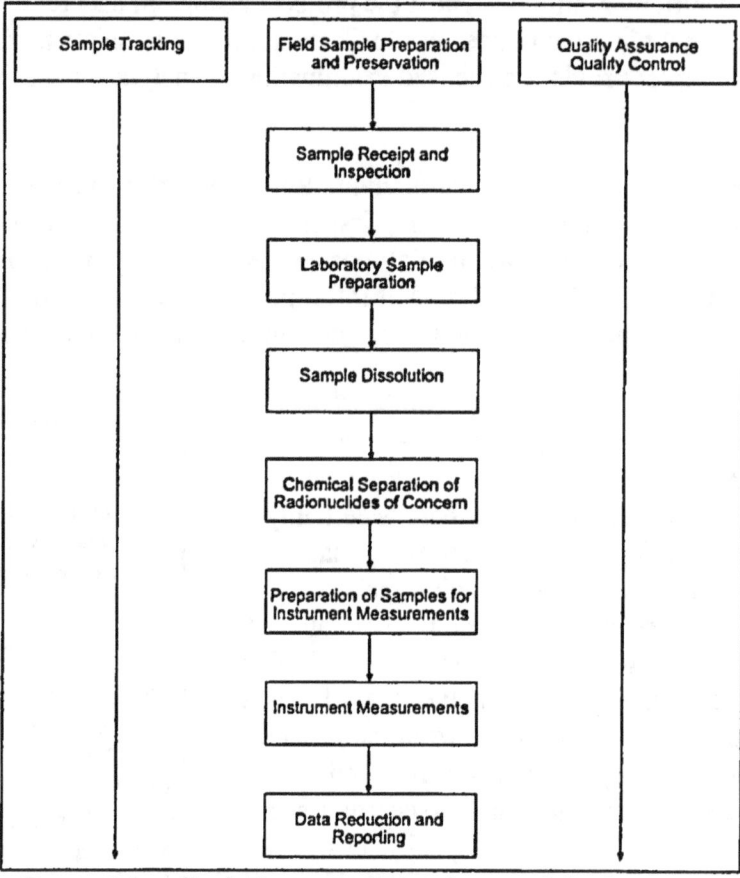

FIGURE 3.1 — Typical components of an analytical process

nature and data requirements of the project, and the resolution of some of these planning issues may affect the selection of methods (see Section 6.5, "Project-Specific Considerations for Method Selection"). This section presents each planning issue as an activity to be accomplished during a directed planning process and also identifies the expected outcome of the activity in general terms. The resolution of these general analytical planning issues, particularly those that are key planning issues for a project, provides the basic framework of the APSs and, therefore, should be identified and resolved before proceeding to matrix-specific planning issues. The resolution of these issues normally results (at a minimum) in an analyte list, identified matrices of concern, measurement quality objectives (MQOs), and established frequencies and acceptance criteria for quality control (QC) samples. The resolution of matrix-specific issues, particularly those that are key issues for a project, normally provides the necessary additions and modifications to the basic framework of the APSs needed to complete and finalize the specifications. MARLAP recommends that any assumptions made during the resolution of key analytical planning issues are documented, and that these assumptions are incorporated into the appropriate narrative sections of project plan documents. Documenting these assumptions may help answer questions or help make decisions during the implementation and assessment phases of the project.

3.3.1 Develop Analyte List

From an analytical perspective, one of the most important planning issues is the target analyte list—the radionuclides of concern for the project. Note that the target analyte list may also include nonradioactive hazardous constituents, which could influence the analytical protocols, including sample collection and waste disposal issues. Although this issue probably would be dealt with by the same planning team, its discussion is outside the scope of MARLAP. For many projects, data are available from previous activities for this purpose. Four possible sources of information are (1) historical data, (2) process knowledge, (3) previous studies, and (4) information obtained from conducting a preliminary survey or characterization study. Although discussed separately in Section 3.3.3, the identification and characterization of matrices of concern often is done concurrently with the development of an analyte list.

Historical data are one source of existing information. Many activities associated with radioactive materials have been well documented. For example, activities licensed by the Nuclear Regulatory Commission (NRC) or NRC Agreement States normally generate much documentation. Chapter 3 of MARSSIM (2000) provides guidance on obtaining and evaluating historical site data.

Another source of existing information is process knowledge. Some sites are associated with a specific activity or process that involved radioactive material, where the process was well defined and the fate of the radioactive material in the process was known or controlled. Examples include uranium and rare earth ore processing, operations at Department of Energy (DOE) weapons facilities, and operations at commercial nuclear power plants (see Section 6.5.2 for additional discussion on process knowledge).

A third source of existing information is previous studies. Similar projects or studies of related topics can provide valuable information during a directed planning process. Previous studies may provide useful information on background radiation. Many radionuclides are present in measurable quantities in the environment. Natural background radiation is due both to primordial and cosmogenic radionuclides. Anthropogenic background includes radionuclides that are ubiquitous in the environment as a result of such human activities as the atmospheric testing of nuclear weapons. Natural and anthropogenic backgrounds can be highly variable even within a given site. It may be important to consider the background and its variability when choosing an action level and when establishing the MQOs. Every effort should be made to obtain as much existing information as possible prior to initiating a directed planning process.

A fourth source of information is generated by conducting a preliminary survey or characterization study. This preliminary analysis may be necessary if there are little or no historical data that can help identify concentrations of radionuclides of potential concern, or if the existing data are of inadequate quality. The design of preliminary surveys or characterization studies should be part of the project planning process. The need for fast turnaround and lower costs at this stage of the project may lead to different data quality objectives (DQOs) and MQOs that are less restrictive than those used for the primary phase of the project. However, it is important that analytical requirements for the survey or study be established during the project planning process. Gross alpha, gross beta, and gamma spectrometry analyses often are used for preliminary survey or characterization studies. The benefits of performing these types of measurements include:

- Rapid analysis and short turnaround time;
- Relatively low analytical costs; and
- Ability to detect the presence of a wide range of radionuclides in a variety of matrices.

There are also limitations on the use of these analyses. These limitations include:

- No specific identification for pure alpha- or pure beta-emitting radionuclides;
- Low-energy gamma- and beta-emitting radionuclides are generally not detected; and
- Failing to detect the presence of several radionuclides (e.g., ^{3}H and other volatile radionuclides; ^{55}Fe and other radionuclides that decay by electron capture).

Output: An initial list of radionuclides of potential concern including a brief narrative explaining why each radionuclide is on the list as well as an explanation of why certain radionuclides were considered but not listed. This list may be modified as more project-specific information becomes available. It is better to include radionuclides on the initial list even if the probability that they significantly contribute to the addressed concerns is small. The consequence of discovering an additional radionuclide of concern late in a project generally outweighs the effort of evaluating its potential during planning.

3.3.2   Identify Concentration Ranges

Once the radionuclides of concern have been identified, the expected concentration range for each radionuclide should be determined. Historical data, process knowledge, previous studies, and preliminary survey or characterization results if available, can be used to determine the expected concentration range for each analyte. While most analytical protocols are applicable over a fairly large concentration range for the radionuclide of concern, performance over a required concentration range can serve as an MQO for the protocol-selection process, thereby eliminating any analytical protocols that cannot accommodate this need. In addition, knowledge of the expected concentration ranges for all of the radionuclides of concern can be used to identify possible chemical or spectral interferences that might lead to the elimination of some of the alternative analytical protocols.

Output: The expected concentration range for each radionuclide of concern and any constituent with the potential for causing chemical or radiological interference.

3.3.3   Identify and Characterize Matrices of Concern

During a directed project planning process, the matrices of concern should be identified clearly. Typical matrices may include surface soil, subsurface soil, sediment, surface water, groundwater, drinking water, air particulates, biota, structural materials, metals, etc. Historical data, process knowledge, previous studies, conceptual site models, transport models, and other such sources generally are used to identify matrices of concern. It is critical to be as specific as possible when identifying a matrix.

Information on the chemical and physical characteristics of a matrix is extremely useful. Therefore, in addition to identifying the matrices of concern, every effort should be made to obtain any information available on the chemical and physical characteristics of the matrices. This information is particularly important when determining the required specificity of the analytical protocol (i.e., the ability to accommodate possible interferences). It is also important to identify any possible hazards associated with the matrix, such as the presence of explosive or other highly reactive chemicals. Issues related to specific matrices, such as filtration of water samples and removal of foreign material, are discussed in more detail in Section 3.4 ("Matrix-Specific Analytical Planning Issues") and Section 6.5.1.1 ("Matrices").

Output: A list of the matrices of concern along with any information on their chemical and physical characteristics and on possible hazards associated with them. The list of matrices of concern and the analyte list often are developed concurrently. In some cases, one analyst list is applicable to all the matrices of concern, and in other cases there are variations in the analyte list for each matrix.

3.3.4   Determine Relationships Among the Radionuclides of Concern

Known or expected relationships among radionuclides can be used to establish "alternate" radionuclides that may be easier and less costly to measure. In most cases, an "easy-to-measure" radionuclide is analyzed, and the result of this analysis is used to estimate the concentration of one or more radionuclides that may be difficult to measure or costly to analyze.

One of the best known and easiest relationships to establish is between a parent radionuclide and its associated progeny. Once equilibrium conditions have been established, the concentration of any member of the decay series can be used to estimate the concentration of any other member of the series (see Attachment 14A, "Radioactive Decay And Equilibrium"). For example, the thorium decay series contains 12 radionuclides. If each radionuclide in this series is analyzed separately, the analytical costs can be very high. However, if equilibrium conditions for the decay series have been established, a single analysis using gamma spectrometry may be adequate for quantifying all of the radionuclides in the series simultaneously.

Similarly, process knowledge can be used to predict relationships between radionuclides. For example, in a nuclear power reactor, steel may become irradiated, producing radioactive isotopes of the elements present in the steel. These isotopes often include ^{60}Co, ^{63}Ni, and ^{55}Fe. Cobalt-60 decays by emission of a beta particle and two high-energy gamma rays, which are easily measured using gamma spectrometry. Nickel-63 also decays by emission of a beta particle but has no associated gamma rays. Iron-55 decays by electron capture and has several associated X-rays with very low energies. Laboratory analysis of ^{63}Ni and ^{55}Fe typically is time-consuming and expensive. However, because all three radionuclides are produced by the same mechanism from the same source material, there is an expected relationship at a given time in their production cycle. Once the relationship between these radionuclides has been established, the ^{60}Co concentration can be used as an alternate radionuclide to estimate the concentration of ^{63}Ni and ^{55}Fe.

The uncertainty in the concentration ratio between radionuclide concentrations used in the alternate analyte approach should be included as part of the combined standard uncertainty of the analytical protocol in the measurement process. Propagation of uncertainties is discussed in Chapter 19, *Measurement Uncertainty*.

Output: A list of known or potential radionuclide relationships, based upon parent-progeny relationships, previous studies, or process knowledge. A preliminary study to determine the project-specific radionuclide relationships may be necessary, and additional measurements may be required to confirm the relationship used during the project. This information may be used to develop a revised analyte list.

3.3.5 Determine Available Project Resources and Deadlines

The available project resources can have a significant impact on the selection or development of analytical protocols, as well as the number and type of samples to be analyzed. In addition, project deadlines and, in particular, required analytical turnaround times (see Section 6.5.3, "Radiological Holding and Turnaround Times") can be important factors in the selection and development of analytical protocols for a particular project. During a directed planning process, radioanalytical specialists can provide valuable information on typical costs and turnaround times for various types of laboratory analyses.

Output: A statement of the required analytical turnaround times for the radionuclides of concern and the anticipated budget for the laboratory analysis of the samples.

3.3.6 Refine Analyte List and Matrix List

As additional information about a project is collected, radionuclides may be added to or removed from the analyte list. There may be one analyte list for all matrices or separate lists for each matrix. Developing an analyte list is an iterative process, however. The list should become more specific during the project planning process.

Radionuclides might be added to the analyte list when subsequent investigations indicate that additional radionuclides were involved in a specific project. In some cases, radionuclides may be removed from the analyte list. When there is significant uncertainty about the presence or absence of specific radionuclides, the most conservative approach is to leave them on the list even when there is only a small probability that they may be present. Subsequent investigations may determine if specific radionuclides are actually present and need to be considered as part of the project. For example, a research laboratory was licensed for a specific level of activity from all radionuclides with atomic numbers between 2 and 87. Even limiting the analyte list to radionuclides with a half-life greater than six months results in a list containing several dozen radionuclides. A study may be designed to identify the actual radionuclides of concern through the use of historical records and limited analyses to justify removing radionuclides from the analyte list.

Output: A revised analyte list. Radionuclides can always be added to or removed from the analyte list, but justification for adding or removing radionuclides should be included in the project documentation.

3.3.7 Method Performance Characteristics and Measurement Quality Objectives

The output of a directed planning process includes DQOs for a project (Section 2.6, "Results of the Directed Planning Process"). DQOs apply to all data collection activities associated with a project, including sampling and analysis. In particular, DQOs for data collection activities describe the overall level of uncertainty that a decisionmaker is willing to accept for project results. This overall level of uncertainty is made up of uncertainties from sampling and analysis activities.

Because DQOs apply to both sampling and analysis activities, what are needed from an analytical perspective are performance objectives specifically for the analytical process of a particular project. MARLAP refers to these performance objectives as MQOs. The MQOs can be viewed as the analytical portion of the overall project DQOs. In a performance-based approach, the MQOs are used initially for the selection and evaluation of analytical protocols and are subsequently used for the ongoing and final evaluation of the analytical data.

In MARLAP, the development of MQOs for a project depends on the selection of an action level and gray region for each analyte during the directed planning process. The term "action level" is used to denote the numerical value that will cause the decisionmaker to choose one of the alternative actions. The "gray region" is a set of concentrations close to the action level where the project planning team is willing to tolerate a relatively high decision error rate (see Chapter 2 and Appendices B and C for a more detailed discussion of action levels and gray region). MARLAP recommends that an action level and gray region be established for each analyte during the directed planning process.

MARLAP provides guidance on developing MQOs for select method performance characteristics such as:

- The method uncertainty at a specified concentration (expressed as an estimated standard deviation);

- The method's detection capability (expressed as the minimum detectable concentration, or MDC);

- The method's quantification capability (expressed as the minimum quantifiable concentration, or MQC);

- The method's range, which defines the method's ability to measure the analyte of concern over some specified range of concentration;

- The method's specificity, which refers to the ability of the method to measure the analyte of concern in the presence of interferences; and

- The method's ruggedness, which refers to the relative stability of method performance for small variations in method parameter values.

An MQO is a quantitative or qualitative statement of a performance objective or requirement for a particular method performance characteristic. An example MQO for the method uncertainty at a specified concentration, such as the action level, would be: "A method uncertainty of 0.01 Bq/g or less is required at the action level of 0.1 Bq/g." A qualitative example of an MQO for method specificity would be "The method must be able to quantify the amount of ^{226}Ra present, given elevated levels of ^{235}U in the samples."

The list provided in this section is not intended to be an exhaustive list of method performance characteristics, and for a particular project, other method performance characteristics may be important and should be addressed during the project planning process. In addition, one or more of the method performance characteristics listed may not be important for a particular project. From an analytical perspective, a key activity during project planning is the identification of important method performance characteristics and the development of MQOs for them.

In addition to developing MQOs for method performance characteristics, MQOs may be established for other parameters, such as data quality indicators (DQIs). DQIs are qualitative and quantitative descriptors used in interpreting the degree of acceptability or utility of data. The principal DQIs are precision, bias, representativeness, comparability, and completeness. These five DQIs are also referred to by the acronym PARCC; the "A" stands for "accuracy" instead of bias, although both indicators are included in discussions of the PARCC parameters (EPA, 2002). Because the distinction between imprecision and bias depends on context, and because a reliable estimate of bias requires a data set that includes many measurements, MARLAP focuses on developing an MQO for method uncertainty. Method uncertainty effectively combines imprecision and bias into a single parameter whose interpretation does not depend on context. This approach assumes that all potential sources of bias present in the analytical process have been considered in the estimation of the measurement uncertainty and, if not, that any appreciable bias would only be detected after a number of measurements of QC and performance-testing samples have been performed. MARLAP provides guidance on the detection of bias, for example, during analytical method validation and evaluation (Chapters 6, *Selection and Application of an Analytical Method*, and 7, *Evaluating Methods and Laboratories*).

While MARLAP does not provide specific guidance on developing MQOs for the DQIs, establishing MQOs for the DQIs may be important for some projects. EPA (2002) contains more information on DQIs. MARLAP provides guidance on developing MQOs for method performance characteristics in the next section.

3.3.7.1 Develop MQOs for Select Method Performance Characteristics

Once the important method performance characteristics for an analytical process have been identified, the next step is to develop MQOs for them. This section provides guidance on developing MQOs for the method performance characteristics listed in the previous section. As noted, other method performance characteristics may be important for a particular analytical process, and

MQOs should be developed for them during project planning. Many of these issues are discussed in Section 6.5 from the laboratory's perspective.

METHOD UNCERTAINTY

While measurement uncertainty is a parameter associated with an individual result and is calculated after a measurement is performed, MARLAP uses the term "method uncertainty" to refer to the predicted uncertainty of a measured value that would likely result from the analysis of a sample at a specified analyte concentration (see Attachment 3A for a general overview of the concept of measurement uncertainty). Method uncertainty is a method performance characteristic much like the detection capability of a method. Reasonable values for both characteristics can be predicted for a particular method based on typical values for certain parameters and on information and assumptions about the samples to be analyzed. These predicted values can be used in the method selection process to identify the most appropriate method based on a project's data requirements. Because of its importance in the selection and evaluation of analytical protocols and its importance in the evaluation of analytical data, MARLAP recommends that the method uncertainty at a specified concentration (typically the action level) always be identified as an important method performance characteristic, and that an MQO be established for it for each analyte/matrix combination.

The MQO for the method uncertainty at a specified concentration plays a key role in MARLAP's performance-based approach. It effectively links the three phases of the data life cycle: planning, implementation, and assessment. This MQO, developed during the planning phase, is used initially in the selection and validation of an analytical method for a project (Chapter 6). This MQO provides criteria for the evaluation of QC samples during the implementation phase (Appendix C, *MQOs for Method Uncertainty and Detection and Quantification Capability*, and Chapter 7, *Evaluating Methods and Laboratories*). It also provides criteria for verification and validation during the assessment phase (Chapter 8, *Radiochemical Data Verification and Validation*). The use of the project-specific MQOs for the method uncertainty of each analyte in the three phases of the life of a project, as opposed to arbitrary non-project-specific criteria, helps to ensure the generation of radioanalytical data of known quality appropriate for its intended use.

The MQO for method uncertainty for an analyte at a specified concentration, normally the action level, is related to the width of the gray region. The gray region has an upper bound and a lower bound. The upper bound typically is the action level. The width of the gray region is represented by the symbol Δ (delta). Given the importance of the gray region in establishing MQOs, the reader is strongly encouraged to review Section B3.7 in Appendix B and Attachment B-1 of Appendix B, which provide detailed guidance on setting up a gray region.

Appendix C provides the rationale and detailed guidance on the development of MQOs for method uncertainty. Outlined below is MARLAP's recommended guideline for developing MQOs for method uncertainty when a decision is to be made about the mean of a population represented by multiple samples. Appendix C provides additional guidelines for developing MQOs for method uncertainty when decisions are to be made about individual items or samples.

If decisions are to be made about the mean of a sampled population, MARLAP recommends that the method uncertainty (u_{MR}) be less than or equal to the width of the gray region divided by 10 for sample concentrations at the upper bound of the gray region (typically the action level). If this method uncertainty cannot be achieved, the project planners should require at least that the method uncertainty be less than or equal to the width of the gray region divided by 3 (Appendix C).

EXAMPLE

Suppose the action level is 0.1 Bq/g and the lower bound of the gray region is 0.02 Bq/g. If decisions are to be made about survey units based on samples, then the required method uncertainty (u_{MR}) at 0.1 Bq/g is

$$\frac{\Delta}{10} = \frac{0.1 - 0.02}{10} = 0.008 \ \text{Bq/g}$$

If this uncertainty cannot be achieved, then a method uncertainty (u_{MR}) as large as $\Delta / 3 =$ 0.027 Bq/g may be allowed if more samples are taken.

In the example above, the required method uncertainty (u_{MR}) is 0.008 Bq/g. In terms of method selection, this particular MQO calls for a method that can ordinarily produce measured results with expected combined standard uncertainties (1σ) of 0.008 Bq/g or less at sample concentrations at the action level (0.1 Bq/g in this example). Although individual measurement uncertainties will vary from one measured result to another, the required method uncertainty is effectively a target value for the individual measurement uncertainties.

Output: MQOs expressed as the required method uncertainty at a specified concentration for each analyte.

DETECTION AND QUANTIFICATION CAPABILITY

For a particular project, the detection capability or the quantification capability may be identified as an important method performance characteristic during project planning (see Attachment 3B of this chapter and Attachment B-2 of Appendix B for a general overview of the concept of detection of an analyte). If the issue is whether an analyte is present in an individual sample and it is therefore important that the method be able to reliably distinguish small amounts of the analyte from zero, then an MQO for the detection capability should be established during project planning. If the emphasis is on being able to make precise measurements of the analyte concentration for comparing the mean of a sampled population to the action level, then an MQO for the quantification capability should be established during project planning.

Detection Capability

When decisions are to be made about *individual items or samples* (e.g., drinking water samples), and the lower bound of the gray region is zero for the analyte of concern, the detection capability of the method is an important method performance characteristic, and an MQO should be developed for it. MARLAP recommends that the MQO for the detection capability be expressed as a required MDC (Chapter 20, *Detection and Quantification Capabilities*).

Outlined below is MARLAP's recommended guideline for developing MQOs for detection capability. Appendix C provides the rationale along with detailed guidance on the development of MQOs for detection capability.

If the lower bound of the gray region is zero, and decisions are to be made about individual items or specimens, choose an analytical method whose MDC is no greater than the action level.

Quantification Capability

When decisions are to be made about a *sampled population* and the lower bound of the gray region is zero for the analyte of concern, the quantification capability of the method is an important method performance characteristic and an MQO should be developed for it. MARLAP recommends that the MQO for the quantification capability be expressed as a required MQC (Chapter 20).

Outlined below is MARLAP's recommended guideline for developing MQOs for quantification capability. The MQC, as used in the guideline, is defined as the analyte concentration at which the relative standard uncertainty is 10 percent (see Chapter 19). Appendix C provides the rationale along with detailed guidance on the development of MQOs for quantification capability.

If the lower bound of the gray region is zero, and decisions are to be made about a sampled population, choose an analytical method whose MQC is no greater than the action level.

If an MQO for method uncertainty has been established, then establishing an MQO for the quantification capability in terms of a required MQC is somewhat redundant because an MQC is defined in terms of a specified relative standard uncertainty. However, this method performance characteristic is included in MARLAP for several reasons. First, it has been included to emphasize the importance of the quantification capability of a method for those instances where the issue is not whether an analyte is present or not—for example measuring ^{238}U in soil where the presence of the analyte is given—but rather how precisely the analyte can be measured. Second, this method performance characteristic has been included so as to promote the MQC as an important method parameter. And last, it has been included as an alternative to the overemphasis on establishing required detection limits in those instances where detection (reliably distinguishing an analyte concentration from zero) is not the key analytical question.

Output: MQOs for each analyte should be expressed as (a) *MQCs* if the lower bound of the gray region is zero and decisions are to be made about a sample population or (b) *MDCs* if the lower bound of the gray region is zero, and decisions are to be made about individual items or samples.

RANGE

Depending on the expected concentration range for an analyte (Section 3.3.2), the method's range may be an important method performance characteristic. Most radioanalytical methods are capable of performing over a fairly large range of analyte concentrations. However, if the expected concentration range is large for an analyte, the method's range should be identified as an important method performance characteristic and an MQO should be developed for it. The radioanalytical specialist on the project planning team will determine when the expected concentration range of an analyte warrants the development of an MQO for the method's range. Because the expected concentration range for an analyte is based on past data, which may or may not be accurate, the MQO for the method's range should require that the method perform over a larger concentration range than the expected range. This precaution will help minimize the potential for selecting methods that cannot accommodate the actual concentration range of the analyte.

Output: MQOs for the method's concentration range for each analyte.

SPECIFICITY

Depending on the chemical and physical characteristics of the matrices, as well as the concentrations of analytes and other chemical constituents, the method's specificity may be an important method performance characteristic for an analytical process. Method specificity refers to the ability of the method to measure the analyte of concern in the presence of interferences. The importance of this characteristic is evaluated by the radioanalytical specialist based upon information about the expected concentration range of the analytes of concern, other chemical and radioactive constituents that may be present, and the chemical and physical characteristics of the matrices (Sections 3.3.2 and 3.3.3). If it is determined that method specificity is an important method performance characteristic, then an MQO should be developed for it. The MQO can be qualitative or quantitative in nature.

Output: MQOs for the method specificity for those analytes likely affected by interferences.

RUGGEDNESS

For a project that involves analyzing samples that are complex in terms of their chemical and physical characteristics, the method's ruggedness may be an important method performance characteristic. Method ruggedness refers to the relative stability of the method's performance when small variations in method parameter values are made, such as a change in pH, a change in amount of reagents used, etc. The importance of this characteristic is evaluated by the radio-analytical specialist based upon detailed information about the chemical and physical characteristics of the sample. If important, then an MOO should be developed for it, which may require performance data demonstrating the method's ruggedness for specified changes in select method parameters. Youden and Steiner (1975) and ASTM E1169 provide guidance on ruggedness testing.

Output: MQOs for method ruggedness for specified changes in select method parameters.

3.3.7.2 The Role of MQOs in the Protocol Selection and Evaluation Process

Once developed, the MQOs become an important part of the project's APSs and are subsequently incorporated into project plan documents (Chapter 4) and into the analytical Statement of Work (Chapter 5). In MARLAP, MQOs are used initially in the selection, validation, and evaluation of analytical protocols (Chapters 6 and 7). In a performance-based approach, analytical protocols are either accepted or rejected largely on their ability or inability to meet the project MQOs.

3.3.7.3 The Role of MQOs in the Project's Data Evaluation Process

Once the analytical protocols have been selected and implemented, the MQOs and—in particular—the MQOs for method uncertainty, are used in the evaluation of the resulting laboratory data relative to the project's analytical requirements. The most important MQO for data evaluation is the one for method uncertainty at a specified concentration. It is expressed as the required method uncertainty (u_{MR}) at some concentration, normally the action level (for this discussion, it is assumed that the action level is the upper bound of the gray region). When the analyte concentration of a laboratory sample is less than the action level, the combined standard uncertainty of the measured result should not exceed the required method uncertainty.

For example, if the required method uncertainty is 0.01 Bq/g or less at an action level of 0.1 Bq/g, then for any measured result less than 0.1 Bq/g, the laboratory's reported combined standard uncertainty should be less than or equal to 0.01 Bq/g. When the concentration is greater than the action level, the combined standard uncertainty of the measured result should not exceed the relative value of the required method uncertainty. If the required method standard uncertainty is 0.01 Bq/g or less at an action level of 0.1 Bq/g (10 percent of the action level), then for any measured result greater than 0.1 Bq/g, the laboratory's reported combined standard uncertainty should be no greater than 10 percent of the measured result. If an expanded uncertainty is reported with each measured value, and the coverage factor also is specified (see Section 19.3.6, "Expanded Uncertainty"), the combined standard uncertainty may be calculated and checked against the required value. The check described relies on the laboratory's estimate of its measurement uncertainty. Additional checks are needed to ensure that the uncertainties are not seriously underestimated.

Appendix C provides guidance on developing criteria for QC samples based on the MQO for method uncertainty. Specifically, Appendix C contains equations for determining warning and control limits for QC sample results based on the project's MQO for method uncertainty.

The following example illustrates the use of the MQO for method uncertainty in evaluating QC sample results. Chapter 8, *Data Verification and Validation*, provides guidance on developing validation criteria based on the MQO for the required method uncertainty.

EXAMPLE

Suppose the upper bound of the gray region (the action level) is 0.1 Bq/g, and the required method uncertainty (u_{MR}) at this concentration is 0.01 Bq/g, or 10 percent. A routine laboratory control sample (LCS) is prepared with an analyte concentration of 0.150 Bq/g. (For the purpose of this example the uncertainty in the spike concentration is assumed to be negligible.) The lab analyzes the LCS with a batch of samples and obtains the measured result 0.140 ± 0.008 Bq/g, where 0.008 Bq/g is the combined standard uncertainty (1σ).

Question: Is this LCS result acceptable?

Answer: The LCS result may be acceptable if it differs from the accepted true value by no more than three times the required method uncertainty at that concentration. In this example the required method uncertainty is 10 percent at 0.150 Bq/g. So, the LCS result is required to be within 30 percent of 0.150 Bq/g, or in the range 0.105–0.195 Bq/g. Because 0.140 Bq/g is clearly in the acceptance range, the data user considers the result acceptable. Note also that the laboratory's reported combined standard uncertainty is less than the required method uncertainty, as expected.

3.3.8 Determine Any Limitations on Analytical Options

With the outputs of the resolution of a number of key analytical planning issues, such as a refined analyte list, MQOs for the analyte list, known relationship between radionuclides of concern, a list of possible alternate analytes, required analytical turnaround times, the analytical budget, etc., the project planning team may choose to limit the analytical options normally available to the laboratory. This decision may be based on information obtained during project planning, such as the absence of equilibrium between the analyte and other radionuclides in its decay chain, the presence of other radionuclides known to cause spectral interferences, the presence of the analyte in a refractory form, etc. However, in the absence of such considerations, the project planning should allow the laboratory the flexibility of selecting any analytical approach that meets the analytical requirements in the APSs.

The role of the radioanalytical specialist is critical in determining if any limitations on analytical options are necessary because of the many laboratory-related issues and factors involved (see Section 2.5, "Directed Planning Process and Role of the Radioanalytical Specialists"). For example, if several of the radionuclides of concern on the target analyte list are gamma-emitters, the radioanalytical specialist can determine if gamma spectrometry is an appropriate technique given the required MQOs, matrices of concern, possible spectral interferences, etc. The radio-analytical specialist may determine that not only is gamma spectrometry an appropriate technique for the gamma-emitting radionuclides of concern, but because there is evidence that equilibrium conditions are present, the results for gamma spectrometry can be used for other radionuclides of concern in the same decay chain as the gamma-emitting radionuclides. In other instances, such as the use of gamma spectrometry to quantify ^{226}Ra in the presence of elevated levels of ^{235}U, the radioanalytical specialist may determine that gamma spectrometry is not an appropriate analysis due to possible spectral interferences. The following sections provide a brief overview of some measurement options.

3.3.8.1 Gamma Spectrometry

In general, gamma spectrometry has many advantages over other choices. It is capable of identifying and quantifying a large number of radionuclides. In comparison with other analyses, it offers a fairly quick turnaround time, and because it is generally a nondestructive technique and limited sample manipulation is involved, it is relatively inexpensive, particularly compared to analyses that require sample dissolution and chemical separations. It also allows for the use of relatively large sample sizes, thereby reducing the measurement uncertainty associated with sub-sampling at the laboratory. However, given its many advantages, gamma spectrometry cannot be used to analyze for all radionuclides. For example, gamma spectrometry may not be able to achieve the project's MQOs, because some or all of the radionuclides of concern may not be gamma-emitters, interfering radionuclides may present problems, etc. The radioanalytical specialist on the planning team can evaluate the appropriateness of the use of gamma spectrom-etry for some or all of the radionuclides on the analyte list or for alternate analytes.

3.3.8.2 Gross Alpha and Beta Analyses

Gross alpha and gross beta analysis provides information on the overall level of alpha- and beta-emitting radionuclides present in a sample. The analysis has the advantage of a relatively quick turnaround time and generally is inexpensive compared to other analyses. The analysis also has significant limitations. It does not identify specific alpha- and beta-emitting radionuclides, so the source of the overall alpha and beta radiation is not determined by the analysis. It does not detect contribution from low-energy beta-emitting radionuclides such as ^{3}H. Volatile radionuclides may be lost during analysis. The measurement uncertainty of the analysis, particularly for matrices other than water, tends to be larger than the measurement uncertainty of other analyses. However, even with these limitations, gross alpha and beta analysis can be an important and appropriate analysis for a project.

3.3.8.3 Radiochemical Nuclide-Specific Analysis

In many instances, due to the project's MQOs, the lack of an appropriate alternate analyte, the lack of equilibrium conditions, etc., radiochemical nuclide-specific analyses are required. This is often true when radionuclides such as ^{3}H, ^{14}C, ^{90}Sr, isotopes of Pu, ^{99}Tc, etc., are on the analyte list. These analyses generally involve more manipulation of the samples than do gamma spectrometry and gross alpha and beta analysis. These analyses often require sample dissolution and chemical separation of the radionuclides of concern. For liquid scintillation counting, distillation is usually required for water samples, and some oxidative/combustion procedure is usually required for solid samples. Because of this, these analyses generally have longer turnaround times and are more expensive than other analyses.

Given the many analytical factors and considerations involved, *the role of the radioanalytical specialist is critical to determining if any limitations on analysis options are necessary.*

Output: Any limitations on analysis options, if appropriate.

3.3.9   Determine Method Availability

After the required analyses have been specified along with their associated sample matrices, the required MQOs, the analytical turnaround times, etc., the radioanalytical specialist should be able to determine if there are analytical methods currently available to meet the project's requirements.

If there are no known analytical methods that would meet the project's analytical requirements, the project planning team must evaluate options. They may decide to reevaluate the analytical data requirements, such as the MQOs, to see if they can be changed to allow the use of existing methods or increase the analytical budget and project timeline to allow for method development.

Output: A statement of method availability.

3.3.10 Determine the Type and Frequency of, and Evaluation Criteria for, Quality Control Samples

There are three main categories of laboratory QC samples—blanks, replicates, and spikes. In addition, there are different types of blanks, replicates, and spikes. For example, spikes can be matrix spikes, laboratory control samples, external performance evaluation samples, etc. Chapter 18 (*Laboratory Quality Control*) contains a detailed discussion of the different types of QC samples and the information they provide. Because the results of the three main types of QC samples often are used to evaluate different aspects of the analytical process, most projects should employ all three types as part of the QC process.

The frequency of laboratory QC sampling for a project essentially represents a compromise between the need to evaluate and control the analytical process and the resources available. In addition, the nature of the project and the intended use of the data will play a role in determining the frequency of QC samples required. For example, the frequency of QC samples for a project involving newly developed methods for analytes in a complex matrix normally should be greater than the frequency of QC samples for a project using more established methods on a simpler matrix, assuming the intended use of the data is the same for both projects. The radioanalytical specialists on the project planning team play a key role in determining the type and frequency of QC samples for a project.

In order to adequately evaluate laboratory data, it is important that the QC samples be clearly linked to a group of project samples. Typically, this is done by analyzing QC samples along with a batch of samples and reporting the results together (see Chapter 18).

In addition to determining the type and frequency of QC samples, evaluation criteria for the QC sample results should be developed during the directed planning process and incorporated into the project's APSs. Appendix C provides guidance on developing criteria for QC samples and contains equations that calculate warning and control limits for QC sample results based on the project's MQO for method uncertainty.

Output: List of type and frequency of QC samples required and the criteria for evaluating QC sample results.

3.3.11 Determine Sample Tracking and Custody Requirements

A procedural method for sample tracking should be in place for all projects so that the proper location and identification of samples is maintained throughout the life of the project. Sample tracking should cover the entire process from sample collection to sample disposal. For some projects, a chain-of-custody (COC) process is needed. COC procedures are particularly important in demonstrating sample control when litigation is involved. In many cases, federal, state, or local agencies may require that COC be maintained for specific samples. Chapter 11, *Sample Receipt, Inspection, and Tracking*, provides guidance on sample tracking and COC. It is important that the requirements for sample tracking be clearly established during project planning.

Output: Project sample tracking requirements.

3.3.12 Determine Data Reporting Requirements

The data reporting requirements should be established during project planning. This involves determining not only what is to be reported but also how it is to be reported. Consideration also should be given to which information should be archived to allow a complete evaluation of the data in the future. Items that are routinely reported are listed below. It should be noted that this is not a comprehensive list, and some projects may require the reporting of more items while other projects may require the reporting of fewer items:

- Field sample identification number
- Laboratory sample identification number
- Sample receipt date
- Analysis date
- Radionuclide
- Radionuclide concentration units
- Sample size (volume, mass)
- Aliquant size (volume, mass)
- Radionuclide concentration at specified date
- Combined standard uncertainty or expanded uncertainty (coverage factor should be indicated)
- Sample-specific minimum detectable concentration
- Analysis batch identification
- Quality control sample results
- Laboratory instrument identification
- Specific analytical parameters (e.g., chemical yields, counting times, etc.)
- Analytical method/procedure reference

It is important that the required units for reporting specific items be determined during project planning. MARLAP recommends that units of the International System of Units (SI) be used whenever possible. However, because regulatory compliance levels are usually quoted in traditional radiation units, it may be appropriate to report in both SI and traditional units, with one being placed in parentheses. *MARLAP also recommends that all measurement results be reported directly as obtained, including negative values, along with the measurement uncertainty—for example 2σ, 3σ, etc.* This recommendation addresses the laboratory's reporting of data to the project planning team or project manager; additional consideration should be given to how data will be reported to the general public. Additional guidance on data reporting, including a discussion of electronic data deliverables, is provided in Chapter 16, *Data Acquisition, Reduction, and Reporting for Nuclear Counting Instrumentation,* and in Chapter 5, *Obtaining Laboratory Services.*

Output: Data reporting requirements for a project.

3.4 Matrix-Specific Analytical Planning Issues

This section discusses a number of matrix-specific analytical planning issues common to many types of projects. For each matrix there is a discussion of several potential key analytical planning issues specific to that matrix. It should be noted that what may be a key analytical planning issue for one project, may not be a key issue for another project. The list of potential matrix-specific key analytical planning issues discussed in this section is summarized in Table 3.1. Table 3.1 is not a comprehensive list, but rather is an overview of some common matrix-specific planning issues. Parenthetical references associated with "potential key issues" in the table identify Part II chapters where these issues are discussed in detail.

This section is divided into solids, liquids, filters and wipes. While filters and wipes are solids, they are discussed separately because of the unique concerns associated with them.

TABLE 3.1 — Common matrix-specific analytical planning issues

MATRIX	RECOMMENDED KEY ISSUES	POTENTIAL KEY ISSUES (Reference Chapters)
Solids (soil, sediment, structural material, biota, metal, etc.)	Homogenization Subsampling Removal of unwanted material	Container type (Chapter 10) Container material (10) Sample preservation (10) Surveying samples for health and safety (11) Volatile compounds (10) Sample identification (10, 11, 12) Cross-contamination (10) Sample size (10, 11, 12) Compliance with radioactive materials license (11) Compliance with shipping regulations (11) Chemical and physical form of the substrate (13, 14)
Liquids (drinking water, groundwater, precipitation, solvents, oils, etc.)	Is filtering required? Sample preservation Should sample be filtered or preserved first?	Sample identification (Chapters 10, 11, 12) Volume of sample (10) Immiscible layers (12) Precipitation (12) Total dissolved solids (12) Reagent background (12) Compliance with radioactive materials license (11) Compliance with shipping regulations (11)
Filters and Wipes	Filter material Pore size Sample volume or area wiped	Sample identification (Chapters 10, 11, 12) Compliance with radioactive materials license (11) Compliance with shipping regulations (11) Subsampling (12) Background from filter material (12)

3.4.1 Solids

Solid samples consist of a wide variety of materials that include soil and sediment, plant and animal tissue, concrete, asphalt; trash, etc. In general, most solid samples do not require preservation (Chapter 10) but do require specific processing both in the field and in the laboratory. In certain instances, some biota samples may require preservation, primarily in the form of lowered temperatures, to prevent sample degradation and loss of water. Some common analytical planning issues for solid samples include the removal of unwanted materials (Section 3.4.1.1), homogenization and subsampling (Section 3.4.1.2), and sample dissolution (Section 3.4.1.3) For certain types of biological samples, removal and analysis of edible portions may be a key analytical planning issue.

3.4.1.1 Removal of Unwanted Materials

When a solid sample is collected in the field, extraneous material may be collected along with the "intended" sample. For example, when collecting a soil sample, rocks, plant matter, debris, etc., may also be collected. Unless instructed otherwise, samples received by the laboratory typically are analyzed exactly as they are received. Therefore, it is important to develop requirements regarding the treatment of extraneous materials. Ultimately, these guidelines should be based on the project's DQOs. The requirements should clearly state what, if anything, is to be removed from the sample and should indicate what is to be done with the removed materials. The guidelines should indicate where the removal process should occur (in the field, in the laboratory or at both locations) and the material to be removed should be clearly identified.

For soil samples, this may involve identifying rock fragments of a certain sieve size, plant matter, debris, etc., as extraneous material to be removed, weighed, and stored at the laboratory. If material is removed from a soil sample, consideration should be given to documenting the nature and weight of the material removed. For sediment samples, requirements for occluded water should be developed. In the case of biological samples, if the entire sample is not to be analyzed, the analytical portion should be identified clearly.

3.4.1.2 Homogenization and Subsampling

For many types of analyses, a portion of the sample sent to the laboratory must be removed for analysis. As with sampling in the field, this portion of the sample should be representative of the entire sample. Adequate homogenization and proper subsampling techniques are critical to obtaining a representative portion of the sample for analysis. Developing requirements for—and measuring the adequacy of—homogenization processes and subsampling techniques can be complicated for various types of solid matrices. General guidance on homogenization and subsampling is provided in Chapter 12 and Appendix F. The input of the radioanalytical specialist as a member of the project planning team is critical to developing requirements for homogenization processes and subsampling techniques.

3.4.1.3 Sample Dissolution

For many analyses, a portion of the solid sample must undergo dissolution before the analyte of interest can be measured. The decision as to which technique to employ for sample dissolution is best left to the laboratory performing the analysis. The radioanalytical specialist can review any information on the chemical and physical characteristics of the matrices of concern and incorporate any relevant information into the APSs.

3.4.2 Liquids

Liquids include aqueous liquids (e.g., surface water, groundwater, drinking water, aqueous process wastes, and effluents), nonaqueous liquids (e.g., oil, solvents, organic liquid process wastes), and mixtures of aqueous and nonaqueous liquids.

A key analytical planning issue for most liquids is whether or not filtering is required or necessary (Section 10.3.2). The question of whether or not to filter a liquid is generally defined by the fundamental analytical question. If the question is related to total exposure from ingestion, the liquids are generally not filtered or the filters are analyzed separately and the results summed. If the question is concerned with mobility of the analyte the concentration in the liquid fraction becomes more important than the concentration in the suspended solids (although some suspended solids may still be important to questions concerning mobility of contamination). In many projects, all of the liquids are filtered and the question becomes which filters need to be analyzed. Issues related to this decision include where and when to filter (Chapter 10); homogenization and subsampling (Chapter 10); volatile compounds (Chapter 10); screening for health and safety (Chapter 11); and cross-contamination (Chapter 10).

Another key analytical planning issue involves preservation of liquid samples, which is also discussed in Chapter 10. Sample preservation involves decisions about the method of preservation (temperature or chemical, Chapter 10), container type and material (Chapter 10), and chemical composition of the sample (Chapters 13 and 14). Preservation of radionuclides in liquids is generally accomplished in the same manner as preservation of metals for chemical analysis. There are of course exceptions such as for ^{3}H and ^{129}I.

A third key analytical issue results from the first two issues and involves the decision of which issue should be resolved first. Should the sample be filtered and then preserved, or preserved first and filtered later? This issue is also discussed in Chapter 10. In general, acid is used to preserve liquid samples. Because acid brings many radionuclides into solution from suspended or undissolved material, filtering is generally performed in the field prior to preserving the sample with acid.

3.4.3 Filters and Wipes

Filters include a wide variety of samples, including liquid filters, air filters for suspended particulates, and air filters for specific compounds. Once the decision to filter has been made, there are at least three key analytical planning issues: filter material, effective pore size, and volume of material to be filtered.

The selection of filter or wipe material can be very important. The wrong filter or wipe can dissolve, break, clog, or tear during sample collection or processing, thus invalidating the sample. Chapter 10 includes a discussion of the various types of filter and wipe materials. Issues influencing this decision include the volume of material to be filtered, the loading expected on the filter, and the chemical composition of the material to be filtered.

The volume of material to be filtered, or area to be wiped, is generally determined by the detection requirements for the project. Lower detection limits require larger samples. Larger samples may, in turn, result in problems with shipping samples or analytical problems where multiple filters were required to meet the requested detection limits.

3.5 Assembling the Analytical Protocol Specifications

After key general and matrix-specific analytical planning issues have been identified and resolved, the next task of the project planning team is to organize and consolidate the results of this process into APSs for the project. In general, there will be an APS for each type of analysis (analyte-matrix combination). At a minimum, the APS should include the analyte list, the sample matrix, possible interferences, the MQOs, any limitations on analysis options, the type and frequency of QC samples along with acceptance criteria, and any analytical process requirements (e.g., sample tracking requirements). The analytical process requirements should be limited to only those requirements that are considered essential to meeting the project's analytical data requirements. For example, if the analyte of concern is known to exist in a refractory form in the samples, then fusion for sample digestion may be included as an analytical process requirement. However, in a performance-based approach, it is important that the *level of specificity in the APSs should be limited to those requirements that are considered essential to meeting the project's analytical data requirements*. The APS should be a one- or two-page form that summarizes the resolution of key analytical planning issues.

Figure 3.2 provides an example form for APSs with references to sections in this chapter as major headers on the form. Figure 3.3 provides for the purpose of an example, an APS for ^{226}Ra in soil for an information gathering project.

3.6 Level of Protocol Performance Demonstration

As discussed in Section 3.3.7.3, during project planning, the project planning team should deter-mine what level of analytical performance demonstration or method validation is appropriate for the project. The question to be answered is how the analytical protocols will be evaluated. There are three parts of this overall evaluation process: (1) the initial evaluation, (2) the ongoing evalu-ation, and (3) the final evaluation. This section briefly discusses the initial evaluation of protocol performance. Chapters 7 and 8 provide guidance on the ongoing and final evaluation of protocol performance, respectively.

The project planning team should determine what level of initial performance demonstration is required from the laboratory to demonstrate that the analytical protocols the laboratory proposes to use will meet the MQOs and other requirements in the APSs. The project planning team should decide the type and amount of performance data required. For example, for the analysis of ^{3}H in drinking water, the project planning team may decide that past performance data from the laboratory, such as the results of internal QC samples for the analysis of ^{3}H in drinking water, are sufficient for the initial demonstration of performance for the laboratory's analytical protocols if they demonstrate the protocol's ability to meet the MQOs. If the analysis is for ^{238}Pu in a sludge, the project planning team may decide that past performance data (if it exists) would not be sufficient for the initial demonstration of performance. The planning team may decide that satisfactory results on performance evaluation samples would be required for the initial demonstration of analytical protocol performance. Section 6.6 ("Method Validation") provides detailed guidance on protocol performance demonstration/method validation, including a tiered approach based on the project analytical needs and available resources.

3.7 Project Plan Documents

Once the APSs have been completed, they should be incorporated into the appropriate project plan documents and, ultimately, into the analytical Statement of Work. Chapters 4 and 5 provide guidance on the development of project plan documents and analytical Statements of Work, respectively. While the APSs are concise compilations of the analytical data requirements, the appropriate plan documents should detail the rationale behind the decisions made in the develop-ment of the APSs.

Analytical Protocol Specifications

Analyte List: (Section 3.3.1, 3.3.7)   Analysis Limitations: (Sections 3.3.9)

Matrix: (Section 3.3.3)   Possible Interferences: (Sections 3.3.2, 3.3.7)

Concentration Range: (Section 3.3.2)   Action Level: (Section 3.3.7)

MQOs:

(Section 3.3.7)   (Section 3.3.7)

(Section 3.3.7)   (Section 3.3.7)

QC Samples		
Type	**Frequency**	**Evaluation Criteria**
(Section 3.3.10)	(Section 3.3.10)	(Section 3.3.10)
(Section 3.3.10)	(Section 3.3.10)	(Section 3.3.10)
(Section 3.3.10)	(Section 3.3.10)	(Section 3.3.10)
(Section 3.3.10)	(Section 3.3.10)	(Section 3.3.10)

Analytical Process Requirements*	
Activity	**Special Requirements**
Field Sample Preparation and Preservation	(Section 3.4)
Sample Receipt and Inspection	(Section 3.3.12)
Laboratory Sample Preparation	(Section 3.4)
Sample Dissolution	(Section 3.4)
Chemical Separations	(Section 3.4)
Preparing Sources for Counting	(Section 3.4)
Nuclear Counting	(Section 3.4)
Data Reduction and Reporting	(Section 3.3.12)
Sample Tracking Requirements	(Section 3.3.11)
Other	

*Consistent with a performance-based approach, analytical process requirements should be kept to a minimum, therefore none or N/A may be appropriate for many of the activities.

FIGURE 3.2 — Analytical protocol specifications

Analytical Protocol Specifications (Example)

Analyte List: ^{226}Ra

Analysis Limitations: Must perform direct measurement of analyte or analysis of progeny allowed if equilibrium established at laboratory

Matrix: Soil

Possible Interferences: Elevated levels of ^{235}U

Concentration Range: 0.01 to 1.50 Bq/g

Action Level: 0.5 Bq/g

MQOs:

A method uncertainty (u_{MR}) of 0.05 Bq/g or less at 0.5 Bq/g

QC Samples		
Type	**Frequency**	**Evaluation Criteria**
Method blank	1 per batch	See attachment B*
Duplicate	1 per batch	See attachment B*
Matrix Spike	1 per batch	See attachment B*

Analytical Process Requirements	
Activity	**Special Requirements**
Field Sample Preparation and Preservation	None
Sample Receipt and Inspection	None
Laboratory Sample Preparation	None
Sample Dissolution	None
Chemical Separations	None
Preparing Sources for Counting	None
Nuclear Counting	None
Data Reduction and Reporting	See Attachment A*
Sample Tracking Requirements	Chain-of-Custody
Other	

* Attachments A and B are not provided in this example

FIGURE 3.3 — Example analytical protocol specifications

3.8   Summary of Recommendations

- MARLAP recommends that any assumptions made during the resolution of key analytical planning issues are documented, and that these assumptions are incorporated into the appropriate narrative sections of project plan documents.

- MARLAP recommends that an action level and gray region be established for each analyte during the directed planning process.

- MARLAP recommends that the method uncertainty at a specified concentration (typically the action level) always be identified as an important method performance characteristic, and that an MQO be established for it for each analyte.

- MARLAP recommends that the MQO for the detection capability be expressed as a required minimum detectable concentration.

- MARLAP recommends that the MQO for the quantification capability be expressed as a required minimum quantifiable concentration.

- MARLAP recommends that if the lower bound of the gray region is zero, and decisions are to be made about individual items or specimens, an analytical method should be chosen whose MDC is no greater than the action level.

- MARLAP recommends that if the lower bound of the gray region is zero, and decisions are to be made about a sampled population, choose an analytical method whose MQC is no greater than the action level.

- MARLAP recommends that units of the International System of Units (SI) be used whenever possible.

- MARLAP recommends that all measurement results be reported directly as obtained, including negative values, along with the measurement uncertainty.

3.9   References

American Society for Testing and Materials (ASTM) E1169. *Standard Guide for Conducting Ruggedness Test.* 1989. West Conshohocken, PA.

U. S. Environmental Protection Agency (EPA). 2002. *Guidance on Developing Quality Assurance Project Plans* (EPA QA/G-5). EPA/240/R-02/009. Office of Environmental Information, Washington, DC. Available at www.epa.gov/quality/qa_docs.html.

MARSSIM. 2000. *Multi-Agency Radiation Survey and Site Investigation Manual, Revision 1*. NUREG-1575 Rev 1, EPA 402-R-97-016 Rev1, DOE/EH-0624 Rev1. August. Available at: www.epa.gov/radiation/marssim/filesfin.htm.

Youden, W.J. and E.H. Steiner. 1975. *Statistical Manual of the Association of Official Analytical Chemists*. Association of Official Analytical Chemists International, Gaithersburg, MD.

ATTACHMENT 3A
Measurement Uncertainty

3A.1 Introduction

No measurement is perfect. If one measures the same quantity more than once, the result generally varies with each repetition of the measurement. Not all the results can be exactly correct. In fact it is generally the case that no result is exactly correct. Each result has an "error," which is the difference between the result and the true value of the measurand (the quantity being measured). Ideally, the error of a measurement should be small, but it is always present and its value is always unknown. (Given the result of a measurement, it is impossible to know the error of the result without knowing the true value of the measurand.)

Since there is an unknown *error* in the result of any measurement, the measurement always leaves one with some *uncertainty* about the value of the measurand. What is needed then is an estimate of the range of values that could reasonably be attributed to the measurand on the basis of the measurement. Determining such a range of reasonable values is the purpose of evaluating the numerical "uncertainty" of the measurement (ISO, 1993).

This attachment gives only a brief overview of the subject of measurement uncertainty. Chapter 19 (*Measurement Uncertainty*) of this manual describes the evaluation and expression of measurement uncertainty in more detail.

3A.2 Analogy: Political Polling

The uncertainty of a laboratory measurement is similar to the "margin of error" reported with the results of polls and other surveys. Note that a political poll is a form of measurement, the measurand in this case being the fraction of likely voters who support a specified candidate. (The fraction is usually reported as a percentage.) The margin of error for the poll result is a kind of measurement uncertainty.

Suppose a poll of 1200 people indicates that 43 percent of the population supports a particular candidate in an election, and the margin of error is reported to be 3 percent. Then if the polling procedure is unbiased, one can be reasonably confident (but not certain) that the actual percentage of people who support that candidate is really between 40 percent and 46 percent.

Political polling results can be wildly inaccurate, and the predicted winner sometimes loses. One reason for this problem is the difficulty of obtaining an unbiased sample of likely voters for the poll. A famous example of this difficulty occurred in the presidential election of 1936, when a polling organization chose its sample from a list of people who owned telephones and automobiles and predicted on the basis of the poll that Alf Landon would defeat Franklin Roosevelt. A significant source of inaccuracy in the result was the fact that many voters during the Great Depression were not affluent enough to own telephones and automobiles, and those voters tended to support FDR, who won the election in a landslide. Another famous example of inaccurate polling occurred in the 1948 presidential election, when polls erroneously predicted that Thomas Dewey would defeat Harry Truman. It seems that the polls in this case were simply taken too early in the campaign. They estimated the fraction of people who supported Dewey at the time the polls were taken, but the fraction who supported him on election day was lower. So, the margin of error in each of these cases was not a good estimate of the total uncertainty of the polling result, because it did not take into account significant sources of inaccuracy. A more complete estimate of the uncertainty would have combined the margin of error with other uncertainty components associated with possible sampling bias or shifts in public opinion. Similar issues may arise when laboratories evaluate measurement uncertainties.

3A.3   Measurement Uncertainty

To obtain a single numerical parameter that describes the uncertainty of a measured result in the laboratory requires one to consider all the significant sources of inaccuracy. An internationally accepted approach to the expression of measurement uncertainty involves evaluating the uncertainty first in the form of an estimated standard deviation, called a *standard uncertainty* (ISO, 1995). A standard uncertainty is sometimes informally called a "one-sigma" uncertainty.

In the political polling example above, the measurand is the fraction, p, of likely voters who support candidate X. The poll is conducted by asking 1,200 likely voters whether they support candidate X, and counting the number of those who say they do. If m is the number who support X, then the pollster estimates p by the quotient $m / 1200$. Pollsters commonly evaluate the standard uncertainty of p as $u(p) = 1 / 2\sqrt{1200}.$

After the standard uncertainty of a result is calculated, finding a range of likely values for the measurand consists of constructing an interval about the result by adding and subtracting a multiple of the standard uncertainty from the measured result. Such a multiple of the uncertainty is called an *expanded uncertainty*. The factor, k, by which the standard uncertainty is multiplied is called a *coverage factor*. Typically the value of k is a small number, such as 2 or 3. If $k = 2$ or 3, the expanded uncertainty is sometimes informally called a "two-sigma" or "three-sigma" uncertainty. An expanded uncertainty based on a coverage factor of 2 provides an interval about the measured result that has a reasonably high probability of containing the true value of the measurand (often assumed to be about 95 percent), and an expanded uncertainty based on a coverage factor of 3 typically provides an interval with a very high probability of containing the true value (often assumed to be more than 99 percent).

In the polling example, the definition of the margin of error is equivalent to that of an expanded uncertainty based on a coverage factor of $k = 2$. Thus, the margin of error equals 2 times $u(p)$, or $1 / \sqrt{1200}$, which is approximately 3 percent.

3A.4 Sources of Measurement Uncertainty

In radiochemistry the most familiar source of measurement uncertainty is counting statistics. Mathematically, the uncertainty of a radiation measurement due to counting statistics is closely related to the uncertainty represented by the margin of error for a political poll. If one prepares a source from a measured amount of radioactive material, places the source in a radiation counter, and makes several 10-minute measurements, the number of counts observed will not always be the same. A typical set of five results might be as follows:

$$101, 115, 88, 111, 103$$

Similarly, if the political poll described above were repeated five times with different groups of likely voters, the number of respondents in each poll who indicate they support the specified candidate might be as follows:

$$523, 506, 520, 516, 508$$

In either case, whether the numbers come from radiation counting or political polling, there is some inherent variability in the results due to random sampling and counting. In radiation counting, the variability exists partly because of the inherently random nature of radioactive decay and partly because the radiation counter is not perfectly efficient at detecting the radiation emitted from the source. In political polling, the variability exists because only a fraction of voters support the candidate and only a limited number of voters are surveyed.

As noted above, there are other potential sources of uncertainty in a political poll. The difficulty in polling is in obtaining a representative sample of likely voters to be surveyed. A similar difficulty is generally present in radiochemical analysis, since many analytical methods require that only a small fraction of the entire laboratory sample be analyzed. The result obtained for that small fraction is used to estimate the concentration of analyte in the entire sample, which may be different if the fraction analyzed is not representative of the rest of the material.

There are many other potential sources of uncertainty in a radiochemical measurement, such as instrument calibration standards, variable background radiation (e.g., cosmic radiation), contaminants in chemical reagents, and even imperfect mathematical models. Some of these errors will vary randomly each time the measurement is performed, and are considered to be "random errors." Others will be fixed or may vary in a nonrandom manner, and are considered to be "systematic errors." However, the distinction between a random error and a systematic error is relatively unimportant when one wants to know the quality of the result of a single measurement.

Generally, the data user wants to know how close the result is to the true value and seldom cares whether the (unknown) error of the result would vary or remain fixed if the measurement were repeated. So, the accepted methods for evaluating and expressing the *uncertainty* of a measurement make no distinction between random and systematic errors. Components of the total uncertainty due to random effects and systematic effects are mathematically combined in a single uncertainty parameter.

3A.5 Uncertainty Propagation

In a radiochemical measurement one typically calculates the final result, y, called the "output estimate," from the observed values of a number of other variables, $x_1, x_2, ..., x_N$, called "input estimates," using a mathematical model of the measurement. The input estimates might include quantities such as the gross sample count, blank count, count times, calibration factor, decay factors, aliquant size, chemical yield, and other variables. The standard uncertainty of y is calculated by combining the standard uncertainties of all these input estimates using a mathematical technique called "uncertainty propagation." The standard uncertainty of y calculated in this manner is called a "combined standard uncertainty" and is denoted by $u_c(y)$.

Radiochemists, like pollsters, have traditionally provided only partial estimates of their measurement uncertainties, because it is easy to evaluate and propagate radiation counting uncertainty — just as it is easy to calculate the margin of error for a political poll. In many cases the counting uncertainty is the largest contributor to the overall uncertainty of the final result, but in some cases other uncertainty components may dominate the counting uncertainty — just as the polling uncertainty due to nonrepresentative sampling may dominate the uncertainty calculated from the simple margin-of-error formula. MARLAP recommends (in Chapter 19) that all of the potentially significant components of uncertainty be evaluated and propagated to obtain the combined standard uncertainty of the final result.

3A.6 References

International Organization for Standardization (ISO). 1993. *International Vocabulary of Basic and General Terms in Metrology.* ISO, Geneva, Switzerland.

International Organization for Standardization (ISO). 1995. *Guide to the Expression of Uncertainty in Measurement.* ISO, Geneva, Switzerland.

ATTACHMENT 3B
Analyte Detection

3B.1 Introduction

In many cases one of the purposes of analyzing a laboratory sample is to determine whether the analyte is present in the sample.[1] If the data provide evidence that the analyte is present, the analyte is *detected*; otherwise, it is *not detected*. The purpose of this attachment is to explain the issues involved in analyte detection decisions, which are often misunderstood. More details are presented in Chapter 20 (*Detection and Quantification Capabilities*).

The result of a laboratory analysis is seldom if ever exactly equal to the true value of the measurand (the quantity being measured), because the result is affected by measurement error (see Attachment 3A). It is also rare for two or more analyses to produce exactly the same result, because some components of the measurement error vary randomly when a measurement is repeated. Typically some sources of error are well understood (e.g., radiation counting statistics) while others (e.g., reagent contamination and interferences) may or may not be. For these reasons, deciding whether an analyte is present in a sample is not always easy.

Acceptable methods for making detection decisions are based on statistical hypothesis testing. In any statistical hypothesis test there are two hypotheses, which are called the *null hypothesis* and the *alternative hypothesis*. Each hypothesis is a statement whose truth is unknown. Only one of the two hypotheses in a hypothesis test can be true in any given situation. The purpose of the test is to choose between the two statements. The null hypothesis is the statement that is presumed to be true unless there is adequate statistical evidence (e.g., analytical data) to the contrary. When the evidence for the alternative hypothesis is strong, the null hypothesis is rejected and the alternative hypothesis is accepted. When the evidence is weak, the null hypothesis is retained and thus must still be assumed to be true, or at least possibly true. In the context of analyte detection, the null hypothesis states that there is *no* analyte in the sample, while the alternative hypothesis states that there is *some* analyte in the sample.

The concept of a null hypothesis is similar to that of a presumption of innocence in a criminal trial, where the defendant is presumed to be innocent (the null hypothesis) unless there is strong legal evidence to the contrary. If the evidence is strong enough to meet the burden of proof, the defendant is found guilty (the alternative hypothesis). The important point here is that an acquit-

---

[1] In other cases, the analyte's presence in a sample may be known or assumed before the analysis. For example, project planners may want to know whether the concentration of a naturally occurring radionuclide, such as ^{238}U, in soil is above or below an action level, although there is little doubt that the analyte is present. In these cases it is usually not necessary to make a detection decision.

tal does not require proof of innocence—only a lack of proof of the defendant's guilt. Analogous rules apply in statistical hypothesis testing.

In the context of analyte detection, the null hypothesis states that there is no analyte in the sample; so, one must presume that no analyte is present unless there is sufficient analytical evidence to the contrary. Therefore, failing to detect an analyte is not the same thing as proving that no analyte is present. Generally, proving that there is no analyte in a sample is *impossible* because of measurement error. No matter how small the result of the measurement is, even if the result is zero or negative, one cannot be certain that there is not at least one atom or molecule of the analyte in the sample.

3B.2   The Critical Value

When a laboratory analyzes a sample, the measuring instrument produces a response, or gross signal, that is related to the quantity of analyte present in the sample, but random measurement errors cause this signal to vary somewhat if the measurement is repeated. A nonzero signal may be (and usually is) produced even when no analyte is present. For this reason the laboratory analyzes a blank (or an instrument background) to determine the signal observed when no analyte is present in the sample, and subtracts this blank signal from the gross signal to obtain the *net signal*. In fact, since the signal varies if the blank measurement is repeated, there is a blank signal *distribution*, whose parameters must be estimated. To determine how large the instrument signal for a sample must be to provide strong evidence for the presence of the analyte, one calculates a threshold value for the net signal, called the *critical value*, which is sometimes denoted by S_C. If the observed net signal for a sample exceeds the critical value, the analyte is considered "detected"; otherwise, it is "not detected."

Since the measurement process is statistical in nature, even when one analyzes an analyte-free sample, it is possible for the net signal to exceed the critical value, leading one to conclude incorrectly that the sample contains a positive amount of the analyte. Such an error is sometimes called a "false positive," although the term "Type I error" is favored by MARLAP. The probability of a Type I error is often denoted by α. Before calculating the critical value one must choose a value for α. The most commonly used value is 0.05, or 5 percent. If $\alpha = 0.05$, then one expects the net instrument signal to exceed the critical value in only about 5 percent of cases (one in twenty) when analyte-free samples are analyzed.

Figure 3B.1 depicts the theoretical distribution of the net instrument signal obtained when analyzing an analyte-free sample and shows how this distribution and the chosen Type I error probability, α, together determine the critical value of the net signal, S_C. The probability α is depicted as the area under the curve to the right of the dashed line. Note that decreasing the value of α, requires increasing the critical value (shifting the dashed line to the right), and increasing the value of α requires decreasing the critical value (shifting the dashed line to the left).

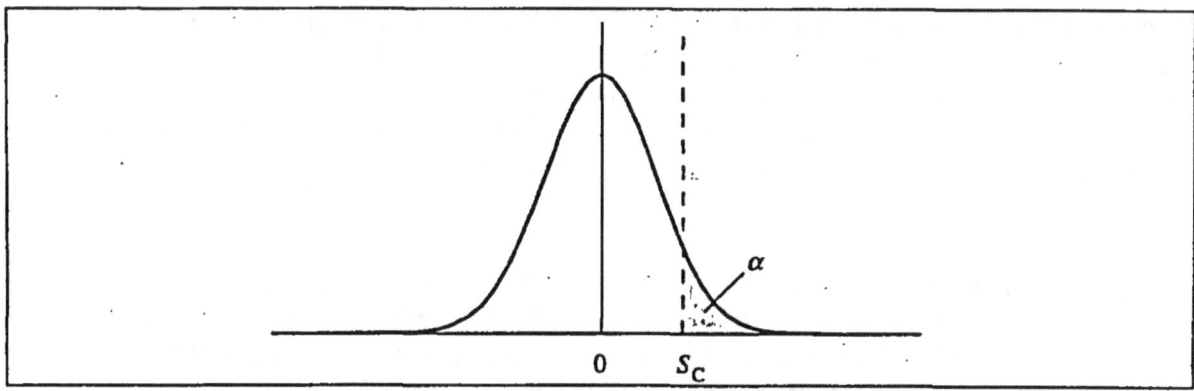

FIGURE 3B.1 — The critical value of the net signal

3B.3   The Minimum Detectable Value

As explained above, the critical value is chosen to limit the probability of a Type I decision error, which means incorrectly concluding that the analyte has been detected when it actually is not present. When the analyte actually *is* present in the sample being analyzed, another kind of decision error is possible: incorrectly failing to detect the analyte. The latter type of error is called a *Type II error*.

The *detection capability* of an analytical measurement process, or its ability to distinguish small positive amounts of analyte from zero, is defined in terms of the probability of a Type II error. The common measure of detection capability is the *minimum detectable value*, which equals the smallest true value (amount, activity, or concentration) of the analyte at which the probability of a Type II error does not exceed a specified value, β.[2] The definition of the minimum detectable value presumes that an appropriate detection criterion (i.e., the critical value) has already been chosen. So, the minimum detectable value is the smallest true value of the analyte that has a specified probability, $1 - \beta$, of generating an instrument signal greater than the critical value. The value of β, like that of α, is often chosen to be 0.05, or 5 percent. (See Figure 20.1 in Chapter 20 for a graphical illustration of the relationship between the critical value and the minimum detectable value.)

In radiochemistry, the minimum detectable value may be called the *minimum detectable concentration* (MDC), *minimum detectable amount* (MDA), or *minimum detectable activity* (also abbreviated as MDA). MARLAP generally uses the term "minimum detectable concentration," or MDC.

---

[2] Although the minimum detectable value is defined theoretically as a "true" value of the analyte, this value, like almost any true value in the laboratory, is not known exactly and can only be estimated. The important point to be made here is that the minimum detectable value should not be used as a detection threshold for the *measured value* of the analyte.

It is common in radiochemistry to report the MDC (or MDA) for the measurement process. Unfortunately, it is also common to use the MDC incorrectly as a critical value, which it is not. It is difficult to imagine a scenario in which any useful purpose is served by comparing a measured result to the MDC. Nevertheless such comparisons are used frequently by many laboratories and data validators to make analyte detection decisions, often at the specific request of project planners.

This common but incorrect practice of comparing the measured result to the MDC to make a detection decision produces the undesirable effect of making detection much harder than it should be, because the MDC is typically at least twice as large as the concentration that corresponds to the critical value of the instrument signal. In principle, a sample that contains an analyte concentration equal to the MDC should have a high probability (usually 95 percent) of producing a detectable result. However, when the MDC is used for the detection decision, the probability of detection is only about 50 percent, because the measured concentration is as likely to be below the MDC as above it. When an analyte-free sample is analyzed, the probability of a Type I error is expected to be low (usually 5 percent), but when the MDC is used for the detection decision, the probability of a Type I error is actually much smaller—perhaps 0.1 percent or less.

Sometimes it may be desirable to have a Type I error rate much less than 5 percent; however, this goal does not justify using the MDC for the detection decision. In this case, the correct approach is to specify the critical value based on a smaller value of α, such as 0.01 instead of 0.05.

MARLAP recommends that when a detection decision is required, the decision should be made by comparing the measured value (e.g., of the net instrument signal) to its critical value—not to the minimum detectable value.

3B.4   Sources of Confusion

There are several potential sources of confusion whenever one deals with the subject of analyte detection in radiochemistry. One source is the lack of standardization of terminology. For example, the term "detection limit" is used with different meanings by different people. In radiochemistry, the detection limit for a measurement process generally means the minimum detectable value. However, in other fields the term may correspond more closely to the critical value. In particular, in the context of hazardous chemical analysis, the term "method detection limit," which is abbreviated as MDL, is defined and correctly used as a critical value (i.e., detection threshold); so, the MDL is not a "detection limit" at all in the sense in which the latter term is commonly used in radiochemistry. Another potential source of confusion is the similarity between the abbreviations MDL and MDC, which represent very different concepts. Anyone who is familiar with only one of these terms is likely to be confused upon first encountering the other.

Another cause of confusion may be the practice of reporting undetectable results as "< MDC." If the measured result is less than the critical value, the practice of reporting "< MDC" may not be ideal, but at least it can be defended on the basis that when the measured value is less than the critical value, the true value is almost certainly less than the MDC. However, if this shorthand reporting format is not explained clearly, a reader may interpret "< MDC" to mean that the *measured* value was less than the MDC and for that reason was considered undetectable. The latter interpretation would be incorrect and might cause the reader to misunderstand the MDC concept. (MARLAP recommends in Chapter 19 that the laboratory always report the measured value and its uncertainty even if the result is considered undetectable.)

3B.5 Implementation Difficulties

Conceptually, the theory of detection decisions and detection limits is straightforward, but the implementation of the theory often presents difficulties. Such difficulties may include:

- Difficulty in preparing and measuring appropriate blanks,
- Variable instrument background,
- Sample-specific interferences, and
- Statistics of low-background radiation counting.

The concept of the "appropriate blank" is that of an artificial sample that is as much like a real sample as practical in all important respects, but which contains none of the analyte being measured. The most appropriate type of blank depends on the analyte and the measurement procedure.

Too often the critical value is based on the distribution of the instrument background, even when it is known that the presence of analyte in reagents and interferences from various sources cause the observed signal for an analyte-free sample to be somewhat elevated and more variable than the instrument background. This practice may produce a high percentage of Type I errors when the critical value is used as a detection threshold. In other cases, the instrument background measurement may overestimate the signal produced by an analyte-free sample and lead to higher Type II error rates. Note that the problem in either of these cases is not the use of the critical value but its incorrect calculation. There is still no justification for using the MDC as a detection threshold. Instead, the critical value should be based on a better evaluation of the distribution of the signal that is observed when analyte-free samples are analyzed.

Even when there are no interferences or reagent contamination, if the instrument background is variable, some of the commonly used expressions for the critical value (which are based on counting statistics only) may be inadequate. Again, the consequence of ignoring such variability when calculating the critical value may be a high percentage of Type I errors. In this case too, the mistake is not in how the critical value is used (as a detection threshold), but in how it is calculated.

A final issue to be discussed is how to calculate an appropriate critical value when the observed blank count is extremely low (e.g., less than 20 counts). Chapter 20 presents expressions for the critical value that should give good results (Type I error rates close to those expected) in these situations when the only variability is that due to counting statistics. However, when the blank count is low and there is additional variability, the usefulness of these expressions cannot be guaranteed, even when they are modified to account for the extra variability.

4 PROJECT PLAN DOCUMENTS

4.1 Introduction

The project plan documents are a blueprint for how a particular project will achieve data of the type and quality needed and expected by the project planning team. In the planning documents, the data user's expectations and requirements, which are developed during the planning process—including the analytical protocol specifications (APSs) and measurement quality objectives (MQOs)—are documented along with the standard operating procedures (SOPs), health and safety protocols, and quality assurance/quality control (QA/QC) procedures for the field and laboratory analytical teams. The objectives of this chapter are to discuss:

- The importance of project plan documents;
- The elements of project plan documents; and
- The link between project planning and project plan documents, in particular the incorporation of the analytical protocols.

The importance of project plan documents is discussed in Section 4.2. Section 4.3 discusses a graded approach to project plan documents. The different types of planning documents and the elements of the project plan documents are discussed in Sections 4.4 and 4.5, respectively. The link between project planning and project plan documents is discussed in Section 4.6.

The project plan documents should be dynamic documents, used and updated over the life of the project. Under a performance-based approach, the analytical protocols requirements in the project plan documents initially may reflect the APSs established by the project planning team and issued in the statement of work (SOW). When the analytical laboratory has been selected, the project plan documents should be updated to reflect the actual protocols to be used. The protocols should be cited, or the SOPs for the protocols should be included as appendices. (APSs and the relation to project measurement quality objectives (MQOs) are discussed in Chapter 3 and represented in Figure 3.2 and 3.3).

While this chapter will address the documentation of QA/QC used in project activities, MARLAP recognizes and fully endorses the need for a quality system as documented in a management plan or quality manual. The development of the project plan documents should be addressed in the quality system requirements documentation. The project plan documents should reflect, and be consistent with, the organization's QA policies and procedures. Guidance on elements of a quality

Contents

4.1	Introduction	4-1
4.2	The Importance of Project Plan Documents	4-2
4.3	A Graded Approach to Project Plan Documents	4-3
4.4	Structure of Project Plan Documents	4-3
4.5	Elements of Project Plan Documents	4-6
4.6	Linking the Project Plan Documents and the Project Planning Process	4-10
4.7	Summary of Recommendations	4-17
4.8	References	4-17

system for environmental data collection activities is available from several sources including ANSI/ASQC E-4 and ISO 9001. QA requirements have been developed by several federal agencies, and they include the following:

- 10 CFR 830.120
- 10 CFR 50, Appendix B
- ANSI N42.23
- ASME NQA-1
- DOE Order 414.1A on QA
- EPA Order 5360.1.A2 on quality systems (2000)
- DOD QA requirement MIL-Q-9858A (1963)

4.2 The Importance of Project Plan Documents

Project plan documents are important in environmental data collection activities to ensure that the type and quantity of data are sufficient for the decision to be made. These documents provide a record of the decisions made during the planning process and integrate the technical operations with the management and quality system practices. Project plans also:

- Support data defensibility for environmental compliance;
- Can be used to defend project objectives and budget; and
- Are a tool for communication with stakeholders.

The development of project plan documents and the implementation of the project plan provide the following benefits:

- Full documentation for legal, regulatory, and historical use of the information;

- Specification of data collection and quality control;

- Documentation of analytical requirements through the incorporation of an APS;

- Implementation of planned data collection activities (through internal and external assessment and oversight activities); and

- Meeting project-specific criteria (i.e., MQOs, DQOs) through data validation and usability assessment.

4.3   A Graded Approach to Project Plan Documents

A graded approach bases the level of management controls applied to an item or work on the intended use of the results and the degree of confidence needed in the quality of the results (ANSI/ASQC E-4). MARLAP recommends a graded approach to project plan development because of the diversity of environmental data collection activities. This diversity in the type of project and the data to be collected impacts the content and extent of the detail to be presented in the plan document. The plan document development team should be flexible in its application of guidance according to the nature of the work being performed and the intended use of the data.

Under a graded approach, a mix of project-specific and site-based quality system documentation may be relied upon to ensure quality. For example, the project specific plan may:

- Address design, work processes, and inspection; and

- Incorporate by citation site-wide plans that address records management, quality improvement, procurement, and assessment.

A comprehensive and detailed project plan is required for some data collection activities because of the need for legal and scientific defensibility of the data. A comprehensive and detailed plan also may be necessary to obtain approval from the Office of Management and Budget for a public survey under the Paperwork Reduction Act to carry out the project (e.g., NRC/EPA survey of Publicly Owned Treatment Works).

Other environmental data collection activities, such as basic studies or small projects, may only require a discussion of the experimental process and its objectives, which is often called a "project narrative statement," "QA narrative statement," or "proposal QA plan" (EPA, 2002). Basic studies and small projects generally are of short duration or limited scope and could include proof of concept studies, exploratory projects, small data collection tasks, feasibility studies, qualitative screens, or initial work to explore assumptions or correlations. Although basic studies and small projects may be used to acquire a better understanding of a phenomenon, they will not by themselves be used to make significant decisions or establish policy. Further discussion on the content of plan documents for basic studies and small projects is provided in Section 4.5.3.

4.4   Structure of Project Plan Documents

The ANSI/ASQC E-4 definition for a QAPP, which is also applicable to other integrated project plan documents, is "a formal document describing in comprehensive detail the necessary QA, QC and other technical activities that must be implemented to ensure that the results of the work performed will satisfy the stated performance criteria." The project plan documents should contain this information in a clear and integrated manner so that all implementation teams can understand their role and the project objectives.

Project plan documents vary in size and format and are referred to by a variety of names. The size of the project plan documents tends to reflect the issuing agency's requirements, complexity, and scope of the project activities. Some projects with multiple phases may have more than one plan document. For example, separate plan documents may be developed for scoping surveys, characterization, and the final status survey for the same site because of the different objectives and data requirements. Available guidance on project plans follows in Section 4.4.1, and a general discussion of various approaches is in Section 4.4.2.

4.4.1   Guidance on Project Plan Documents

National standards guidance on project plan documents is available in:

- ASTM D5283, *Standard Practice for Generation of Environmental Data Related to Waste Management Activities: Quality Assurance and Quality Control Planning and Implementation*;

- ASTM D5612, *Standard Guide for Quality Planning and Field Implementation of a Water Quality Measurements Program*; and

- ASTM PS85, *Standard Provisional Guidance for Expedited Site Characterization of Hazardous Waste Contaminated Sites*.

Guidance on project plans for federal environmental data collection activities is also available (EPA, 2002; 40 CFR 300.430; NRC, 1989; and USACE, 1997 and 2001). Other federal agencies may follow or adapt EPA guidance for quality assurance project plans (QAPPs) (EPA, 2002).

The Intergovernmental Data Quality Task Force (IDQTF) has developed a *Uniform Federal Policy for Quality Assurance Project Plans* (UFP-QAPP) and a *Uniform Federal Policy for Implementing Environmental Quality Systems* (UFP-QS). Agencies participating in IDQTF are the Environmental Protection Agency, Department of Defense, and Department of Energy. The *UFP-QAPP Manual* is a consensus document prepared by the IDQTF work group, and it provides instructions for preparing QAPPs for any environmental data collection operation. Information on IDQTF, including these policies, may be found at www.epa.gov/swerffrr/ documents/data_quality/ufp_sep00_intro.htm#quality.

4.4.2 Approaches to Project Plan Documents

The approach and naming of project plan documents is usually a function of the authoring organization's experience, any controlling federal or state regulations, or the controlling Agency. Project plan, work plan, QAPP, field sampling plan, sampling and analysis plan, and dynamic work plan are some of the names commonly used for project plan documents. The names can however often represent different documents to different agencies, states, companies and even to different people within the same organization.

A work plan is often the primary and integrating plan document when the data collection activity is a smaller supportive component of a more comprehensive project (for example, data collection activity in support of an aspect of an environmental impact statement for a large multi-year project). The QAPP is often the primary document when the data collection activity is a major portion of the project (for example, data collection activity in support of an initial site investigation). A National Contingency Plan (NCP) format (40 CFR 300.430) is appropriate when data collection activities are in support of National Priorities List (NPL) Superfund site projects. The NCP format has a sampling and analysis plan as the primary plan document. The project documentation consists of two integrated documents: a field sampling plan and a QAPP. Stand-alone health and safety plans are also developed.

Traditional site investigations are generally based on a phased engineering approach, which collects samples based on a pre-specified grid pattern and does not provide the framework for making changes in the plan in the field. The work plan (the project plan document) for the site investigation typically will specify the number of samples to be collected, the location of each sample and the analyses to be performed. A newer concept is to develop a dynamic work plan (the project plan document), which, rather than specifying the number of samples to be collected and the location of each sample, would specify the decisionmaking logic that will be used in the field to determine where the samples will be collected, when the sampling will stop, and what analyses will be performed. Guidance on dynamic work plans is available in ASTM PS85 and EPA (2003).

MARLAP does not recommend a particular project plan document approach, title or arrangement. Federal and state agencies have different requirements for the various environmental data collection activities. In certain cases there are regulatory requirements. If an organization has successful experience addressing the essential content of plan documents (Section 4.5) in a well-integrated, documented format, it is usually unnecessary and wasteful of time and monies to change a proven approach. The project plan document should reflect, and be consistent with, the organization's QA policies and procedures.

MARLAP recommends a primary project plan document that includes other documents by citation or as appendices. The primary project plan document serves to integrate the multi-disciplinary sections, other management plans, and stand alone documents into a coherent plan. Appropriate management plans may include the Health and Safety Plan, Waste Management Plan, Risk Analysis Plan, Community Relations Plan, or Records Management Plan. If a detailed discussion of the project already exists in another document, which is available to project participants, then a brief description of site history and incorporation of the document into the project plan document by reference may be appropriate. Incorporation by citation may also be appropriate when the complexity of the project requires an extensive discussion of background issues. Other documents that should be integrated, if available, are the report on the planning process, the data validation plan (Chapter 8), and the DQA plan (Chapter 9). If stand alone documents are not immediately available to project participants, they should be appended to the (primary) project plan document.

4.5   Elements of Project Plan Documents

A project plan document must address a range of issues. The extent of the detail is dependent on the type of project and the intended use of the results as previously discussed in applying a graded approach to plan documents (Section 4.3). For all projects, the project plan document must provide the project information and decisions developed during the project planning process. Project plan documents should address:

- The project's DQOs and MQOs;

- The sampling and analytical protocols that will be used to achieve the project objectives; and

- The assessment procedures and documentation that are sufficient to confirm that the data are of the type and quality needed.

Content of plan documents is discussed in Section 4.5.1. The integration of project plan documents is discussed in Section 4.5.2. Special consideration of project documentation for small projects is discussed in Section 4.5.3.

4.5.1   Content of Project Plan Documents

The plan document development team should remain flexible with regard to format and should focus instead on the appropriate content of plan documents needed to address the elements listed above. The content of plan documents, regardless of the title or format, will include similar information, including:

- The project description and objectives;

- Identification of those involved in the data collection and their responsibilities and authorities;

- Enumeration of the QC procedures to be followed;

- Reference to specific SOPs that will be followed for all aspects of the projects; and

- Health and safety protocols.

The project plan document(s) should present the document elements as integrated chapters, appendices, and stand alone documents, and plans should be included by citation. Table 4.1 provides summary information on project plan elements for three different plan documents: project plans, dynamic work plans, and QAPPs as provided in ASTM and EPA guidance. The table also illustrates the similarity of project plan content.

TABLE 4.1 — Elements of project plan documents

Project Plan (ASTM D5283 and D5612)	Dynamic Work Plan (ASTM PS85)	QAPP (EPA, 2002)
Project Management Identify individuals with designated responsibility and authority to: (1) develop project documents; (2) select organizations to perform the work; (3) coordinate communications; and (4) review and assess final data.		**A. Project Management** A1 Approval Sheet A2 Table of Contents A3 Distribution List A4 Project Organization
Background Information Reasons for data collection. Identify regulatory programs governing data collection.	1. **Regulatory Framework** 2. **Site Descriptions and History of Analyte Use and Discovery** 3. **Analysis of Prior Data and Preliminary Conceptual Site Model**	A5 Problem Definition and Background
Project Objectives • Clearly define objectives of field and laboratory work. • Define specific objectives for the sampling location. • Describe intended use of data.	**Dynamic Technical Program** • Essential questions to be answered or specific objectives. • Identify the investigation methods and the areas in which they may be applied. • Provide clear criteria for determining when the project objectives have been met.	A6 Project Description. A7 Quality Objectives and Criteria for Measurement Data. A8 Special Training Requirements/Certifications. A9 Documentation and Records.

Project Plan (ASTM D5283 and D5612)	Dynamic Work Plan (ASTM PS85)	QAPP (EPA, 2002)
Sampling Requirements Sample requirements are specified, including: • Sampling locations. • Equipment and Procedures (SOPs). • Sample preservation and handling.	**Field Protocols and Standard Operating Procedures** (this section may be attached as a separate document) [* see footnote]	**B. Measurement/Data Acquisition** B1 Sampling Process Designs. B2 Sampling Method Requirements. B3 Sample Handling and Custody Requirements.
Analytical Requirements The analytical requirements are specified, including: • Analytical procedures (SOPs). • Analyte list. • Required method uncertainty. • Required detection limits. • Regulatory requirements and DQO specifications are considered.		B4 Analytical Methods Requirements.
Quality Assurance and Quality Control Requirements • QA/QC requirements are addressed for both field and laboratory activities. • Type and frequency of QC samples will be specified. • Control parameters for field activities will be described. • Performance criteria for laboratory analysis will be specified. • Data validation criteria (for laboratory analysis) will be specified.	**Quality Assurance and Quality Control Plan**	B5 Quality Control Requirements. B6 Instrument/Equipment Testing Inspection and Maintenance Requirements. B7 Instrument Calibration and frequency. B8 Inspection/Acceptance Requirements for Supplies and Consumables. B9 Data Acquisition Requirements for Non-direct Measurements. B10 Data Management.
Project Documentation All documents required for planning, implementing, and evaluating the data collection efforts are specified, may include: • SOW, Work Plan, SAP, QAPP, H&S Plan, Community Relations Plan. • Technical reports assessing data. • Requirements for field and analytical records.	1. **Data Management Plan** 2. **Health and Safety Plan** 3. **Community Relations Plan**	**C. Assessment/Oversight** C1 Assessments and response Actions. C2 Reports to Management. **D. Data Validation and Usability** D1 Data Review, Verifications and Validation Requirements. D2 Verification and Validation Methods. D3 Reconciliation with DQO.

* The combined Dynamic Technical Program section and Field Protocols and SOPs section is the functional equivalent of a Field Sampling and Analysis Plan.

Appendix D (*Content of Project Plan Documents*) provides more detailed guidance on the content of project plan documents following the outline developed by EPA requirements (EPA, 2001) and guidance (EPA, 2002) for QAPPs for environmental data operations. The EPA element identifiers (A1, A2, etc.) and element titles are used in the tables and text of this chapter for ease of cross reference to the appropriate section in Appendix D. The EPA elements for a QAPP are used to facilitate the presentation and do not represent a recommendation by MARLAP on the use of a QAPP as the project plan document format.

4.5.2   Plan Documents Integration

MARLAP strongly discourages the use of a number of stand-alone plan components of equivalent status without integrating information and without a document being identified as a primary document. For large project plan compilations, it is appropriate to issue stand-alone portions of the plan that focus on certain activities such as sampling, analysis or data validation, since it can be cumbersome for sampling and laboratory personnel to keep the entire volume(s) of the project plan document readily available. However, each stand-alone component should contain consistent project information, in addition to the component specific plan information, such as the following:

- A brief description of the project including pertinent history;
- A brief discussion of the problem to be solved or the question to be answered (DQO);
- An organizational chart or list of key contact persons and means of contact;
- The analyte(s) of interest; and
- The appropriate health and safety protocols and documentation requirements.

In addition, a cross-referenced table is helpful in the primary document, which identifies where project plan elements are located in the integrated plan document.

4.5.3   Plan Content for Small Projects

The project plan documents for small projects and basic studies (Section 4.3) generally consist of three elements: the title and approval sheet, the distribution list, and a project narrative. The project narrative should discuss in a concise manner the majority of issues that are normally addressed in a project plan document, such as a QAPP. A typical project narrative may be a concise and brief description of the following list (keyed to the QAPP column in Table 4.1, [EPA, 2001]):

- Problem and site history (A5)
- Project/task organization (A4)
- Project tasks, including a schedule and key deliverables (A6)
- Anticipated use of the data (A5, A6)
- MQOs (A7)
- Sampling process design requirements and description (B1)
- Sample type and sampling location requirements (B2)
- Sample handling and custody requirements (B3)
- Analytical protocols (B4)

- QC and calibration requirements for sampling and analysis (B5, B7)
- Inspection and maintenance of analytical instrumentation (B6)
- Plans for peer or readiness reviews prior to data collection (C1)
- Assessments to be conducted during actual operation (C1)
- Procedure for data review (D2)
- Identification of any special reports on QA/QC activities, as appropriate (C2)
- Reconciliation with DQOs or other objectives (D3)

4.6 Linking the Project Plan Documents and the Project Planning Process

Directed planning processes (Chapter 2 and Appendix B) yield many outputs, such as the APSs (Chapter 3), which must be captured in project plan documents to ensure that data collection activities are implemented properly. MARLAP recommends that the project plan documents integrate all technical and quality aspects for the life cycle of the project, including planning, implementation, and assessment.

The project plan should be a dynamic document, used and updated over the life of the project. For example, the analytical methods requirements in the project plan documents (B4) will initially reflect the APSs established by the project planning team (Chapter 3) and issued in the SOW or BOA task order (Chapter 5). When the analytical laboratory has been selected (Chapter 7), the project plan document should be updated to reflect the specific analytical protocols: the actual protocols to be used, which should be included by citation or inclusion of the SOPs as appendices. MARLAP also recommends using a formal process to control and document changes if updates of the original project plan document are needed.

Table 4.2 presents a crosswalk of the elements of the EPA QAPP guidance with outputs of a directed planning process to illustrate how to capture and integrate the outputs of the planning process into the plan document(s).

TABLE 4.2 — Crosswalk between project plan document elements and directed planning process

ID	Project Plan Document Elements (QAPP—EPA, 2001)*	Content	Directed Planning Process Input
PROJECT MANAGEMENT			
A1	Title and Approval Sheet	Title and approval sheet.	
A2	Table of Contents	Document control format.	
A3	Distribution List	Distribution list for the plan document revisions and final guidance.	Include the members of the project planning team and stakeholders.
A4	Project/Task Organization	1) Identify individuals or organizations participating in the project and discuss their roles and responsibilities.	The directed planning process: • Identifies the stakeholders, data users, decisionmakers.

ID	Project Plan Document Elements (QAPP—EPA, 2001)*	Content	Directed Planning Process Input
		2) Provide an organizational chart showing relationships and communication lines.	• Identifies the core planning team and the technical planning team members responsible for technical oversight. • Identifies the specific people/organizations responsible for project implementation (sampling and analysis).
A5	Problem Definition/ Background	1) State the specific problem to be solved and decision to be made. 2) Include enough background to provide a historical perspective.	Project planning team: • Documents the problem, site history, existing data, regulatory concerns, background levels and thresholds. • Develops a decision statement.
A6	Project/Task Description	Identify measurements, special requirements, sampling and analytical methods, action levels, regulatory standards, required data and reports, quality assessment techniques, and schedules.	Project planning team identifies: • Deadlines and other constraints that can impact scheduling. • Existing and needed data inputs. Project planning team establishes: • Action levels and tolerable decision error rates that will be the basis for the decision rule. • The optimized sampling and analytical design as well as quality criteria.
A7	Quality Objectives and Criteria for Measurement Data	1) Identify DQOs, data use, type of data needed, domain, matrices, constraints, action levels, statistical parameters, and acceptable decision errors. 2) Establish MQOs that link analysis to the user's quality objectives.	Project planning team: • Identifies the regulatory standards and the action level(s). • Establishes the decision rule. • Describes the existing and needed data inputs. • Describes practical constraints and the domain. • Establishes the statistical parameter that is compared to the action level. • Establishes tolerable decision error rates used to choose quality criteria. • Establishes quality criteria linked to the optimized design. • Establishes data verification, validation and assessment criteria and procedures. • Establishes APSs and MQOs.
A8	Special Training Requirements/ Certification	Identify and discuss special training/certificates required to perform work.	Project planning team: • Identifies training, certification, accreditation requirements for field and laboratory.

ID	Project Plan Document Elements (QAPP—EPA, 2001)*	Content	Directed Planning Process Input
			• Identifies federal and state requirements for certification for laboratories. • Identifies federal and state requirements for activities, such as disposal of field-generated residuals.
A9	Documentation and Record	Itemize the information and records, which must be included in a data report package including report format and requirements for storage etc.	Project planning team: • Indicates whether documents will be controlled and the distribution list incomplete. • Identifies documents that must be archived. • Specifies period of time that documents must be archived. • Specifies procedures for error corrections (for hard copy and electronic files).
MEASUREMENT/DATA ACQUISITION			
B1	Sampling Process Designs (Experimental Designs)	(1) Outline the experimental design, including sampling design and rationale, sampling frequencies, matrices, and measurement parameter of interest. (2) Identify non-standard methods and validation process.	Project planning team establishes the rationale for and details of the sampling design.
B2	Sampling Methods Requirements	Describe sampling procedures, needed materials and facilities, decontamination procedures, waste handling and disposal procedures, and include a tabular description of sample containers, sample volumes, preservation and holding time requirements.	Project planning team specifies the preliminary details of the optimized sampling method.
B3	Sample Handling and Custody Requirements	Describe the provisions for sample labeling, shipment, sample tracking forms, procedures for transferring and maintaining custody of samples.	Project planning team describes the regulatory situation and site history, which can be used to identify the appropriate sample tracking level.
B4	Analytical Methods Requirements	Identify analytical methods and procedures including needed materials, waste disposal and corrective action procedures.	Project planning team: • Identifies inputs to the decision (nuclide of interest, matrix, etc.). • Establishes the allowable measurement uncertainty that will drive choice of the analytical protocols. • Specifies the optimized sampling and analytical design.

ID	Project Plan Document Elements (QAPP—EPA, 2001)	Content	Directed Planning Process Input
B5	Quality Control Requirements	(1) Describe QC procedures and associated acceptance criteria and corrective actions for each sampling and analytical technique. (2) Define the types and frequency of QC samples should be defined along with the equations for calculating QC statistics.	Project planning team: • Establishes the allowable measurement uncertainty, which will drive QC acceptance criteria. • Establishes the optimized analytical protocols and desired MQOs.
B6	Instrument/Equipment Testing Inspection and Maintenance Requirements	1) Discuss determination of acceptable instrumentation performance. 2) Discuss the procedures for periodic, preventive and corrective maintenance.	
B7	Instrument Calibration and Frequency	(1) Identify tools, gauges and instruments, and other sampling or measurement devices that need calibration. (2) Describe how the calibration should be done.	Project planning team establishes the desired MQOs, which drive acceptance criteria for instrumentation performance.
B8	Inspection/Acceptance Requirements for Supplies and Consumables	Define how and by whom the sampling supplies and other consumables will be accepted for use in the project.	
B9	Data Acquisition Requirements (Non-direct Measurements)	Define criteria for the use of non-direct measurement data such as data that come from databases or literature.	Project planning team: • Identifies the types of existing data that are needed or would be useful. • Establishes the desired MQOs that would also be applicable to archived data.
B10	Data Management	(1) Outline of data management scheme including path of data, use of storage and& record keeping system.(2) Identify all data handling equipment and procedures that will be used to process, compile, analyze the data, and correct errors.	
ASSESSMENT/OVERSIGHT			
C1	Assessments and Response Actions	(1) Describe the number, frequency and type of assessments needed for the project. (2) For each assessment: list participants and their authority, the schedule, expected information, criteria for success and unsatisfactory conditions and those who	Project planning team establishes the MQOs and develops statements of the APSs, which are used in the selection of the analytical protocols and in the ongoing evaluation of the protocols.

ID	Project Plan Document Elements (QAPP—EPA, 2001)*	Content	Directed Planning Process Input
		will receive reports and procedures for corrective actions.	
C2	Reports to Management	Identify the frequency, content and distribution of reports issued to keep management informed.	
DATA VALIDATION AND USABILITY			
D1	Data Review, Verification and Validation Requirements	State the criteria including specific statistics and equations, which will be used to accept or reject data based on quality.	Project planning team: • Establishes the MQOs for the sample analysis, and may also discuss completeness and representativeness requirements that will be the basis of validation. • Establishes the action level(s) relevant to the project DQOs. • Establishes the data validation criteria.
D2	Verification and Validation Methods	Describe the process to be used for validating and verifying data, including COC for data throughout the lifetime of the project.	Project planning team: • Determines appropriate level of custody. • May develop a validation plan.
D3	Reconciliation With Data Quality Objectives	Describe how results will be evaluated to determine if DQOs are satisfied.	Project planning team: • Defines the necessary data input needs. • Defines the constraints and boundaries with which the project has to comply. • Defines the decision rule. • Identifies the hypothesis and tolerable decision error rates. • Defines MQOs for achieving the project DQOs.

[Adapted from: EPA, 2002]
* EPA QAPP elements are discussed in Appendix D

4.6.1 Planning Process Report

MARLAP recommends including, by citation or as an appendix, the report on the directed planning process in the project plan documents. If the planning process was not documented in a report, MARLAP recommends that a summary of the planning process be included in the project plan document section on Problem Definition/Background (A5) that addresses assumptions and decisions, established action levels, the DQO statement, and APSs (which include the established MQOs and any specific analytical process requirements). Additional detailed information on the APSs including the MQOs will be presented in the project plan document sections on project/

task description (A6), quality objectives and criteria for measurement data (A7), and analytical methods requirements (B4). MARLAP views the project plan documents as the principal product of the planning process.

4.6.2 Data Assessment

Assessment (verification, validation, and DQA) is the last step in the project's data life cycle and precedes the use of data. Assessment, and in particular DQA, is designed to evaluate the suitability of project data to answer the underlying project question or the suitability of project data to support the project decision. The project planners should define the assessment process in enough detail that achievement or failure to meet goals can be established upon project completion. An important output of the directed planning process to be captured in the project plan document is the data verification, validation and assessment criteria and procedures.

4.6.2.1 Data Verification

Analytical data verification assures that laboratory conditions and operations were compliant with the contractual SOW and the project plan. Verification compares the data package to these requirements (contract compliance) and checks for consistency and comparability of the data throughout the data package and completeness of the results to ensure all necessary documentation is available. Performance criteria for verification should be documented in the contract and in the project plan document in the sections that address data review, verification, and validation requirements (D1), and verification and validation methods (D2).

4.6.2.2 Data Validation

Validation addresses the reliability of the data. During validation, the technical reliability and the degree of confidence in reported analytical data are considered. Data validation criteria and procedures should be established during the planning process and captured in the project plan document (and the SOW for the validation contractor). Performance criteria for data validation can be documented directly in the project plan document in data review, verifications, and validation requirements (D1) and verifications and validation methods (D2) or in a separate plan, which is included by citation or as an appendix in the project plan document.

Guidance on data validation plans is provided in Section 8.3. The data validation plan should contain the following information:

- A summary of the project, which provides sufficient detail about the project's APSs, including the MQOs;

- The set of data to be validated and whether all the raw data will be reviewed and in what detail;

- The necessary validation criteria and the MQOs deemed appropriate for achieving project DQOs;

- Specifications on what qualifiers are to be used and how final qualifiers are to be assigned; and

- Information on the content of the validation report.

4.6.2.3 Data Quality Assessment

Data quality assessment consists of a scientific and statistical evaluation of project-wide knowledge to determine if the data set is of the right type, quality and quantity to support its intended use. The data quality assessor integrates the data validation report, field information, assessment reports and historical project data and compares the findings to the original project objectives and criteria (DQOs).

Performance criteria for data usability for the project should be documented in the project plan documents in a section on DQA or reconciliation of the data results with DQOs (D3) or in a separate plan, which is included by citation or as an appendix in the project plan document. Guidance on DQA plans is provided in Section 9.5. The DQA plan should contain the following information:

- A summary of the project, which provides sufficient detail about the project's DQOs and tolerable decision error rates;

- Identification of what issues will be addressed by the DQA;

- Identification of any statistical tests that will be used to evaluate the data;

- Description of how the representativeness of the data will be evaluated (for example, review the sampling strategy, the suitability of sampling devices, subsampling procedures, assessment findings);

- Description of how the accuracy of the data, including potential impact of non-measurable factors (for example, subsampling bias) will be considered (for example, review the APSs and the analytical plan, the suitability of analytical protocols, subsampling procedures, assessment findings);

- Description of how the MQOs will be used to determine the usability of measurement data (that is, did the uncertainty in the data significantly affect confidence in the decision);

- Identification of what will be included in the DQA report; and

- Identification of who will receive the report and the mechanism for its archival.

4.7 Summary of Recommendations

- MARLAP recommends using a graded approach to project plan writing because of the diversity of environmental data collection activities.

- MARLAP recommends developing a primary integrating project plan that includes other documents by citation or as appendices.

- MARLAP recommends developing project plan documents that integrate all technical and quality aspects for the life-cycle of the project, including planning, implementation, and assessment.

- MARLAP recommends including, by citation or as an appendix, the report on the directed planning process in the project plan documents.

- If the planning process was not documented in a report, MARLAP recommends that a summary of the planning process addressing assumptions and decisions, established action levels, the DQO statement, and APSs (which include the established MQOs and any specific analytical process requirements) be included in the project plan documents.

- MARLAP recommends using a formal process to control and document changes if updates of the original project plan document are needed.

4.8 References

American Society for Quality Control (ANSI/ASQC) E-4. *Specifications and Guidelines for Quality Systems for Environmental Data Collection and Environmental Technology Programs.* 1995. American Society for Quality Control, Milwaukee, Wisconsin.

American Society of Mechanical Engineers (ASME) NQA-1. *Quality Assurance Program Requirements for Nuclear Facilities.* New York, New York. 1989.

American Society of Testing and Materials (ASTM) D5283. *Standard Practice for Generation of Environmental Data Related to Waste Management Activities: Quality Assurance and Quality Control Planning and Implementation.* 1992. West Conshohocken, PA.

American Society of Testing and Materials (ASTM) D5612. *Standard Guide for Quality Planning and Field Implementation of a Water Quality Measurements Program*, 1994. West Conshohocken, PA.

American Society of Testing and Materials (ASTM) PS85. *Standard Provisional Guidance for Expedited Site Characterization of Hazardous Waste Contaminated Sites.* 1996. West Conshohocken, PA.

Code of Federal Regulations (CFR). 1999. 10 CFR 50 Appendix B, "Quality Assurance Criteria for Nuclear Power Plants and Fuel Reprocessing Plants."

Code of Federal Regulations (CFR). 1994. 10 CFR 830.120, "Nuclear Safety Management - Quality Assurance Requirements."

Code of Federal Regulations (CFR). 1997. 40 CFR 300.430, "National Oil and Hazardous Substance Pollution Contingency Plan – Remedial Investigation/Feasibility Study and Selection of Remedy."

U.S. Department of Defense (DOD). 1963. *Quality Program Requirements.* Military Specification MIL-Q-9858A. Washington, DC.

U.S. Department of Energy (DOE). 1991. *Quality Assurance.* DOE Order 414.1A (Replaced DOE Order 5700.6C), Washington, DC.

U.S. Environmental Protection Agency (EPA). 2000. *EPA Policy and Program Requirements for the Mandatory Agency-Wide Quality System.* EPA Order 5360.1A2, Revised. Washington, DC. Available at www.epa.gov/quality/qa_docs.html.

U.S. Environmental Protection Agency (EPA). 2001. *EPA Requirements for Quality Assurance Project Plans* (EPA QA/R-5). EPA/240/B-01/003. Office of Environmental Information, Washington, DC. Available at www.epa.gov/quality/qa_docs.html.

U. S. Environmental Protection Agency (EPA). 2002. *Guidance on Developing Quality Assurance Project Plans* (EPA QA/G-5). EPA/240/R-02/009. Office of Environmental Information, Washington, DC. Available at www.epa.gov/quality/qa_docs.html.

U.S. Environmental Protection Agency (EPA). 2003. *Using Dynamic Field Activities for On-Site Decision Making: A Guide for Project Managers.* EPA/540/R-03/002. Office of Solid Waste and Emergency Response, Washington, DC. Available at www.epa.gov/superfund/programs/dfa/index.htm.

International Organization for Standardization (ISO) 9001. *Quality Systems - Model for Quality Assurance in Design, Development, Installation and Servicing.* 1994.

U.S. Army Corps of Engineers (USACE). 1997. *Chemical Quality Assurance for Hazardous, Toxic and Radioactive Waste Projects.* Engineer Manual EM 200-1-6. Available at www.environmental.usace.army.mil/info/technical/chem/chemguide/chemusac/ chemusac.html.

U.S. Army Corps of Engineers (USACE). 2001. *Requirements for the Preparation of Sampling and Analysis Plans.* Engineer Manual EM 200-1-3. Available at www.environmental.usace. army.mil/info/technical/chem/chemguide/chemusac/chemusac.html.

U.S. Nuclear Regulatory Commission (NRC). 1989. *Standard Format and Content of Decommissioning Plans for Licensees Under 10 CFR Parts 30, 40, and 70.* Regulatory Guide 3.65.

5 OBTAINING LABORATORY SERVICES

5.1 Introduction

This chapter provides guidance on obtaining radioanalytical laboratory services. In particular, this chapter discusses the broad items that should be considered in the development of a procurement for laboratory services. Throughout this chapter, MARLAP uses the request for proposal (RFP) as an example of a procurement mechanism. Agencies and other organizations may use a variety of procurement mechanisms, depending upon circumstances and policies. The RFP typically includes a statement of work (SOW), generic contract requirements, and the description of the laboratory qualification and selection process. It should be noted that for some agencies or organizations, not all technical, quality, and administrative aspects of a contract are specified in a SOW; many are in the procurement document (RFP) or resulting contract. More detailed guidance and discussion on the content of the SOW and other contracting issues can be found in Appendix E (*Contracting Laboratory Services*). Appendix E includes types of procurement mechanisms (with emphasis on the request for proposal), typical proposal requirements, proposal evaluation and scoring, pre-award proficiency samples and audits, and post-award contract management. This chapter is written for contracting outside laboratory services, but the principal items and information provided would apply equally to similar services not requiring a formal contract, such as a service agreement within an Agency or organization. It should be noted that the information and specifications of a SOW may appear in many procurement documents other than a contract resulting from a RFP. These include purchase and work orders, task orders under existing Basic ordering agreements, or Government-wide Acquisition Contracts and Multiple Acquisition Schedule contracts, such as those offered by the U.S. General Services Administration. MARLAP recommends that technical specifications be prepared in writing in a single document, designated a SOW, for all radioanalytical laboratory services, regardless of whether the services are to be contracted out or performed by an Agency's laboratory.

Analytical protocol specifications (APSs) should be compiled in the SOW in order for the laboratory to propose the analytical protocols that the laboratory wishes to use for the project (Chapter 6). The development of APSs, which includes the measurement quality objectives (MQOs), is described in detail in Chapter 3, and the incorporation of these protocols into the relevant project plan documents is covered in Chapter 4. These specifications should include such items as the MQOs, the type and frequency of quality control (QC) samples, the level of performance demonstration needed, number and type of samples, turnaround times, and type of data package.

Contents	
5.1 Introduction	5-1
5.2 Importance of Writing a Technical and Contractual Specification Document	5-2
5.3 Statement of Work—Technical Requirements	5-2
5.4 Request for Proposal—Generic Contractual Requirements	5-7
5.5 Laboratory Selection and Qualification Criteria	5-11
5.6 Summary of Recommendations	5-15
5.7 References	5-16

Section 5.3 discusses the technical requirements of a SOW, Section 5.4 provides guidance on generic contractual requirements, and Section 5.5 discusses various elements of the laboratory selection and qualification criteria.

5.2 Importance of Writing a Technical and Contractual Specification Document

One objective of the SOW and contract documents is to provide the analytical requirements in a concise format that will facilitate the laboratory's selection of the appropriate analytical protocols. The authors of the SOW may be able to extract most, if not all, of the necessary technical information from properly prepared project plan documents (Chapter 4). If specific information is not available, the author should contact the planning team. The preparation of a SOW can be viewed as a check to make sure that the project planning documents contain all the information required for the selection and implementation of the appropriate analytical protocols. One important aspect of writing the SOW is that it should clearly identify the project laboratory's responsibility for documentation to be provided for subsequent data verification, validation, and quality assessment. These project laboratory requirements should be addressed in the assessment plans developed during directed planning (Chapter 2).

5.3 Statement of Work—Technical Requirements

A review of the project plan documents (Chapter 4) should result in a summary list of the technical requirements needed to develop a SOW. Much of this information, including the project MQOs and any unique analytical process requirements, will be contained in the APSs. When possible, a project summary of sufficient detail (i.e., process knowledge) to be useful to the laboratory should be included in the SOW. The project planning team is responsible for identifying and resolving key analytical planning issues and for ensuring that the resolutions of these issues are captured in the APSs. Consistent with a performance-based approach, the level of specificity in the APSs is limited to those requirements that are essential to meeting the project's analytical data requirements. In response to such project management decisions, the laboratory may propose for consideration several alternative validated methods that meet the MQOs under the performance-based approach (such as measurement of a decay progeny as an alternate radionuclide; see Section 6.6, "Method Validation"). Chapter 7 provides guidance on the evaluation of a laboratory and analytical methods.

The SOW should specify what the laboratory needs to provide in order to demonstrate its ability to meet the technical specifications in the RFP. This should include documentation relative to the method validation process to demonstrate compliance with the MQOs and information on previous contracts for similar analytical work as well as performance in performance evaluation (PE) programs using the proposed method. Any specific requirements on sample delivery

(Section 5.3.7) should also be made clear to the laboratory. In addition, the requirements for the laboratory's quality system should be discussed.

5.3.1 Analytes

Each APS should state the analyte of concern. The SOW should specify all analytes of concern and, when possible, an analyte's expected chemical form and anticipated concentration range (useful information for separating high activity samples from low activity samples) and potential chemical or radiometric interferences (Sections 3.3.1, "Develop Analyte List," and 3.3.2, "Identify Concentration Ranges"). In some instances, because of process knowledge and information on the absence of equilibrium between analytes and their parents and progeny, the SOW may require the direct measurement of an analyte rather than allowing for the measurement of other radionuclides in the analyte's decay chain. In these cases, the SOW should indicate the analyses to be performed. Examples of analyses include gross alpha and beta, gamma spectrometry, and radionuclide/matrix-specific combinations such as ^{3}H in water and ^{238}Pu in soil.

5.3.2 Matrix

Each APS should state the sample matrix to be analyzed. The sample matrix for each radionuclide or analysis type (e.g., gamma-ray spectrometry) should be listed and described in detail where necessary. The matrix categories may include surface soil, sub-surface soil, sediment, sludge, concrete, surface water, ground water, salt water, aquatic and terrestrial biota, air, air sample filters, building materials, etc. Additional information should be provided for certain matrices (e.g., the chemical form of the matrix for solid matrices) in order for the laboratory to select the appropriate sample preparation or dissolution method (Section 3.3.3, "Identify and Characterize Matrices of Concern").

5.3.3 Measurement Quality Objectives

The APSs should provide the MQOs for each analyte-matrix combination. The MQOs can be viewed as the analytical portion of the overall project data quality objectives (DQOs). An MQO is a statement of a performance objective or requirement for a particular method performance characteristic. Examples of method performance characteristics include the method's uncertainty at some concentration, detection capability, quantification capability, specificity, analyte concentration range, and ruggedness. An example MQO for the method uncertainty at some analyte concentration such as the action level would be, "A method uncertainty of 0.5 Bq/g or less is required at the action level of 5.0 Bq/g" (Chapters 1, 3, and 19). The MQOs are a key part of a project's APSs. Chapter 3 provides guidance on developing MQOs for select method performance characteristics.

5.3.4 Unique Analytical Process Requirements

The APS should state any unique analytical processing requirement. The SOW should give any matrix-specific details necessary for the laboratory to process the sample, such as type of soil, type of debris to be removed, whether or not filtering a sample at the laboratory is required, processing whole fish versus edible parts, drying of soils, information on any known or suspected interferences, hazards associated with the sample, etc. (see Section 3.4, "Matrix-Specific Analytical Planning Issues"). In some cases, unique analytical process requirements or instructions should be specified that further delineate actions to be taken in case problems occur during sample processing. For example, the SOW may require that the laboratory reprocess another aliquant of the sample by a more robust technique when a chemical yield drops below a stated value.

If necessary, special instructions should be provided as to how or when the analytical results are to be corrected for radioactive decay or ingrowth. In some cases, the sample collection date may not be the appropriate date to use in the decay or ingrowth equations.

5.3.5 Quality Control Samples and Participation in External Performance Evaluation Programs

The SOW should state the type and frequency of internal QC samples needed as well as whether they are to be included on a batch or some other basis (see Chapter 18, *Laboratory Quality Control*). The batch size may be defined in the SOW and may vary depending on the analysis type. The quality acceptance limits for all types of QC samples should be stated (see Appendix C for guidance on developing acceptance limits for QC samples based on the MQO for method uncertainty). In addition, the SOW should state when and how the project manager or the contracting officer's representative should be notified about any nonconformity. In addition, the SOW should spell out the conditions under which the laboratory will have to reanalyze samples due to a nonconformance.

The evaluation of the laboratory's ability to perform the required radiochemical analyses should be based on the acceptability of the method validation documentation submitted by the laboratory. The evaluation should also include the laboratory's performance in various external PE programs administered by government agencies or commercial radioactive source suppliers that are traceable to a national standards laboratory or organization, such as the National Institute of Standards and Technology (NIST). The source supplier's measurement capabilities and manufacturing processes should be linked to NIST according to ANSI N42.22 (additional information on evaluating a laboratory's performance is provided in Chapter 7). As such, the RFP should request the laboratory's participation in a PE program, traceable to a national standards organization, appropriate for the analytes and matrices under consideration. In addition, the weighting factor (Appendix C) given to scoring the laboratory's performance in such a program should be provided to the laboratory. Some examples of government programs include

DOE's Quality Assessment Program (QAP) and the Mixed Analyte Performance Evaluation Program (MAPEP) and the NIST-administered National Voluntary Laboratory Accreditation Program (NVLAP) Performance Testing (PT) providers.

5.3.6 Laboratory Radiological Holding and Turnaround Times

The SOW should include specifications on the required laboratory radiological holding time (i.e., the time between the date of sample collection and the date of analysis) and the sample processing turnaround time (i.e., the time between the receipt of the sample at the laboratory to the reporting of the analytical results). Such radiological holding and turnaround times, which are usually determined by specific project requirements, are typically specified in terms of calendar or working days. The SOW should state whether the laboratory may be requested to handle expedited or rush samples. In some cases, time constraints become an important aspect of sample processing (e.g., in the case of radionuclides that have short half-lives). Some analyses will call for specific steps that take a prescribed amount of time. Requesting an analytical protocol that requires several days to complete is obviously not compatible with a 24-hour turnaround time. This highlights the need for input from radioanalytical specialists during the planning process.

In some cases, the required sample-processing turnaround times are categorized according to generic headings such as routine, expedited or rush, and emergency sample processing. Under these circumstances, the SOW should specify the appropriate category for the samples and analyses.

5.3.7 Number of Samples and Schedule

Estimating the volume of work for a laboratory is commonly considered part of the planning process that precedes the initiation of a project. Thus, the SOW should estimate the anticipated amount of work and should spell out the conditions under which the laboratory will have to reanalyze samples due to some non-conformance. Similarly, the estimate should allow the laboratory to judge if its facility has the capacity to compete for the work. The estimate for the number of samples is a starting point, and some revision to the volume of work may occur, unless the laboratory sets specific limits on the number of samples to be processed.

The SOW should indicate whether samples will be provided on a regular basis, seasonally, or on some other known or unknown schedule. It should also be specified if some samples may be sent by overnight carrier for immediate analysis. Holidays may be listed when samples will not be sent to the laboratory. The SOW should state if Saturday deliveries may be required. Furthermore, it should specify whether samples will be sent in batches or individually, and from one location or different locations.

The carrier used to ship samples to the laboratory should be experienced in the delivery of field samples, provide next day and Saturday deliveries, have a package tracking system and be familiar with hazardous materials shipping regulations.

5.3.8 Quality System

The RFP should require that a copy of the laboratory's Quality System documentation (such as a Quality Manual), related standard operating procedures (including appropriate methods) and documentation (such as a summary of the internal QC and external PE sample results) be included with the proposal, as necessary. Only those radioanalytical laboratories that adhere to a well-defined quality system can ensure the appropriate quality of scientifically valid and defensible data. The laboratory's Quality System (NELAC, 2002; ANSI N42.23; ISO/IEC 17025) for a radioanalytical laboratory should address at a minimum the following items:

- Organization and management;
- Quality system establishment, audits, essential quality controls and evaluation and data verification;
- Personnel (qualifications and resumes);
- Physical facilities—accommodations and environment;
- Equipment and reference materials;
- Measurement traceability and calibration;
- Test methods and standard operating procedures (methods);
- Sample handling, sample acceptance policy and sample receipt;
- Records;
- Subcontracting analytical samples;
- Outside support services and supplies; and
- Complaints.

The Intergovernmental Data Quality Task Force (IDQTF) has developed a *Uniform Federal Policy for Implementing Environmental Quality Systems*. Agencies participating in the IDQTF are the Environmental Protection Agency, Department of Defense, and Department of Energy. The *Uniform Federal Policy* is a consensus document prepared by the IDQTF work group and it provides recommendations and guidelines for documentation and implementation of acceptable quality systems for federal agencies. Information on IDQTF and this policy may be found at www.epa.gov/swerffrr/documents/data_quality/ufp_sep00_intro.htm#quality.

5.3.9 Laboratory's Proposed Methods

Under the performance-based approach to method selection, the laboratory will select and identify radioanalytical methods (Chapter 6) that will meet the MQOs and other performance specifications of the SOW. MARLAP recommends that the laboratory submit the proposed methods and required method validation documentation with the formal response. The SOW should state that the proposed methods and method validation documentation will be evaluated in accordance with established procedures by a technical evaluation committee (TEC) based on experience, expertise, and professional judgement. MARLAP uses the term TEC for the group that performs this function. Agencies and other organizations may use various terms and procedures for this process.

The TEC should provide their findings and recommendations to the organization's contracting officer for further disposition. In some cases, the organization may inform a laboratory that the proposed methods were deemed inadequate, and, if appropriate, request that the laboratory submit alternative methods with method validation documentation within a certain time period.

When the methods proposed by the laboratories have been deemed adequate to meet the technical specifications of the SOW, the TEC may want to rank the proposed methods (and laboratories) according to various factors (e.g., robustness, performance in PE programs or qualifying samples, etc.) as part of the contract scoring process.

5.4   Request for Proposal—Generic Contractual Requirements

Not all quality and administration aspects of a contract are specified in a SOW. Many quality (e.g., requirement for a quality system), administrative, legal, and regulatory items need to be specified in a RFP and eventually in the contract. Although not inclusive, the items or categories discussed in the following sections should be considered as part of the contractual requirements and specifications of a RFP.

5.4.1 Sample Management

The RFP should require the laboratory to have an appropriate sample management program that includes those administrative and quality assurance aspects covering sample receipt, control, storage and disposition. The RFP should require the laboratory to have adequate facilities, procedures, and personnel in place for the following actions (see Chapter 11, *Sample Receipt, Inspection, and Tracking*; and Chapter 17, *Waste Management in a Radioanalytical Laboratory*):

- Receive, log-in, and store samples in a proper fashion to prevent deterioration, cross-contamination, and analyte losses;

- Verify the receipt of each sample shipment: compare shipping documentation with samples actually received; notify the point of contact or designee by telephone within a prescribed number of business days and subsequently provide details in all case narratives of any discrepancies in the documentation;

- Sign, upon receipt of the samples, the sample receipt form or, if required, chain of custody (COC) form(s) submitted with each sample release. Only authorized laboratory personnel should sign the forms. The signature date on the COC form, if required, is normally the official sample receipt date. All sample containers should be sealed prior to their removal from the site; and

- Store unused portions of samples in such a manner that the analyses could be repeated or new analyses requested, if required, for a certain specified time period following the submission of an acceptable data package. Unused sample portions should be stored with the same sample handling requirements that apply to samples awaiting analysis. Documentation should be maintained pertaining to storage conditions and sample archival or disposal.

- Treat, store, or dispose of sample processing wastes, test and calibration sources, and samples (see also Section E.4.4.5, "Sample Storage and Disposal," and Chapter 17)

5.4.2   Licenses, Permits and Environmental Regulations

Various federal, state, and local permits, licences and certificates (accreditation) may be necessary for the operation of a radioanalytical laboratory. The RFP should require the laboratory to have the necessary government permits, licenses, and certificates in place before the commencement of any laboratory work for an awarded contract. The following sections provide a partial list of those provisions that may be necessary. Some projects may require special government permits in order to conduct the work and transport and analyze related samples. For these cases, the necessary regulations or permits should be cited in the RFP.

5.4.2.1 Licenses

When required, the laboratory will be responsible for maintaining a relevant Nuclear Regulatory Commission (NRC) or Agreement State License to accept low-level radioactive samples for analyses. In certain circumstances, the laboratory may have to meet host nation requirements if operating outside the United States (e.g., military fixed or deployed laboratories located overseas).

When necessary, the laboratory should submit a current copy of the laboratory's radioactive materials license with their proposal. Some circumstances may require a copy of the original radioactive materials license. For more complete information on license requirements, refer to either the NRC or state government offices in which the laboratory resides, or to 10 CFR 30.

5.4.2.2 Environmental and Transportation Regulations

Performance under a contract or subcontract must be in compliance with all applicable local, state, federal, and international laws and regulations. Such consideration must not only include relevant laws and regulations currently in effect, but also revisions thereto or public notice that has been given that may reasonably be anticipated to be effective during the term of the contract.

The laboratory may be required to receive (and in some cases ship) samples according to international, federal, state, and local regulations (see Section 10.2.10, "Packaging and Shipping," for details). In particular, the laboratory should be aware of U.S. Postal Service and Department of Transportation (DOT) hazardous materials regulations applicable to the requirements specified in the SOW and that appropriate personnel should be trained in these regulations. International shipping also is subject to International Air Transport Association Dangerous Goods Regulations.

5.4.3 Data Reporting and Communications

The type of information, schedules and data reports required to be delivered by the laboratory, as well as the expected communications between the appropriate staff or organizations, should be delineated in the RFP. The required schedule and content of the various reports, including sample receipt acknowledgment, chain of custody, final data results, data packages, QA/QC project summaries, status reports, sample disposition, and invoices should be provided in the RFP. In addition, the expected frequency and lines of communications should be specified.

In some cases, the RFP may request relevant information relative to the point-of-contact for certain key laboratory positions such as the Laboratory Director, Project Manager, QA Officer, Sample Manager, Record Keeping Supervisor, Radiation Safety or Safety Officer and Contracting Officer. Contact persons should be identified along with appropriate telephone numbers (office, FAX, pager), e-mail, and postal and courier addresses.

5.4.3.1 Data Deliverables

The SOW should specify what data are required for data verification, validation, and quality assessment. A data package, the pages of which should be sequentially numbered, may include a project narrative, the results in a specified format including units, a data review checklist, any non-conformance memos resulting from the work, sample receipt acknowledgment or chain of custody form (if required), sample and quality control sample data, calibration verification data, and standard and tracer information. In addition, the date and time of analysis, instrument identification, and analyst performing the analysis should be included on the appropriate paperwork. At the inception of the project, initial calibration data may be required for the detectors used for the work. When a detector is recalibrated, or a new detector is placed in service, updated calibration data should be required whenever those changes could affect the analyses in question. In some cases, only the summary or final data report may be requested. In these cases, the name of the data reviewer, the sample identification information, reference and analysis dates, and the analytical results along with the reported measurement uncertainties should be reported.

The SOW should specify the acceptable formats for electronic and hard copy records. The SOW also should state at what intervals the data will be delivered (batch, monthly, etc.).

5.4.3.2 Software Verification and Control

The policy for computer software verification, validation and documentation typically are included in the laboratory's Quality Manual. If there are specific software verification and validation requirements germane to the project, the RFP should instruct or specify such requirements. ASTM E919, "Standard Specification for Software Documentation for a Computerized System," describes computer program documentation that should be provided by a software supplier. Other sources for software QC are ANSI ANS 10.3 "Documentation of Computer Software" and IEEE Standard 1063, "IEEE Standard for Software User Documentation."

5.4.3.3 Problem Notification and Communication

Communication is key to the successful management and execution of the contract. Problems, schedule delays, potential overruns, etc., can be resolved quickly only if communication between the laboratory and organization's representative is conducted promptly. The RFP should state explicitly when, how, and in what time frame communication or notification is required by the laboratory for special technical events, such as the inability to meet MQO specifications for a sample or analyte, when a QC sample result is outside of an acceptance limit or some other non-conformance and when—if required by the project manager—the laboratory fails to meet its internal QC specifications.

The laboratory should document and report all deviations from the method and unexpected observations that may be of significance to the data reviewer or user. Such deviations should be documented in the narrative section of the data package produced by the contract laboratory. Each narrative should be monitored closely to assure that the laboratory is documenting departures from contract requirements or acceptable practice.

Communication from the organization's representative to the laboratory is also important. A key element in managing a contract is the timely review of the data packages provided by the laboratory. Early identification of problems allows for corrective actions to improve laboratory performance and, if necessary, the cessation of laboratory analyses until solutions can be instituted to prevent the production of large amounts of data that are unusable. Note that some sample matrices and processing methods can be problematic for even the best laboratories. Thus, the organization's technical representative must be able to discern between failures due to legitimate reasons and poor laboratory performance.

5.4.3.4 Status Reports

The SOW may require the laboratory to submit, on a specified frequency, sample processing status reports that include such information as the sample identification number, receipt date, analyses required, expected analytical completion date and report date. Depending on the project's needs, a status report may include the disposition of remaining portions of samples following sample processing or sample processing wastes.

5.4.4 Sample Re-Analysis Requirements

There may be circumstances when samples should be reanalyzed due to questionable analytical results or suspected poor quality as reflected by the laboratory's batch QC or external PT samples. Specific instructions and contractual language should be included in the RFP that address such circumstances and the resultant fiscal responsibilities (Appendix E).

5.4.5 Subcontracted Analyses

MARLAP recommends that the RFP state that subcontracting will be permitted only with the contracting organization's approval. In addition, contract language should be included giving the contracting organization the authority to approve proposed subcontract laboratories. For continuity or for quality assurance, the contract may require one laboratory to handle the entire analytical work load. However, the need may arise to subcontract work to another laboratory facility if the project calls for a large number of samples requiring quick turnaround times or specific methodologies that are not part of the primary laboratory's support services. The use of multiple service providers adds complexity to the organization's tasks of auditing, evaluating and tracking services.

Any intent to use a subcontracted laboratory should be specified in the response to the RFP or specific task orders. The primary laboratory should specify which laboratory(ies) are to be used, require that these laboratories comply with all contract or task order requirements, and verify that their operations can and will provide data quality meeting or exceeding the SOW requirements. Subcontract laboratories should be required to allow the contracting organization full access to inspect their operations, although it should be understood that the primary laboratory should maintain full responsibility for the performance of subcontract laboratories.

5.5 Laboratory Selection and Qualification Criteria

A description of the laboratory qualification and selection process should be stated in the RFP. The initial stages of the evaluation process focus on the technical considerations only. Cost will enter the selection process later. The organization's TEC considers all proposals and then makes an initial selection (see Figures E.6a and E.6b in Appendix E), at which time some laboratories may be eliminated based on the screening process. The laboratory selection process is based on predetermined criteria that are related to the RFP and how a laboratory is technically able to support the contract. A laboratory that is obviously not equipped to perform work according to the RFP is certain to be dropped early in the selection process. In some cases, the stated ability to meet the analysis request may be verified by the organization, through pre-award audits and proficiency testing as described below. Letters notifying unsuccessful bidders may be sent at this time.

5.5.1 Technical Proposal Evaluation

The RFP requires each bidding contractor laboratory to submit a technical proposal and a copy of its Quality Manual. This Quality Manual is intended to address all of the technical and general laboratory requirements. As noted previously, the proposal and Quality Manual are reviewed by members of the TEC who are both familiar with the proposed project and are clearly knowledgeable in the field of radiochemistry and laboratory management.

5.5.1.1 Scoring and Evaluation Scheme

The RFP should include information concerning scoring of proposals or weighting factors for areas of evaluation. This helps a laboratory to understand the relative importance of specific sections in a proposal and how a proposal will be evaluated or scored. This allows the laboratory to focus on those areas of greater importance. If the laboratory submits a proposal that lacks sufficient information to demonstrate support in a specific area, the organization can then indicate how the proposal does not fulfill the need as stated in the request. Because evaluation formats differ from organization to organization, laboratories may wish to contact the organization for additional organization-specific details concerning this process. A technical evaluation sheet (TES) may be used in conjunction with the proposal evaluation plan as outlined in the next section (see Figures E.6a and E.6b in Appendix E) to list the total weight for each factor and to provide a space for the evaluator's assigned rating. In the event of a protest, the TES can be used to substantiate the selection process. The TES also provides areas to record the RFP number, identity of the proposer, and spaces for total score, remarks, and evaluator's signature. The scoring and evaluation scheme is based on additional, more detailed, considerations which are discussed briefly in the Sections E.4 and E.5 of Appendix E.

Once all proposals are accepted by the organization, the TEC scores the technical portion of the proposal. MARLAP recommends that all members of the TEC have a technical understanding of the subject matter related to the proposed work. These individuals are also responsible for responding to any challenge to the organization's selection for the award of the contract. Their answers to such challenges are based on technical merit in relation to the proposed work.

5.5.1.2 Scoring Elements

Although each organization may have a different scoring process to evaluate a laboratory's response to a RFP, there are various broad categories or common elements that are typically evaluated. For example, these may include the following:

- Technical merit;
- Adequacy and suitability of laboratory resources and equipment;
- Staff qualifications;
- Related experience and record of past performance; and
- Other RFP requirements.

Although each organization may score or weight these items differently, performance-based contracting requires the weighting of past performance of the contractor as a significant technical element. Each of these elements is considered in the following paragraphs. Outlined below are the key elements that are discussed in more detail in Appendix E.

TECHNICAL MERIT

The response to the RFP should include details of the laboratory's quality system and all the analytical methods to be employed by the laboratory as well as the method validation documentation (Section 6.6). The information provided should outline or demonstrate that the methods proposed are likely to be suitable and meet the APSs. The methods should be evaluated against the APSs and MQOs provided in the SOW. Chapter 7 provides guidance on the evaluation of methods and laboratories. The laboratory's Quality Manual should be reviewed for adequacy and completeness to ensure the required data quality.

ADEQUACY AND SUITABILITY OF LABORATORY RESOURCES AND EQUIPMENT

When requested, the laboratory will provide a listing of the available instrumentation or equipment by analytical method category. In addition, the RFP may request information on the available sample processing capacity and the workload for other clients during the proposed contract period. The information provided should be evaluated by the TEC to determine if the laboratory has the sample processing capacity to perform the work. The instrumentation and equipment must be purchased, set-up, calibrated, and on-line before award of contract. In addition, the laboratory should provide information relative to the adequacy and suitability of the laboratory space available for the analysis of samples.

STAFF QUALIFICATIONS

The RFP should require the identification of the technical staff and their duties, along with their educational background and experience in radiochemistry, radiometrology or laboratory operations. The laboratory staff that will perform the radiochemical analyses should be employed and trained prior to the award of the contract. Appendix E provides guidance on staff qualifications.

RELATED EXPERIENCE AND RECORD OF PAST PERFORMANCE

The RFP should require the laboratory to furnish references in relation to its past or present work. To the extent possible, this should be done with regard to contracts or projects similar in composition, duration and number of samples to the proposed project. In some cases, the laboratory's previous performance for the same Agency may be given special consideration.

OTHER RFP REQUIREMENTS

Within the response to the RFP, the laboratory should outline the various programs and commitments (QA, safety, waste management, etc.) as well as submit various certifications, licences, and permits to ensure the requirements of the RFP will be met. The reasonableness of the proposed work schedule, program, and commitments should be evaluated by the TEC. In addition, if accreditation is required in the RFP, the TEC should confirm the laboratory's accreditation for radioanalytical services by contacting the organization that provided the certification. The National Environmental Laboratory Accreditation Conference (NELAC) is an organization formed to establish and promote performance standards for the inspection and operation of environmental laboratories in support of the National Environmental Laboratory Program (NELAP). States and federal agencies serve as the accrediting authorities within NELAP. If state-accredited, a laboratory typically is accredited by the state in which it resides, and if the state is a NELAP-recognized accrediting authority, the accreditation is recognized by other states and federal agencies approved under NELAP. If the state is not a NELAP-recognized accrediting authority, and an organization expects a laboratory to process samples from other states or the federal government, then additional accreditations may be required. The TEC should review and confirm the applicability and status of the licenses and permits with respect to the technical scope and duration of the project.

5.5.2 Pre-Award Proficiency Evaluation

Some organizations may elect to send proficiency or PT samples (sometimes referred to as "performance evaluation" or "PE" samples) to the laboratories that meet a certain scoring criteria in order to demonstrate the laboratory's analytical capability. The composition and number of samples should be determined by the nature of the proposed project. The PT sample matrix should be composed of well-characterized materials. It is recommended that site specific PT matrix samples or method validation reference material (MVRM; see Section 6.5.1, "Matrix and Analyte Identification") be used when available.

Each competing lab should receive an identical set of PT samples. The RFP should specify who will bear the cost of analyzing these samples as well as the scoring scheme (e.g., "pass/fail" or a sliding scale). Any laboratory failing to submit results should be disqualified. The results should be evaluated and each laboratory given a score. This allows the organization to make a second cut—after which only two or three candidate laboratories are considered.

5.5.3 Pre-Award Assessments and Audits

The RFP should indicate that the laboratories with the highest combined scores for technical proposals and proficiency samples may be given an on-site audit. A pre-award assessment or audit may be performed to provide assurance that a selected laboratory is capable of fulfilling the contract in accordance with the RFP. In other words, is the laboratory's representation of itself accurate? To answer this question, auditors should be looking to see that a candidate laboratory appears to have all the required elements to meet the proposed contract's needs. Refer to Appendix E for details on the pre-award assessments and audits.

5.6 Summary of Recommendations

- MARLAP recommends that technical specifications be prepared in writing in a single document, designated a SOW, for all radioanalytical laboratory services, regardless of whether the services are to be contracted out or performed by an Agency's laboratory.

- MARLAP recommends that the MQOs and analytical process requirements contained in the SOW be provided to the laboratory.

- MARLAP recommends that the SOW include the specifications for the action level and the required method uncertainty for the analyte concentration at the action level for each analyte/matrix.

- MARLAP recommends that the laboratory submit the proposed methods and required method validation documentation with the formal response.

- MARLAP recommends that the RFP state that subcontracting will be permitted only with the contracting organization's approval.

- MARLAP recommends that all members of the TEC have a technical understanding of the subject matter related to the proposed work.

5.7 References

5.7.1 Cited References

American National Standard Institute (ANSI) N42.23. *Measurement and Associated Instrumentation Quality Assurance for Radioassay Laboratories.* 2003.

American National Standard Institute (ANSI) ANS 10.3. *Documentation of Computer Software.*

American Society for Testing and Materials (ASTM) E919. Standard Test Methods for *Software Documentation for a Computerized System.* West Conshohocken, PA.

International Electrical and Electronics Engineers (IEEE). Standard 1063. *Software User Documentation.*

International Standards Organization/International Electrotechnical Commission (ISO/IEC) 17025. *General Requirements for the Competence of Testing and Calibration Laboratories,* International Organization for Standardization, Geneva, Switzerland. December 1999, 26 pp.

National Environmental Laboratory Accreditation Conference (NELAC). 2002. *NELAC Standards.* NELAC, Washington, DC. Available at: www.epa.gov/ttn/nelac/2002standards. html.

5.7.2 Other Sources

U.S. Department of Energy (DOE). 2001. *Management Assessment and Independent Assessment Guide.* DOE G 414.1-1A. May. Available at: www.directives.doe.gov/pdfs/doe/doetext/ neword/414/g4141-1a.html.

U.S. Department of Energy (DOE). 1997. *Model Statement of Work for Analytical Laboratories.* Albuquerque Operations Office. Prepared by AGRA Earth and Environmental, Inc., Albuquerque, NM. March.

U.S. Department of Energy (DOE). 1999. *Quality Assurance Management System Guide for use with 10 CFR 830.120 and DOE O 414.1.* DOE G 414.1-2. June. Available at: www. directives.doe.gov/cgi-bin/explhcgi?qry1666663647;doe-120#HL002.

U.S. Department of Energy (DOE). 2001. *Management Assessment And Independent Assessment Guide For Use With 10 CFR, Part 830, Subpart A, and DOE O 414.1A, Quality Assurance; DOE P 450.4, Safety Management System Policy; and DOE P 450.5, Line ES&H Oversight*

Policy. DOE G 414.1-1A. May. Available at: www.directives.doe.gov/pdfs/doe/doetext/ neword/414/g4141-1a.html.

U. S. Environmental Protection Agency (EPA). 2002. *Guidance for Quality Assurance Project Plans.* EPA QA/G-5. EPA/240/R-02/009. Office of Environmental Information, Washington, DC. Available at www.epa.gov/quality/qa_docs.html.

U.S. Nuclear Regulatory Commission (NRC). 1994. *NRC Procedures for Placement and Monitoring of Work With the U.S. Department of Energy (DOE).* Directive 11.7, May 3.

U.S. Nuclear Regulatory Commission (NRC). 1996. *NRC Acquisition of Supplies and Services.* Directive 11.1, July 23.

Office of Federal Procurement Policy (OFPP) 1997. *Performance-Based Service Contracting (PBSC) Solicitation/Contract/Task Order Review Checklist.* August 8. Available at: www.arnet.gov/Library/OFPP/PolicyDocs/pbscckls.html.

6 SELECTION AND APPLICATION OF AN ANALYTICAL METHOD

6.1 Introduction

This chapter provides guidance to both the project manager and the laboratory on the selection and application of analytical method. It offers guidance to the project manager on the development of the analytical protocol specifications (APSs) from the laboratory's perspective on method appropriateness and availability. It offers guidance to the laboratory on the key elements to consider when selecting an analytical method (Section 1.4.5, "Analytical Protocol") to meet the objectives of the APSs contained in the statement of work (SOW). Assuming that the laboratory has received a SOW, certain subsections within Section 6.5 provide guidance on how to review and properly evaluate the APSs therein. However, Section 6.5 also provides guidance for the project planning team on the important laboratory considerations needed to develop the measurement quality objectives (MQOs). Section 6.6 deals with method validation requirements and has been written for both the project planners and the laboratory.

Because the method constitutes the major part of the analytical protocol (Chapter 1), this chapter focuses on the selection of a method. However, other parts of the protocol should be evaluated for consistency with the method (Figure 6.1). MARLAP recommends the performance-based approach for method selection. Thus, the laboratory should be able to propose whichever method meets the project's analytical data requirements (MQOs), within constraints of other factors such as regulatory requirements, cost, and project deadlines. The selection of a method by the laboratory is in response to the APSs (Chapter 3) that were formulated during the directed planning process (Chapter 2) and documented in the SOW (Chapter 5, *Obtaining Laboratory Services*). In most project plan documents, the project manager or the project planning team has the authority and responsibility for approving the methods proposed by the laboratory. The APSs will, at a minimum, document the analytes, sample matrices, and the MQOs. A MQO is a statement of a performance objective or requirement for a particular method performance characteristic. The MQOs can be viewed as the analytical portion of the data quality objectives (DQOs; see Chapter 3).

Background material in Section 6.2.1 provides the reader with the subtleties of the performance-based approach to method selection, contrasted with the use of prescribed methods

Contents

6.1	Introduction	6-1
6.2	Method Definition	6-3
6.3	Life Cycle of Method Application	6-5
6.4	Generic Considerations for Method Development and Selection	6-9
6.5	Project-Specific Considerations for Method Selection	6-11
6.6	Method Validation	6-22
6.7	Analyst Qualifications and Demonstrated Proficiency	6-32
6.8	Method Control	6-33
6.9	Continued Performance Assessment	6-34
6.10	Documentation To Be Sent to the Project Manager	6-35
6.11	Summary of Recommendations	6-36
6.12	References	6-36
Attachment 6A: Bias-Testing Procedure		6-39

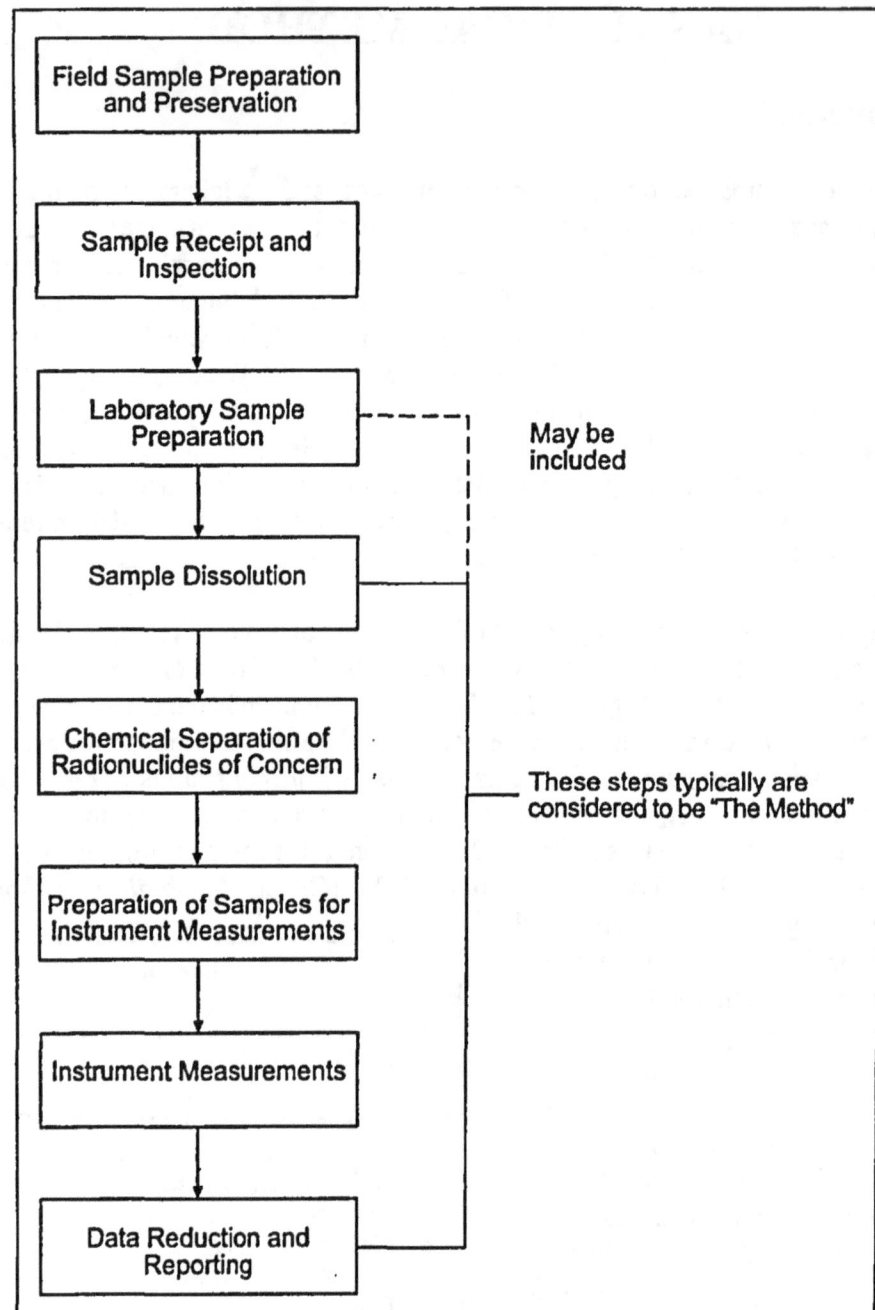

FIGURE 6.1 — Analytical process

and the importance of the directed planning process and MQOs in the selection of the method.
This chapter does not provide a listing of existing methods with various attributes indexed to
certain applications. Analytical methods may be obtained from national standards bodies,
government laboratories and publications, and the open literature.

In this chapter, project method validation is defined as the demonstration of method applicability for a particular project. MARLAP recommends that only methods validated for a project's application be used. This recommendation should not be confused with the general method validation that all methods should undergo during method development. The laboratory should validate the method to the APS requirements of a SOW for the analyte/matrix combination and provide the project method validation documentation to the project manager prior to the implementation of routine sample processing (Section 6.6.2). If applicable, consideration should be given to the uncertainty of the laboratory's protocol for subsampling (heterogeneity) of the received field sample when selecting a method. Appendix F provides guidance on the minimization of subsampling uncertainty.

Section 6.3 provides an overview of the generic application of a method for a project and how a laboratory meets the recommendations of the guidance provided in this and other chapters. Generic considerations for the method selection process that a laboratory should evaluate are provided in Section 6.4. Project-specific considerations for method selection relevant to APSs are discussed in Section 6.5. Recommendations on the degree of project method validation specified by the project planning team are outlined in Section 6.6. Sections 6.7, 6.8, and 6.9 provide guidance on analyst qualifications, method control, and continued laboratory performance assessment, respectively. Section 6.10 outlines recommendations for the method proposal and method validation documentation that a laboratory should send to the project manager.

6.2   Method Definition

For this chapter, a laboratory "method" includes all physical, chemical, and radiometric processes conducted at a laboratory in order to provide an analytical result. These processes, depicted in Figure 6.1, may include sample preparation, dissolution, chemical separation, mounting for counting, nuclear instrumentation counting, and analytical calculations. This chapter will emphasize the laboratory's selection of the radioanalytical method that will be proposed in response to a SOW. Each method is assumed to address a particular analyte in a specified matrix or, in some cases, a group of analytes having the same decay emission category that can be identified through spectrometric means (e.g., gamma-ray spectrometry). However, it should be emphasized that the project planning team should have evaluated every component of the APSs for compatibility with respect to all analytes in a sample and the foreseen use of multiple analytical methods by the laboratory. For example, samples containing multiple analytes must be of sufficient size (volume or mass) to ensure proper analysis and to meet detection and quantification requirements. Multiple analytes in a sample will require multiple analyses for which a laboratory may use a sequential method that addresses multiple analytes or stand-alone individual methods for each analyte. The analytical protocol must ensure that the samples are properly preserved for each analyte and sufficient sample is collected in the field to accommodate the analytical requirements.

Certain aspects of a method are defined in this chapter in order to facilitate the method selection process. The following subsections describe the underlying basis of a performance-based approach to method selection and provide a functional definition related to MARLAP.

Performance-Based Approach and Prescriptive Method Application

MARLAP uses a performance-based approach to selecting a method, which is based on a demonstrated capability to meet defined project performance criteria (e.g., MQOs). With a properly implemented quality system, a validated method should produce appropriate and technically defensible results under the applicable conditions. The selection of any new method usually requires additional planning and, in some cases, may result in additional method development or validation. The selection of a method under the performance-based approach involves numerous technical, operational, quality, and economic considerations. However, the most important consideration in the selection of a method under the performance-based approach is compliance with the required MQOs for the analytical data. These requirements should be defined in the SOW or appropriate project plan document.

When developing the MQOs, the project planning team should have evaluated all processes that have a potential to affect the analytical data. Those involved in the directed planning process should understand and communicate the needs of the project. They should also understand how the sampling (field, process, system, etc.) and analytical activities will interact and the ramifications that the data may have on the decisionmaking process. These interactive analysis and communication techniques should be applied in all areas where analytical data are produced. As new projects are implemented, it should not be assumed that the current methods are necessarily the most appropriate and accurate; they should be reevaluated based on project objectives. The application of a performance-based approach to method selection requires the quantitative evaluation of all aspects of the analytical process. Once the MQOs for a project have been determined and incorporated into the APSs, under the performance-based approach, the laboratory will evaluate its existing methods and propose one or more methods that meet each APS. This chapter contains guidance on how to use the APSs in the laboratory's method evaluation process.

The objective of a performance-based approach to method selection is to facilitate the selection, modification, or development of a method that will reliably produce quality analytical data as defined by the MQOs. Under the performance-based approach, a laboratory, responding to a SOW, will propose a method that best satisfies the requirements of the MQO and the laboratory's operations.

In certain instances, the requirement to use prescribed methods may be included in the SOW. The term "prescribed methods" has been associated with those methods that have been selected by industry for internal use or selected by a regulatory agency, such as the U.S. Environmental Protection Agency (EPA), for specific programs. The methods for analyzing radionuclides in drinking water prescribed by EPA (1980) provides an example of applying a limited number of methods to a well-defined matrix. In many companies or organizations, prescribed methods are widely used. Methods that have been validated for a specific application by national standard setting organizations such as the American Society for Testing and Materials (ASTM), American National Standards Institute (ANSI), American Public Health Association (APHA), etc., may also be used as prescribed methods by industry and government agencies.

Typically, the prescribed methods were selected by an organization to meet specific objectives for a regulation under consideration or for a program need. In most cases, the prescribed methods had undergone some degree of method validation, and the responsible organization had required a quality system to demonstrate continued applicability and quality, as well as laboratory proficiency. The use of any analytical method, whether prescribed or from the performance-based approach, has a life cycle that can be organized into the major categories of selection, validation, and continued demonstrated capability and applicability. This chapter will cover in detail only the first two of these categories. A discussion on ongoing laboratory evaluations is presented in Chapter 7 (*Evaluating Methods and Laboratories*) and Appendix C (*MQOs for Method Uncertainty and Detection and Quantification Capability*).

A final note should be made relative to prescribed methods and the performance-based approach to method selection. The performance-based approach for method selection allows more latitude in dealing with the potential diversity of matrices (such as waste-, sea-, ground- or surface water; biota; air filters; waste streams; swipes; soil; sediment; and sludge) from a variety of projects, or in dealing with different levels of data quality requirements or a laboratory's analytical proficiency. Even though the prescribed method approach may initially appear suitable and cost effective, it does not allow a laboratory to select a method from the many possible methods that will meet the MQOs.

Many individuals have the wrong impression that prescribed methods do not need to be validated by a laboratory. However, as discussed in this chapter, all methods should be validated to some level of performance for a particular project by the laboratory prior to their use. In addition, the laboratory should demonstrate continued proficiency in using the method through internal QC and external performance evaluation (PE) programs that use performance testing (PT) samples (Chapter 18, *Laboratory Quality Control*).

6.3 Life Cycle of Method Application

In responding to a SOW for a given analyte/matrix combination, a laboratory may have one or more methods that may be appropriate for meeting the MQOs. The final method selected from a set of methods may be influenced by many other technical, operational, or quality considerations. Figure 6.2 provides an overview of the life cycle of the method application. Figure 6.3 expands the life cycle into a series of flow diagrams.

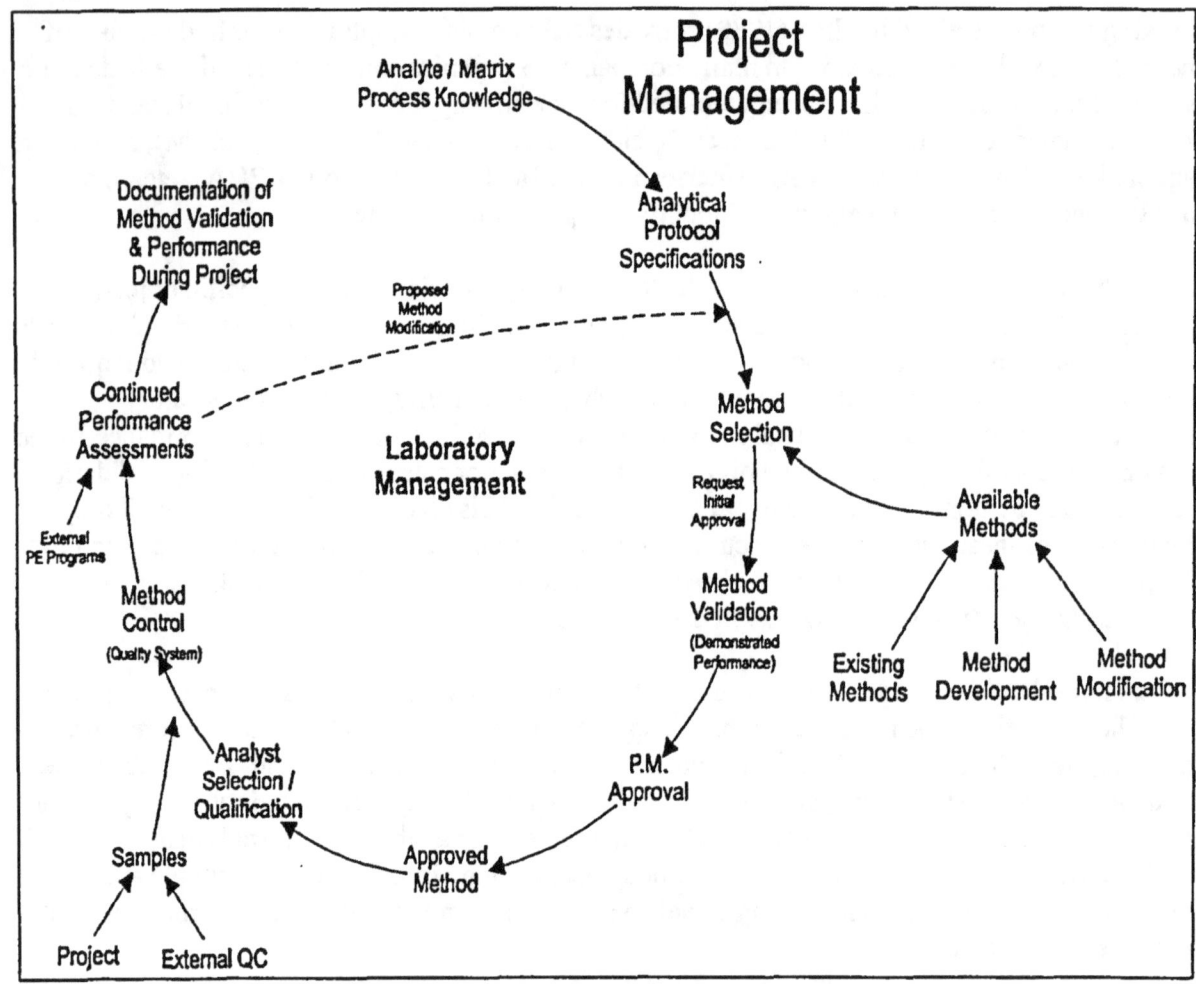

FIGURE 6.2 — Method application life cycle

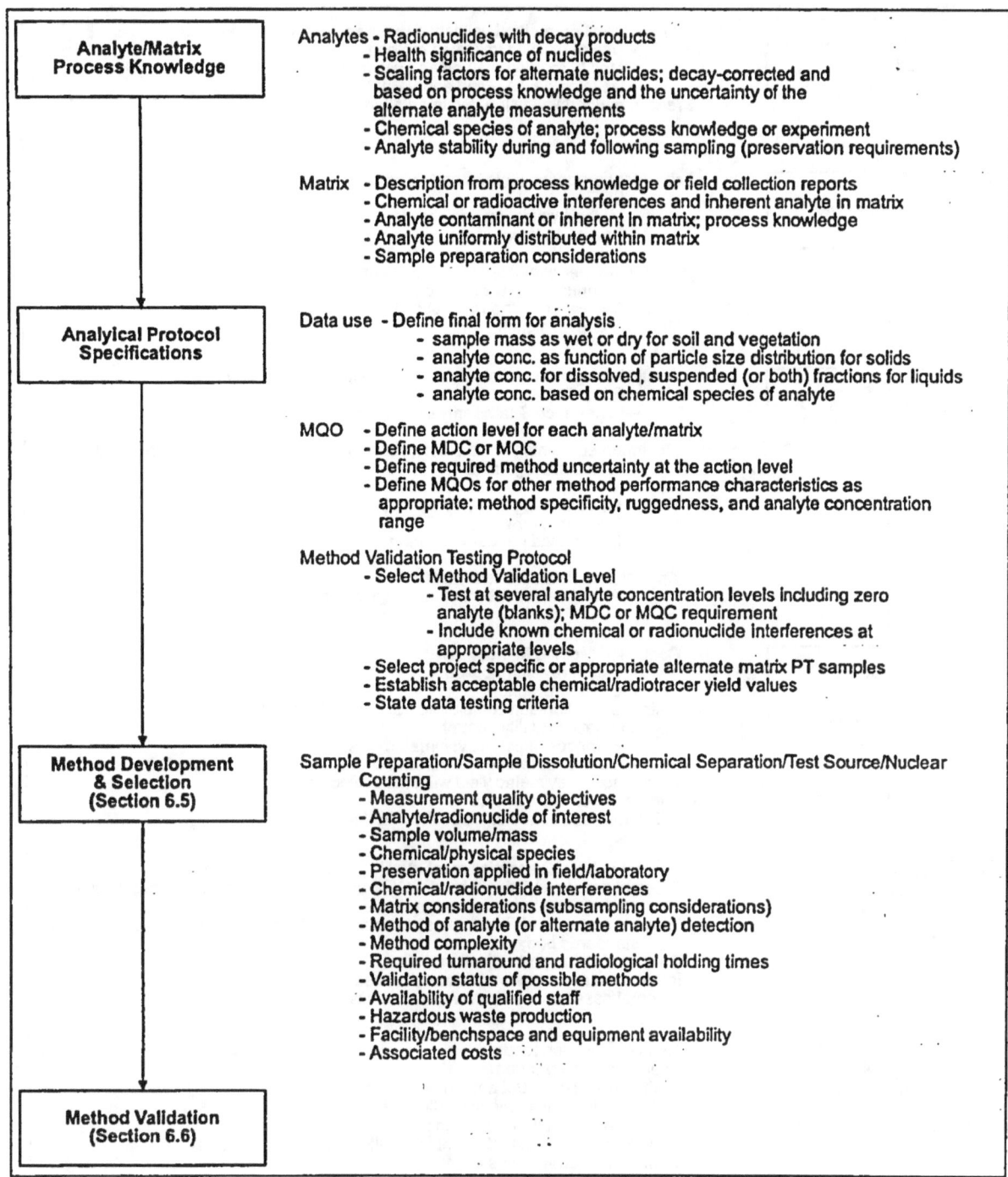

Figure 6.3 — Expanded Figure 6.2 addressing the laboratory's method evaluation process

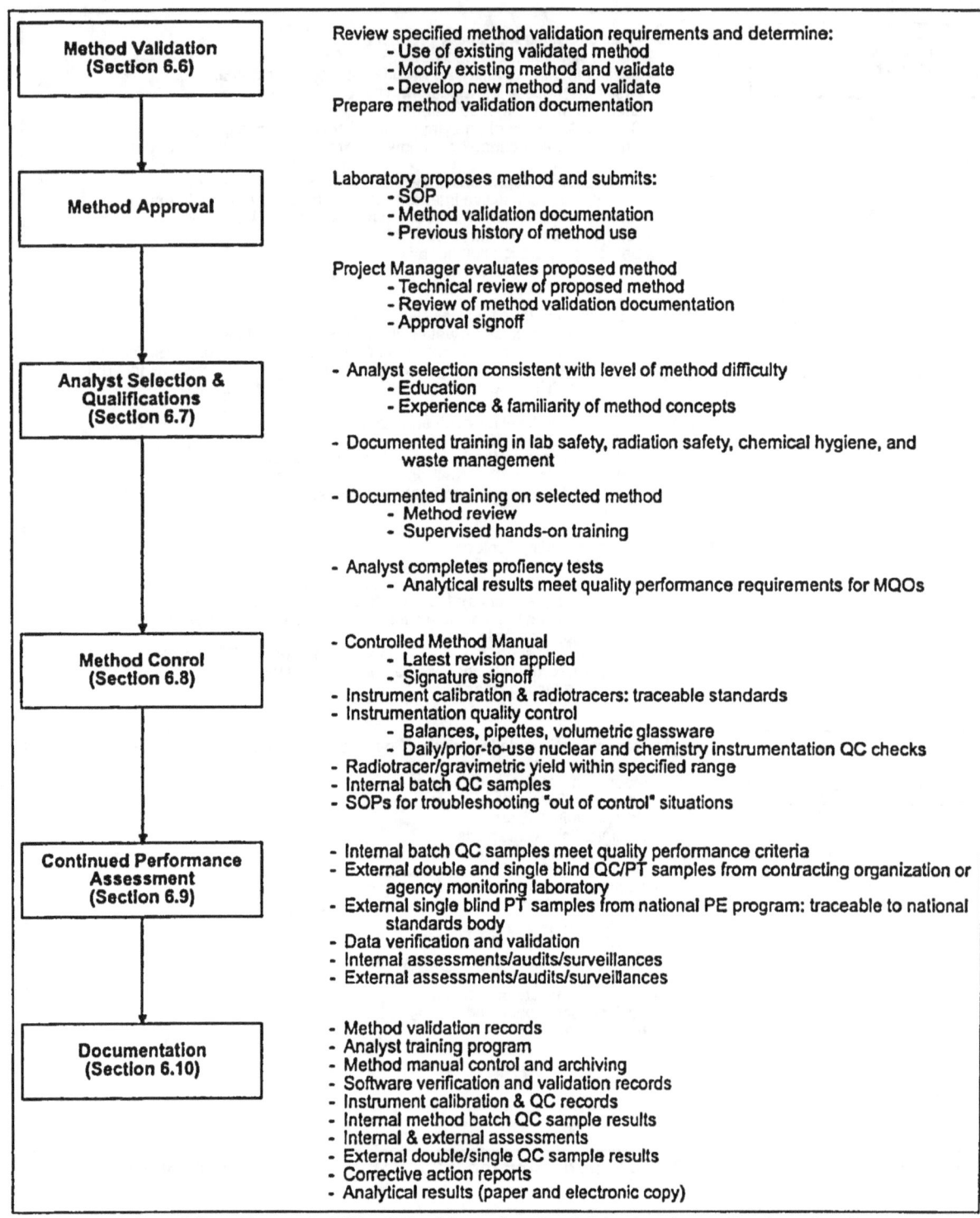

FIGURE 6.3 (cont'd) — Expanded Fig. 6.2 addressing the laboratory's method evaluation process

6.4   Generic Considerations for Method Development and Selection

This section provides guidance on the technical, quality, and operational considerations for the development of a new method or the selection of an existing radioanalytical method. Unless required by a regulatory or internal policy, rarely should a method be specified in an APS or a SOW. MARLAP recommends that a SOW containing the MQOs and analytical process requirements be provided to the laboratory.

If the nature of the samples and analytes are known in advance, and variations in a sample matrix and analyte concentration are within a relatively small range, the development or selection of analytical methods is easier. In most situations, however, the number of samples, sample matrices, analyte interferences, chemical form of analytes, and variations among and within samples may influence the selection of a method for a given analyte. A number of radioanalytical methods are available, but no single method provides a general solution (all have advantages and disadvantages). The method selection process should consider not only the classical radiochemical methods involving decay emission detection (alpha, beta or gamma) but also non-nuclear methods, such as mass spectrometric and kinetic phosphorescence analysis.

In the performance-based approach to method selection, the laboratory may select and propose a gross measurement (alpha, beta, or gamma) method that can be applied to analyte concentrations well below the action level for the analyte, as well as an analyte specific method for analyte levels exceeding a proposed "screening level" that is a fraction of the action level. For example, it may be acceptable to propose a gross measurement method when its method uncertainty meets the method uncertainty (absolute or u_{MR}) requirement at concentration levels much below the action level. A gross measurement method may be employed initially for some projects. Such an approach would have to be agreed to by the laboratory and project manager. The project method validation, discussed in Section 6.6.2, should demonstrate that the gross measurement method can measure the analyte of interest (directly or indirectly) at a proposed analyte screening level concentration and meet the method uncertainty requirement (u_{MR}) in the presence of other radionuclides. Appendix C provides guidance on how to determine an acceptable method uncertainty at an analyte concentration relative to the action level.

In general, the development or selection of a method follows several broad considerations. These include analyte and matrix characteristics, technical complexity and practicality of methods, quality requirements, availability of equipment, facility and staff resources, regulatory concerns, and economic considerations. Each of the broad considerations can be detailed. The following list, although not inclusive, provides insight into the selection of an appropriate method. Many of these categories are discussed in subsequent MARLAP Part II chapters.

- Analyte/radionuclide/isotope of interest
  - Decay emission (particle or photon), atom detection, or chemical (photon detection)
  - Half-life of analyte

- Decay products (progeny); principal detection method or interference
- Chemical/physical forms (e.g., gas, volatile)
- Use of nondestructive or destructive sample analysis

- Level of other radionuclides or chemical interference
  - Level of decontamination or selectivity required, e.g., a decontamination factor of 10^3 for an interfering nuclide (^{60}Co) present with the analyte of interest (^{241}Pu)
  - Resolution of measurement technique
  - Ruggedness of technique for handling large fluctuations in interference levels and variations in a matrix
  - Radionuclides inherent in background

- Matrix
  - Destructive testing
    - Stable elemental interferences
    - Difficulty in dissolution of a matrix
    - Difficulty in ensuring homogeneity of aliquant
    - Inconsistency in chemical forms and oxidation states of the analyte versus the tracer
  - Non-destructive testing
    - Heterogeneity of final sample for analysis
    - Self absorption of particle/photon emissions within a matrix

- Degree of method complexity
  - Level of technical ability required of analysts
  - Reproducibility of quality results between analysts
  - Method applicability to sample batch processing
  - Extensive front-end chemical-processing technique (sample dissolution, analyte concentration and purification/isolation, preparation for final form for radiometrics)
  - Nuclear instrumentation oriented technique (minimal chemical processing)

- Required sample turnaround time
  - Half-life of analyte
  - Sample preparation or chemical method processing time
  - Nuclear instrumentation measurement/analysis time
  - Chemical or sample matrix preservation time
  - Batch processing
  - Degree of automation available/possible

- Status of possible methods and applications
  - Validated for the intended application
  - Staff qualified and trained to use method(s)
  - Existing method QC

   o Specialized equipment, tracers, reagents, or materials available

- Hazardous or mixed-waste production
  o Older classical techniques versus new advanced chemical technologies
  o Availability and expense of waste disposal

- Associated costs
  o Labor, instrumentation usage, facilities, radiological waste costs
  o Method applicability to portable or mobile laboratory facilities
  o Availability of service hookups
  o Need for facility environmental controls
  o Need for regulatory permitting of mobile laboratory facility

6.5   Project-Specific Considerations for Method Selection

Certain parameters of the APSs (see Chapter 3 and the example in Figure 3.2) within the SOW are important to the method selection process. These include the analytes, matrix type, matrix characterization, analyte and matrix interferences, analyte speciation information gathered from process knowledge, sample process specifications (such as radiological holding times and sample processing turnaround times), and the MQOs. While these issues should be resolved during project planning, they are presented here as guidance to the laboratory for their review and evaluation of the technical adequacy of the SOW and to provide context for the method evaluation and selection process. Many of the issues from the project planning point of view are discussed in Section 3.3.

6.5.1   Matrix and Analyte Identification

The first step in selecting a method is knowing what analytes and sample matrices are involved. The following sections discuss what important information should accompany analyte and matrix identification.

6.5.1.1 Matrices

A detailed identification and description of the sample matrix are important aspects in the selection of an analytical method to meet the MQOs. The SOW should provide the necessary detailed sample matrix description, including those important matrix characteristics gathered from process knowledge. The laboratory should evaluate whether the existing sample preparation and dissolution steps of a method (Chapters 10 and 12 through 15) will be sufficient to meet the MQOs or the general or project method validation requirements. The matrix will also determine, to a certain extent, waste handling and disposal at the laboratory. If the matrix description is too vague or generic, the laboratory should contact the technical representative named in the SOW and request additional information.

The laboratory should ensure that the sample matrix description in the SOW reflects what is considered to be the "sample" by the project manager and the description is of sufficient detail to select the method preparation or analyte isolation steps that will meet the MQOs for the matrix. The laboratory should not accept generic sample matrix descriptions such as liquids or solids. For example, the differences between potable water and motor oil are obvious, but both may be described as a "liquid sample." However, there may be only subtle differences between potable surface water and groundwater but major differences between potable and process effluent waters. The laboratory should consider how much method ruggedness is needed in order to address the varied amounts of possible stable elements or compounds within a non-specified water matrix. Furthermore, when water from a standing pool is received in the laboratory, it may contain some insoluble matter. Now the questions arise whether the sample is the entire contents of the container, what remains in the container, the insoluble material, or just the water? A clay will act as an ion exchange substrate, while a sand may have entirely different retention properties. Both can be described as a soil or sediment, but the properties with which they retain a radionuclide are substantially different; thus, the method to properly isolate a particular radionuclide will vary. The laboratory should ensure that the selected method is consistent with the intended sample matrix, and the analytical results convey analyte concentration related to the proper matrix (i.e., Bq/L dissolved, Bq/L suspended, or Bq/L total). For such cases, the laboratory should request the project manager to clarify the "matrix" or "sample" definition.

Matrices generically identified as "solid" require additional clarification or information in order to select and validate a method properly. For example, sludges from a sewage treatment facility may be classified as a solid, but the suspended and aqueous portions (and possibly the dried residual material) of the sample may have to be analyzed. Normally, the radionuclide concentration in soils and sediments is reported in terms of becquerels per dry weight. However, certain projects may require additional sample process specifications (Section 6.5.4) related to the soil or sediment matrix identification that will affect the method selection process and the reporting of the data. This may involve sectioning of core samples, specified drying temperature of the sample, determining wet-to-dry weight ratio, removing organic material or detritus, homogenizing and pulverizing, sieving and sizing samples, etc. In order to determine the average analyte concentration of a sample of a given size containing radioactive particles, proper sample preparation and subsampling coupled with the applicable analytical methods are required (Chapter 12, *Laboratory Sample Preparation*, and Appendix F, *Laboratory Subsampling*). For alpha-emitting radionuclides, the method selected may only be suitable to analyze a few grams of soil or sediment, depending on the organic content. The laboratory should identify to the project manager the typical subsample or aliquant size that is used for the proposed method. Information should be provided to the laboratory on process knowledge. This information should indicate when sample inhomogeneities may exist due to:

- Radioactive particles;
- Selected analyte adsorption onto soil or sediment particles;
- Special chemical forms of the analyte; or

- Any other special analyte circumstances.

Based on this information, the laboratory should propose sample preparation and analytical methods that will address these matrix characteristics. Information on the solubility of the analyte can be used to select the dissolution method employed (see Chapter 13, *Sample Dissolution*). The laboratory should submit the proposed methods annotated with the suspected matrix characterization issues.

When selecting the methods for the analysis of flora (terrestrial vegetation, vegetables, aquatic plants, algae, etc.) or fauna (terrestrial or aquatic animals) samples, the detailed information on the matrix or the unique process specifications should be used by the laboratory to select or validate the method, or both. The laboratory should ensure that the specific units for the analytical results are consistent with the matrix identification and unique process specifications stated in the SOW. Most flora and fauna results are typically reported in concentrations of wet weight. However, for dosimetric pathway analyses, some projects may want only the edible portion of the sample processed and the results to reflect this portion, e.g., fillet of sport fish, meat and fluid of clams, etc. For the alpha- and beta-emitting radionuclides, aquatic vegetation normally is analyzed in the dry form, but the analyte concentration is reported as wet weight. The laboratory should ensure that the sample preparation method (Chapter 12) includes the determination of the necessary wet and dry weights.

These considerations bear not only on the method selected but also on how the sample should be collected and preserved during shipment. When possible, the laboratory should evaluate the proposed sample collection and preservation methods, as well as timeliness of shipping, for consistency with the available analytical methods. Discrepancies noted in the SOW for such collateral areas should be brought to the attention of the project manager. For example, sediment samples that have been cored to evaluate the radionuclide depth profile should have been collected and treated in a fashion to retain the depth profile. A common method is to freeze the core samples in the original plastic coring sleeves and ship the samples on ice. The SOW should define the specifics on how to treat the core samples and the method of sectioning the samples (e.g., cutting the cores into the desired lengths or flash heating the sleeves with subsequent sectioning).

The SOW should have properly delineated the proper matrix specifications required for project method validation. The purpose of the method validation reference material (MVRM) is to provide a matrix, which closely approximates that of the project samples to be analyzed (Section 6.6). The sample matrix must be characterized to the extent that the pertinent parameters are used to prepare the MVRM for the project method validation (Section 6.6.2). The laboratory should ensure that sufficient information and clarity have been provided on the matrix to conduct a proper method validation.

6.5.1.2. Analytes and Potential Interferences

The SOW should describe the analytes of interest and the presence of any other chemical and radionuclide contaminants (potential method interferences and their anticipated concentration) that may be in the samples. This information should be provided in the SOW to allow the laboratory's radiochemist to determine the specificity and ruggedness of a method that will address the multiple analytes and their interferences. The delineation of other possible interfering radionuclides is extremely important in the selection of a method to ensure that the necessary decontamination factors and purification steps are considered.

The size of the sample needed by the laboratory will depend on the number of analytes and whether the laboratory will select individual methods for each analyte or a possible "sequential" analytical method, where several analytes can be isolated from the same sample and analyzed. If a sample size is listed in the SOW, the laboratory should determine if there will be sufficient sample available to analyze all analytes, the associated QC samples, and any backup sample for re-analyses. Other aspects, such as the presence of short-lived analytes or analytes requiring very low detection limits, may complicate the determination of a proper sample size.

The laboratory should ensure that the project method validation requirements in the SOW are consistent with the analytes and matrix. The project method validation protocols defined in Section 6.6.2 are applicable to methods for single analyte analyses or to a "sequential method" where several analytes are isolated and analyzed. The laboratory should develop a well-planned protocol for project method validation that considers the method(s), analyte(s), matrix and validation criteria.

6.5.2 Process Knowledge

Process knowledge typically is related to facility effluent and environmental surveillance programs, facility decommissioning, and site remediation activities. Important process knowledge may be found in operational history or regulatory reports associated with these functions or activities. It is imperative that the laboratory review the information provided in the SOW to determine whether the anticipated analyte concentration and matrix are consistent with the scope of the laboratory operations. Process knowledge contained in the SOW should provide sufficient detail for the laboratory to determine, quickly and decisively, whether or not to pursue the work. If sufficient detail is not provided in the SOW, the laboratory should request the project planning documents. Laboratories having specialized sample preparation facilities that screen the samples upon arrival can make the necessary aliquanting or dilutions to permit the processing of all low-level samples in the laboratories. Laboratories that have targeted certain sectors of the nuclear industry or a particular nuclear facility may be very knowledgeable in the typical chemical and physical forms of the analytes of a given sample matrix and may not require detailed process knowledge information. However, under these circumstances, the laboratory's method should be robust and rugged enough to handle the expected range of analyte concentrations, ratios of radionuclide and chemical interferences, and variations in the sample matrix.

Process knowledge may provide valuable information on the possible major matrix constituents, including major analytes, chemical/physical composition, hazardous components, radiation levels, and biological growth (e.g., bacteria, algae, plankton, etc.) activities. When provided, the laboratory should use this information to determine if the sample collection and preservation methodologies are consistent with the proposed radioanalytical method chosen. In addition, the information also should be reviewed to ensure that the proposed sample transportation or shipping protocols comply with regulations governing the laboratory operation.

Process knowledge information in the SOW may be used by the laboratory to refine method selection from possible radiometric/chemical interferences, chemical properties of the analytes or matrix, and hazardous components, among others. Chapter 14 describes the various generic chemical processes that may be used to ensure proper decontamination or isolation of the analyte from other interferences in the sample. These include ion exchange, co-precipitation, oxidation/reduction, and solvent extraction among others. The process knowledge information provided in the SOW should be reviewed to determine whether substantial amounts of a radionuclide that normally would be used as a radiotracer will be present in the sample. Similarly, information on the levels of any stable isotope of the analyte being evaluated is equally important. Substantial ambient or background amounts of either a stable isotope of the radionuclide or the radiotracer in the sample may produce elevated and false chemical yield factors. In addition, substantial amounts of a stable isotope of the analyte being evaluated may render certain purification techniques inadequate (e.g., ion exchange or solid extractants).

6.5.3 Radiological Holding and Turnaround Times

The SOW should contain the requirements for the analyte's radiological holding and sample turnaround times. MARLAP defines radiological holding time as the time differential between the date of sample collection and the date of analysis. It is important that the laboratory review the specifications for radionuclides that have short half-lives (less than 30 days), because the method proposed by the laboratory may depend on the required radiological holding time. For very short-lived radionuclides, such as ^{131}I or ^{224}Ra, it is very important to analyze the samples within the first two half-lives in order to meet the MQOs conveniently. A laboratory may have several methods for the analysis of an analyte, each having a different analyte detection and quantification capability. Of the possible methods available, the method(s) selected and proposed by the laboratory should address the time-related constraints of the radioanalytical process, such as the radiological holding time requirement, half-life of the analyte, and the time available after sample receipt at the laboratory. When a laboratory has several methods to address variations in these constraints, it is recommended that the laboratory propose more than one method with a clarification that addresses the radiological holding time and MQOs. In some cases, circumstances arise which require the classification of sample processing into several time-related categories (Chapter 5). For example, the determination of ^{131}I in water can be achieved readily within a reasonable counting time through direct gamma-ray spectrometry (no chemistry) using a Marinelli beaker counting geometry, when the detection requirement is 0.4 Bq/L and the radiological holding time is short. However, when the anticipated radiological holding time is in the order of weeks, then a radiochemistry method using beta detection or beta-gamma coincidence counting would be more appropriate to meet the detection requirement. The more sensitive method also may be used when there is insufficient sample size or when the analyte has decayed to the point where the less sensitive method cannot meet the required MQOs. Another example would be the analysis of ^{226}Ra in soil, where the laboratory could determine the ^{226}Ra soil concentration through the quantification of a ^{226}Ra decay product by gamma-ray spectrometry after a certain ingrowth period, instead of direct counting of the alpha particle originating from the final radiochemical product (micro-precipitate) using alpha spectrometry.

Sample (processing) turnaround time normally means the time differential from the receipt of the sample at the laboratory to the reporting of the analytical results. As such, the laboratory should evaluate the SOW to ensure that the sample turnaround time, radiological holding time, data reduction and reporting times, and project needs for rapid data evaluation are consistent and reasonable. Method selection should take into consideration the time-related SOW requirements and operational aspects. When discrepancies are found in the SOW, the laboratory should communicate with the project manager and resolve any issue. Additionally, the response to the SOW should include any clarifications needed for sample turnaround time and/or radiological holding time issues.

6.5.4 Unique Process Specifications

Some projects may incorporate detailed sample processing parameters, specifications, or both within the SOW. Specifications for parameters related to sample preparation may include the degree of radionuclide heterogeneity in the final sample matrix prepared at the laboratory, the length of the sections of a soil or sediment core for processing, analysis of dry versus wet weight material, partitioning of meat and fluid of bivalves for analyses, and reporting of results for certain media as a dry or wet weight. Specifications related to method analysis could include radionuclide chemical speciation in the sample matrix. The laboratory must evaluate these specifications carefully, since various parameters may affect the method proposed by the laboratory. When necessary, the laboratory should request clarification of the specifications in order to determine a compatible method. In addition, the laboratory should ensure that the project method validation process is consistent with the unique process requirements. In some cases, not all special process specifications must be validated and, in other cases, site-specific materials (also referred to as MVRM) will be required for method validation. When necessary, the laboratory also should request site-specific reference materials having the matrix characteristics needed for proper method validation consistent with the special process requirements. It is incumbent upon the laboratory to understand clearly the intent of the special process specifications and how they will be addressed.

6.5.5 Measurement Quality Objectives

The specific method performance characteristics having a measurement quality objective may include:

- Method uncertainty at a specified analyte concentration level;
- Quantification capability (minimum quantifiable concentration);
- Detection capability (minimum detectable concentration);
- Applicable analyte concentration range;
- Method specificity; and
- Method ruggedness.

How each of these characteristics affect the method selection process will be discussed in detail in the subsequent paragraphs.

6.5.5.1 Method Uncertainty

From the directed planning process, the required method uncertainty at a stated analyte concentration should have been determined for each analyte/matrix combination. The method uncertainty requirement may be linked to the width of the gray region (Appendices B and C). MARLAP recommends that the SOW include the specifications for the action level and the required method uncertainty for the analyte concentration at the action level for each combination of analyte and matrix. For research and baseline monitoring programs, the action level and gray region concepts may not be applicable. However, for these applications, the project manager should establish a concentration level of interest and a required method uncertainty at that level. The laboratory should ensure that this method uncertainty requirement is clearly stated in the SOW.

The laboratory should select a method that will satisfy the method uncertainty requirement at the action level or other required analyte level. MARLAP uses the term "method uncertainty" to refer to the predicted uncertainty of a result that would be measured if a method were applied to a hypothetical laboratory sample with a specified analyte concentration. The uncertainty of each input quantity (method parameter) that may contribute significantly to the total uncertainty should be evaluated. For some methods, the uncertainty of an input quantity may vary by analyst or spectral unfolding software. Chapter 19 provides guidance on how to calculate the combined standard uncertainty of the analyte concentration, and Section 19.6.12 shows how to predict the uncertainty for a hypothetical measurement. For most basic methods, uncertainty values for the following input quantities (parameters) may be necessary when assessing the total uncertainty:

- Counting statistics (net count rate);
- Detector efficiency, if applicable;
- Chemical yield (when applicable) or tracer yield;

- Sample volume/weight;
- Decay/ingrowth factor; and
- Radiometric interference correction factor.

Typically, for low-level environmental remediation or surveillance activities, only those input quantities having an uncertainty greater than one percent significantly contribute to the combined standard uncertainty. Other than the radiometric interference correction factor and counting uncertainties, most input quantity uncertainties normally do not vary as a function of analyte concentration. At analyte levels near or below the detection limit, the counting uncertainty may dominate the method's uncertainty. However, at the action level or above, the counting uncertainty may not dominate.

When appropriate, the laboratory should determine the method uncertainty over the MQO analyte concentration range (Section 6.5.5.4), including the action level or other specified analyte concentration. The laboratory's project method validation (Section 6.6.2) should demonstrate or show through extrapolation or inference (e.g., from a lower or higher range of concentrations) that this method uncertainty requirement can be met at the action level or specified analyte concentration value. Method validation documentation should be provided in the response to the SOW.

6.5.5.2 Quantification Capability

For certain projects or programs, the project planning team may develop an MQO for the quantification capability of a method. The quantification capability, expressed as the minimum quantifiable concentration (MQC), is the smallest concentration of the analyte that ensures a result whose relative standard deviation is not greater than a specified value, usually 10 percent. Chapter 19 provides additional information on the minimum quantifiable concentration.

For example, if the MQC requirement for ^{89}Sr is 1.0 Bq/g (with a 10 percent relative standard deviation), the laboratory should select a method that has sufficient chemical yield, beta detection efficiency, low background, and sample (processing) turnaround time for a given sample mass to achieve a nominal measurement uncertainty of 0.1 Bq/g. The same forethought that a laboratory gives to estimating a method's minimum detectable concentration (MDC) for an analyte should be given to the MQC requirement. The laboratory should consider the uncertainties of all input quantities (detector efficiency, chemical yields, interferences, etc.), including the counting uncertainty when selecting a method. This is an important consideration, because for some methods, the counting uncertainty at the MQC level may contribute only 50 percent of the combined standard uncertainty. Therefore, the laboratory may have to select a method that will meet the MQC requirement for a variety of circumstances, including variations in matrix constituents and chemical yields, radionuclide and chemical interferences, and radioactive decay. In addition, sufficient sample size for processing may be critical to achieving the MQC specification.

During the project method validation process, the ability of the method to meet the required MQC specification should be tested. The method validation acceptance criteria presented in Section 6.6 have been formulated to evaluate the MQC requirement at the proper analyte concentration level, i.e., action level or other specified analyte concentration.

Since the laboratory is to report the analyte concentration value and its measurement uncertainty for each sample, the project manager or data validator easily can evaluate the reported data to determine compliance with the MQC requirement. Some projects may send PT material spiked at the MQC level as a more in-depth verification of the compliance with this requirement.

6.5.5.3 Detection Capability

For certain projects or programs, the method selected and proposed by the laboratory should be capable of meeting a required MDC for the analyte/matrix combination for each sample analyzed. For certain monitoring or research projects, the required analyte MDC may be the most important MQO to be specified in the SOW. For such projects, the MDC specification may be based on the analyte concentration of interest or the state-of-the-art capability of the employed technology or method. No matter what premise is used to set the value by the project planning team, the definition of, or the equation used to calculate, the analyte MDC should be provided in the SOW (Chapter 20). Furthermore, the SOW should specify how to treat appropriate blanks or the detector background when calculating the MDC. The laboratory should be aware that not all agencies or organizations define or calculate the MDC in the same manner. It is important for the laboratory to check that the SOW clearly defines the analyte detection requirements. In most cases, it would be prudent for the laboratory to use a method that has a lower analyte MDC than the SOW required MDC.

In some situations, a radiochemical method may not be robust or specific enough to address interferences from other radionuclides in the sample. The interferences may come from the incomplete isolation of the analyte of interest resulting in the detection of the decay emissions from these interfering nuclides. These interferences would increase the background of the measurement for the analyte of interest and, thus, increase the uncertainty of the measurement background. Consequently, an *a priori* MDC that is calculated without prior sample knowledge or inclusion of the interference uncertainties would underestimate the actual detection limit for the sample under analysis. Another example of such interferences or increase in an analyte's background uncertainty can be cited when using gamma-ray spectrometry to determine ^{144}Ce in the presence of ^{137}Cs. The gamma energy usually associated with the identification and quantification of ^{144}Ce is 133.5 keV. The gamma energy for ^{137}Cs is 661.6 keV. If a high concentration of ^{137}Cs is present in the sample, the Compton scattering from the 661.6 keV into the 133.5 keV region may decrease the ability to detect ^{144}Ce by one to two orders of magnitude over an *a priori* calculation that uses a nominal non-sample specific background uncertainty. Another example can be cited for alpha-spectrometry and the determination of isotopic uranium. If some interfering metal is present in unexpected quantities and carries onto the final filter mount or electrodeposited plate, a substantial decrease in the peak resolution may occur (resulting in an increased width of the alpha peak). Depending on the severity of the problem, there may be overlapping alpha peaks resulting in additional interference terms that should be incorporated into the MDC equation. In order to avoid subsequent analyte detection issues, it is important for the laboratory to inquire whether or not the project manager has considered all the constituents (analytes and interferences) present in the sample when specifying a detection limit for an analyte.

The laboratory should include documentation in the response to the SOW that the method proposed can meet the analyte's MDC requirements for the method parameters (e.g., sample size processed, chemical yield, detector efficiency, counting times, decay/ingrowth correction factors, etc.). When practicable, care should be given to ensure the blank or detector background uncertainty includes contributions from possible anthropogenic and natural radionuclide interferences. In addition, any proposed screening method should meet the detection limit requirement in the presence of other radionuclide interferences or natural background radioactivity. When appropriate or required, the laboratory should test the method's capability of meeting the required MDC using MVRMs that have analytes and interferences in the expected analyte concentration range. Upon request, the project manager should arrange to provide MVRMs to the laboratory.

6.5.5.4 Applicable Analyte Concentration Range

The SOW should state the action level for the analyte and the expected analyte concentration range. The proposed method should provide acceptable analytical results over the expected analyte concentration range for the project. Acceptable analytical results used in this context means consistent method precision (at a given analyte concentration) and without significant bias. The applicable analyte concentration range may be three or four orders of magnitude. However, most radioanalytical methods, with proper analyte isolation and interference-decontamination steps, will have a linear relationship between the analytical result and the analyte concentration. For certain environmental monitoring or research projects, the laboratory should ensure that there are no instrument or analytical blank background problems. If the background is not well-defined, there may be an inordinate number of false positive and false negative results.

In its response to the SOW, the laboratory should include method validation documentation that demonstrates the method's capability over the expected range. The laboratory's project method validation (Section 6.6) should demonstrate or show through extrapolation or inference (e.g., from a different range of concentrations) that the method is capable of meeting the analyte concentration range requirement.

6.5.5.5 Method Specificity

The proposed method should have the necessary specificity for the analyte/matrix combination. Method specificity refers to the method's capability, through the necessary decontamination or separation steps, to remove interferences or to isolate the analyte of interest from the sample over the expected analyte concentration range. Method specificity is applicable to both stable and radioactive constituents inherent in the sample. Certain matrices, such as soil and sediments, typically require selective isolation of femtogram amounts of the analyte from milligrams to gram quantities of matrix material. In these circumstances, the method requires both specificity and ruggedness to handle variations in the sample constituents.

If other radionuclide interferences are known or expected to be present, the SOW should provide a list of the radionuclides and their expected concentration ranges. This information enables the laboratory to select and propose a method that has the necessary specificity to meet the MQOs. As an alternative, the project manager may specify in the SOW the degree of decontamination a method needs for the interferences present in the samples. If the laboratory is not provided this information, method specificity cannot be addressed properly. The laboratory should ensure that related information on the matrix characteristics, radiometric or chemical interferences, and chemical speciation is provided to properly select a method.

6.5.5.6 Method Ruggedness

Ruggedness is the ability of the method to provide accurate analytical results over a range of possible sample constituents, interferences, and analyte concentrations, as well as to tolerate subtle variations in the application of the method by various chemists (EPA, 2002; APHA, 1998). Ruggedness is somewhat qualitative (Chapter 7). Therefore, the desirable parameters of a rugged method are difficult to specify quantitatively. A ruggedness test usually is conducted by systematically altering the critical variables (or quantities) associated with the method and observing the magnitude of the associated changes in the analytical results. ASTM E1169 provides generic guidance on how to conduct method ruggedness tests under short-term, high-precision conditions. In many cases, a rugged method may be developed over time (typically when difficulty is experienced applying an existing method to variations in the sample matrix or when two analysts have difficulty achieving the same level of analytical quality).

A laboratory may have several methods for an analyte/matrix combination. Samples from different geographical locations or different processes may have completely different characteristics. Therefore, the laboratory should select a method that is rugged enough to meet the APSs in the SOW. As indicated in Section 6.6.2, the prospective client may send site-specific MVRM samples for the method validation process or for PT samples (Chapter 7).

6.5.5.7 Bias Considerations

As discussed earlier, the proposed method should provide acceptable analytical results over the expected analyte concentration range for the project. Acceptable results used in this context means consistent method precision (at a given analyte concentration) and without significant bias. According to ASTM (E177, E1488, D2777, D4855), "bias of a measurement process is a generic concept related to a constant or systematic difference between a set of test results from the process and an accepted reference value of the property being measured," or "the difference between a population mean of the measurements or test results and the accepted reference or true value." ASTM (D2777) defines precision as "the degree of agreement of repeated measurements of the same property, expressed in terms of dispersion of test results (measurements) about the arithmetical mean result obtained by repetitive testing of a homogeneous sample under specified conditions." MARLAP considers bias to be a persistent difference of the measured result from the true value of the quantity being measured, which does not vary if the measurement is repeated. Normally, bias cannot be determined from a single result or a few results (unless the bias is large) because of the analytical uncertainty component in the measurement. Bias may be expressed as the percent deviation from a "known" analyte concentration. Note that the estimated bias, like any estimated value, has an uncertainty—it is not known exactly.

If bias is detected in the method validation process (see Section 6.6.4, "Testing for Bias") or from other QA processes, the laboratory should make every effort to eliminate it when practical. Implicitly, bias should be corrected before using the method for routine sample processing. However, in some cases, the bias may be very small and not affect the overall data quality. The project manager should review the method validation documentation and results from internal QC and external PE programs obtained during the laboratory review process (Chapter 7) and determine if there is a bias and its possible impact on data usability.

6.6 Method Validation

Without reliable analytical methods, all the efforts of the project may be jeopardized. Financial resources, timeliness, and public perception and confidence are at risk, should the data later be called into question. Proof that the method used is applicable to the analyte and sample matrix of concern is paramount for defensibility. The project manager should ensure the methods used in the analyses of the material are technically sound and legally defensible.

The method selected and proposed by the laboratory must be based on sound scientific principles and must be demonstrated to produce repeatable results under a variety of sample variations. Each step of the method should have been evaluated and tested by a qualified expert (radio-analytical specialist) in order to understand the limits of each step and the overall method in terms of the MQOs. These steps may involve well-known and characterized sample digestion, analyte purification and decontamination steps that use ion exchange, solvent extraction, precipitation and/or oxidation /reduction applications. Method validation will independently test the scientific basis of the method selected for a given analyte and sample matrix.

EURACHEM (1998) interprets method validation as "being the process of defining an analytical requirement, and confirming that the method under consideration has performance capabilities consistent with what the application requires. Implicit in this is that it will be necessary to evaluate the method's performance capabilities." As such, the laboratory is responsible for ensuring that a method is validated adequately. MARLAP distinguishes between general method validation and project method validation. During the development of an analytical method or prior to the first use of a recognized industry or government method, laboratories typically perform a general method validation. General method validation is normally conducted to determine the capability of the method for a single analyte/matrix combination to meet internal laboratory quality requirements.

For the purposes of MARLAP, project method validation is the demonstration that the radioanalytical method selected by the laboratory for the analysis of a particular radionuclide in a given matrix is capable of providing analytical results that meet a project's MQOs and any other requirements in the APS. A proposed method for a specific combination of analyte and matrix should be validated in response to the requirements within a SOW. Demonstration of method performance to meet project-specific MQOs prior to analyzing project samples is a critical part of the MARLAP process.

Methods obtained from recognized industry standards (ASTM, ANSI, APHA) or government method manuals may have been validated for certain general applications by the developing or issuing laboratory. However, prior to their use, other laboratories planning to use these methods need to perform general and project method validations to ensure that the method meets labora-tory performance criteria (for generic applications) and project method validation criteria, respectively. In some cases, the laboratory's quality requirements and method attributes for general method validation may be less stringent compared to a project method validation. For example, a method's precision or chemical yield range requirement may be less stringent for general method validation than for project method validation requirements. MARLAP recommends that a method undergo some basic general validation prior to project method validation.

In the discussion on general and project method validation, certain terms related to test samples are used. These include method validation reference materials (MVRMs) and internal and external PT materials. MVRM refers to site-specific materials that have the same or similar chemical, physical, and nuclear properties as the proposed project samples. Normally, MVRMs can be prepared by at least two mechanisms:

- Spiking background or blank material from a site with the radionuclides of interest; or
- Characterizing the site material containing the radionuclides of interest to a high degree of accuracy.

Although MVRM is the most appropriate material for testing a laboratory's project-specific performance or for validating a method for a particular project, its availability may be limited depending on the project manager's ability to supply such material. Internal PT materials (samples) are materials prepared by the laboratory, typically as part of a laboratory's QC program and method validation process. A matrix spike (internal batch QC sample) may be considered an internal PT material. External PT materials are materials prepared for use in an external government or commercial PE program. When available and applicable, external PT samples may be used for validating methods. PT and MVRM samples should be traceable to a national standards body, such as the National Institute of Standards and Technology in the United States.

An analytical laboratory's quality system should address the requirements and attributes for general method development, including some level of validation. However, general validation will not address the specific requirements of project method validation. MARLAP recommends that when a method is applied to a specific project, the method should then undergo validation for that specific application.

6.6.1 General Method Validation

A general method validation process should be a basic element in a laboratory's quality system. General method validation is applied to an analyte(s)/matrix combination, such as ^{90}Sr in water, but can be applied to a "sequential" method to determine multiple analytes. In most cases, a matrix of typical constituents will be used when evaluating the method. A general method validation protocol should address the important aspects of the methods that influence the results (e.g., inclusion of radiotracers, standard addition, alternate analyte analyses, etc.) and the basic quality requirements of a laboratory's quality system. General guidance on single laboratory method validation can be found in IUPAC (2002) and EURACHEM (1998). For most applications, the method should be evaluated for precision and relative bias for several analyte concentration levels. In addition, the absolute bias, critical level and the *a priori* minimum detectable concentration of the method, as determined from appropriate blanks, should be estimated. (See Section 6.6.4 for a discussion on testing for absolute and relative bias.) There should be a sufficient number of test level concentrations and replicate PT samples to make realistic estimates of the quality parameters. During validation, the method should also be evaluated in terms of factors most likely to influence a result (e.g., ruggedness) so that the method can handle minor deviations to the method and precautions may be written into the method (Youden and Steiner, 1975). In addition, IUPAC (2002) recommends that method validation evaluate the following parameters: applicability, selectivity, calibration and linearity, range of applicable analyte concentrations, detection and determination (quantification capability) limit, sensitivity, fitness of purpose, matrix variation and measurement uncertainty.

Laboratories that have participated in an interlaboratory collaborative study whose data were included in a published method (having an appropriate number of test levels and replicate samples, e.g., ASTM D2777 and Youden and Steiner, 1975) would be considered to have an acceptable general validated method for the analyte/matrix combination under study. These collaborative studies have at a minimum three or four different analyte concentration levels (excluding blanks) with three replicates or Youden pairs per analyte concentration level. A well-planned collaborative study will include expected interferences and matrix variations.

6.6.2 Project Method Validation Protocol

A laboratory's project method validation protocol should include the evaluation of the method for project-specific MQOs for an analyte and internal quality performance criteria as well as other generic parameters. With a properly designed method validation protocol, important information may be ascertained from the analytical results generated by the method validation process.

The parameters that should be specified, evaluated, or may be ascertained from the analytical results generated by the project method validation process are listed below:

- Defined Method Validation Level (Table 6.1)
- APSs including MQOs for each analyte/matrix
  - Chemical or physical characteristics of analyte when appropriate
  - Action level (if applicable)
  - Method uncertainty at a specific concentration
  - MDC or MQC
  - Bias (if applicable)
  - Applicable analyte concentration range
  - Method blanks
  - Other qualitative parameters to measure the degree of method ruggedness or specificity
- Defined matrix for testing, including chemical and physical characteristics that approximate project samples
- Selected project-specific or appropriate alternative matrix PT samples, including known chemical or radionuclide interferences at appropriate levels
- Defined sample preservation
- Stated additional data testing criteria (such as acceptable chemical/radiotracer yield values)

In order to demonstrate properly that a method will meet project MQOs, the method should be evaluated over a range of analyte concentrations that cover the expected analyte concentration range for the project (Section 6.5.5.4). The middle of the concentration range should be set near the action level. The preparation and analysis of the test samples should result in a measurement uncertainty that is equal to or less than the required method uncertainty. In addition, anticipated or known chemical and radionuclide interferences should be added in the appropriate "interference to analyte" activity or concentration ratio. As a requirement of the project method validation process, appropriate method blanks (containing similar interferences when practical) should be analyzed concurrently with the matrix spikes to determine analyte interferences and to estimate the absolute bias near the detection limit (Section 6.6.4, "Testing for Bias").

The number of validation samples requires is a function of the validation level sought. As shown in Table 6.1, the number of samples may vary from 9 to 21.

TABLE 6.1 — Tiered project method validation approach

Validation Level	Application	Sample Type*	Acceptance Criteria§	Levels† (Concentrations)	Replicates	No. of Analyses
A Without Additional Validation	Existing Validated Method	—	Method Previously Validated (By One of the Validation Levels B through E)	—	—	—
B	Same or Similar Matrix	Internal PT	Measured Value Within $\pm 2.8 u_{MR}$ or $\pm 2.8 \varphi_{MR}$ of Known Value	3	3	9
C	Similar Matrix/New Application	Internal or External PT	Measured Value Within $\pm 2.9 u_{MR}$ or $\pm 2.9 \varphi_{MR}$ of Known Value	3	5	15
D	Newly Developed or Adapted Method	Internal or External PT	Measured Value Within $\pm 3.0 u_{MR}$ or $\pm 3.0 \varphi_{MR}$ of Known Value	3	7	21
E	Newly Developed or Adapted Method	MVRM Samples	Measured Value Within $\pm 3.0 u_{MR}$ or $\pm 3.0 \varphi_{MR}$ of Known Value	3	7	21

* PT and MVRM samples should be traceable to a national standards body, such as NIST in the United States. Internal PT samples are prepared by the laboratory. External PT samples may be obtained from a performance evaluation program or from a commercial radioactive source producer that has traceability to a national standards body. Blank samples should be representative of the matrix type being validated.

§ The acceptance criterion is applied to each analysis in the method validation, not to the mean of the analyses. u_{MR} is the required absolute method uncertainty for analyte concentrations at or below the action level and φ_{MR} is the required relative method uncertainty for analyte concentrations above the action level (see Figure C.1 in Appendix C). The acceptance criteria are chosen to give a false rejection rate of ~5% when the measurement process is unbiased, with a standard deviation equal to the required method uncertainty (u_{MR} or φ_{MR}). The stated multiplier, k, for the required method uncertainty was calculated using the formula $k = z_{0.5 + 0.5(1-\alpha)^{1/N}}$ where N is the number of measurements, α is the desired false rejection rate, and, for any p, z_p denotes the p-quantile ($0 < p < 1$) of the standard normal distribution.

† Concentration levels should cover the expected analyte concentration range for a project including the action level concentration. A set of five appropriate blanks (not considered a level) should be analyzed during the method validation process. The blank data and the estimated absolute bias in the mean blank concentration value (see Attachment 6A in this chapter for applicable statistical tests) shall be reported as part of the method validation documentation.

6.6.3 Tiered Approach to Project Method Validation

While MARLAP recommends that as each new project is implemented, the methods used in the analysis of the associated samples undergo some level of validation, it is the project manager's responsibility to assess the level of method validation necessary. Although the end result of method validation is to ensure that the selected method meets the MQOs for an analyte/matrix, the level of validation depends on the extent of method development. Therefore, MARLAP recommends a tiered approach for project method validation. The recommended level of validation for new or existing methods are provided in the next four sections, based on level of effort: no additional validation, modification of a method for a similar matrix, new application of a method, and newly developed or adapted methods. The suggested levels of validation are indicative of the modification required of the method. It should be noted that the method validation requirements of Table 6.1 permit the laboratory to use internal and external PT or site-specific MVRM samples and also permit the project manager to provide PT or site-specific MVRM samples for the laboratory to use or analyze. As part of the qualifying process, a project manager may provide PT samples. In this case, the project manager should ensure consistency with the method validation requirements of Table 6.1. Most laboratories normally have documentation on the general or overall performance of a method. This documentation may supplement, or occasionally may be sufficient to meet the project method validation criteria.

The tiered approach to project method validation outlined in Table 6.1 was developed to give the project manager flexibility in the method validation process according to the project MQOs The degree of method validation increases from the lowest (Level A) to the highest (Level E). Figure 6.4 illustrates that—for a given validation level—the relative assurance in a method meeting the MQOs and the relative effort for method validation required by the laboratory are directly related. For certain projects, achieving the highest degree of assurance in method suitability (e.g., for a difficult sample matrix with interferences) would require validation using site-specific PT samples (Level E). This validation level also requires 21 samples: more laboratory effort compared to the other levels.

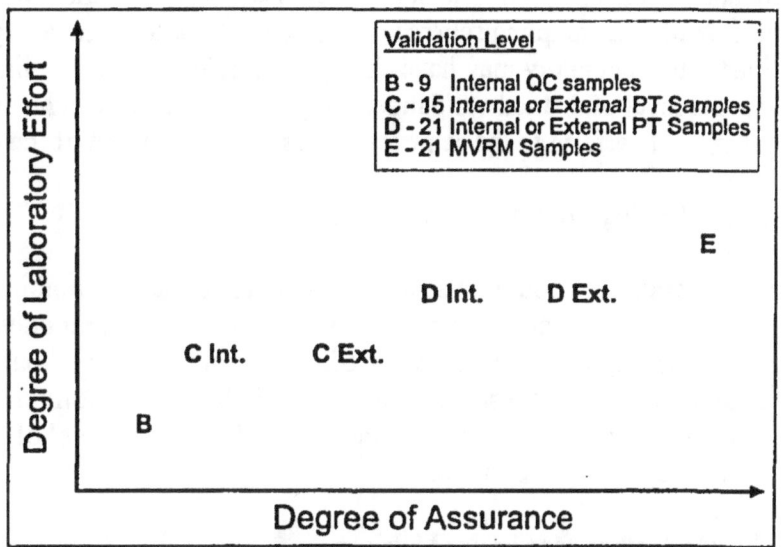

FIGURE 6.4 — Relationship between level of laboratory effort, method validation level, and degree of assurance of method performance under the tiered approach to method validation

Each of the validation levels evaluates the proposed method over the expected concentration range of the analytes and interferences. Requiring that each analytical result be within the interval of the known value ± ~3 times the required method uncertainty (u_{MR} or φ_{MR}) at the action level ensures a high degree of confidence that a method will meet the MQO. (See Appendix C for the definition of the required method uncertainty at the action level or other stated concentration, u_{MR}.) In addition to evaluating the method uncertainty, the method should be evaluated for bias (Section 6.6.4).

During the method validation process, the laboratory should ensure that the standard deviation for the samples analyzed is consistent with the estimated individual sample measurement uncertainty. An evaluation should be conducted for replicate sample analyses that have the same approximate relative measurement uncertainties. If the estimated measurement uncertainty of a given sample is much different than the observed method precision for the replicate analyzes, then the laboratory may not have properly estimated the uncertainty of one of the input parameters used to calculate the combined standard uncertainty.

6.6.3.1 Existing Methods Requiring No Additional Validation

For completeness, it is necessary to consider the possibility that a previously validated method requires no additional validation (Level A of Table 6.1) for a specific project. As noted in the table, the method should have previously undergone some level (Level B through E) of validation. It may be that the samples (matrix and analyte specific) associated with a new project are sufficiently similar to past samples analyzed by the same laboratory that the project manager feels additional validation is unwarranted. The decision to use Level A method validation should be made with caution. While the sampling scheme may be a continuation, the analytical processing capabilities at the laboratory may have changed sufficiently to merit limited method validation. Without some level of method validation, the project manager has no assurance that the analytical laboratory will perform to the same standards as an extension of the earlier work.

6.6.3.2 Routine Methods Having No Project Method Validation

When a laboratory has a routine method for a specific radionuclide/matrix combination that has had no previous project method validation, a project manager may select method validation Level B to validate the method for project sample analyses. Since the routine method has been used on a regular basis for client and PE program samples, there should be sufficient information on the performance of the method. As such, the minimum method validation protocol of Level B should be adequate to verify the method's performance.

6.6.3.3 Use of a Validated Method for Similar Matrices

When a previously validated method is to be used in the analysis of samples that are similar to the matrix and analyte for which the method was developed, MARLAP recommends that validation of the method be implemented according to Level B or C of Table 6.1. These levels will provide a reasonable assurance to both the laboratory and the project manager that the method will meet the required MQOs. Level B requires the least amount of effort for the laboratory but may not satisfy the level of method validation required by the project. When the laboratory does not have the capability to produce internal QC samples, the Level C validation protocol should be used.

Since a method inherently includes initial sample preparation, projects that have severe differences in analyte heterogeneity may require a moderate change in a radiochemical method's initial sample treatment. A change in the method to address the increased heterogeneity of the analyte distribution within the sample may require another method validation depending on the ruggedness of the method and the degree of analyte heterogeneity. In this case, Level C validation would be appropriate.

6.6.3.4 New Application of a Validated Method

Methods that have been validated for one application normally require another validation for a different application, such as a different sample matrix. In addition, the MQOs may change from one project to another or from one sample matrix to another. The validation process for an existing validated method should be reviewed to ensure applicability of the new (which can be more or less restrictive) MQOs. Applying an existing method to another matrix is not recommended without further method validation. MARLAP recommends, based on the extent of the modification and the difficulty of the matrix, that Level C of Table 6.1 be used to validate the performance of the modified method.

Both internal and external PT samples may be used for Level C validation. However, the project manager should specify the PT matrix. It should be recognized that national or commercial PE programs may not provide the necessary matrices or the required analyte concentrations needed for the Level C validation protocol. However, some radioactive source suppliers have the capability to produce high quality PT materials for method validation.

Validation of an existing method for a different application depends on the extent of the departure from the original method application, in terms of:

• Dissimilarity of matrices;
• Chemical speciation of the analyte or possible other chemical interference;
• Analyte, chemical or radiometric interferences;
• Complete solubilization of the analyte and sample matrix; and
• Degree of analyte or sample matrix heterogeneity.

When the chemical separation of the analyte varies from that for which the method was originally validated, the method should be so modified and subsequent validation performed. For example, if the original method was developed and validated to extract iodide using ion exchange chromatography, and a new application requires that iodine and iodate be quantified as well as iodide, then the method should be validated for the new analytes. Another example would be the initial development of a method for soluble plutonium in soil using acid dissolution and then applying the same method to high-fired plutonium oxide in soil. For these two examples, if the original methods were to undergo the validation process for the new application, definite deficiencies and poor results would be evident. Portions of the original method would have to be modified to address the chemical speciation problems. The modified method requires validation to ensure that the MQOs for the new application can be met.

When additional analyte, chemical, or sample matrix interferences are known to exist for a new application, the previously validated method should undergo further validation. For example, applying a method developed for the analysis of an analyte in an environmental matrix containing few interfering radionuclides would be inappropriate for the analysis of process waste waters containing many interfering radionuclides at high concentrations. In essence, the degree of decontamination (degree of interference removal) or analyte purification (isolation of the analyte from other radionuclides) necessary for one application may be completely inadequate or inappropriate for another application (an indication of method specificity).

Another example would be the use of a method for soil analysis employing ^{234}Th as a radiotracer for chemical yield for the isotopic analysis of thorium when the soil also has a high concentration of uranium. Thorium-234 is a decay product of ^{238}U and will exist in the sample as a natural analyte, thus creating an erroneous chemical yield. A third example is the application of a ^{90}Sr method developed for freshwater to seawater samples for which the amount of chemical interferences and ambient strontium levels are extensive.

Some matrices and analytes may be solubilized easily through acid dissolution or digestion. For some applications, the analyte of interest may be solubilized from the sample matrix through an acid extraction process. The applicability of such methods should be carefully chosen and, most important, the method must be validated for each application. Definite problems and misapplication can be the result of using an acid extraction process when a more robust complete sample dissolution is necessary. These examples illustrate the deficiencies of the initial method validation when applied to the modified sample parameters.

6.6.3.5 Newly Developed or Adapted Methods

MARLAP recommends that methods developed by the laboratory or adapted from the literature that have not been previously validated for a project be validated according to Levels D or E of Table 6.1. These levels provide the most comprehensive testing of method performance. Levels D and E have an increased number of replicates and the data obtained should provide the best estimate of a method's precision and bias. When the matrix under consideration is unique, the method should be validated using the same matrix (e.g., MVRM) as determined in Level E. This is extremely important for process/effluent waters versus laboratory deionized water and for various heavy metal radionuclides in soils or sediments when compared to spiked sand or commercial topsoil. For site-specific materials containing severe chemical and radionuclides interferences, many methods have been unable to properly address the magnitude of interferences.

6.6.4 Testing for Bias

The laboratory should test the method for bias.[1] In fact, the laboratory should check for at least two types of bias: *absolute* and *relative*. Attachment 6A describes a statistical hypothesis test that may be used to check for each type.

It is assumed here that the mean response of the method is an essentially linear function of analyte concentration over the range of the method. This function can be characterized by its y-intercept, which equals the mean response at zero concentration, and its slope, which equals the ratio of the change in the mean response to a change in sample analyte concentration. The absolute bias of the method is equated here with the y-intercept, and the relative bias is equated with the difference between the slope and 1.

Detecting and quantifying an absolute or relative bias in a measurement process may be difficult if the bias is small in relation to the uncertainty of a measurement. Typically, an absolute bias is most easily observed by analyzing blank samples, and a relative bias is most easily observed by analyzing high-activity certified reference materials (CRMs); however, if the bias is very small, the number of sample measurements required to detect it may make the effort impractical.

6.6.4.1 Absolute Bias

Testing for absolute bias is most important when one of the purposes of analysis is to determine whether the analyte is present either in individual laboratory samples or in a sampled population. An absolute bias in the measurement process can lead to incorrect detection decisions. Likely causes of such a bias include inadequate corrections made by the laboratory for instrument background, laboratory reagent contamination, and other interferences.

It is presumed here that the laboratory attempts to eliminate any absolute bias in the measurement process by blank- or background-correcting all measured results. For example, such a correction may be based on measurements of *instrument background* or analyses of *reagent blank* samples. To test whether the corrections are adequate, the laboratory should analyze a series of *method blank* samples, applying all appropriate corrections exactly as for ordinary samples, and perform a *t*-test on the results. To avoid the appearance of a false bias, the determinations of the correction terms (e.g., background or reagent blank) should be repeated for each method blank sample analyzed.

---

[1] Technically, the laboratory tests the *measurement process* for bias. The measurement process represents the laboratory's implementation of the *method* using particular instruments, analysts, quality control, etc.

6.6.4.2 Relative Bias

Testing the method for relative bias is most important when one of the purposes of analysis is to quantify the amount of analyte present either in a sample or in a sampled population, and perhaps to determine whether the analyte concentration is above or below some positive action level.

To test for relative bias, the laboratory may analyze an appropriate CRM (or spiked sample) a number of times. To avoid the appearance of a false bias, the laboratory should replicate as many steps in the measurement process as possible for each analysis.

6.6.5   Project Method Validation Documentation

Project method validation, depending on the required level of validation, can be accomplished by the project manager sending PT samples to the laboratory or by the laboratory using internal or external PT samples. When PT samples are sent to a laboratory to evaluate or validate the laboratory's method and capabilities, the appropriate technical representative should retain all records dealing with applicable method validation protocols (Section 6.6.2), PT sample preparation certification, level of validation (from Table 6.1), results, and evaluations. Evaluations include comparison of individual results to the validation acceptance criterion, absolute bias in blanks and, if available, statistical analyses of the data for method precision and bias. The laboratory should provide the necessary documentation to the project manager for these PT samples as required by the SOW. The laboratory should request feedback from the project manager as to the method performance. This information, along with the sample analytical results documentation, should be retained by the laboratory for future method validation documentation.

When the laboratory conducts its own project method validation, all records, laboratory workbooks, and matrix spike data used to validate an analytical method should be retained on file and retrievable for a specified length of time after the method has been discontinued. Data evaluations such as comparison of individual results to the validation acceptance criterion and absolute bias in blanks and, when available, method precision and bias, should be part of the data validation package sent to the project manager. All method validation documentation should be retained as part of the documentation related to the laboratory's quality system.

6.7   Analyst Qualifications and Demonstrated Proficiency

The required level of qualification of an analyst is commensurate with the degree of difficulty and sophistication of the method in use. The selection of the analyst for the method application is typically determined initially on experience, education and proven proficiency in similar methods. Basic guidance for the minimum education and experience for radioassay laboratory technicians and analysts has been provided in Appendix E (*Contracting Laboratory Services*) and ANSI N42.23.

For radiochemical methods, there may be several analysts involved. At most major laboratories, different individuals may be involved in the sample preparation, radiochemistry, and radiation detection aspects of the method. In these cases, the entire staff involved in the method should undergo method proficiency tests to demonstrate their ability to meet quality requirements and performance goals. The staff involved in the initial validation of an acceptable method would be considered proficient in their particular role in the method application and the results of their performance should be documented in their training records.

Successful proficiency is established when the performance of the analyst or staff meet predefined quality requirements defined in the laboratory's quality system or a SOW, as well as processing goals. Parameters involved in operational processing goals are typically turnaround time, chemical yields, frequency of re-analyses (percent failure rate), and frequency of errors.

The continued demonstrated analyst proficiency in the method is usually measured through the acceptable performance in internal QC and external PE programs associated with routine sample processing.

6.8   Method Control

Method control is an inherent element of a laboratory's quality system. Simply stated, method control is the ongoing process used to ensure that a validated method continues to meet the expected requirements as the method is routinely used. Method control is synonymous with process control in most quality systems. For a laboratory operation, method control can be achieved by the application of the following:

- Controlled method manual (latest revision and signature sign-off);

- Calibration standards and radiotracers that are traceable to a national standards body such as the National Institute of Science and Technology (NIST) in the United States;

- An instrument QC program that properly evaluates the important method parameters on an appropriate frequency;

- Radiotracers should be evaluated routinely for consistent concentration;

- Chemical yields should be evaluated for trends or deficiencies;

- Internal QC and external PT samples to determine deviations from expected quality performance ranges;

- Standard operating procedures for troubleshooting "out of control" situations; and

- Problem reporting, corrective action, and quality improvement process.

The method control elements described above typically are addressed in the quality manual of the laboratory or the project plan document for the project under consideration. Refer to Chapter 18 for additional information.

6.9   Continued Performance Assessment

The assessment of a laboratory's continued performance is covered in detail in Chapter 7. However, it is important to discuss briefly certain aspects of evaluating a method's continued performance from the perspective of a laboratory.

A performance indicator system should be in place that assesses and provides feedback on the quality of the routine processing. The most useful and cost-effective means of assessing a method's performance is through the implementation of internal QC or external performance evaluation programs or both. Of course, it can be argued that method assessment through a QC or PE program evaluates the combined performance of the method and the analyst. However, statistical and inferential interpretation of the QC/PE data can provide insight into whether the method is failing or whether an analyst is underperforming. Chapters 7 and 18 and Appendix C provides guidance on quality control programs and the use of the internal laboratory QC or external PE data to assess the laboratory's performance in meeting performance criteria.

The laboratory management should use the internal QC program to detect and address radioanalytical issues before the client does. Many SOWs require the use of internal QC samples for every batch of project samples (Chapter 18). In effect, the client is essentially setting the level of internal quality control and the frequency of method performance evaluation. It should be recognized that an internal QC program evaluates method performance related to the initial calibrations or internal "known values." An external NIST-traceable PE program will detect method biases relative to the national standard or to the agency's PE program.

Some users of laboratory services have developed "monitoring" laboratory programs (ANSI N42.23). For these programs, the user engages a recognized independent monitoring laboratory to intersperse double- and single-blind external PT materials into batches of normal samples submitted to a laboratory. The complexity and frequency of the monitoring laboratory PT samples vary among programs, projects, and Federal and state agencies. An external double-blind PE program conducted by a monitoring laboratory using site-specific matrices probably provides the most realistic estimate of the method's or laboratory's true performance. When the monitoring laboratory is traceable to a national standards body (such as NIST in the United States), either directly or through an authorized reference laboratory (ANSI N42.23), the monitoring laboratory program will provide an estimate of any method bias as related to the national standard.

Method performance can also be determined, although on a less frequent basis, through the laboratory's participation in the various PE programs. For a laboratory providing services to government agencies, the participation in such programs is typically a requirement. The PE programs commonly send out non site-specific PT materials on a quarterly or semiannual basis.

The laboratory's performance in certain PE programs is public knowledge. Such information is useful to project managers in selecting a laboratory during the laboratory selection and qualifying processes. Similar to the monitoring laboratory, when the laboratory conducting the PE program is traceable to NIST, either directly or through a NIST reference laboratory (ANSI N42.23), the PE program may provide an estimate of the bias as related to the national standard as well as the precision of the method, depending on the distribution of replicate samples.

Some projects require that all analytical results received from a laboratory undergo a data verification and validation process. Chapter 8 provides more detail on these processes. When properly conducted, certain aspects and parameters of the method can be assessed during the data verification and validation process.

Internal and external audits/assessments are also key elements in a laboratory's quality system to assess the continuing performance of a method (Chapter 7). The level and frequency of the audits and assessments typically vary according to the magnitude and importance of the project and on the performance of the laboratory. Another quality system element that is very effective is a self-assessment program. A functioning and effective self-assessment program may identify weaknesses or performance issues more readily and timely than formal internal and external audits.

6.10 Documentation To Be Sent to the Project Manager

The documentation related to the life cycle of a method application is essentially the information gathered during the use of the method. A formal method documentation program is unnecessary since the information should be part of the quality system documentation. Documented information available from the quality system, related to a method's development, validation, and control, include the following:

- Method validation protocol and results;
- Analyst training and proficiency tests;
- Method manual control program;
- Instrument calibration and QC results;
- Internal QC and external PT sample results;
- Internal and external assessments; and
- Corrective actions.

Data verification and validation information should be kept available and retained for those projects requiring such processes. In addition to QA documentation, the analytical results, either in hard copy or electronic form, should be available from the laboratory for a specified length of time after the completion of a project.

6.11 Summary of Recommendations

- MARLAP recommends the performance-based approach for method selection.

- MARLAP recommends that only methods validated for a project's application be used.

- MARLAP recommends that a SOW containing the MQOs and analytical process requirements be provided to the laboratory.

- MARLAP recommends that the SOW include the specifications for the action level and the required method uncertainty for the analyte concentration at the action level for each combination of analyte and matrix.

- MARLAP recommends that a method undergo some basic general validation prior to project method validation.

- MARLAP recommends that when a method is applied to a specific project, the method should then undergo validation for that specific application.

- MARLAP recommends that as each new project is implemented, the methods used in the analysis of the associated samples undergo some level of validation. However, it is the project manager's responsibility to assess the level of method validation necessary.

- MARLAP recommends a tiered approach for project method validation.

6.12 References

American National Standards Institute (ANSI) N42.23. *Measurement and Associated Instrumentation Quality Assurance for Radioassay Laboratories*. 2003.

American Public Health Association (APHA). 1998. *Standard Methods for the Examination of Water and Waste Water, 20ʰ Edition*. Washington, DC. Available at: www.standardmethods.org.

American Society for Testing and Materials (ASTM) D2777. *Standard Practice for Determination of Precision and Bias of Applicable Test Methods of Committee D-19 on Water*. West Conshohocken, PA.

American Society for Testing and Materials (ASTM) D4855. *Standard Practice for Comparing Methods*. West Conshohocken, PA.

American Society for Testing and Materials (ASTM) E177. *Standard Practice for Use of the Terms Precision and Bias in ASTM Test Methods*. West Conshohocken, PA.

American Society for Testing and Materials (ASTM) E1169. *Standard Guide for Conducting Ruggedness Tests*. West Conshohocken, PA.

American Society for Testing and Materials (ASTM) E1488. *Standard Guide for Statistical Procedures to Use in Developing and Applying ASTM Test Methods*. West Conshohocken, PA.

U.S. Environmental Protection Agency (EPA). 1980. *Prescribed Procedures for Measurement of Radioactivity in Drinking Water*. Environmental Monitoring and Support Laboratory, Cincinnati, OH. EPA 600-4-80-032, August.

U. S. Environmental Protection Agency (EPA). 2002. *Guidance for Quality Assurance Project Plans*. EPA QA/G-5. EPA/240/R-02/009. Office of Environmental Information, Washington, DC. Available at: www.epa.gov/quality/qa_docs.html.

EURACHEM. 1998. *The Fitness for Purpose of Analytical Methods, A Laboratory Guide to Method Validation and Related Topics*. ISBN 0-948926-12-0. Available at: www.eurachem.ul.pt/index.htm.

International Union of Pure and Applied Chemistry (IUPAC). 2002. "Harmonized Guidelines for Single-Laboratory Validation of Methods of Analysis." *Pure Appl. Chem.*, 74:5, pp. 835-855.

Youden, W.J. and E.H. Steiner. 1975. *Statistical Manual of the AOAC*. Association of Official Analytical Chemists International, Gaithersburg, MD. Available at: www.aoac.org/pubs/pubcat81.htm.

ATTACHMENT 6A
Bias-Testing Procedure

6A.1 Introduction

This attachment describes a statistical test that may be used to determine whether a laboratory measurement process is biased. The laboratory should check for both "absolute bias" and "relative bias," as defined in Section 6.6.4, "Testing for Bias."

Testing for absolute bias involves repeated analyses of method blank samples. Testing for relative bias requires repeated testing of spiked samples, such as certified reference materials (CRMs) or standard reference materials (SRMs). In either case, it is assumed here that replicate analyses are done at one concentration level, the estimate of which is called the *reference value* and denoted by K. When method blanks are analyzed, the reference value is zero.

6A.2 The Test

Whenever one performs a hypothesis test, one must choose the significance level of the test, which is denoted by α. Most often α is chosen to be 0.05, or 5 percent, but other values are possible. The significance level is the specified maximum acceptable probability of incorrectly rejecting the null hypothesis when it is actually true.

The hypothesis test described below is a t-test, modified if necessary to account for the uncertainty of the reference value. The test statistic is denoted by $|T|$ and is calculated by the equation

$$|T| = \frac{|\overline{X} - K|}{\sqrt{s_X^2 / N + u^2(K)}} \tag{6.1}$$

where
- $\overline{X}$ is the average measured value
- s_X is the experimental standard deviation of the measured values
- N is the number of measurements
- K is the reference value (typically $K = 0$ for method blanks)
- $u(K)$ is the standard uncertainty of the reference value (typically $u(K) = 0$ for method blanks)

When method blanks are analyzed, $K = u(K) = 0$, and the statistic may be calculated as

$$|T| = \frac{|\overline{X}|}{s_X / \sqrt{N}} \tag{6.2}$$

The number of *effective degrees of freedom* for the T statistic is calculated as follows:

$$\nu_{eff} = (N - 1)\left(1 + \frac{u^2(K)}{s_X^2 / N}\right)^2 \tag{6.3}$$

When $K = u(K) = 0$, the number of effective degrees of freedom is $N - 1$, which is an integer. However, if $u(K) > 0$, then ν_{eff} generally is not an integer[2]; so ν_{eff} should be truncated (rounded down) to an integer. Then, given the chosen significance level, α, the critical value for $|T|$ is defined to be $t_{1-\alpha/2}(\nu_{eff})$, the $(1 - \alpha/2)$-quantile of the t-distribution with ν_{eff} degrees of freedom (e.g., see Table G.2 in Appendix G). So, a bias in the measurement process is indicated if

$$|T| > t_{1-\alpha/2}(\nu_{eff}) \tag{6.4}$$

A measure of the power of this t-test for bias is the *minimum detectable bias* (MDB), which may be defined as the smallest bias ($\pm$) that can be detected with a specified probability, $1 - \beta$. The MDB is a function of α, β, N, and the standard deviation of the measured results, σ_X, at the given concentration level. Achieving a small value for the MDB may require the analysis of many replicate samples. If $\alpha = \beta = 0.05$, then at least 16 analyses are needed to ensure the MDB is less than the measurement standard deviation. Fifty-four measurements would be necessary to ensure MDB $\leq \sigma_X / 2$.

EXAMPLE 6.1

Suppose a laboratory analyzes a series of 9 method blanks and obtains the following results (Bq):

| 0.714 | 2.453 | −1.159 | 0.845 | 0.495 | 0.993 | 0.472 | −0.994 | 0.673 |

Determine whether the data indicate an absolute bias. Use a significance level of $\alpha = 0.05$.

Calculate the average of the measured results.

$$\overline{X} = \frac{1}{N}\sum_{i=1}^{N} X_i = \frac{4.492}{9} = 0.49911$$

Note that $\overline{X}$ is the best available estimate of the bias, but it has not yet been determined to be statistically significant.

---

[2] When the value of ν_{eff} is > 20, one may assume a ν_{eff} value of infinity.

Next calculate the experimental standard deviation.

$$s_X = \sqrt{\frac{1}{N-1}\sum_{i=1}^{N}(X_i - \bar{X})^2} = \sqrt{\frac{1}{9-1}\sum_{i=1}^{9}(X_i - 0.49911)^2} = \sqrt{1.15455} = 1.0745$$

In this example, the reference value is $K = 0$, with a standard uncertainty of $u(K) = 0$. So, the value of the test statistic, $|T|$, is found as follows.

$$|T| = \frac{|\bar{X}|}{s_X/\sqrt{N}} = \frac{0.49911}{1.0745/\sqrt{9}} = 1.3935$$

Since $u(K) = 0$, the number of effective degrees of freedom is

$$\nu_{eff} = N - 1 = 8$$

So, the critical value for the statistic is

$$t_{1-\alpha/2}(\nu_{eff}) = t_{0.975}(8) = 2.306$$

Since $1.3935 \le 2.306$, no bias is detected.

EXAMPLE 6.2

Suppose a laboratory performs 7 replicate analyses of a standard reference material and obtains the following results (Bq/L):

50.74 53.08 50.73 50.92 51.50 51.11 52.61

Suppose also that the reference value for the SRM is 49.77 Bq/L with a combined standard uncertainty of 0.25 Bq/L.

Determine whether the data indicate a relative bias. Use a significance level of $\alpha = 0.05$.

Calculate the average of the measured results.

$$\bar{X} = \frac{1}{N}\sum_{i=1}^{N}X_i = \frac{360.69}{7} = 51.527$$

Note that the best estimate of the relative bias is $\bar{X}/K - 1$, which equals +0.0353.

Calculate the experimental standard deviation.

$$s_X = \sqrt{\frac{1}{N-1}\sum_{i=1}^{N}(X_i - \overline{X})^2} = \sqrt{\frac{1}{7-1}\sum_{i=1}^{7}(X_i - 51.527)^2} = 0.94713$$

Calculate the value of the test statistic, $|T|$.

$$|T| = \frac{|\overline{X} - K|}{\sqrt{s_X^2/N + u^2(K)}} = \frac{|51.527 - 49.77|}{\sqrt{0.94713^2/7 + 0.25^2}} = 4.024$$

The number of effective degrees of freedom for the statistic is calculated as follows.

$$v_{eff} = (N-1)\left(1 + N\frac{u^2(K)}{s_X^2}\right)^2 = (7-1)\left(1 + 7\frac{0.25^2}{0.94713^2}\right)^2 = 13.28$$

Note that v_{eff} is then truncated to 13. Next calculate the critical value for $|T|$.

$$t_{1-\alpha/2}(v_{eff}) = t_{0.975}(13) = 2.160$$

Since $|T| = 4.024 > 2.160 = t_{1-\alpha/2}(v_{eff})$, a bias is detected.

6A.3 Bias Tests at Multiple Concentrations

The discussion above describes a test for bias based on replicate measurements at one concentration level. If replicate measurements are done at each of several concentration levels, the bias test should be performed for each level to evaluate whether there is an "overall" method bias for the entire concentration range based on an α false rejection rate. For this test, the value of α used for each concentration level should be replaced by a smaller value, α', given by

$$\alpha' = 1 - (1 - \alpha)^{1/m} \tag{6.5}$$

where m denotes the number of concentration levels. For example, if the desired overall method false rejection rate is $\alpha = 0.05$ and the number of levels is three ($m = 3$), the value of α' for a given test level is 0.01695. When the bias test, using the α' value, for every concentration level indicates no bias, then the method would be considered free of bias based on an α false rejection rate over the concentration range evaluated. However, this overall method bias test should not be misused or misinterpreted. In some cases, a project manager or laboratory may be more interested to know if a bias exists at one specific test concentration and not at others. For example, the evaluation of the rate of false- or non-detection for blanks (zero radionuclide concentration) may be more important for a particular project than evaluating the overall method bias for all test levels.

A possible alternative when testing is done at several concentration levels is to use weighted linear regression to fit a straight line to the data and perform hypothesis tests to determine whether the intercept is 0 and the slope is 1. However, determining the most appropriate numerical weights may not be straightforward.

7 EVALUATING METHODS AND LABORATORIES

7.1 Introduction

This chapter provides guidance for the initial and ongoing evaluation of radioanalytical laboratories and methods proposed by laboratories. Appendix E, *Contracting Laboratory Services*, provides additional guidance on the initial laboratory evaluation. More details about evaluating and overseeing a laboratory's performance can be found in ASTM E1691 and ASTM E548.

The performance-based approach to method selection allows a laboratory the freedom to propose one or several methods for a specific analyte/matrix combination that will meet the needs of the analytical protocol specifications (APSs) and measurement quality objectives (MQOs) delineated in the statement of work (SOW). However, the laboratory should demonstrate, through a method validation process, that the method is capable of producing analytical results of quality that meet the needs of the SOW (Chapter 5, *Obtaining Laboratory Services*). Guidance and recommendations on the selection of an analytical method based on the performance-based approach is presented in Chapter 6 (*Selection and Application of an Analytical Method*). Section 7.2 provides guidance on how to evaluate the methods proposed by a laboratory. Section 7.3 provides guidance on the initial evaluation of a laboratory, and Section 7.4 discusses the continual evaluation of the quantitative measures of quality and operational aspects of the laboratory once sample processing has commenced.

Method applicability and performance compliance should be demonstrated prior to the initiation of the sample analyses, as well as during the project period. A defined logical process for demonstrating and documenting that the analytical method selected meets the project's data needs and requirements may involve, for example, a review of the method validation documentation, an evaluation of past performance data from other projects (if available), the analysis of external performance evaluation (PE) program results, the analysis of matrix-specific standard reference materials (or method validation reference materials) sent during the initial work period and throughout the project, and the final evaluation of the performance during the data verification and validation process (see Chapter 8, *Radiochemical Data Verification and Validation*).

In addition to the evaluation of the analytical methods, the capability of the laboratory to meet all SOW requirements needs to be reviewed and evaluated. Supporting information, such as method validation documentation, safety manuals, licenses and certificates, and quality manual are typically submitted with the response to the request for proposals (RFP). A generic evaluation of the laboratory operation may be conducted during the initial laboratory audit or assessment. This may be an initial

Contents		
7.1	Introduction	7-1
7.2	Evaluation of Proposed Analytical Methods .	7-2
7.3	Initial Evaluation of a Laboratory	7-15
7.4	Ongoing Evaluation of the Laboratory's Performance	7-20
7.5	Summary of Recommendations	7-32
7.6	References	7-33

onsite audit. This first evaluation covers those generic SOW requirements dealing with the laboratory's capability and operation, including verification of adequate facilities, instrumentation, and staffing and staff training and qualifications. Following the first audit, emphasis should be on ensuring the laboratory continues to meet the APSs through a continuous or ongoing evaluation effort.

7.2 Evaluation of Proposed Analytical Methods

A laboratory may submit several methods for a particular APS contained in the SOW, but each method should be evaluated separately and, if appropriate, approved by the project manager or designee. The method should be evaluated to be consistent with the overall analytical process that includes the proposed field sampling and preservation protocols (Chapter 1). The project manager may delegate the method review process to a technical evaluation committee (TEC) that has a radioanalytical specialist. MARLAP recommends that a radioanalytical specialist review the methods for technical adequacy. The acceptance, especially of a new method, may be the most critical aspect of the performance-based approach for method selection. Acceptance of the method requires the project manager to verify that the method is scientifically sound.

Each step of the method should be evaluated by a radioanalytical specialist in order to understand how the results are derived. These steps may involve sample digestion, analyte purification and decontamination steps that use ion exchange, solvent extraction, precipitation or oxidation/ reduction applications. Once these steps have been reviewed, and the method evaluation data (e.g., from method validation documentation or various performance evaluation results) confirm that the proposed method is acceptable, the project manager should have the confidence necessary to endorse and verify the use of the method in the analysis of the routine samples.

As discussed in Chapter 6, the laboratory should provide method validation and analytical data that demonstrates method performance. The data should show conclusively that the proposed method meets the requirements as defined by the APSs. If method performance is questionable, additional data may be required. For such cases, the project manager may decide to send performance testing (PT) materials to the laboratory in order to evaluate or validate the method. The preparation of the PT material used to evaluate the method should be based on sound scientific principles and representative of the expected sample matrix (see Chapter 6 on method validation options using site-specific materials). If there is sufficient reason to believe that the PT material is an adequate substitute for the sample matrix and that the laboratory will follow the same method, then the need to justify each step in the method may be drastically reduced.

7.2.1 Documentation of Required Method Performance

Certain documentation submitted by the laboratory with the proposed methods, as well as available external information on the laboratory's analytical performance, should be reviewed and evaluated by the radioanalytical specialist. Table 7.1 outlines where such information typically can be found by the TEC. This section will discuss various information categories that may be available during the method evaluation process.

TABLE 7.1 — Cross reference of information available for method evaluation

Evaluation Element Addressed	Method Validation	Internal and External QC Reports	External PE Programs	Internal/ External QA Assessments	Information from RFP and Other Sources
Analyte/Matrix	●				●
Process Knowledge					●
Previous Experience					●
Radiological Holding Time	●	O	O	●	●
Turnaround Time		O	O	●	●
Unique Process Specifications	●				●
Bias	●	●	●	O	●
Method Uncertainty (MQC/MDC)	●	●	●	O	●
Analyte/Interference Range	●	●	●		●
Method Ruggedness	●	O	●	●	●
Method Specificity	●	O	●	●	●

● Information relevant to method evaluation should be present.
O Information relevant to method evaluation may be present.

7.2.1.1 Method Validation Documentation

Chapter 6 outlines the various method validation options that can be specified by the project manager. In the MARLAP process, the method validation requirements will be contained in the SOW. The laboratory must submit the necessary method validation documentation consistent with the SOW specification. The laboratory may choose to validate a method to a higher degree of validation or to submit method validation documentation for a higher degree of validation than that specified by the SOW. The radioanalytical specialist or project manager should review the documentation to ensure that validation criteria for the number of analyte concentration levels and replicates meet or exceed the required validation criteria (Chapter 6, Table 6.1). Although not specified in the method validation protocol, some laboratories may include chemical and analytical interferences in their method validation plan to gain a perspective on the method's specificity and ruggedness. However, it should be noted that the graded approach to method validation presented in Chapter 6 does inherently increase the degree of ruggedness in terms of having the method address site-specific materials which may include chemical and radionuclide interferences.

In addition to reviewing the documentation for compliance with the method validation protocol, the results of the method validation process should be evaluated to determine if the project specific MQOs will be met. The method validation may or may not have been specifically conducted for the project at hand. When the method has been validated (Chapter 6, Section 6.6) to the SOW specifications (validation level and MQOs), then evaluation of the documentation can be straight forward. If the method has been previously validated for the MQOs of other projects, then the laboratory should provide a justification and calculations to show that the method validation results will meet the MQOs for the new project. The TEC should verify these calculations and review the assumptions and justifications for reasonableness and technical correctness.

7.2.1.2 Internal Quality Control or External PE Program Reports

The documentation of internal QC and external PE program results should be reviewed relative to the MQOs. Method uncertainty and internal biases can be estimated from the information available in the laboratory's internal quality control reports, summaries of batch QC results that may be submitted with the RFP response and external PE program reports. The TEC should review these documents and, when possible, estimate the method uncertainty and bias for various analyte concentration levels. However, it is imperative that no confusion exists in terms of what method produced the results: the proposed method or another method available to the laboratory. This is especially important when reviewing external PE program results. It should also be noted that although a laboratory may meet performance acceptance criteria for an external PE program, this fact may have no bearing on whether the method will meet the MQOs of the SOW.

Review of the internal batch QC data can provide additional information on typical sample analysis times and rates of blank contamination and sample reanalysis. This information is important when comparing methods (from the same or between laboratories) in terms of APS characteristics. The frequency of blank contamination would be very important to national characterization studies (groundwater or soil analyses) for the determination of ambient analyte levels. Method evaluation for these projects may weight the blank contamination rate more heavily than other SOW parameters. The rate of sample reanalysis would be important to projects having pending operations that are conducted based on a short sample processing turnaround time (TAT). In some site remediation projects, the contractor may remain onsite pending analytical results. A delay in reporting data or not meeting a TAT due to sample reanalysis may be costly. Projects of this nature may weight TAT and low sample reanalyses more heavily than other SOW parameters.

7.2.1.3 Method Experience, Previous Projects, and Clients

When permitted by former clients, the laboratory may submit information relative to the previous or ongoing clients and projects for which the proposed method has been used. The TEC should verify with the laboratory's clients that the laboratory has previous experience using the method. When available and allowed, the information should also include the analyte(s) and interferences and their applicable concentration range, matrix type, and project size in terms of the number of samples per week or other time periods. From this information, the TEC can evaluate whether or not to contact the laboratory's client for further information on the operational adequacy of the method. The client may offer some information on the quality of the results based on their external single- or double-blind QC program, percent completion of reports, TAT, and sample re-analysis frequency. The sharing of laboratory assessment reports may be advantageous when reviewing the performance of the laboratory during its employment of the method.

7.2.1.4 Internal and External Quality Assurance Assessments

When available, internal and external quality assurance assessment reports should be evaluated to determine the adequacy of the method performance based on previous projects. Problems with the conduct of the method due to procedural and technical issues may be readily evident. These issues may include an ineffective corrective action program creating delayed remedies to problems, insufficient understanding of the method, inadequate training of staff, internal and project-specific QC issues, and higher-than-expected failure rates for sample TATs and re-analyses. Information in these reports may disclose problems with a particular method that are not common to another proposed method. As such, the TEC may give one method a higher weighting factor than another method.

7.2.2 Performance Requirements of the SOW—Analytical Protocol Specifications

Under the performance-based approach to method selection, a laboratory will propose one or several analytical methods that can meet the stated APSs and MQOs in the SOW for a given analyte and matrix combination. Chapters 3, 5, and 6 discuss the APSs and MQOs in detail in terms of their basic description, their inclusion in a SOW, and as key considerations for identifying existing validated methods or developing new methods. The purpose of this section is to provide guidance on what available information should be evaluated in order to approve the various proposed methods.

The radioanalytical specialist should review the process-knowledge information and determine if the proposed method is adequately specific, rugged, and applicable to address these issues. Discussions on method specificity and ruggedness may be found on in subsections on pages 7-12 and 7-13, respectively.

As discussed in Section 6.5.2 and above, process knowledge is extremely important for identifying potential radioanalytical problems on some projects. Historical information or process knowledge may identify chemical and radionuclide interferences, expected analyte and interfering radionuclide concentration ranges, sample analyte heterogeneity issues, and the physiochemical form of the analyte, and the sample matrix substrate. In some special cases, it may be necessary to determine if the radiological holding time will be an issue if the laboratory must analyze an alternative nuclide to determine supported and unsupported radionuclides (decay progeny nuclides) in the matrix.

The following subsections cover key aspects of the SOW that should be addressed during the method evaluation and approval process.

7.2.2.1 Matrix and Analyte Identification

The TEC should review the method(s) proposed by the laboratory to determine if the method under evaluation is applicable for the analyte/matrix combination specified in the SOW. In some cases, several methods may be proposed, including gross screening methods and specific radionuclide or isotopic methods having high specificity and ruggedness (Section 6.5.1.1 has additional guidance). Each method should be evaluated on its own application and merit. When methods are proposed by the laboratory that use alternative nuclides (such as decay products) to determine the analyte of interest, the TEC should carefully review the objective or summary of the method to determine if the proposed method is truly applicable for the analyte of interest given the radiological holding time and MQOs (i.e., can it properly quantify the analyte of interest through decay progeny measurements?). For gross screening techniques, the TEC should evaluate the analyte's decay scheme to determine the underlying gross radiation category (beta, alpha, X-ray, or gamma-ray emitting) and the applicability of the proposed method's radiation detection methodology.

Each proposed method should be evaluated to determine if the method can analyze the sample matrix identified in the SOW. A method validated for water cannot be applied to soil samples without modification and validation (Section 6.5). The planning team should have made—through historical process knowledge, previous matrix characterization studies or common experience—a determination on the uniqueness of the site-specific matrices compared to typical matrices and provided guidance in the SOW as to the level of method validation. In addition, if the radioanalytical specialist of the project planing team is concerned that the physiochemical form of the analyte or the sample matrix substrate may present special problems to the radioanalytical process, a detailed description of the analyte and matrix should have been included in the SOW. Chapters 12 (*Laboratory Sample Preparation*) and 13 (*Sample Dissolution*) discuss possible sample matrix problems and Section 6.5 provides guidance on the need for method validation. The radioanalytical specialist should carefully review the summary of the method to determine if the proposed method is applicable for the sample matrix.

At this point, if it is determined that the proposed method(s) is not applicable and cannot meet the SOW specifications, there is no need to continue the method evaluation process.

7.2.2.2 Radiological Holding and Turnaround Times

The radioanalytical specialist also should review the proposed method in light of the radiological holding time, analyte's half-life and typical sample delivery options and determine if the method is capable of meeting the MQOs in a reasonable counting period given the typical method parameters (such as sample weight processed, chemical yields, radiation detection efficiency, branching ratio and background, ingrowth periods for decay progeny analysis, etc.). Radiological holding time is defined as the time between the sample collection and the end of the sample-counting interval, while sample processing TAT refers to the time between sample receipt at the laboratory and the issuance of an analytical report. The physical (analyte's half-life) and chemical (stability or preservation concerns) characteristics of the analyte, as well as biological degradation for some matrices, usually will dictate the radiological holding time. Project-specific schedules and practicalities related to project and laboratory processing capacities normally enter into establishing TATs. If the radiological holding time appears to be a critical issue, then the client should request information on the typical batch size being processed by the laboratory for that method. This information is needed in the method evaluation and review process. For very short-lived analytes, too large a batch size may result in the later samples having much larger uncertainties than the earlier samples. In these cases, the laboratory will count the sample (or final processing products) longer in order to achieve client-requested minimum detectable concentrations. This is not often practical because the loss of counts due to analyte decay is more significant than any gain achieved by counting the sample longer.

In some cases, the laboratory may want to propose two methods for a short-lived analyte: one for normal delivery and processing schedules and another method for situations when lower detection limits are needed. An example of such a situation is the analysis of ^{131}I in environmental media. A method with adequate detection limits for reasonable radiological holding times is gamma spectrometry. Another method that can be applied for lower detection limits or longer radiological holding times is radiochemical separation followed by beta-gamma coincidence counting.

Certain projects may be concerned with the chemical speciation of the analyte in the sample. For these projects, the radiological holding time should have been specified to ensure that the chemical species are not altered prior to processing. The project normally should specify chemical preservation specifications applicable at the time of sample collection.

Preservation techniques should be used when deterioration of biological samples may become a problem (Chapter 10, *Field and Sampling Issues that Affect Laboratory Measurements*). However, the radiological holding time should be specified to limit problems with sample degradation. The radioanalytical specialist should evaluate the method in light of the foregoing information and determine its adequacy to meet the radiological holding time and the pertinent MQOs

A laboratory's sample (processing) TAT for a method typically is not related to the method's technical basis unless the radiological holding time and the TAT are nearly equal for a short-lived analyte. However, sufficient time should be available between the completion of sample analysis and the delivery of the analytical report. Meeting the radiological holding time but failure to meet the TAT will not affect the quality of the analytical results but may place a hardship on the project to meet schedules. The TEC should review the proposed method, the radiological holding time and the TAT to determine if the method can process the samples in a reasonable time period to meet the TAT. The sample delivery rate, sample batch size, level of data automation and the laboratory's existing sample processing capacity will affect the laboratory's ability to meet the TAT requirement.

7.2.2.3 Unique Processing Specifications

The TEC should review the proposed methods for compliance or applicability to unique sample processing specifications stated in the SOW. Chapter 6 provides a limited discussion on what a project may identify as unique or special sample process specifications. Examples may include chemical speciation, analyte depth profiles, analyte particle size distribution, analyte heterogeneity within the sample, wet-to-dry analyte concentration ratios in biologicals, and possible scaling factors between radionuclides in the sample. In some cases, the proposed method(s) for the analyte(s) may have to be evaluated with respect to all analytes or other sample preparation specifications in order to determine method applicability and adequacy.

7.2.2.4 Measurement Quality Objectives

Method performance characteristics (method uncertainty, quantification capability, detection capability, applicable analyte concentration range, method specificity, and method ruggedness) will be discussed in the following subsections. For a particular project, MQOs normally will be developed for several (but not all) of the performance characteristics discussed below.

METHOD UNCERTAINTY

The SOW should specify the required method uncertainty at a stated analyte concentration (or activity level) for each sample matrix and the level of method validation (Section 6.6) needed to qualify the method at the stated analyte concentration.

MARLAP uses the term "method uncertainty" to refer to the predicted uncertainty of a result that would be measured if a method were applied to a hypothetical laboratory sample with a specified analyte concentration. As presented in Chapter 6 and formulated in Chapter 20 (*Detection and Quantification Capabilities*), the method uncertainty of the analyte concentration for a given method is determined by mathematically combining the standard uncertainties of the many input quantities (parameters), involved in the entire radioanalytical process. This will involve making some assumptions and normally involve using typical or worst case values for a conservative estimate of the method uncertainty. Some of these input quantities, and thus the method uncertainty, vary according to analyte level or concentration in the final measured product; others do not. In some cases, the magnitude of the method uncertainty for an analyte may increase in proportion to the magnitude (concentration/activity) of any interfering radionuclide present in the final measurement product. Therefore, it is imperative that the TEC evaluate the laboratory's submitted documentation relative to this requirement, especially the information provided on method specificity, given the historical or expected interfering nuclides and the needed decontamination factors (chemical separation factors) to render a good measurement for the analyte of interest.

In evaluating the documentation relevant to meeting the method uncertainty requirement, it is important to determine if the method validation requirements stated in the SOW have been met. The TEC should review the submitted method validation documentation and verify that the method's performance meets the requirements of Table 6.1 (Chapter 6) for the specified validation level. It is important that the laboratory submit definitive documentation of method validation compliance for the method uncertainty requirement.

The method performance documentation may include documentation or data from method validation, internal or external (organization sending QC samples) QC data, external PE program data, and results of prequalifying laboratories by sample analyses. By evaluating the actual QC and PE program performance data, it can be determined if the quoted measurement uncertainty for a reported QC sample result (calculated by the laboratory) truly reflects the method uncertainty under routine processing of samples. The required method uncertainty can be viewed as a target value for the overall average measurement uncertainty for the samples at a specified analyte concentration. It is important that the precision, as calculated from repeated measurements, is consistent with the laboratory's stated measurement uncertainty for a given sample result whose analyte concentration is near the specified concentration. If the quoted measurement uncertainty of a QC or test measurement is quoted to be $\pm$ 10 percent and QC or PE program data indicates a data set standard deviation of $\pm$ 20 percent, then the laboratory may not have identified all possible uncertainty components or may have underestimated the magnitude of a component.

QUANTIFICATION CAPABILITY

A requirement for the quantification capability of a method and the required method validation criteria may be specified in a SOW. The quantification capability, expressed as the minimum quantifiable concentration (MQC), is the smallest concentration of the analyte that ensures a result whose relative standard deviation is not greater than a specified value, usually 10 percent.

The project manager or TEC should review available documentation on the method to determine if the laboratory can meet the method quantification requirement. Method validation documentation sent by the laboratory should demonstrate explicitly, or by extrapolation, that the method, using certain input quantities and their uncertainties, can meet the quantification requirement. The method validation acceptance criteria presented in Section 6.6 have been formulated to evaluate the MQC requirement at the proper analyte concentration level, i.e., action level or other specified analyte concentration.

Some projects may send performance testing material spiked at the MQC level as a more in-depth verification of the compliance with this requirement. Laboratories may also submit documentation for internal QC or external PE program results that cover the MQC value. The TEC should evaluate the reported results to determine if the MQC requirement can be met.

DETECTION CAPABILITY

A radiochemical method's detection capability for an analyte is usually expressed in terms of minimum detectable concentration (MDC) or activity (MDA). Chapter 19 provides the definition and mathematical equations for the MDC[1] and MDA. A MDC requirement for each analyte/matrix combination may be stated in a SOW. Any proposed method should document the basis and equation for calculating the MDC. The supporting documentation on the method should contain the input quantity values that may be entered into the MDC equation to calculate the detection capability under a variety of assumptions. The TEC should evaluate the assumptions and parameter values for reasonableness and practicality. This evaluation is especially important for recently validated methods that have a limited routine processing history. MARLAP recommends that the TEC perform an independent calculation of the method's MDC using laboratory-stated typical or sample-specific parameters.

When the proposed method has been validated recently or previously used on similar projects, sufficient data should exist that either are directly related to testing the method's detection capability or can be used to estimate the method's detection capability. Any data submitted that document direct testing of the method's detection capability should be reviewed for appropriateness or applicability, reasonableness, and accuracy. If method detection testing is performed, it normally will be for one analyte concentration level or value. It should not be expected that the MDC testing process included varying the magnitude of the method's many parameters over a wide range.

The reported quantitative results of the blanks can be used to estimate the MDC to within a certain degree of confidence (for most methods). At or below the MDC value, the majority of the measurement uncertainty typically is due to the Poisson counting uncertainty. For well-controlled

---

[1] The MDC should not be confused with the concept of the critical value (Chapter 20).

methods, the uncertainties of the other method parameters (input quantities), such as sample weight, detection efficiency, and chemical yield, may range up to 10 percent. Therefore, a simple rule of thumb to estimate the MDC for most methods involves reviewing the measurement uncertainty for the reported blank results. If the blanks were analyzed to meet the MDC requirement, then the reported MDC (based on blank and sample paired observations) for most methods should be between 3 and 4 times the measurement uncertainty of the blank when the background counts (per measurement interval) are greater than 10. It is more complicated to estimate the MDC for methods that use low background detectors (such as alpha spectrometry) having background counts less than 10 per counting interval. The TEC should evaluate the blank data to determine the reasonableness of the quoted MDC values. These rules of thumb can be applied to actual samples when the quoted analyte concentration value is less than two times its associated combined standard uncertainty value.

APPLICABLE ANALYTE CONCENTRATION RANGE

The applicable analyte concentration range can vary substantially depending on whether the project deals with process waste streams, environmental remediation or monitoring, or environmental or waste tank characterization research. The proposed method being evaluated should provide accurate results over the analyte concentration range stated in the SOW. Acceptable analytical results used in this context means consistent method uncertainty (at a given analyte concentration) and without significant bias. The range may be over several decades, from a minimum value (the MDC for some projects) to 100 times the action level or MQC.

Due to the effects of the Poisson counting uncertainty, most methods will provide more precise results at higher analyte concentration levels compared to those concentration levels near zero. At concentration levels near zero, background effects will render the results less precise. If the background (instrument or ambient levels of analyte in the matrix) is not well characterized, a bias may also exist. For projects or programs (environmental characterization research) that have no action level requirement, the lower portion of the required concentration range or the MDC requirement may be most important. For those situations, particular emphasis should be placed on evaluating method and reagent blank data (i.e., net results that take into account inherent analyte content in the reagents or tracers) to ensure that a bias does not exist. Refer to Section 7.2.2.5, "Bias Considerations," on page 7-13 for additional guidance.

Typically, radiation detection systems are linear in signal response over a very large range of count rates. However, depending on the magnitude of the chemical or radionuclide interferences in the sample, the method may not produce linear results over the entire application range. Therefore, it is critical that when a mixture of radionuclides is present in a sample, the method must provide sufficient "analyte selectivity/isolation or impurity decontamination" to ensure valid results and "method linearity." In some cases, such as that for pure beta-emitting analytes, the degree of needed decontamination from other interfering nuclides may be as much as six orders of magnitude.

There are several sources of information available from the laboratory that should be reviewed and possibly evaluated to ensure the method is capable of meeting this MQO. These include method validation documentation, previous projects or experience using the method, PE program results, internal and external QC sample results, and prequalifying test samples. When evaluating the data, the TEC should evaluate the method's performance as a function of analyte concentration with and without interferences. However, this evaluation would be most valid when the samples were processed to the same MQO (especially MDC or MQC), a situation that may not be realistic for different projects. If the MDC requirement results in a longer counting time from one project to another, there may be an impact on the method's uncertainty for a given analyte concentration due to difference in the Poisson counting uncertainty. Bias typically is not affected by increasing the counting time. A graphical plot of this data would be visually helpful and may be used to determine if the method uncertainty requirement would be met at the action level (extrapolation may be necessary).

METHOD SPECIFICITY

Method specificity refers to the ability of the method to measure the analyte of concern in the presence of other radionuclide or chemical interferences. The need for or degree of method specificity depends on the degree or magnitude of the interferences and their effect on the ability to measure the analyte of interest. Gross alpha, beta, and gamma-ray methods (which do not have fine resolution) are considered to be methods of low specificity and are used when individual nuclide specificity is not possible or needed. Radiochemical methods involving sample digestion, purification and decontamination steps followed by alpha spectrometry, such as for ^{239}Pu in soil, are considered methods of high specificity. However, the relative degree of specificity of these nuclide specific methods depends on the number of analyte isolation and interference decontamination steps. High-resolution gamma-ray spectrometry employing a germanium (Ge) detector is considered to have better specificity than the lower resolution sodium iodide (NaI) gamma-ray spectrometry.

The TEC should evaluate the proposed methods for adequacy to meet the specificity requirements stated in the SOW. As mentioned in Chapter 6, methods of low specificity, such as gross radiation detection methods, may be proposed if the methods meet the MQOs. For example, when a single analyte having a relatively elevated action level needs to be evaluated, such as ^{137}Cs in soil at an action level of 222 Bq/kg (6 pCi/g), then a method with less specificity (gross counting methods for gamma-ray or beta emitting nuclides) may be sufficient to meet the MQOs. For this example, a less expensive NaI gamma-ray spectrometric analysis with a lower resolution capability may be more desirable compared to a more costly high resolution germanium gamma-ray spectrometric analysis. If greater method specificity for a certain analyte/matrix combination has been required in the SOW, then a high resolution non-destructive sample analysis method (such as high resolution gamma-ray spectrometry) or a destructive sample analysis by a detailed radiochemical method would be appropriate. For proposed methods of high specificity, it is important that the TEC review and evaluate the basic purification and decontamination steps of the method, or the resolution of the radiation detection system, for adequacy in relation to the expected mixture of analytes and interferences. For radiochemical methods, the TEC may be able to estimate the needed distribution/partition coefficients, extraction and solubility factors, etc., of the various purification steps and compare the values against the needed decontamination factors for the interfering chemical or radionuclide interferences.

The adequacy of method specificity can be evaluated by the analytical results from the analysis of site-specific PT materials during method validation and/or laboratory prequalifying tests. A further discussion on the use of these materials is presented below.

METHOD RUGGEDNESS

Method ruggedness refers to the ability of the method to produce accurate results over wide variations in sample matrix composition and chemical and radionuclide interferences, as well as when steps (such as pH adjustments) in the method are varied slightly by the analyst. For some projects, the matrix composition and level of analyte or interferences may very dramatically in a given project.

Ruggedness studies have been defined by EPA (2002). A testing protocol for method ruggedness has been outlined by the American Public Health Association (APHA). Some laboratories may have developed methods according to the APHA protocol for method ruggedness or are using methods contained in standards methods (APHA, 1998). Documentation on any internal ruggedness study may be available from the laboratory.

As mentioned in Chapter 5 and 6, the use of site-specific PT materials is a means of testing the ruggedness of a method for a defined project. If ruggedness and method specificity are concerns due to the sample matrix of a defined project, then a variety of site-specific performance testing materials should be sent to the laboratory as part of the prequalification process or as a method validation requirement. National PE programs, such as DOE's Multiple Analyte Performance Evaluation Program (MAPEP) and Quality Assessment Program (QAP), use generic PT materials and may not be applicable or representative of the matrices for a defined project. The results of the prequalifying or method validation processes using site-specific PT materials should be evaluated by the TEC to determine the adequacy of the method to meet this MQO parameter. If the sample matrix and analytes are fairly standard, then no other evaluation of the available information may be necessary.

7.2.2.5 Bias Considerations

The method proposed by the laboratory should produce analytical results that are unbiased. MARLAP considers bias to be a persistent difference of the measured result from the true value of the quantity being measured, which does not vary if the measurement is repeated. Normally, bias cannot be determined from a single result or a few results (unless the bias is large). Bias may be expressed as the percent deviation in (or deviation from) the "known" analyte concentration. Since bias is estimated by repeated measurements, there will be an uncertainty in the calculated value. It is incumbent upon the project manager or TEC to evaluate the proposed methods for possible bias over the applicable analyte concentration range. A laboratory should eliminate all known biases before using a method. However, there may be circumstances, such as the processing of site-specific sample matrices, that may produce some inherent bias that is difficult to assess or correct in a reasonable time or economical fashion. For the methods proposed, the project manager must determine if the magnitude of the bias will significantly affect the data quality.

A bias can be positive or negative. Methods may have a bias at all analyte concentration levels due to the improper determinations of chemical yield, detector efficiency or resolution, subtraction of interferences, and improper assumptions for the analyte's half-life or an emission branching ratio. When reporting an analyte concentration based on a decay progeny analysis, improper ingrowth assumptions may lead to a bias.

MARLAP recommends that the project manager or TEC evaluate the available data provided by the laboratory or from performance evaluations for bias, based on multiple analyses covering the applicable analyte concentration range. One means of estimating a bias is through the evaluation of external PE program data.[2] For proper evaluation of the PE program sample results, it is essential that the PE program provider use sample preparation techniques that will produce performance testing (PT) samples (or a sample distribution) having insignificant "within or between" sample analyte heterogeneity and whose analyte concentrations are accurately known.

For the purpose of evaluating whether a laboratory method has an observable bias based on multiple laboratory internal QC samples (matrix or method spikes) or external PE program samples, the following equations can be used:

$$D_i = 100 * \left(\frac{X_i - Y_{i_{Known}}}{Y_{i_{Known}}} \right) \tag{7.1}$$

where D_i is the percent deviation, X_i is an individual analytical result and $Y_{i\,known}$ is the "known" value for the sample analyzed. The D_i should be determined for each test sample in the data set. The mean percent deviation for the method for a series of analyses in the data set can be estimated by the equation:

---

[2] In order to standardize against the national standard (NIST), an external performance evaluation program should be implemented by a well-qualified provider that has standardized its reference materials to NIST or is participating in a NIST traceability program

$$\overline{D} = \frac{\sum_{i=1}^{N} (D_i)}{N} \tag{7.2}$$

Refer to various references (ASTM D2777, NBS 1963, Taylor 1990) for applicable tests that may be performed to determine if there is a statistical difference at a given significance level.

There may be a negative or positive bias at low analyte concentrations due to the improper determination of the appropriate detector background or analytical blank value. For an individual blank result, the result (net activity or concentration value) would be considered to be a statistically positive value if the magnitude of its value is greater than 1.65 times the quoted measurement uncertainty. An older, much less conservative approach was to consider a reported value as a positive value when the magnitude of a result was greater than 3 times the measurement uncertainty.

Since the measurement process is statistical in nature and involves the subtraction of an appropriate background or blank which also has an uncertainty, there is a 50 percent probability (half of the results) that the analytical result for a blank sample will have a negative magnitude, e.g., -1.5 ± 2.0. For an individual blank measurement, the measurement may be considered to be problematic when the negative magnitude is greater than 2 or 3 times the measurement uncertainty.

For most radionuclides, other than those that are naturally occurring, the major source of a positive blank is from contamination, either cross-contamination from other samples or dirty glassware during sample processing or from tracer impurities. A poor estimate of the instrument background or ambient analyte levels in the matrix/reagent can lead to results being too negative in magnitude. A statistical test should be performed on a series of the data results to determine if there is a negative bias. The relative importance of the negative bias depends on the magnitude of the negative bias, magnitude of the action level and type of project.

7.3 Initial Evaluation of a Laboratory

The basic information to be considered in the initial evaluation of a laboratory has been summarized according to major categories in Figure 7.1. Not all categories will be discussed in detail as subsections. Some categories may be grouped and discussed under a single generic subsection heading. In order to allow for flexibility, no definitive guidance or detailed acceptance criteria for the parameters under discussion will be provided.

7.3.1 Review of Quality System Documents

A radiochemical laboratory providing usable analytical data should have a quality manual. A review of this document by a knowledgeable evaluator can reveal a great deal about the quality

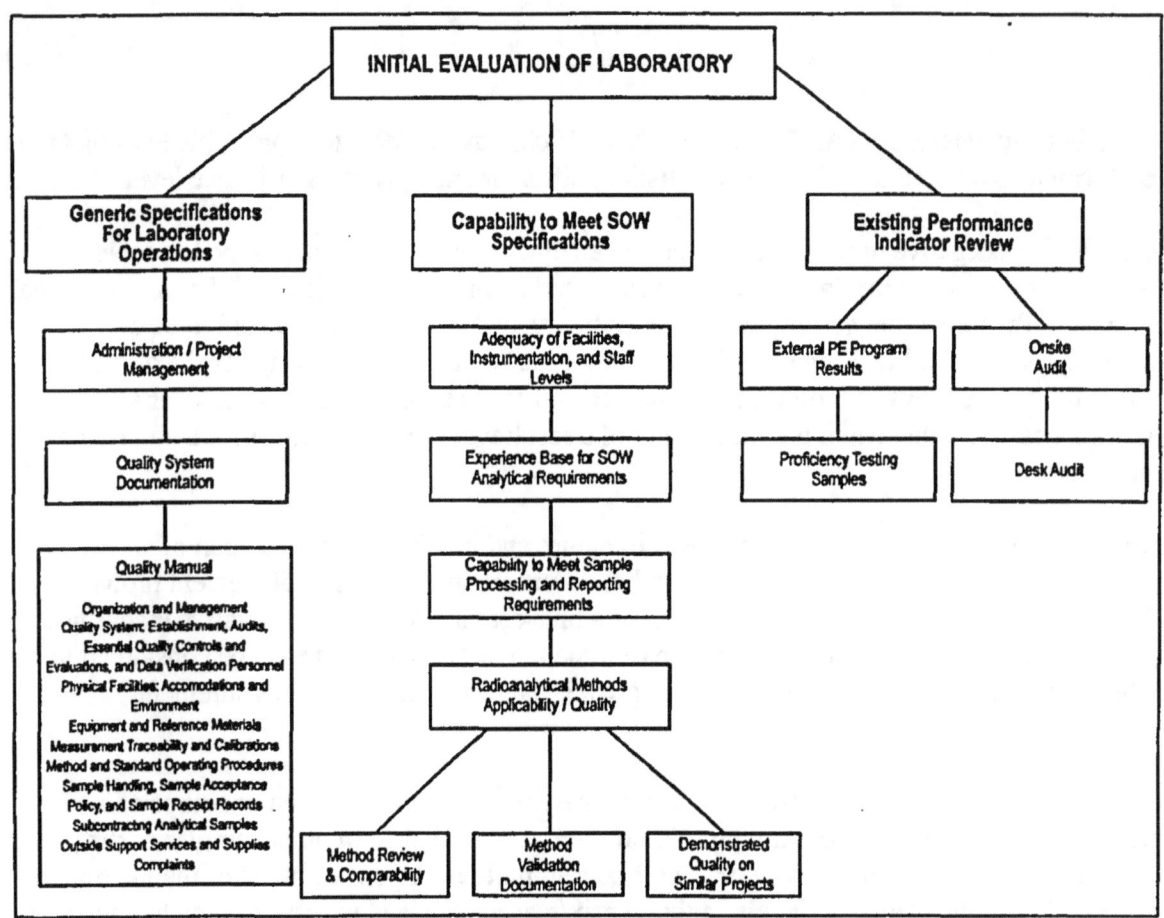

FIGURE 7.1 — Considerations for the Initial evaluation of a laboratory

and acceptability of the laboratory relative to the work to be performed. A well-developed quality manual contains a description of the quality system and descriptive material covering most other aspects of a laboratory's operation. The standard operating procedures, method documentation, list of instrumentation, and personnel resumes should be reviewed. For some projects, the project manager may require the laboratory to develop a specific project quality plan, system, and manual. The following items, taken from the NELAC *Quality Systems* (NELAC 2002), should be discussed at a minimum:

- Organization and management
- Quality system establishment, audits, essential quality controls and evaluation, and data verification
- Personnel (qualifications and resumes)
- Physical facilities (accommodations and environment)
- Equipment and reference materials
- Measurement traceability and calibration
- Test methods and standard operating procedures (methods)

- Sample handling, sample acceptance policy and sample receipt
- Records
- Subcontracting analytical samples
- Outside support services and supplies
- Complaints

The laboratory evaluation should involve a review of the quality system documents for completeness, thoroughness, and clarity.

7.3.2 Adequacy of Facilities, Instrumentation, and Staff Levels

Many factors enter into a laboratory's ability to meet the analytical requirements of a SOW. The resources and facilities of a laboratory may become stretched depending on the number of clients, the analytical services needed, and the deadlines of the committed work activities. Some SOWs may request information about the current workload of the laboratory and available facilities, staff and nuclear instrumentation for the specified work scope. The resources needed will vary considerably depending on the analysis and number of samples: from minimal bench space, hoods, and nuclear instrumentation for fairly simple gross analyses to maximum bench space, hoods, staff, and nuclear instrumentation for low-level analyses of soil. In addition, the laboratory capacity also depends on the number of samples that are routinely processed in a batch. Various factors may control the batch size, including the hood processing area, bench space, and equipment setup, available number of radiation detectors, counting time, and half-life of radionuclide, among others.

The adequacy of the facilities, instrumentation, and staff levels can be estimated by two general mechanisms: detailed supporting information provided by the laboratory in response to the SOW and an initial onsite audit. Information received from the prospective laboratory may provide an estimate of the laboratory's resources, but an initial onsite audit verifies the actual existence and maintenance of the resources.

7.3.3 Review of Applicable Prior Work

If required in a SOW, a laboratory will provide a list of clients for whom radioanalytical services had been performed that are considered comparable in terms of work scope, DQOs, MQOs, APSs, and project type. A written or oral verification of the client list should be performed. As part of the verification process, the following items related to adherence to contract or project requirements should be discussed and documented:

- Radionuclides analyzed;
- Sample matrices types;
- Laboratory capacity (number of samples per week or another time period);
- MQO for method uncertainty, detection and quantification capability;

- Radiological holding times;
- Sample turnaround times;
- Corrective actions; and
- Communications related to schedule, capacity, or quality issues.

It should be noted that under performance-based contracting, a laboratory's prior work for an agency should be considered, either as a positive or negative performance weighting factor, when scoring a laboratory's performance during the technical evaluation process.

7.3.4 Review of General Laboratory Performance

Some laboratories compile a semiannual or annual QA report summarizing the internal QC sample results for the methods used during a given time period, as well as an internal quality assessment report summarizing the internal and external audit findings and corrective actions taken. Although the laboratory's internal quality criteria for a given radionuclide/matrix may be different from the project MQOs, the internal QC sample results can be used to gauge the laboratory's performance capabilities. If these documents are available, they should be reviewed for documentation of process control and pertinent quality parameters such as bias, precision, unusually high number of positive blank detection, chemical recoveries, turnaround times, number of recurring deficiencies or findings, and corrective action effectiveness.

7.3.4.1 Review of Internal QC Results

A quality assessment report may contain a summary of various QA-related activities, including internal audits and surveillance, report of conditions adverse to quality, investigation requests, corrective actions, and the results of external PE programs and internal QC samples. The content and frequency of the reports normally are outlined in the laboratory's quality manual. Frequently, this type of quality assessment report may be submitted with the laboratory's response to the RFP without request. The TEC may want to specifically request such a report when available.

When the laboratory's quality system is effectively implemented, the information contained in these QA reports can be used not only to gauge the quality of the analyses but also the effective-ness and timeliness of such quality system activities as identifying conditions adverse to quality, controlling and monitoring the radioanalytical quality using internal QC samples, and corrective actions. The internal QC sample results can be used to gauge the laboratory's performance capability. Results of the QC samples for a radionuclide and sample matrix should be reviewed for both the batch QC samples and single- or double-blind samples submitted by the QA officer. Batch QC samples typically include laboratory control samples, method blanks, matrix spikes, splits, and duplicates. Such parameters as acceptable percent deviation for spiked samples, acceptable precision as measured by duplicate sample analyses, false nuclide detection, positive blanks, and compliance to internal quality requirements should be reviewed, depending on the type of QC sample. The single- and double-blind samples submitted independently by the QA officer are considered more operationally independent than the batch QC samples.

When quality problems are observed by the reviewer, it is important to check if the laboratory's quality system also has found and reported the same problem and whether an investigation or corrective action has been undertaken.

Additional specific guidance is provided in Chapter 18 (*Laboratory Quality Control*) on evaluating internal QC samples to meet internal laboratory QC performance criteria. It is recommended that the project managers review this chapter to gain a perspective on how to use reported internal QC results to gauge a laboratory's potential to meet project MQOs.

7.3.4.2 External PE Program Results

Typically, a laboratory's performance or capability to perform high quality radiochemical analyses can be evaluated through two external PE program mechanisms. The first mechanism, which may not be available for all projects, is the submittal, as an initial laboratory evaluation process, of project-specific PT samples prepared by the organization or a contracted source manufacturer. When previous knowledge or experience exists, well-characterized site-specific matrix samples containing the nuclides of interest can be used. This approach can use site-specific matrix materials for background samples or for samples spiked with target analytes. For this evaluation mechanism, and depending on the number and type of samples, the laboratory's capability to meet all proposed project MQOs and quality performance specifications may be evaluated.

The second mechanism, available to most projects, is the laboratory's participation in government or commercial PE programs for radiochemical analyses. Each PE program has its own acceptable performance criteria related to a laboratory's bias with respect to the PE program's "known" analyte concentration value. Acceptable performance criteria are established for each nuclide/matrix combination. A PE program may also evaluate a laboratory based on a false positive analyte detection criterion. Typically, the laboratory's performance data in government PE programs are provided in reports available to the public.

The project manager should be aware that the acceptable performance criteria used by the PE programs may be inconsistent with or more lenient than the MQOs of the project. The laboratory's performance should be evaluated in terms of the established MQOs of the project rather than a PE program's acceptable performance criteria. In some cases, the laboratories could be ranked as to their level of performance in these programs.

7.3.4.3 Internal and External Quality Assessment Reports

Most laboratories undergo several external and internal QA audits per year, with resultant audit reports. Typically, a summary of the findings and commitments of internal and external quality audits or assessments are tracked on some type of QA database as part of the laboratory's corrective action process. Access to the audit reports or database information may be limited. This information is not normally requested as part of the RFP process, nor do most laboratories submit such information with their response to an RFP. Therefore, obtaining previous QA audit information from a laboratory outside a formal, external, onsite audit process may be limited.

7.3.5 Initial Audit

An initial assessment or audit may be performed to provide assurance that a potentially selected laboratory is capable of fulfilling the project requirements in accordance with the SOW. Essentially, the objectives of an initial audit are twofold. The first objective is to verify that what the laboratory claims in response to the SOW or RFP, such as the various quality and safety programs, are being correctly and fully implemented, and when used during the project period, will ensure that stipulated requirements will be met. The second objective is to determine if the laboratory has the instruments, facilities, staffing levels and other operational requirements available to handle the anticipated volume of work. In other words, is the laboratory's proposal realistic when compared to the actual facilities? To answer this question, auditors will be looking to see whether a candidate laboratory has all the required elements to meet the project needs.

Detailed guidance and information on what should be evaluated in an initial audit has been provided in Appendix E, Section E5.5 and Table E7. This section also contains recommendations on the key items or parameters that should be reviewed during the initial audit. Depending on the project, other quality or operational parameters/requirements (such as requirements related to chemical speciation or subsampling at the laboratory) not covered in Appendix E should be included in the initial audit plan.

7.4 Ongoing Evaluation of the Laboratory's Performance

The evaluation framework presented here is intended to be sufficiently generic to cover the operations of a laboratory performing work according to a SOW as recommended in Chapter 5. As described in MARLAP, MQOs are a key component of the SOW. Therefore, the sample schedule, analyses to be performed, MQOs, and other analytical requirements have been defined. The methods selected by the laboratory have been demonstrated to meet the MQOs and have been approved by the project manager. In addition, the laboratory and its programs should have undergone an initial audit to ensure that the laboratory has met or is capable of meeting project requirements, including sample processing capacity, sample TATs, deliverables for analytical reports, etc. This would include maintaining a satisfactory quality system that includes monitoring and controlling the radioanalytical processes through an instrument and internal sample QC program and the acceptable performance in an external PE program.

The ongoing evaluation of a laboratory's performance includes the evaluation of the method applicability or the quality of the data produced, and assessing the laboratory's quality system and operations through onsite or desk audits or assessments. The continued method performance can be evaluated through the laboratory's internal sample QC program, a possible external QC program maintained by the project manager, or an external PE program. It should be noted that samples used to control and monitor the quality of laboratory analyses have been defined according to their use. For example, batch or external QC samples are used to control as well as monitor the quality of the analytical process (the process can be stopped immediately if the QC sample results indicate that the process is outside appropriate SOW specifications or laboratory control limits). As defined previously, PT samples are used to compare the performance of the radioanalytical processing to some acceptance criteria but are not used to control the process.

The ongoing evaluation of the laboratory quality system and operations is accomplished through a visit to the laboratory or by a desk audit (the review of records and data from the laboratory). These audits or assessments are more focused on whether the laboratory is meeting project specifications rather than whether the laboratory has the capability to meet project or SOW requirements.

Once a laboratory has initiated work on a project, the laboratory's performance should be evaluated for the duration of the project. The quality of the radioanalytical measurements, as well as the pertinent key operational aspects of the laboratory, should be evaluated against the requirements of the MQOs and SOW. Both the quantitative and qualitative measures of laboratory performance should be evaluated on a continual basis. In addition, the operational aspects of the laboratory germane to the effective implementation of the project requirements should be evaluated/monitored on a continual basis.

7.4.1 Quantitative Measures of Quality

The laboratory's ongoing demonstrated ability to meet the MQOs and other APS requirements can be evaluated through various quantitative measures using internal QC data and external PE program QC data. From these data, quantitative tests, as outlined in Appendix C can be used to measure and monitor the MQO parameters on a short-term basis. Also, the QC and PE program data can be used to evaluate the laboratory's performance, on a long-term trending basis, in meeting other quality related parameters such as bias and precision, unusually high number of positive blank detection, false nuclide detection, MDC or MQC adherence, radiological holding times, etc. The following subsections will discuss the use of data from these samples to evaluate the laboratory's radioanalytical quality with respect to the requirements.

7.4.1.1 MQO Compliance

MARLAP recommends that project-specific MQOs be established and incorporated into the SOW for laboratory radioanalytical services. Appendix C provides guidance on developing the MQOs for method uncertainty, detection capability, and quantification capability. Establishing a gray region and action level are important to the development of the MQOs. For certain research programs and characterization studies, the concept of an action level may not be applicable. For these studies or programs, the MDC requirement and restrictions on the frequency of false positive detections may be more important. As such, the project planning team for these programs should establish the basis for their own MQOs and develop tests to evaluate a laboratory's performance to meet the requirements. These tests may be different from those presented below.

MARLAP recommends that a MQO for method uncertainty be established for each analyte/ matrix combination. The method uncertainty is affected by laboratory sample preparation, sub-sampling, and the analytical method. In the absence of other information, the required method uncertainty (u_{MR}) at the upper bound of the gray region (UBGR) may be defined as:

$$u_{MR} = \frac{\Delta}{10} \tag{7.3}$$

where u_{MR} is the method uncertainty and Δ is the width of the gray region (difference between the upper and lower bounds of the gray region) as defined in Appendix C. In terms of the relative fraction of the upper bound of the gray region (action level), φ_{MR}, is defined:

$$\varphi_{MR} = \frac{u_{MR}}{\text{UBGR}} \tag{7.4}$$

The following subsections describe methods to quantitatively monitor a laboratory's performance relative to meeting this principal MQO through the use of internal or external batch QC samples. In some cases, the laboratory's internal quality program may have more restrictive quality control limitations for method performance compared to the proposed control limits used by the project manager to monitor adherence to the MQO for method uncertainty. Evaluation of the labora-tory's performance in NIST-traceable external PE programs will determine the degree of bias of the laboratory's method with respect to the national standard, as opposed to the determination of the laboratory's internal bias through the use of internal QC samples. The tests presented assume that all known internal (related to QC values and calibrations) and external (calibration differ-ences with respect to the national standard) biases have been defined and eliminated and, as such, the difference between the measured result and the "expected known" value is a result of the method uncertainty only.

USE OF INTERNAL QC SAMPLE RESULTS

For most projects, the SOW will specify that the laboratory incorporate internal QC samples within a defined batch of samples. The QC samples may include a laboratory control sample, sample duplicates, a matrix spike sample and a method or reagent blank, or both. Appendix C provides examples on the use of the following quantitative tests to measure a laboratory's performance in meeting the MQO for method uncertainty.

Quality Performance Tests and Acceptance Criteria for Quality Control Samples

Laboratory Control Sample (LCS). The analyte concentration of an LCS should be high enough so that the resulting Poisson counting uncertainty is small and the relative uncertainty limit φ_{MR} is appropriate with respect to the action level and the spike concentration chosen. The percent deviation (%D) for the LCS analysis is defined as

$$\%D = \frac{SSR-SA}{SA} \times 100\%$$

(7.5)

where
SSR   is the measured result (spiked sample result) and
SA   is the spike activity (or concentration) added.

It is assumed that the uncertainty of SA is negligible with respect to the uncertainty of SSR. Refer to Appendix C for the basic assumption and limitation of this test. For long-term trending, the %D results should be plotted graphically in terms of a quality control chart as described in Chapter 18. The warning and control limits on %D are summarized below:

Laboratory Control Samples	
Statistic:	%D
Warning limits:	$(\pm 2\varphi_{MR}) \times 100\%$
Control limits:	$(\pm 3\varphi_{MR}) \times 100\%$

Duplicate Analyses. The acceptance criteria for duplicate analysis results depend on the analyte concentration of the sample, which is estimated by the average $\bar{x}$ of the two measured results x_1 and x_2.

$$\bar{x} = \frac{x_1 + x_2}{2}$$

(7.6)

When $\bar{x} < UBGR$, the absolute difference $|x_1 - x_2|$ of the two measurements is used in the testing protocol. For these tests, only upper warning and control limits are used, because the absolute value $|x_1 - x_2|$ is being tested.

When $\bar{x} \geq UBGR$, the acceptance criteria may be expressed in terms of the *relative percent difference* (RPD) defined as

$$RPD = \frac{|x_1 - x_2|}{\bar{x}} \times 100\% \tag{7.7}$$

The requirements for duplicate analyses are summarized below.

Duplicate Analyses

If $\bar{x} < UBGR$:
    Statistic: $|x_1 - x_2|$
    Warning limit: $2.83 \, u_{MR}$
    Control limit: $4.24 \, u_{MR}$

If $\bar{x} \geq UBGR$:

    Statistic: $RPD = \dfrac{|x_1 - x_2|}{\bar{x}} \times 100\%$

    Warning limit: $2.83 \, \varphi_{MR} \times 100\%$
    Control limit: $4.24 \, \varphi_{MR} \times 100\%$

Method Blanks. When an aliquant of a blank material is analyzed, the target value is zero. However, the measured value may be either positive or negative. The applicable warning and control uncertainty limits for blank samples are defined as:

Method Blanks

    Statistic: Measured Concentration Value
    Warning limits: $\pm 2u_{MR}$
    Control limits: $\pm 3u_{MR}$

Matrix Spikes. The acceptance criteria for matrix spikes are more complicated than those described above for the other laboratory QC samples because of the pre-existing activity that is inherent to the unspiked sample. The pre-existing activity (or concentration) must be measured and subtracted from the activity measured after spiking.

MARLAP recommends the "Z score," defined below, as the test for matrix spikes.

$$Z = \frac{\text{SSR} - \text{SR} - \text{SA}}{\varphi_{MR}\sqrt{\text{SSR}^2 + \max(\text{SR, UBGR})^2}} \qquad (7.8)$$

where:
- SSR   is the spiked sample result,
- SR   is the unspiked sample result,
- SA   is the spike concentration added (total activity divided by aliquant mass), and
         $\max(\text{SR,UBGR})$ denotes the maximum of SR and UBGR.

The warning and control limits for Z are set at ± 2 and ± 3, respectively. It is assumed that the uncertainty of SA is negligible with respect to the uncertainty of SSR. For long-term trending, the Z results should be plotted graphically in terms of a quality control chart, as described in Chapter 18.

The requirements for matrix spikes are summarized below.

Matrix Spikes

Statistic: $Z = \dfrac{\text{SSR} - \text{SR} - \text{SA}}{\varphi_{MR}\sqrt{\text{SSR}^2 + \max(\text{SR, UBGR})^2}}$

Warning limits: ± 2
Control limits: ± 3

USE OF EXTERNAL PE PROGRAM AND QC SAMPLE RESULTS

Information on a laboratory's performance in an external PE program or from double-blind QC samples is very useful in monitoring a laboratory's ability to meet MQOs. A PE program will provide a snapshot in time whereas external QC samples included with samples submitted to the laboratory permit a continuous evaluation of the method's performance. When traceable to NIST, the PE program will elucidate any measurement or instrument calibration biases as related to the national standard. An external QC program may not have NIST traceability, and thus calibration biases to the national standard would not be determined.

For monitoring the performance of a laboratory using external PE program and QC sample results, the tests provided in the previous subsection ("Use of Internal QC Sample Results," page 7-25) may be used when there are sufficient data. The test equations assume that the project has an MQO for method uncertainty at a specific concentration. In addition, it is assumed that the Poisson counting uncertainty for the radioanalysis of these samples is minimal.

Results from PE Programs

In many SOWs, the laboratory is required to participate in a recognized PE program for the nuclides and media of interest. In some cases, a certificate of participation may be needed as part of response to the RFP. However, it also should be noted that although a laboratory may meet performance acceptance criteria for an external PE program, this fact may have no bearing on whether the method will meet the MQOs of the SOW.

Monitoring ongoing laboratory performance is limited due to the minimum frequency of testing of the PE program, i.e., usually quarterly or semiannually. Some PE programs require multiple measurements to estimate precision but most only request a single result be reported. In addition, the concentration of the analyte typically never approaches an action level value and the media used are not site specific. For PE program samples, when possible, the laboratory should analyze a sample to reach a 1σ Poisson counting uncertainty that is less than five percent.

Multiple Analyses and Results

When a PE program requires the analysis of multiple samples, the laboratory's measurement precision and bias (to a "known value") at the analyte concentration may be estimated and reported by the PE program provider. When only duplicates sample results are reported, then the tests for laboratory control samples and duplicate analyses given in the previous section should be used. The duplicate analysis test can be used as is, but the laboratory control sample test should be evaluated based on the mean of the duplicate results. By using the mean of the two results, the LCS test provides a better estimate of any laboratory measurement bias with respect to the PE program provider. As discussed in Appendix C, the measurement (combined standard) uncertainty of each measured result value should be smaller than the required u_{MR} or φ_{MR}.

Results from External QC Samples

The project manager may elect to establish an external QC program wherein QC samples are submitted to the laboratory with each batch of routine samples for the purpose of "controlling," rather than monitoring, the quality of the analytical processes. The types of QC samples may include matrix spikes, blanks, and possibly duplicates if prepared under controlled and exacting protocols. An agency may use a qualified reference or monitoring laboratory (ANSI N42.23) to prepare the performance testing materials. When available, these QC samples may be prepared from site-specific materials.

When acceptance criteria are not met, the organization may issue a stop-work order and request corrective actions and reanalysis before routine processing can resume. In order to do this, the SOW must define the performance acceptance criteria and stipulate that the agency or organization has the right to stop laboratory processing when the performance requirements are not met. This application is not widespread but may have merit for certain project types. For example, research or national monitoring programs may monitor groundwater for specific naturally occurring radionuclides at state-of-art detection levels. For these programs, frequent false positive results, due to the application of incorrect instrument background or an analytical blank to the analytical result, would be unacceptable. Rather than permit a high rate of false positive results to continue, the agency can use the external batch QC samples to detect problems early and have the laboratory discontinue sample processing until a root cause is discovered and a corrective action undertaken. Non-conformance of a single analysis to performance criteria would not warrant the issuance of a stop work order unless a severe blunder has occurred. Typically, a certain amount of statistical trending of the data is in order to truly elucidate deficiencies.

Since the number of QC samples is similar to the recommendations for the laboratory's internal batch QC samples, there should be sufficient data for trending. The statistical tests provided in the section on "Use of Internal QC Sample Results," beginning on page 7-25, may be applied to these QC samples.

7.4.1.2 Other Parameters

The laboratory's performance in meeting the requirements for the other APSs that are listed in the SOW should be evaluated quantitatively when possible. In some cases, the information needed to perform the evaluations may be found in the final analytical results data package. For certain types of evaluations, a follow-up onsite or desk audit may be needed to complete the evaluation, e.g., a review of logbooks on unique processes or software algorithms and the analytical data base for proper spectral resolution.

RADIOLOGICAL HOLDING AND TURNAROUND TIMES

The data packages or analytical results report should contain the sample collection (reference), sample analysis, and reporting dates. From this information, the radiological holding and sample processing TATs can be calculated and compared against requirements. When a method uses a decay progeny to measure the analyte of interest (^{222}Rn to measure ^{226}Ra), the decay of the parent nuclide and ingrowth of the decay progeny are important parameters for evaluation. Unless requested in the SOW, most laboratories do not report the ingrowth factor as a standard output. Therefore, the information on the sample-specific ingrowth factor may not be readily available and reported only on the data sheets or during audits. When required, these time related requirements will be evaluated for compliance during data verification and validation.

CHEMICAL YIELD

When appropriate, the SOW may specify limits on the chemical yield for each analyte. For radionuclides, this requirement typically is related to the provision of robust or rugged methods so that extreme yields become flags indicating potential problems. Wide swings in the chemical yield may be indicative of method's difficulty handling matrix or radionuclide interferences. The data packages or analytical results report should contain the chemical yield for each analyte listed. This reported value can be compared to the SOW yield limit. When required, these requirements will be evaluated for compliance during data verification and validation.

SPECTRAL RESOLUTION

Problems with spectral resolution of gamma-ray and alpha spectra cannot be evaluated through a review of the analytical results report. If spectral resolution limits have been stated in the SOW, the evaluator should review and evaluate each sample spectrum against the SOW limit. Spectral information may be available in data packages when required or may be obtained during audits.

During an initial audit, a preliminary evaluation of the method's SOP and review of past performance data for spectral resolution should be undertaken. The TEC may want to determine the baseline or typical spectral resolution for the radiation detection systems that will be used in the analysis of project samples. Trends of the spectral resolution of each detection system during the conduct of the project may be used to determine compliance with a spectral resolution specification.

7.4.2 Operational Aspects

Once a laboratory begins providing radioanalytical services, certain operational aspects need to be reviewed and evaluated periodically to determine if the laboratory is maintaining project requirements or if new problems have occurred. It is also important to ensure that the laboratory has been properly maintained and is operated and managed in a manner that will not create a liability to any client. Many of the operational areas that were discussed in Sections 7.3.1 and 7.3.2 for the initial evaluation of a laboratory also should be evaluated periodically to ensure commitments are being met. The audit frequency varies according to the organization and the extent of the project or contract. Desk audits can be conducted more frequently than onsite audits because they require fewer resources. However, not all operational aspects may be reviewed during desk audits. The operational aspects that may be considered during desk and onsite audits are presented below.

7.4.2.1 Desk Audits

A desk audit is conducted as an off-site activity, usually by a technical representative of the project manager. A radioanalytical specialist should review all technical aspects of the desk audit, including method and calculation (data reduction) changes, method performance, instrument recalibrations, corrective actions, and case narratives. The desk audit is most useful when performed periodically to monitor certain activities or programs following an extensive onsite laboratory audit. However, for some smaller projects, the desk audit may be the only assessment mechanism used to monitor the laboratory's operations. The desk audit may be used to review or monitor the following operational aspects or items:

- Organization and Management
  - Changes in key personnel
  - Reassignments

- Quality System
  - Internal and external audits conducted, including laboratory certification audits
  - Corrective action implementations
  - Quality control and performance evaluations
    - Instrument and batch sample QC results
    - External PE program results
  - Laboratory data verification (narrative status reports)
  - Additional method validation studies

- Certificates, licenses, equipment, and reference materials
  - Standard and tracer certificates
  - New and updates to instrument calibrations
  - Instrument repairs and new instruments put into service
  - NRC/State radioactive materials licence updates
  - State or EPA drinking water certification status changes

- Personnel
  - Updates to staff qualification/proficiency for methods
  - Updates to staff training files
    - Radiation and chemical safety
    - Quality assurance
    - Technical principles
    - Hands-on training records

- Radioanalytical Methods and Standard Operating Procedures
  - Updates to methods and SOPs
  - Technical basis for updates
  - Detection limits or method uncertainty studies

- Sample Receipt, Handling and Disposal
  - Sample receipt acknowledgment
  - Chain-of-custody
  - Sample- and waste-disposal tracking logs and manifests

Desk audits may also be used to review the data packages provided by the laboratory and, periodically, to verify certain method results by hand calculations. In addition, verification of compliance to radiological holding and turnaround times may be performed during the desk audit. In the absence of a full data verification and validation program (Chapter 8), the desk audit may be used to periodically evaluate the detailed instrument and data reduction reports of the data packages for method adherence, technical correctness and valid application.

7.4.2.2 Onsite Audits

The onsite laboratory audit is more comprehensive and resource intensive than a desk audit. An onsite audit typically is conducted to assess, periodically and in depth, a laboratory's capability to meet project requirements. Section E.5.5 of Appendix E provides guidance on the conduct of an initial onsite audit during a contract award process. EPA (1997) provides limited guidance on the conduct of an audit for a radiological laboratory. NELAC (2002) provides some generic guidance on laboratory assessments, although not specifically for a radiological laboratory.

Onsite audits usually cover the operational aspects delineated in Section 7.4.2.1 and also provide an opportunity to evaluate the physical conditions at the laboratory, in terms of adequacy and upkeep of the facilities, and the full application or conduct of programs and resources. Information sent in data packages or submitted for desk audits can be confirmed or verified during an onsite audit. Furthermore, an onsite audit permits the tracking of a sample from receipt through processing to sample storage and disposition and can verify the related instrument and batch QC samples specific to the sample being tracked. During an onsite audit, the auditors may have interviews with the staff to gauge their technical proficiency and familiarity with methods.

For large projects, onsite audits may be formal in nature and have a predefined audit plan, which has been developed by a designated audit team, for a specific project or program. The audit team typically is comprised of qualified QA representatives and technical experts. MARLAP recommends that the audit team include a radioanalytical specialist familiar with the project's or program's technical aspects and requirements.

In addition to the items in Section 7.4.2.1 ("Desk Audits"), the following items and programs should be assessed during an onsite laboratory audit:

- Organization and Management
  - Qualifications of assigned laboratory project manager
  - Implementation of management's policy on quality
  - Timeliness of addressing client complaints
  - Timeliness of implementing corrective actions

- Physical Facilities

o Adequacy of facilities (sample receipt, processing, instrumentation and storage areas, waste processing and storage, offices, etc.)
o Physical conditions of facilities including laboratories, hoods, bench tops, floors, offices, etc.
o Environmental controls, such as climate control (heating, ventilation, air conditioning) and electrical power regulation
o Sample processing capacity
o Sample storage conditions including chain-of-custody lockup areas and cross contamination control (separation of samples by project and from radioactive sources or wastes)

- Instrumentation and Equipment
  o Age of nuclear instrumentation and equipment
  o Functionality of nuclear instrumentation and equipment
  o Calibrations and QC logs
  o Maintenance and repair logs
  o Sample throughput capacity
  o Contamination control for radiation detectors
  o Background spectra of radiation detectors

- Methods and Standard Operating Procedures
  o Use of latest revisions of methods and SOPs (spot check method manuals used by technical staff)
  o Conformance to method application (surveillance of method implementation)
  o Effectiveness of administering the controlled method manual

- Certifications, Licenses and Certificates of Traceability
  o Ensure existence and applicability of, and conformance to, certifications and licenses
  o Noted citations during audits related to certifications and licenses
  o Ensure use of NIST-traceable materials (calibration standards)/review of vendors' report of NIST traceability

- Waste Management Practices
  o Adherence to waste management SOPs
  o Proper packaging, labeling, manifests, etc.
  o Sample storage and records
  o Training and qualification records

- Radiological Controls
  o Adherence to radiological safety SOPs
  o Contamination control effectiveness (spill control, survey requirements and adherence, posted or restricted areas, proper ventilation, cleaning policies, etc.)

o Badging and survey adherence

- Personnel
  - o Number and technical depth of processing staff
  - o Training files
  - o Testing/qualifications
  - o Personal interviews to determine familiarity of methods and safety SOPs

- Quality Systems
  - o Performance indicator program (feedback from program)Quality assurance reports (QC and audits) for all laboratory processing
  - o Ongoing method evaluations and validations
  - o Corrective action program (effectiveness and outstanding issues for all processing; spot check for implementation of corrective actions)
  - o Records/reports related to audits of vendors used by laboratory
  - o Reagent control program (spot check conformance for effectiveness)
  - o Audits of laboratories that are subcontracted
  - o Laboratory's data verification and validation processes

- Software Verification and Validation
  - o Spot review of key method calculation and data reduction programs that include MDC, MQC, and measurement uncertainty; spectral unfolding routines or crosstalk factors; application of instrument background and analytical blanks; etc.
  - o Spot verification of consistency between electronic data deliverable and data packages

- Radiological Holding and Sample Turnaround Times
  - o Verification of compliance to radiological holding and sample TAT specifications (spot check samples and confirm paperwork)

7.5   Summary of Recommendations

- MARLAP recommends that a radioanalytical specialist review the methods for technical adequacy.

- MARLAP recommends that the TEC perform an independent calculation of the method's MDC using laboratory-stated typical or sample-specific parameters.

- MARLAP recommends that the project manager or TEC evaluate the available data provided by the laboratory or from performance evaluations for bias, based on multiple analyses covering the applicable analyte concentration range.

- MARLAP recommends that project-specific MQOs be established and incorporated into the SOW for laboratory radioanalytical services.

- MARLAP recommends that a MQO for method uncertainty be established for each analyte/matrix combination.

- MARLAP recommends the "Z score" as the test for matrix spikes.

- MARLAP recommends that an audit team include a radioanalytical specialist familiar with the project's or program's technical aspects and requirements.

7.6 References

American National Standards Institute (ANSI) N42.23. Measurement and Associated Instrumentation Quality Assurance for Radioassay Laboratories. 2003.

American Public Health Association (APHA). 1998. *Standard Methods for the Examination of Water and Waste Water, 20ʰ Edition*. Washington, DC. Available at: www.standardmethods. org.

American Society for Testing and Materials (ASTM) D2777. *Standard Practice for Determination of Precision and Bias of Applicable Test Methods of Committee D-19 on Water.*

American Society for Testing and Materials (ASTM) E177. *Standard Practice for Use of the Terms Precision and Bias in ASTM Test Methods.* West Conshohocken, PA.

American Society for Testing and Materials (ASTM) E548. *Standard Guide for General Criteria Used for Evaluating Laboratory Competence.* West Conshohocken, PA.

American Society for Testing and Materials (ASTM) E1580. *Standard Guide for Surveillance of Accredited Laboratories.* West Conshohocken, PA.

American Society for Testing and Materials (ASTM) E1691. *Standard Guide for Evaluation and Assessment of Analytical Chemistry Laboratories.* West Conshohocken, PA.

U.S. Environmental Protection Agency (EPA). 1997. *Manual for the Certification of Laboratories Analyzing Drinking Water*. EPA 815-B-97-001. Office of Ground Water and Drinking Water, Washington, DC. Available at: www.epa.gov/safewater/certlab/labfront.html.

U. S. Environmental Protection Agency (EPA). 2002. *Guidance for Quality Assurance Project Plans*. EPA QA/G-5. EPA/240/R-02/009. Office of Environmental Information, Washington, DC. Available at www.epa.gov/quality/qa_docs.html.

International Organization for Standardization (ISO) 17025. *General Requirements for the Competence of Testing and Calibration Laboratories*. International Organization for Standardization, Geneva, Switzerland. 1999.

National Bureau of Standards (NBS). 1963. *Experimental Statistics*. NBS Handbook 91, National Bureau of Standards, Gaithersburg, MD.

National Environmental Laboratory Accreditation Conference (NELAC). 2002. *NELAC Standards*. Chapter 5, *Quality Systems*. NELAC, Washington, DC. Available at: www.epa.gov/ttn/nelac/2002standards.htm.

Taylor, John K. 1990. *Statistical Techniques for Data Analysis*. Lewis Publishers, Chelsea, MI.

8 RADIOCHEMICAL DATA VERIFICATION AND VALIDATION

8.1 Introduction

The goal of the data collection process is to produce credible and cost-effective data to meet the needs of a particular project. The process can be divided into several stages, as illustrated in the data life cycle (Chapter 1). This chapter is the first of two chapters that address the assessment phase of the project. Because the efficiency and success of these assessment activities are heavily dependent on the completion of the preceding steps in the data collection process, especially the initial planning activity (Chapter 2), the integration of planning and assessment is discussed in Section 8.2 prior to presenting material on data verification and validation.

Data verification compares the material delivered by the laboratory to the requirements in the statement of work (SOW) and identifies problems, if present, that should be investigated during data validation. Data validation uses the outputs from data verification and compares the data produced with the measurement quality objectives (MQOs) and any other analytical process requirements contained in the analytical protocol specifications (APSs) developed in the planning process. The main focus of data validation is determining data quality relative to the project-specific MQOs. It may not be necessary in all instances to validate all project data. This chapter outlines a validation plan that specifies the data deliverables and data qualifiers to be assigned that will facilitate the data quality assessment. The project-specific data validation plan should establish a protocol that prioritizes the data to be validated. This is to eliminate unnecessarily strict requirements that commit scarce resources to the in-depth evaluation of data points with high levels of acceptable uncertainty. For example, results very much above or below an action level may not require rigorous validation, since relatively large measurement uncertainty would not affect the ultimate decision or action. Planners should also identify those samples or data sets that have less rigorous standards for data quality and defensibility.

This chapter presents suggested criteria to evaluate data and addresses the appropriate function and limits of radiochemical techniques and measurements. Since calibration is more efficiently evaluated as part of an audit, this chapter does not recommend that the complete calibration-support documentation be included as part of the data package. MARLAP recommends that calibration be addressed in a quality system and through an audit (Chapter 18, *Laboratory Quality Control*), although demonstration of calibration may be required as part of a project's deliverables. Detector calibration, self-absorption curves, and efficiencies should be addressed as part of the evaluation of

Contents

8.1	Introduction	8-1
8.2	Data Assessment Process	8-2
8.3	Validation Plan	8-7
8.4	Other Essential Elements for Data Validation	8-11
8.5	Data Verification and Validation Process	8-12
8.6	Validation Report	8-29
8.7	Summary of Recommendations	8-31
8.8	Bibliography	8-31

laboratories during the procurement process and continued during subsequent assessments (Chapter 7, *Evaluating Methods and Laboratories*). Availability and retention of calibration records are decisions that are project-specific, but should be clearly identified for contract clarity and to assure project completeness (i.e., customer needs met). External sources of information, such as performance evaluation sample results and internal laboratory control samples, provide useful interim information on calibration status and accuracy.

8.2 Data Assessment Process

Figure 1.1 in Chapter 1 graphically depicts the three phases of the data life cycle—planning, implementation, and assessment—and the associated activities and products of each phase. *While these activities are addressed in separate chapters in MARLAP, it should be emphasized that integration of planning, sampling, and analysis with subsequent data verification, data validation, and data quality assessment (DQA) is essential.*

This section reviews the data life cycle from the perspective of the assessment phase and focuses on those issues that have the potential to impact the quality and usability of the data. Section 8.2.1 addresses the development of the assessment procedures during project planning. Section 8.2.2 considers assessment needs for documentation and a quality system during implementation. Section 8.2.3 focuses on the assessment phase and addresses the interrelationship of the three assessment processes. This introduction to the data life cycle process emphasizes the importance of linkages among planning, implementation, and assessment.

8.2.1 Planning Phase of the Data Life Cycle

Directed project planning and the development of the associated data quality objectives (DQOs), MQOs, and other specifications for the project are reviewed in Chapters 2 and 3. *These chapters emphasize the need for planners to thoroughly define the assessment processes (i.e, verification, validation and data quality assessment) in sufficient detail that success or failure in meeting goals can be determined upon project completion.* MARLAP recommends that the assessment criteria of a project be established during the directed planning process and documented in the respective plans as part of the project plan documents. This requires the project planning team to develop detailed procedures for data verification, data validation, and data quality assessment, as well as identify the actual personnel who will perform assessment or the required qualifications and expertise of the assessors.

The development of these procedures during the directed planning process will increase the likelihood that the appropriate documentation will be available for assessment, and that those generating and assessing data will be aware of how the data will be assessed. A secondary advantage, which assessment plans have, is that prior to their completion, they often result in the detection of design flaws (e.g., lack of proper quality control [QC] samples, lack of a field audit) that upon correction will result in the complete information necessary for the proper assessment of data usability.

The culmination of the planning process is documentation of the outputs of the directed planning process in the project plan documents. The project plan documents should capture the DQOs, MQOs, and the optimized data collection design (i.e., analytical protocol specifications, sampling and analysis plans, and standard operating procedures [SOPs]). The project plans should also include the assessment plans as discussed above, and describe the field, laboratory, safety, and quality assurance (QA) activities in sufficient detail that the project can be implemented as designed. Chapter 4 discusses guidance for writing project plan documents.

If the directed planning process, its outputs (DQOs, MQOs, optimized sampling and analysis designs), and associated assumptions are not documented well in project plan documents, the assessment phase will have difficulties evaluating the resulting data in terms of the project's objectives.

8.2.2 Implementation Phase of the Data Life Cycle

The project plans are executed during the implementation phase. Ideally, the plans would be implemented as designed, but due to errors, misunderstandings, the uncontrolled environments under which sampling is implemented, and matrix-specific issues that complicate sample handling and analysis, most project plans are not implemented without some deviation.

Understanding the realities of implementation, the assessment process, in particular the DQA process, will evaluate the project's implementation by considering: (a) if the plans were adequate to meet the project's DQOs, (b) if the plans were implemented as designed, and (c) if the plans as implemented were adequate to meet the project DQOs. MARLAP recommends that project objectives, implementation activities and QA/QC data be well documented in project plans, reports, and records, since the success of the assessment phase is highly dependent upon the availability of such information.

Documentation and record keeping during the planning and implementation phase of the data life cycle are essential to subsequent data verification, data validation, and data quality assessment. Thorough documentation will allow for a determination of data quality and data usability. *Missing documentation can result in uncertainty, and a lack of critical documentation (e.g., critical quality control results) can result in unusable data. The quality and usability of data can not be assessed if the supporting documentation is not available.*

8.2.2.1 Project Objectives

The DQOs, MQOs, and other specifications, requirements, and assumptions developed during the planning phase will influence the outcomes during the subsequent implementation and assessment phases of the data life cycle. It is important that these objectives, specifications, requirements, and assumptions are well documented and available to those implementing the program so they can make informed decisions. This documentation is reviewed during the DQA process (see discussions of the reviews of project DQOs in Section 9.6.1.1, sampling plans in Section 9.6.2.1, and analysis plans in Section 9.6.3.1).

8.2.2.2 Documenting Project Activities

The assessment of data in terms of sampling and analytical MQOs requires an accurate record of QC sample data and compliance with specifications and requirements. If these records are missing or inadequate, then compliance with APSs, including the MQOs that were identified during the planning phase, will not be ascertainable and will raise questions regarding quality.

Additional documentation is required to assess compliance with plans and contracts, and to assess field and laboratory activities (e.g., compliance with SOPs) and the associated organizational systems (e.g., laboratory quality manual). This information is gleaned from the review of field and laboratory notebooks, deviation reports, chain-of-custody forms, verification reports, audit reports, surveillance reports, performance evaluation sample analyses, corrective action reports and reports to management that may identify deviations, contingencies, and quality problems. Assessment of these types of contemporaneous records allow for the assessment of data in the context of pertinent issues that may have arisen during project implementation.

Project records should be maintained for an agreed upon period of time, which should be specified in project plan documents. Record maintenance should comply with all regulatory requirements and parallel the useful life of the data for purposes of re-assessment as questions arise or for purposes of secondary data uses that were not originally anticipated.

8.2.2.3 Quality Assurance/Quality Control

To ensure that the data collection activity generates data of known quality, it is essential that the project plan documents specify the requirements for an appropriate quality system that is capable of implementing the quality controls and the quality assurance necessary for success.

The quality system will oversee the implementation of QC samples, documentation of QC sample compliance or noncompliance with MQOs, audits, surveillances, performance-testing sample analyses, corrective actions, quality improvement and reports to management. The documentation generated by these quality assurance activities and their outputs during project implementation will be a key basis for subsequent assessments and data usability decisions.

8.2.3 Assessment Phase of the Data Life Cycle

Assessment of environmental data currently consists of three separate and identifiable phases: data verification, data validation, and DQA. Verification and validation pertain to evaluation of analytical data. *Verification and validation are considered as two separate processes, but as the MARLAP recommended planning process is implemented, they may be combined.* DQA considers all sampling, analytical, and data handling details, external QA assessments, and other historical project data to determine the usability of data for decision-making.

Figure 8.1 is a graphical depiction of the assessment phase. Although the figure portrays a linear progression through the various steps, and from verification and validation to data quality

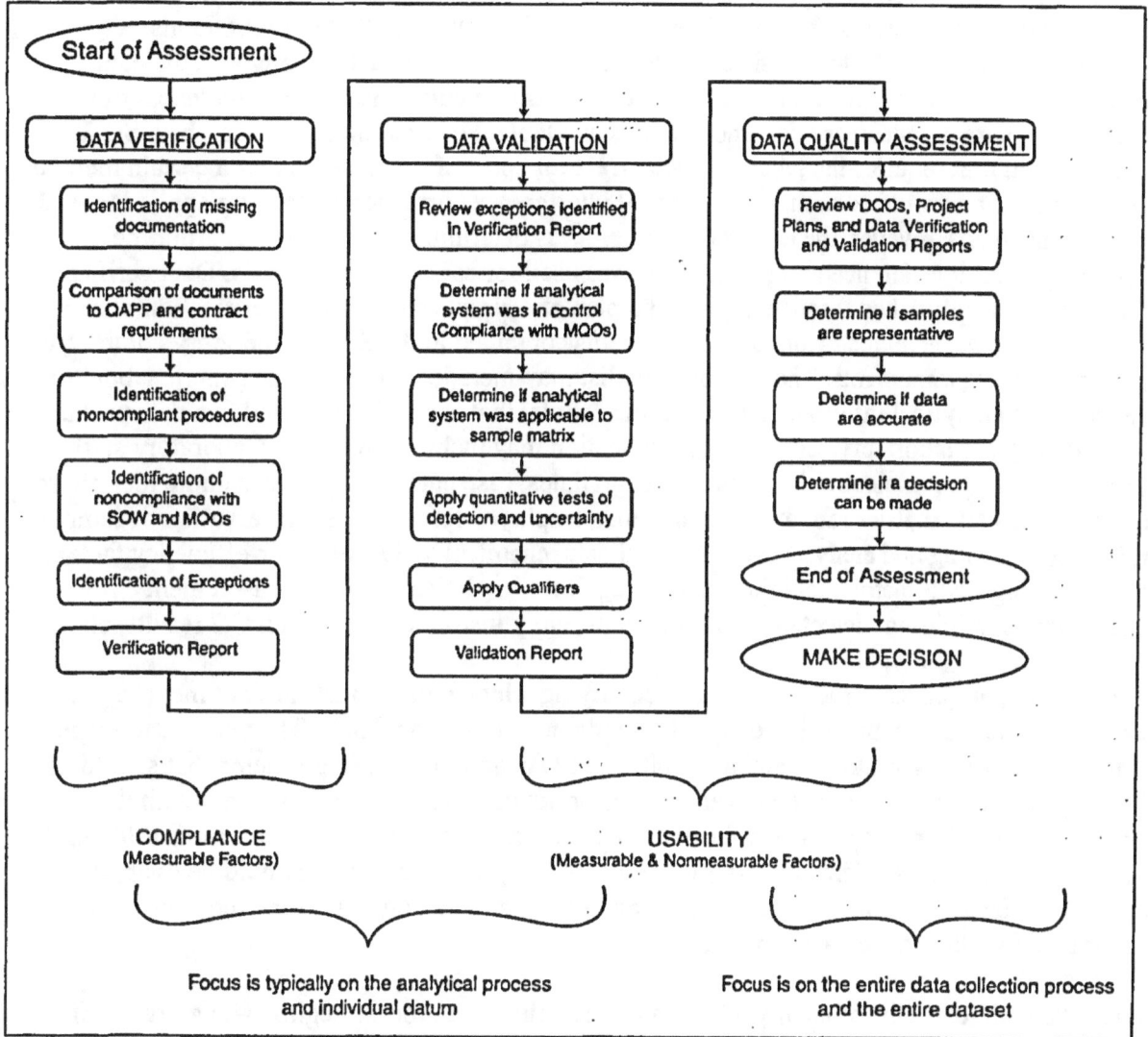

FIGURE 8.1 — The assessment process

assessment, this linear advancement is not entirely necessary. It is possible for parallel progress within an assessment process (e.g., existing documents are verified while waiting for the production of others) and between assessment processes (e.g., analysis of the DQOs for data quality assessment while data validation is being completed). *Typically, the focus of verification and validation is on the analytical process and on a data point by data point review, while data quality assessment considers the entire data collection process and the entire data set as it assesses data quality.*

Analytical data *verification* assures laboratory conditions and operations were compliant with the SOW based on project plan documents. The updated project plan documents specify the analytical protocols the laboratory should use to produce data of acceptable quality and the content of the analytical data package (see Section 1.5, "The MARLAP Process"). Verification compares the analytical data package delivered by the laboratory to these requirements (compliance), and checks for consistency and comparability of the data throughout the data package, correctness of basic calculations, data for basic calculations, and completeness of the results to ensure all necessary documentation is available. For example, there may be a SOW specification that requires the laboratory to correct for spectral interferences for a certain method. Data verification would confirm that a spectral interference correction factor was applied to each sample analysis. However, the data validation process determines whether the appropriate or correct spectral interference correction factor was used for each sample analysis. Data verification can be accomplished through the use of a plan or a simple checklist. A verification plan or checklist may be developed from the requirements contained in the SOW, laboratory contract, or project planning documents. The plan or checklist may include verification of generic laboratory and specific analytical information (outputs or records) that should be reported in a data package generated by the laboratory. Compliance verification may include a review of laboratory staff signatures (written or electronic), data and report dates, case narrative reports, sample identifiers, radionuclides and matrices for analyses, methods employed for analyses, preservation of samples, reference/sampling and analysis dates, spectral data, chemical yields, detector-efficiency factors, decay and ingrowth factors, radiological holding times, analytical results, measurement uncertainties, minimum detectable concentrations, daily instrument and batch QC results, etc.

The verification process produces a report identifying which requirements are not met (i.e., exceptions qualified with an "E" to alert the validator; see Section 8.3.3). The verification report is used to confirm laboratory compliance with the SOW and to identify problems that should be investigated during data validation. Verification works iteratively and interactively with the generator (i.e., laboratory) to assure receipt of all necessary data in the correct format. Although the verification process identifies specific problems, the primary function should be to apply appropriate feedback to the laboratory resulting in corrective action improving the analytical services before the project is completed.

Validation addresses the reliability of the data. The validation process begins with a review of the verification report and laboratory data package to identify its areas of strength and weakness.

This process involves the application of qualifiers that reflect the impact of not meeting the MQOs and any other analytical process requirements. Validation then evaluates the data to determine the presence or absence of an analyte, and the uncertainty of the measurement process. During validation, the technical reliability and the degree of confidence in reported analytical data are considered. The data validator should be a scientist with radiochemistry experience.

Validation flags (i.e., qualifiers) are applied to data that do not meet the performance acceptance criteria established in the SOW and the project plan documents. The products of the validation process are validated data and a validation report stating which data are acceptable, which data are sufficiently inconsistent with the validation acceptance criteria in the expert opinion of the validator, and a summary of the QC sample performance. The appropriate data validation tests should be established during the project planning phase. The point of validation is to perform a systematic check on a set of data being used to meet the project MQOs and any other analytical process requirements. Documenting that such a check cannot be done is an appropriate and essential validation activity. (For example, applying numerical tests to data already determined to be unreliable is of no value.)

Data Quality Assessment is the last phase of the data collection process, and consists of a scientific and statistical evaluation of project-wide knowledge to assess the usability of data sets. To assess and document overall data quality and usability, the data quality assessor integrates the data validation report, field information, assessment reports, and historical project data, and compares the findings to the original project DQOs. The DQA process uses the combined findings of these multi-disciplinary assessments to determine data usability for the intended decisions, and to generate a report documenting that usability and the causes of any deficiencies. It may be useful for a validator to work with the assessor to assure the value of the validation process (e.g., appropriateness of rejection decision) and to make the process more efficient. DQA will be covered in Chapter 9.

8.3   Validation Plan

The validation plan should integrate the contributions and requirements of all stakeholders and present this information in a clear, concise format. To achieve this goal, validation planning should be part of initial planning (e.g., directed planning process) to assure that the data will be validated efficiently to determine its reliability and technical defensibility in an appropriate context and to an appropriate degree.

The validation plan is an integral part of the project plan documents (Chapter 4), and should be included as either a section within the plan or as a stand-alone document attached as an appendix. The validation plan should be approved by an authorized representative of the project, the validation group performing the validation, and any other stakeholder whose agreement is needed.

The information and documentation identified in the validation plans should be communicated to the laboratory as part of the SOW. Integration of validation plan specifications, contractual requirements, and validator instructions/contracts is essential to ensure data collection process efficiency. Implementation of the data validation plan ensures that proper laboratory procedures are followed and that data are reported in a format useful for validation and assessment. This also improves the cost-effectiveness of the data-collection process.

The data validation plan should contain the following information:

- A summary of the project's technical and quality objectives in terms of sample and analyte lists, required measurement uncertainty, and required detection limit and action level on a sample/analyte-specific basis. It should specify the scope of validation, e.g., whether all the raw data will be reviewed and in what detail (Section 8.3.1).

- The necessary validation criteria (derived from the MQOs) and performance objectives deemed appropriate for achieving project objectives (Section 8.3.2).

- Direction to the validator on what qualifiers are to be used and how final qualifiers are assigned (Section 8.3.3).

- Direction to the validator on the content of the validation report (Section 8.3.4).

8.3.1 Technical and Quality Objectives of the Project

The identity of key analytes and how the sample results drive project decisions should be specified in the validation plan. In addition, the plan should define the association of required quality control samples with project samples.

This section of the validation plan should specify the following:

- Quality control acceptance criteria;

- Level of measurement uncertainty considered unusually high and unacceptable (tests of unusual uncertainty and rejection); and

- Action level and MQOs for detection and quantification capability (e.g., required detection and quantification limit).

The *quality control acceptance criteria* serve two purposes: (1) to establish if the analytical process was in control; and (2) to determine if project requirements were met. If the analytical process is in control, the assumption was that the analysis was performing within established limits and indicates a reasonable match among matrix/analyte/method. Generally, this means that routine data quality expectations are appropriate. The *tests of unusual* (i.e., analysis not in control) *uncertainty* should verify the data meet the statistical confidence limits for uncertainty associated with the planning process. During validation, the uncertainty associated with sampling cannot be estimated. The *tests of detection* determine the presence or absence of analytes.

8.3.2 Validation Tests

Validating data requires three specific decisions that will allow the validator to qualify the data. The project planning team should determine:

- Which QC samples should be employed and how do they relate to the samples?

- Which validation tests are appropriate?

- What validation limits should be used for the specific tests?

The answers to these questions are driven by the need to know whether the data meets the MQOs for the project, and the allocation of resources between planning and implementation (i.e., conservative review may be more costly than real or perceived value in the decision). This section of the validation plan should address the following:

- Specific validation tests to be used, and

- Statistical confidence intervals or fixed limit intervals applied to each of the validation tests and criteria based on the MQOs for the project (see Appendix C, *Measurement Quality Objectives For Method Uncertainty and Detection And Quantification Capability*).

8.3.3 Data Qualifiers

Data qualifiers are codes placed on an analytical result that alert data users to the validator's or verifier's concern about the result. This section of the validation plan should outline:

- The basis for rejection or qualification of data; and
- The qualification codes that will be assigned.

These issues are discussed in detail in Section 8.5, which provides guidance for assigning data qualifiers.

The verification process uses a qualifier ("E") to alert the validator to noncompliance, including missing documentation, contract compliance, etc. This qualifier may be removed or replaced during validation, based on the validator's interpretation of the effect of the noncompliance on the data's integrity.

   E   A notice to the validator that something was noncompliant.

The validation process uses the qualifiers listed below to identify data points that do not meet the project MQOs or other analytical process requirements listed in the SOW or appropriate project plan document. The assignment of the "J" and "R" qualifiers relies heavily on the judgement and expertise of the reviewer and therefore, these qualifiers should be assigned as appropriate at the end of data validation.

   U   A normal, not detected (< critical value) result.

   Q   A reported combined standard uncertainty, which exceeds the project's required method uncertainty.

   J   An unusually uncertain or estimated result.

   R   A rejected result: the problems (quantitative or qualitative) are so severe that the data can not be used.

The data validator should be aware that a data qualifier or a set of qualifiers does not apply to all similar data. The data validator should incorporate the project MQOs into the testing and qualifying decision-making process.

During the data validation process the data validator may use additional qualifiers based on QC sample results and acceptance criteria. These qualifiers may be summarized as "U," "J," "R," or "Q" in the final validation report. The final validation reports should also include a summary of QC sample performance for use by the data assessor.

   S   A result with a related spike result (laboratory control sample [LCS], matrix spike [MS] or matrix spike duplicate [MSD]) that is outside the control limit for recovery (%R); "S+" or "S-" used to indicate high or low recovery.

   P   A result with an associated replicate result that exceeds the control limit.

   B   A result with associated blank result, which is outside the control limit, "B+" or "B-"used to indicate high or low results.

8.3.4   Reporting and Documentation

The purpose of this section is to define the format and program needs for validation reports and supporting documentation. This section should include:

- Documentation and records that should be included in a validation report;

- Disposition requirements for records and documents from the project;

- Report format, i.e., a summary table with results, uncertainties and qualifiers; and

- Procedures for non-conformance reporting, which detail the means by which the laboratory communicates nonconformances against the validation plan. The procedures should include all instances where the analytical data requirements and validation requirements established by the planning process and validation plan, respectively, cannot be met due to sample matrix problems or unanticipated laboratory issues (loss of critical personnel or equipment).

Detailed information about the validation report is presented in Section 8.6.

8.4   Other Essential Elements for Data Validation

Effective data validation is dependent on:

- A SOW and project plan documents that clearly define the data needs and the data quality requirements (i.e., MQOs); and

- A data package that has been verified for completeness, consistency, compliance, and correctness.

8.4.1   Statement of Work

The analytical services procurement options should be considered during the planning process. The SOW should specify the QC requirements that will be evaluated by the validator (see Chapter 5, *Obtaining Laboratory Services*). The elements that should be specified include, but are not limited to:

- External performance evaluation (PE) participation and acceptance criteria;
- Replicate sample frequency and acceptance criteria;
- LCS and acceptance criteria;
- Blank requirements and acceptance criteria;
- MS and MSD samples and acceptance criteria;
- Uncertainty calculations; and
- Sample result equations and calculations including corrections for yield, percent moisture, efficiencies and blank, if applied.

Section 8.5.2 provides guidance on evaluating QC sample results based on the project's MQO for measurement uncertainty.

8.4.2 Verified Data Deliverables

Verification compares the sample receipt information and the sample report delivered by the laboratory against the SOW and produces a report that identifies those requirements that were not met (called exceptions). Verification can be accomplished using a plan or checklist, which doesn't necessarily need to be project-specific. Verification exceptions normally identify:

- Required steps not carried out by the laboratory (i.e., correction for yield, proper signatures);

- Method QC not conducted at the required frequency (i.e., blanks, duplicates); and

- Method QC not meeting pre-set acceptance criteria (i.e., noncompliant laboratory control sample analysis).

The verifier checks the data package (paper or electronic) for completeness, consistency, correctness, and compliance. Completeness means all required information is present. Consistency means values are the same when reported redundantly on different reports, or transcribed from one report to another. Correctness means the reported results are based on properly documented and correctly applied algorithms. Compliance means the data pass numerical QC tests based on parameters or limits derived from the MQOs specified in the SOW.

The verifier should provide, within the verification package, checklists for contract or SOW specifications, noted deficiencies related to contract compliance, noted discrepancies or obvious quality related problems, and pertinent external QC results. *The verification package notes the deficiencies, discrepancies, and quality-related problems that could not be resolved with the laboratory.* The validator should take this information into consideration during the data validation process.

8.5 Data Verification and Validation Process

In its most basic form, data validation focuses on the reliability of each data point. After each point is evaluated, summary conclusions concerning the validity of groups of data (sets) are drawn and finally, after the reliability of all data sets has been established, an overall conclusion about the quality and defensibility of a project's analytical database is reached (DQA).

The first step in establishing the reliability of an analytical measurement is to determine that the measurement analytical process used in making the measurement is in control. That is, the sample handling and analysis system is performing within an accepted operating range (established by instrument manufacturer, method, contract specifications, or long-term historical laboratory performance). After it has been determined that the measurement analytical process is in control, it is necessary to demonstrate that the sample is responding as expected when introduced into the measurement system.

The measurement process includes devices such as detectors for measuring radioactive decay emissions and balances for determining the mass of materials. The measurement process also includes the software that takes the output from the measurement device and calculates the result as a quantity of target radionuclide (activity/mass or activity/volume). The measurement process performance normally is specified by the SOW and appropriate project plan documents, and monitored by routine laboratory quality control procedures. Laboratory performance against these requirements is determined by the verification process.

When a sample is analyzed, new sources of variability are encountered in addition to those associated with the measurement process. These sources include laboratory subsampling, sample preparation (e.g., digestion, leaching, etc.), and sample matrix effects, to list a few. These processes, taken together with the previously discussed measurement process, comprise the analytical process.

The performance of the analysis can be predicted based on previous experience with similar materials. Analysis performance is monitored by laboratory quality control procedures specified in the SOW and appropriate project plan documents. Since each sample matrix, analyte, and method set is unique, the evaluation of overall analysis performance and resulting data is the role of a knowledgeable validator.

Using the validation plan, which specifies QC samples, validation tests, and validation limits, validation occurs in four stages:

- Determine whether the sample handling and analysis system is in control (Section 8.5.1);
- Determine whether QC sample analyses meet specified MQOs (Section 8.5.2);
- Apply validation tests of detection and unusual uncertainty (Section 8.5.3); and
- Determine final data qualifiers and document the results (Section 8.5.4).

For some methods (e.g., gamma spectrometry), identification of the analyte is also a primary decision. The laboratory's ability to identify analytes reliably is best checked by auditors and verified by reviewing the instrument's energy-calibration file.

8.5.1   The Sample Handling and Analysis System

As described in earlier sections of this guidance, it is necessary to know the extent to which the data delivered for validation meet the requirements of the SOW and appropriate project plan documents. These documents normally specify the minimum acceptable performance of the analytical process. These specifications are the basis of the tests of quality control (QC tests) that establish that the sample handling and analysis system is in control at the time the analyses were performed. It is also necessary to know that all reporting requirements are complete. Normally, this evaluation against the requirements is made during the data verification process. If the data do not conform to the requirements, notification should be provided in the verification report.

The review of the verification package (and data package) by the validator determines if sufficient information is provided to proceed with data validation. The outcome of the verification process is the designation of exceptions to the quality control tests. These exceptions should be flagged with a qualifier (re-evaluated by the validator), which is appended to a data or report requirement that does not meet specifications to alert the validator of potential problems. The validator should then determine if sufficient reliable data are available to proceed with validation. The validator should use the data requirements and criteria developed in the validation plan to determine if the quality control exceptions have an adverse impact on one or more of the data points being validated.

Rarely, if ever, should quality control exceptions result in the decision to reject a complete data set. Those types of situations should have been detected by the laboratory during the analytical process and the samples reanalyzed. The validator should not reject (assign an "R" code) single data points based on a single QC test exception. Normally, only numerous QC exceptions *and* failures in one or more of the tests of detection and uncertainty are sufficient reason to reject data. The validation report should fully explain the assignment of all qualifiers as previously discussed.

The following paragraphs discuss some of the more important evaluations that should be applied to the sample handling and analysis system. *Some of these items (e.g., calibration, verification of self-absorption curves, and efficiency) may be checked during an audit instead of during data verification and validation.* Limited guidance is provided on how the QC test may impact data quality and defensibility.

8.5.1.1 Sample Descriptors

Sample descriptors include sample identification number, analytical method, analyte, and matrix, among others.

Criteria. Each sample should have a unique identifier code that can be cross-referenced to a unique sample or an internally generated laboratory sample. This unique identifier and associated sample descriptors should be included in all analytical reports to properly document the sample and requested analysis (Chapters 10, *Field and Sampling Issues that Affect Laboratory Measurements*, and 11, *Sample Receipt, Inspection, and Tracking*).

The matrix and other characteristics of the sample that affect method selection and performance should be clearly identified. The method(s) used in sample preparation and analysis should be identified. If laboratory replicate analyses are reported for a sample, they should be distinguishable by a laboratory-assigned code.

Verification. Check that criteria related to the sample description (e.g., stated description of sample type) have been addressed and pertinent documentation is included in the analytical data package. If necessary documentation is missing, the data should be flagged with an "E" code.

Validation. Missing information will decrease the confidence in any result reported on a sample(s) and justify the assignment of a "J" code. Missing information may be inferred from other information in the data package. For example, if the sample matrix is not provided, it may be inferred from:

- The aliquant units are expressed in units of mass or volume;
- The sample preparation method is specific for soils;
- The final results are expressed in units of mass; and
- The sampling report describes sampling soil.

The majority of related information should support the decision that the exception does not decrease the confidence in the result. If the supporting information is incomplete or conflicting, the assignment of a "J" code to data points is warranted. If documentation is inadequate to support the reporting of a data point, the data point should be qualified with an "R" code.

8.5.1.2 Aliquant Size

Criteria. The aliquant or sample size used for analysis should be documented so that it can be checked when reviewing calculations, examining dilution factors or analyzing any data that requires aliquant as an input. It is also imperative that the appropriate unit (liter, kilogram, etc.) is assigned to the aliquant.

Verification. Check that criteria related to sample aliquanting (e.g., stated aliquant size) have been addressed, and pertinent documentation is included in the analytical data package. If aliquant size documentation is missing, the data should be flagged with an "E" code.

Validation. The missing information will increase the uncertainty on any result reported on a sample(s) and justify the assignment of a "J" code.

8.5.1.3 Dates of Sample Collection, Preparation, and Analysis

Criteria. The analytical data package should report date of sampling, preparation, and analysis. These data are used to calculate radiological holding times, some of which may be specified in the sampling and analysis plan.

There are few circumstances where radiological holding times are significant for radionuclides. The best approach to minimize the impact of holding time on analysis is to analyze the samples as quickly as possible. Holding times may be applied to samples that contain radionuclides with short half-lives. Holding times would apply to these radionuclides to prevent reporting of high measurement uncertainties and MDCs, and to detect the radionuclide, if present at low concentration, before it decays to undetectable levels.

Verification. Check that criteria related to sample radiological holding time (e.g., stated date of sample collection and analysis) have been addressed, and pertinent documentation is included in the analytical data package. If information on radiological holding time is missing, the data should be flagged with an "E" code.

If a holding time is specified in the project plan documents or validation plan, the reported values should be compared to this specification. If the holding time is exceeded, the affected criteria (holding time) should be flagged with an "E" code.

Validation. The data points impacted by the missed holding time should be flagged with a "J" code by the validator or the justification for discounting the holding time impact described in the narrative section of the validation report.

8.5.1.4 Preservation

Criteria. Appropriate preservation is dependent upon analyte and matrix and should be defined in sampling and analysis documentation. Generally, radiochemical samples are preserved to prevent precipitation, adsorption to container walls, etc. The criteria (required presence or absence) for this QC process should be provided in the sampling and analysis plan (Chapter 10).

Verification. Check that criteria related to sample preservation (e.g., stated preservation technique or verification thereof) have been addressed, and pertinent documentation is included in the analytical data package. If information on sample preparation is missing, the data should be flagged with an "E" code.

Validation. If exceptions to the preservation criteria are noted, the validator should decide if a "J" code should be assigned to data points because the improper preservation increased the overall uncertainty in the data. In some cases where improper preservation severely impacts data quality or defensibility (e.g., the use of acid preservation in water samples being analyzed for ^{14}C), the validator should assign an "R" qualifier. The assessor may elect to use the data, but they have the responsibility of addressing the data quality and defensibility in the assessment report.

8.5.1.5 Tracking

Criteria. Each analytical result should be linked to the instrument or detector on which it was counted. The requirement for this linkage normally is found in the project plan documents. The analytical sequence log (or some other suitable record) should be available in the data package submitted by the laboratory.

Verification. Check to see that the criteria related to instrument or detector linkage are found in the analytical data package. If the data are not linked to a counting instrument or detector, the data should be flagged with an "E" code.

Validation. The validator may consider the absence of linking a sample to a detection system into their evaluation of data quality and usability. At most, this should result in increasing the uncertainty of the determination and possibly assigning a "J" code to the data. This would not occur normally unless one or more of the detectors used in analyzing the samples was shown to be unreliable. Then, the inability to link a reliable detector to a sample increases the uncertainty of the data point(s).

8.5.1.6 Traceability

Criteria. The traceability of standards and reference materials to be used during the analysis should be specified in the sampling and analysis plan.

Verification. Check that criteria related to traceability of reference materials and standards (completed source manufacturer and internal calibration certificates) have been addressed, and pertinent documentation is included in the analytical data package or has been verified during an audit. If documentation on the traceability of reference materials and standards is missing, the data should be flagged with an "E" code.

Validation. The validator may factor the absence of the traceability into their evaluation of data quality and usability. At most, this should result in increasing the uncertainty of the determination and the possible assignment of a "J" code to the data. This would not occur normally unless one or more of the standards used in analyzing the samples was shown to be unreliable. Then, the inability to trace a reliable standard to a sample increases the uncertainty of the data point(s).

8.5.1.7 QC Types and Linkages

Criteria. The type and quantity of QC samples should be identified and listed in the SOW and the results provided by the laboratory in a summary report. Replicates and matrix spike results should be linked to the original sample results. The approximate level of matrix spike concentrations should be specified in the SOW, but the actual levels should be reported by the laboratory. The QC analyses should be linked to the original sample.

Verification. Check that there is linkage of QC samples to project samples and pertinent documentation is included in the analytical data package. If linkage information is missing, the data should be flagged with an "E" code.

Validation. The validator should compare any QC sample exceptions to similar ones that precede and follow the nonconforming QC sample. If these are in control, the validator can discount the impact of the single QC sample exception on the data results (i.e., analytical blunder). If a trend of failing values is found, the validator should consider if they affected a group of data points to the extent that the level of uncertainty was increased. This may warrant the assignment of a "J" code to the data.

8.5.1.8 Chemical Separation (Yield)

Criteria. Yield assesses the effects of the sample matrix and the chemical separation steps on the analytical result and estimates the analyte loss throughout the total analytical process. Yield is typically measured gravimetrically (with a carrier) or radiometrically (with a radiotracer). All the components in the calculation of the yield should be identified in a defined sequence. These specifications are found in the project plan documents.

Criteria for acceptable chemical yields may be given in the project plan documents. The criteria should be based on historical data for the method and matrix. In that case, yield is determined on both quality control samples and actual samples.

The most important yield-related question is whether the yield has been determined accurately. Typically, a yield estimate that is much greater than 100 percent cannot be accurate, but an estimate may also be questionable if the yield is far outside its historical range. Extremely low yields may lead to large measurement uncertainties.

Verification. Check that criteria related to chemical yield (e.g., calculated gravimetric or radiotracer yield determinations) have been addressed, and pertinent documentation is included in the analytical data package. If information on chemical yield is missing, the data should be flagged with an "E" code.

Validation. The experimentally determined yield is used to normalize the observed sample results to 100 percent yield. Exceptions to the yield value outside the range specified in the project plan documents may result in the validator assigning a "J" qualifier to otherwise acceptable data.

8.5.1.9 Self-Absorption

Criteria. For some radiochemical analytical methods, the SOW may specify the generation of a self-absorption curve, which correlates mass of sample deposited in a known geometry to detector efficiency.

Verification. For certain radionuclides, check that criteria related to self absorption curves (e.g., verification or copies of curves (or point-source values) covering the required weight ranges, emission type and energy) have been addressed and pertinent documentation is included in the analytical data package or has been verified during an audit. If the documentation is missing, the data should be flagged with an "E" code.

Validation. If required self-absorption curves are missing, the validator may qualify affected data with a "J" qualifier to signify an increased level of uncertainty in the measurement because of the inability to correct the measured value for self-absorption.

8.5.1.10 Efficiency, Calibration Curves, and Instrument Background

Criteria. For some methods based on decay-emission counting, efficiency is reported as measured count rate divided by the disintegration rate. For several methods, these efficiency determinations will depend on the energy of the emitted particle. For example, in gamma-ray spectrometry, a curve is fitted (measured activity/absolute disintegration rate) as a function of the gamma-ray energy for which this ratio is determined. A method like alpha spectrometry employs tracer radionuclides to determine a sample-specific "effective" efficiency factor, which is a product of the chemical yield and the detector efficiency. The specific efficiency criterion required for the project may be specified in the SOW. The determination of detector efficiency is a detailed process that is best checked during an audit of the laboratory's capabilities and is usually not part of the verification and validation process. Instrument background count rate is determined for each detector for each region of interest and subtracted from the sample count rate.

Verification. Check that each efficiency determination, efficiency calibration curve, and instrument background called for in the project plan documents is included in the analytical data package, if required. In many cases, this means assessing whether the proper calibration or efficiency was applied for the sample analyzed. If the documentation is missing, the data should be flagged with an "E" code. Each background subtracted should be appropriate for the radionuclide of interest. The proper use of crosstalk factors is one example of proper background subtraction.

Validation. If required factors are missing, the validator may select to qualify affected data with a "J" qualifier to signify an increased level of uncertainty in the measurement because of the inability to correct the measured value for efficiency.

8.5.1.11 Spectrometry Resolution

Criteria. The measured resolution of alpha and gamma-ray spectrometers, in terms of the full width of a peak at half maximum (FWHM), can be used to assess the adequacy of instrument setup, detector selectivity, and chemical separation technique that may affect the identification and quantification of the analyte. When sufficient peak definition (i.e., sufficient number of counts to provide an adequate Gaussian peak shape) has been reached for a sample, the resolution of the analyte peak should be evaluated to determine if proper peak identification and separation or deconvolution was made. Spectral information should be provided in the data packages to accomplish this evaluation.

Verification. There are no established acceptance criteria, but resolution data (e.g., FWHM) should be provided in the package or available during an audit.

Validation. If required calculations (multiplet analysis) are missing, the validator may elect to qualify affected data with a "J" code to signify an increased level of uncertainty in the measurement. In addition, if severe peak interference has not been corrected properly, the data should be qualified with a "J" code. An "R" code may be applied if there is no separation of the analyte peaks.

8.5.1.12 Dilution and Correction Factors

Criteria. Samples for radiochemistry are usually not diluted, but a larger sample may be digested, taking an aliquant for analysis to obtain a more representative subsample. The dilution factors are normally used for tracers and carriers. Dilutions of the stock standards are prepared and added to the samples. This dilution normally affects yield calculations, laboratory control samples, and matrix spikes. This data should be provided in the data package so that the final calculations of all data affected by dilution factors can be recalculated and confirmed, if required.

Other correction factors that may be applied to the data are dry weight correction, ashed weight correction, and correction for a two-phased sample analyzed as separate phases.

Verification. Check that criteria related to sample dilution and correction factors (e.g., factors have been stated) have been addressed and pertinent documentation is included in the analytical data package. If sample dilution information is missing, the data should be flagged with an "E" code.

Validation. Those results impacted by missing dilution factors should be flagged with a "J" or "R" qualifier, reflecting increased uncertainty in the data point(s). "R" may be warranted if the calculation cannot be confirmed due to missing data.

8.5.1.13 Counts and Count Time (Duration)

Criteria. The count time for each sample, QC analysis, and instrument background should be recorded in the data package. The ability to detect radionuclides is directly related to the count time.

Verification. Check that criteria related to instrument counting times (e.g., stated data about test source and background counting times) have been addressed and pertinent documentation is included in the analytical data package. If instrument or detector counting times are missing, the data should be flagged with an "E" code.

Validation. The validator should estimate the impact of the actual count times on the ability to detect the target analyte and the impact on the uncertainty of the measurement. *If the MQOs are met, the sample should not be qualified for count time.* The qualifiers should be adjusted accordingly and the justification provided in the validation report.

8.5.1.14 Result of Measurement, Uncertainty, Minimum Detectable Concentration, and Units

Criteria. MARLAP recommends that the result of each measurement, its expanded measurement uncertainty, and the estimated sample- or analyte-specific MDC be reported for each sample in the appropriate units. These values, when compared with each other, provide information about programmatic problems with the calculations, interference of other substances, and bias. The report should state the coverage factor used if calculating expanded measurement uncertainties, and the Type I and Type II error probabilities used to calculate MDCs.

Verification. Check to see that all criteria relating to linkages among result, measurement uncertainties, MDC, and sample identification are found in the analytical data package. If any of the criteria or actual linkages are missing, they should be flagged with an "E" code.

Validation. The validator should assign data qualifiers to those data points for which they feel sufficient justification exists. Each qualifier should be discussed in the validation report.

8.5.2   Quality Control Samples

Historically, data validation has placed a strong emphasis on review of QC sample data (laboratory control samples, duplicates, etc). The assumption is that if the analytical process was in control and the QC samples responded properly, then the samples would respond properly. It is possible to have excellent performance on simple matrices (e.g., QC samples), but unacceptable performance on complex matrices reported in the same batch as the QC samples. Directly evaluating the sample performance is essential to determine measurement uncertainty and the likelihood of false positive and negative detection of the target analyte.

Method blanks and laboratory control samples relate to the analytical batch (a series of similar samples prepared and analyzed together as a group) quality control function. They are required by most analytical service contracts, sampling and analysis plans, and project plan documents. They serve a useful function as monitoring tools that track the continuing analytical process during extended analytical sequences. They are the most ideal samples analyzed as part of a project. Normally, their performance is compared to fixed limits derived from historical performance or additionally project specific limits derived from the MQOs.

Laboratory duplicates and matrix spikes are quality control samples that directly monitor sample system performance. The laboratory duplicates (two equal-sized samples of the material being analyzed, prepared, and analyzed separately as part of the same batch) measure the overall precision of the sample measurement process beginning with laboratory sub-sampling of the sample. Matrix spikes (a sample containing a known amount of target analyte added to the sample) provide a direct measure of how the target analyte responds when the sample is prepared and measured, thereby estimating a possible bias introduced by the sample matrix.

Other QC tests can be applied to determine how the analytical process performs during the analysis of samples. These are yield, detector efficiency, test-source self-absorption, resolution, and drift. They are the same QC tests that were applied to routine QC samples (blanks and laboratory control samples) in the previous discussion of the analytical process, but now are applied to samples. The difference lies in how performance is measured. Fixed limits based on historical performance or statistics are usually the basis for evaluating the results of routine QC samples.

The following paragraphs discuss how QC tests should be used to determine if the results for QC samples meet the project MQOs. Guidance is provided on how to relate QC sample *and* sample performance to determine sample data quality and defensibility. Direction is also given about how to assign data qualifiers to sample data based on the tests of quality control. Appendix C provides guidance on developing criteria for evaluating QC sample results. Specifically, Appendix C contains equations that allow for the determination of warning and control limits for QC sample results based on the project's MQO for measurement uncertainty.

8.5.2.1 Method Blank

The method blank (Section 18.4.1) is generated by carrying all reagents and added materials normally used to prepare a sample through the same preparation process. It establishes how much, if any, of the measured analyte is contributed by the reagents and equipment used in the preparation process. For an ideal system, there will be no detected concentration or activity.

Measured results are usually corrected for instrument background and may be corrected for reagent background. Therefore, it is possible to obtain final results that are less than zero.

Criteria. The requirement for a method blank is usually established in the SOW and appropriate plan documents. The objective is to establish the target analyte concentration or activity introduced by the sample preparation sequence. Method blanks are normally analyzed once per analytical batch.

Other types of blanks, such as field blanks and trip blanks, are used to evaluate aspects of the data collection effort and laboratory operations that are not directly related to the validation of environmental analytical data quality or technical defensibility. They can be important to the overall data assessment effort, but are beyond the scope of this guidance (Chapter 10).

See Appendix C (*Measurement Quality Objectives for Method Uncertainty and Detection and Quantification Capability*) for guidance on developing criteria for evaluating blanks based on the project's MQO for method uncertainty.

Verification. If a method blank was required but not performed, or if the required data are missing, the verifier flags the data with an "E" code.

Validation. If a blank result does not comply with the established criteria, the associated samples are flagged "B+" to indicate that the blank result is greater than the upper limit, or "B-" to indicate that the blank result is less than the lower limit.

8.5.2.2 Laboratory Control Samples

The laboratory control sample (LCS) is a QC sample of known composition or an artificial sample (created by spiking a clean material similar in chemical and physical properties to the sample), which is prepared and analyzed in the sample manner as the sample (see Section 18.4.3, "Laboratory Control Samples, Matrix Spikes, and Matrix Spike Duplicates"). In an ideal situation, the LCS would give 100 percent of the concentration or activity known to be present in the fortified sample or standard material. Acceptance criteria for the LCS sample are based on the complexity of the matrix and the historical capability of the laboratory and method to recover the activity. The result normally is expressed as percent recovery.

Criteria. The objective of the LCS is to measure the response of the analytical process to a QC sample with a matrix similar to the sample. This will allow inferences to be drawn about the reliability of the analytical process.

See Appendix C for guidance on developing control limits for LCS results based on the project's MQO for method uncertainty.

Verification. If a required LCS is not analyzed, or if required information is missing, the verifier flags the data with an "E" code.

Validation. When the measured result for the LCS is outside the control limits, the associated samples are flagged with the "S" qualifier (S+ or S-).

8.5.2.3 Laboratory Replicates

Replicates are used to determine the precision of laboratory preparation and analytical procedures. Laboratory replicates are two aliquants selected from the laboratory sample and carried through preparation and analysis as part of the same batch.

The discussion of field replicates is beyond the scope of this chapter.

Criteria. The objective of replicate analyses is to measure laboratory precision based on each sample matrix. The variability of the samples due to the analyte's heterogeneity in the sample is also reflected in the replicate result. The laboratory may not be in control of the precision. Therefore, replicate results are used to evaluate reproducibility of the complete laboratory process that includes subsampling, preparation, and analytical process.

See Appendix C for guidance on developing control limits for replicate results based on the project's MQO for method uncertainty.

Verification. If replicate analyses are required but not performed, or if the required data are not present in the report, the verifier flags the data with an "E" code.

Validation. When the replicate analysis is outside the control limit, the associated samples are flagged with the "P" qualifier.

8.5.2.4 Matrix Spikes and Matrix Spike Duplicates

A matrix spike is typically an aliquant of a sample fortified (spiked) with known quantities of target radionuclides and subjected to the entire analytical procedure to establish if the method or procedure is appropriate for the analysis of a particular matrix. In some cases, specifically prepared samples of characterized materials that contain or are spiked with the target radionuclide and are consistent with the sample matrix may be used as matrix spikes. Matrix spike duplicates are used in a similar fashion as laboratory sample replicates, but in cases where there are insufficient quantities of target radionuclides in the laboratory sample replicates to provide statistically meaningful results.

Criteria. Matrix spike samples provide information about the effect of each sample matrix on the preparation and measurement methodology. The test uncovers the possible existence of recovery problems, based on either a statistical test or a specified fixed control limit.

See Appendix C for guidance on developing criteria for evaluating matrix spikes based on the project's MQO for method uncertainty.

Verification. If a required matrix spike analysis was not performed, or if the required information is missing, the data should be flagged with an "E" code.

Validation. If the results of the matrix spike analysis do not meet the established criteria, the samples should be qualified with an "S+" or "S-" indicating unacceptable spike recoveries.

8.5.3   Tests of Detection and Unusual Uncertainty

8.5.3.1 Detection

The purpose of a test of detection is to decide if each result for a regular sample is significantly different from zero. Since most radiochemistry methods always produce a result, even if a very uncertain or negative one, some notion of a non-detected but measured result may be needed for some projects. A nondetected result is generally as valid as any other measured result, but it is too small relative to its measurement uncertainty to give high confidence that a positive amount of analyte was actually present in the sample. Ordinarily, if the material being analyzed is actually analyte-free, most results should be "nondetected."

For some projects, detection may not be an important issue. For example, it may be known that all the samples contain a particular analyte, and the only question to be answered is whether the mean concentration is less than an action level. However, all laboratories should be able to perform a test of detection routinely for each analyte in each sample.

Criteria. An analyte is considered detected when the measured analyte concentration exceeds the critical value (see Chapter 20, _Detection and Quantification Capabilities_). Both values are calculated by the laboratory performing the measurement; so, the detection decision can be made at the laboratory and indicated in its report. If there is no evidence of additional unquantified uncertainty in the result (e.g., lack of statistical control or blank contamination), the laboratory's decision may be taken to be final.

Verification. Typically, the role of the verifier is limited to checking that required information, such as the critical value, is present in the report. If information is missing, the result should be flagged with an "E" code.

Validation. The validator examines the result of the measurement, its critical value, and other information associated with the sample and the batch in which it was analyzed, including method blank results in particular, to make a final determination of whether the analyte has been detected with confidence. If the data indicates the analyte has been detected in both the sample and the method blank, its presence in the sample may be questionable. A quantitative comparison of the total amounts of analyte in the sample and method blank, which takes into account the associated measurement uncertainties, may be needed to resolve the question.

8.5.3.2 Detection Capability

Criteria. If the project requires a certain detection capability, the requirement should be expressed as a required minimum detectable concentration (RMDC). The data report should indicate the RMDC and the sample-specific estimate of the actual minimum detectable concentration (MDC) for each analyte in each sample.

In some situations, it may not be necessary or even possible for a laboratory to meet the MDC requirement for all analytes in all samples. In particular, if the analyte is present and quantifiable at a concentration much greater than the action level, a failure to meet a contract-required detection limit is usually not a cause for concern. A failure to meet the RMDC is more often an important issue when the analyte is not detected.

Verification. The RMDC specified in the contract is compared to the sample-specific MDC achieved by the method. The analytes that do not meet the RMDC are flagged with an "E" code.

Validation. If the sample-specific MDC estimate exceeds the RMDC, the data user may be unable to make a decision about the sample with the required degree of certainty. A "UJ" qualifier is warranted if the estimated MDC exceeds the RMDC and the analyte was not detected by the analysis. A final decision about the usability of the data should be made during the data assessment phase of the data collection process.

An assignment of "R" to the data points affected by this type of exception may be appropriate in some cases, but the narrative report may classify the data as acceptable (no qualifier), "U," or "J," based on the results of the tests of detection and uncertainty. This allows the assessor to make an informed judgement about the usability of the data point(s) and allows them the opportunity to provide a rationale of why the data can be used in the decision process.

8.5.3.3 Large or Unusual Uncertainty

When project planners follow MARLAP's recommendations for developing MQOs, they determine a required method uncertainty at a specified analyte concentration. The required method uncertainty is normally expressed in concentration units, but it may be expressed as a relative method uncertainty (percent based on the upper bound of the gray region, which is normally the action level). It is reasonable to expect the laboratory's combined standard uncertainty at concentrations lower than the action level to be no greater than the required method uncertainty (expressed in concentration units) and to expect the laboratory's relative combined standard uncertainty at concentrations above the action level to be no greater than the required relative method uncertainty (expressed as a percent). Each measured result should be checked against these expectations (see Appendix C).

Criteria. The reported combined standard uncertainty is compared to the maximum allowable standard uncertainty. Either absolute (in concentration units) or relative uncertainties (expressed as a percent) are used in the comparison, depending on the reported concentration. The result is qualified with a "Q" if the reported uncertainty is larger than the requirement allows.

Verification. The test for large uncertainty is straightforward enough to be performed during either verification or validation. If there is a contractual requirement for measurement uncertainty, the verifier should perform the test and assign the "E" qualifier to results that do not meet the requirement. Note that it may sometimes happen that circumstances beyond the control of the laboratory make it impossible to meet the requirement.

Validation. If a "Q" qualifier is assigned, the validator may consider any special circumstances that tend to explain it, such as interferences, small sample sizes, or long decay times, which were beyond the control of the laboratory. He or she may choose to remove the qualifier, particularly if it is apparent that the original uncertainty requirement was too restrictive.

8.5.4 Final Qualification and Reporting

The final step of the validation process is to assign and report final qualifiers for all regular sample results. The basis for assignment of final qualifiers is qualifiers and reasons from all previous tests, patterns of problems in batches of samples, and validator judgement.

The difficult issue during final qualifier assignment is rejecting data. What follows summarizes some of the issues to consider when thinking about rejecting data.

Rejecting a result is an unconditional statement that it is not useable for the intended purpose. A result should only be rejected when the risks of using it are significant relative to the benefits of using whatever information it carries. If the DQA team or users feel data is being rejected for reasons that don't affect usability, they may disregard all validation conclusions. Rejected results should be discarded and not used in the DQA phase of the data life cycle.

There are three bases on which to reject data:

1. Insufficient or only incorrect data are available to make fundamental decisions about data quality. For example, if correctly computed uncertainty estimates are not available, it is not possible to do most of the suggested tests. If the intended use depends on a consistent, high level of validation, it may be proper to reject such data.

   The missing data should be fundamental. For example, missing certificates for standards are unlikely to be fundamental if laboratory performance on spiked samples is acceptable. In contrast, if no spiked sample data is available, it may be impossible to determine if a method gives even roughly correct results, and rejection may be appropriate.

2. Available data indicate that the assumptions underlying the method are not true. For example, QC samples may demonstrate that the laboratory's processes are out of control. Method performance data may indicate that the method simply does not work for particular samples. These problems should be so severe that is not possible to make quantitative estimates of their effects.

3. A result is "very unusually uncertain." It is difficult to say what degree of uncertainty makes a result unusable. Whenever possible, uncertain data should be rejected based on multiple problems with one result, patterns in related data, and the validator's judgement, not the outcome of a single test. This requires radiochemistry expertise and knowledge of the intended use.

Based on an evaluation of the tentative qualifiers, final qualifiers are assigned to each regular sample result.

After all necessary validation tests have been completed and a series of qualifiers assigned to each data point based on the results of the tests, a final judgment to determine which, if any, final qualifiers will be attached to the data should be made. The individual sample data from the laboratory should retain all the qualifiers. The basic decision making process for each result is always subject to validator judgement:

- As appropriate, assign a final "R";

- If "S", "P", or "B" were assigned, determine whether the qualifiers warrant the assignment of an "R";

- If "R" is not assigned, but some test assigned a tentative S, P, B, Q, or J, or a pattern exists that makes it appropriate, assign a final S, P, B, Q, or J and summarize QC sample performance;

- If a final S, B, or J was assigned, + or -, but not both, was tentatively assigned, and the potential bias is not outweighed by other sources of uncertainty, make the + or - final; and

- For non-R results, if any test assigned a tentative "U," make it final.

The final validation decision should address the fact that the broader purpose of validation is to contribute to the total data collection process, i.e., effectively translate and interpret analytical results for efficient use by an assessor. This means the validator should examine the full range of data available to search for and utilize relationships among the data elements to support the acceptance and use of data that falls outside method or contract specifications and data validation plan guidance.

8.6 Validation Report

The final product of validation is a package that summaries the validation process and its conclusions in an orderly fashion. This package should include:

- *A narrative or summary table written by the validator that summarizes exceptional circumstances*: In particular, it should document anything that prevented executing the planned validation tests. Further, the narrative should include an explicit statement explaining why data has been rejected or qualified based on the findings of the validation tests and the validator's judgment.

- *A list of validated samples that provides a cross-reference of laboratory and client sample identifiers*: This report should also include other identifiers useful in the context of the project, such as reporting batch, chain of custody, or other sample management system sample information.

- *A summary of all validated results with associated uncertainty for each regular sample with final qualifiers*: Unless specified in the sampling and analysis plan, non-detects are reported as measured, not replaced by a detection limit or other "less than" value.

- A summary of QC sample performance and the potential effect on the data both qualified and not qualified.

Assuming the client wants additional information, the following, more detailed reports can be included in the validation package. Otherwise, they are simply part of the validation process and the verification contract compliance:

- A detailed report of all tentative qualifiers and associated reasons for their assignment;

- QC sample reports that document analytical process problems; and

- Reports that summarize performance by method—these should support looking across related analyses at values such as yields and result ratios.

The data in the summary reports should be available in a computer-readable format. If no result was obtained for a particular analyte, the result field should be left blank. The validation report should package analytical results as effectively as possible for application and use by the individual assembling and assessing all project data.

The validation report should contain a discussion describing the problem(s) found during the validation process. For the validation codes, the discussion summarizes the performance criteria established in the validation plan. If the validation test performance criteria were changed (e.g., increased or decreased level of unusual uncertainty) because the nature of the sample matrix or analyte was different than expected, the new criteria should be explained in the report and the qualifiers applied using the new criteria. The approval of the project manager should be obtained (and documented) before the new criteria are applied. The project manager should communicate the changes to the project planning team to maintain the consensus reached and documented during validation planning.

Well-planned and executed analytical activities can be expected to meet reasonable expectations for data reliability. This means that for most data points or data sets, the results of the tests of quality control, detection, and unusual uncertainty will show that the data are of sufficient quality and defensibility to be forwarded to the assessor with little or no qualification for final assessment. A small number of points will be rejected because random errors in the analytical process or unanticipated matrix problems resulted in massive failure of several key validation tests.

A smaller number of data points will show conflicting results from the validation tests and present the greatest challenge to the validator. The more important the decision and the lower the required detection limit, the more common this conflict will become, and the more critical it is that the data validation plan provide guidance to the validator about how to balance the conflicting results. Is the ability to detect the analyte more important than the associated statistical unusual uncertainty, or is the presence of the analyte relatively definite but the unusual uncertainty around the project decision point critical to major decisions? The necessary guidance should be developed during the planning phase to guide the final judgment of the validator.

8.7 Summary of Recommendations

- MARLAP recommends that project objectives, implementation activities and QA/QC data be well documented in project plans, reports, and records, since the success of the assessment phase is highly dependent upon the availability of such information.

- MARLAP recommends that calibration be addressed in a quality system and through an audit, although demonstration of calibration may be required as part of a project's deliverables.

- MARLAP recommends that the assessment criteria of a project be established during the directed planning process and documented in the respective plans as part of the project plan documents.

- MARLAP recommends that the result of each measurement, its expanded measurement uncertainty, and the estimated sample- or analyte-specific MDC be reported for each sample in the appropriate units.

8.8 Bibliography

American National Standards Institute (ANSI) N13.30. *Performance Criteria for Radiobioassay.* 1996.

U.S. Environmental Protection Agency (EPA). 1994. *Contract Laboratory Program National Functional Guidelines for Inorganic Data Review.* EPA-540/R-94-013 (PB94-963502). February. Available from www.epa.gov/oerrpage/superfund/programs/clp/download/fginorg.pdf.

U.S. Environmental Protection Agency (EPA). 2002. *Guidance on Environmental Data Verification and Data Validation* (EPA QA/G-8). EPA/240/R-02/004. Office of Environmental Information, Washington, DC. Available at www.epa.gov/quality/qa_docs.html.

9 DATA QUALITY ASSESSMENT

9.1 Introduction

This chapter provides an overview of the data quality assessment (DQA) process, the third and final process of the overall data assessment phase of a project. Assessment is the last phase in the data life cycle and precedes the use of data. Assessment—in particular DQA—is intended to evaluate the suitability of project data to answer the underlying project questions or the suitability of project data to support the project decisions. The output of this final assessment process is a determination as to whether a decision can or cannot be made within the project-specified data quality objectives (DQOs).

The discussions in this chapter assume that prior to the DQA process, the individual data elements have been subjected to the first two assessment processes, "data verification" and "data validation" (see Chapter 8, *Radiochemical Data Verification and Validation*). The line between these three processes has been blurred for some time and varies from guidance to guidance and practitioner to practitioner. Although the content of the various processes is the most critical issue, a common terminology is necessary to minimize confusion and to improve communication among planning team members, those who will implement the plans, and those responsible for assessment. MARLAP defines these terms in Section 1.4 ("Key MARLAP Concepts and Terminology") and the Glossary and discusses assessment in Section 8.2 ("Data Assessment Process").

This chapter is not intended to address the detailed and specific technical issues needed to assess the data from a specific project but rather to impart a general understanding of the DQA process and its relationship to the other assessment processes, as well as of the planning and implementation phases of the project's data life cycle. The target audience for this chapter is the project planner, project manager, or other member of the planning team who wants to acquire a general understanding of the DQA process; not the statistician, engineer, or radiochemist who is seeking detailed guidance for the planning or implementation of the assessment phase. Guidance on specific technical issues is available (EPA, 2000a and b; MARSSIM, 2000; NRC, 1998).

This chapter emphasizes that assessment, although represented as the last phase of the project's data life cycle, should be planned during the directed planning process, and the needed documentation should be provided during the implementation phase of the project.

Section 9.2 reviews the role of DQA in the assessment phase. Section 9.3 discusses the graded approach to DQA. The role of the DQA

Contents		
9.1	Introduction	9-1
9.2	Assessment Phase	9-2
9.3	Graded Approach to Assessment	9-3
9.4	The Data Quality Assessment Team	9-4
9.5	Data Quality Assessment Plan	9-4
9.6	Data Quality Assessment Process	9-5
9.7	Data Quality Assessment Report	9-25
9.8	Summary of Recommendations	9-26
9.9	References	9-27

team is discussed in Section 9.4. Section 9.5 describes the content of DQA plans. Section 9.6 details the activities that are involved in the DQA process.

9.2   Assessment Phase

The assessment phase is discussed in Section 8.2. This present section provides a brief overview of the individual assessment processes, their distinctions, and how they interrelate.

"Data verification" generally evaluates compliance of the analytical process with project-plan and other project-requirement documents, and the statement of work (SOW), and documents compliance and noncompliance in a data verification report. Data verification is a separate activity in addition to the checks and review done by field and laboratory personnel during implementation. Documentation generated during the implementation phase will be used to determine if the proper procedures were employed and to determine compliance with project plan documents (e.g., QAPP), contract-specified requirements, and measurement quality objectives (MQOs). Any data associated with noncompliance will be identified as an "exception," which should elicit further investigation during data validation.

Compliance, exceptions, missing documentation, and the resulting inability to verify compliance should be recorded in the data verification report. Validation and DQA employ the verification report as they address the usability of data in terms of the project DQOs.

"Data validation" qualifies the usability of each datum after interpreting the impacts of exceptions identified during verification. *The validation process should be well defined in a validation plan that was completed during the planning phase.* The validation plan, as with the verification plan or checklist, can range from sections of a project plan to large and detailed stand-alone documents. Regardless of its size or format, the validation plan should address the issues presented in Section 8.3, "Validation Plan." Data validation begins with a review of project objectives and requirements, the data verification report, and the identified exceptions. The data validator determines if the analytical process was in statistical control (Section 8.5.2, "Quality Control Samples") at the time of sample analysis, and whether the analytical process as implemented was appropriate for the sample matrix and analytes of interest(Section 8.5.1, "The Sample Handling and Analysis System"). If the system being validated is found to be under control and applicable to the analyte and matrix, then the individual data points can be evaluated in terms of detection (Section 8.5.3.1), detection capability (Section 8.5.3.2), and unusual uncertainty (Section 8.5.3.3). Following these determinations, the data are assigned qualifiers (Section 8.5.4) and a data validation report is completed (Section 8.6). Validated data are rejected only when the impact of an exception is so significant that the datum is unreliable.

While both data validation and DQA processes address usability, the processes address usability from different perspectives. "Data validation" attempts to interpret the *impacts of exceptions* identified during verification and the impact of project activities on the usability of an individual datum. In contrast, "data quality assessment" considers the *results of data validation* while evaluating the usability of the entire data set.

During data validation, MARLAP strongly advises against the rejection of data unless there is a significant argument to do so (Chapter 8). As opposed to rejecting data, it is generally preferable that data are qualified and that the data validator details the concerns in the data validation report. However, there are times when data should be rejected, and the rationale for the rejection should be explained in the data validation report. There are times when the data validator may have believed data should be rejected based on a viable concern, yet during DQA, a decision could be made to employ the rejected data.

In summary, data validation is a transition from the compliance testing of data verification to usability determinations. The results of data validation, as captured in the qualified data and validation reports, will greatly influence the decisions made during the final assessment process, which is discussed in Section 9.6 ("Data Quality Assessment Process).

9.3   Graded Approach to Assessment

The sophistication of the assessment phase—and in particular DQA and the resources applied— should be appropriate for the project (i.e., a "graded approach"). Directed planning for small or less complex projects usually requires fewer resources and typically involves fewer people and proceeds faster. This graded approach to plan design is also applied to the assessment phase. Generally, the greater the importance of a project, the more complex a project, or the greater the ramifications of an incorrect decision, the more resources will be expended on assessment in general and DQA in particular.

It is important to note that the depth and thoroughness of a DQA will be affected by the thoroughness of the preceding verification and validation processes. Quality control or statement of work (SOW) compliance issues that are not identified as an "exception" during verification, or qualified during validation, will result in potential error sources not being reviewed and their potential impact on data quality will not be evaluated. Thus, while the graded approach to assessment is a valid and necessary management tool, it is necessary to consider all assessment phase processes (data verification, data validation, and data quality assessment) when assigning resources to assessment.

9.4   The Data Quality Assessment Team

The project planning team is responsible for ensuring that its decisions are scientifically sound and comply with the tolerable decision-error rates established during planning. MARLAP recommends the involvement of the data assessment specialist(s) on the project planning team during the directed planning process. This should result in a more efficient assessment plan and should increase the likelihood that flaws in the design of the assessment processes will be detected and corrected during planning. Section 2.4 ("The Project Planning Team") notes that it is important to have an integrated team of operational and technical experts. The data assessment specialist(s) who participated as members of the planning team need not be the final assessors. However, using the same assessors who participated in the directed planning process is advantageous, since they will be aware of the complexities of the project's goals and activities.

The actual personnel who will perform data quality assessment, or their requisite qualifications and expertise, should be specified in the project plan documents. The project planning team should choose a qualified data assessor (or team of data assessors) who is technically competent to evaluate the project's activities and the impact of these activities on the quality and usability of data. Multi-disciplinary projects may require a team of assessors (e.g., radiochemist, engineer, statistician) to address the diverse types of expertise needed to assess properly the representativeness of samples, the accuracy of data, and whether decisions can be made within the specified levels of confidence. Throughout this manual, the term "assessment team" will be used to refer to the assessor expertise needed.

9.5 Data Quality Assessment Plan

To implement the assessment phase as designed and ensure that the usability of data is assessed in terms of the project objectives, a detailed DQA plan should be completed during the planning phase of the data life cycle. This section focuses on the development of the DQA plan and its relation to DQOs and MQOs.

The DQA plan should address the concerns and requirements of all stakeholders and present this information in a clear, concise format. Documentation of these DQA specifications, requirements, instructions, and procedures are essential to assure process efficiency and that proper procedures are followed. Since the success of a DQA depends upon the prior two processes of the assessment phase, it is key that the verification and validation processes also be designed and documented in respective plans during the planning phase. Chapter 8 lists the types of guidance and information that should be included in data verification and validation plans.

MARLAP recommends that the DQA process should be designed during the directed planning process and documented in a DQA plan. The DQA plan is an integral part of the project plan documents and can be included as either a section or appendix to the project plan or as a cited stand-alone document. If a stand-alone DQA plan is employed, it should be referenced by the project plan and subjected to a similar approval process.

The DQA plan should contain the following information:

- A short summary and citation to the project documentation that provides sufficient detail about the project objectives (DQOs), sample and analyte lists, required detection limit, action level, and level of acceptable uncertainty on a sample- or analyte-specific basis;

- Specification of the necessary sampling and analytical assessment criteria (typically expressed as MQOs for selected parameters such as method uncertainty) that are appropriate for measuring the achievement of project objectives and constitute a basis for usability decisions;

- Identification of the actual assessors or the required qualifications and expertise that are required for the assessment team performing the DQA (Section 9.4);

- A description of the steps and procedures (including statistical tests) that will constitute the DQA, from reviewing plans and implementation to authoring a DQA report;

- Specification of the documentation and information to be collected during the project's implementation;

- A description for any project-specific notification or procedures for documenting the usability or non-usability of data for the project's decisionmaking;

- A description of the content of the DQA report;

- A list of recipients for the DQA report; and

- Disposition and record maintenance requirements.

9.6 Data Quality Assessment Process

MARLAP's guidance on the DQA process has the same content as other DQA guidance (ASTM D6233; EPA, 2000a and b; MARSSIM, 2000; NRC, 1998; USACE, 1998), however, MARLAP presents these issues in an order that parallels project implementation more closely. The MARLAP guidance on the DQA process can be summarized as an assessment process that—following the review of pertinent documents (Section 9.6.1)—answers the following questions:

- Are the samples representative? (Section 9.6.2)
- Are the analytical data accurate? (Section 9.6.3)
- Can a decision be made? (Section 9.6.4)

Each of these questions is answered first by reviewing the plan and then evaluating the implementation. The process concludes with the documentation of the evaluation of the data usability in a DQA Report (Section 9.7).

The DQA Process is more global in its purview than the previous verification and validation processes. The DQA process should consider the combined impact of all project activities in making a data usability determination. The DQA process, in addition to reviewing the issues raised during verification and validation, may be the first opportunity to review other issues, such as field activities and their impact on data quality and usability. A summary of the DQA steps and their respective output is presented in Table 9.1.

TABLE 9.1 — Summary of the DQA process

DQA PROCESS	Input	Output for DQA Report
1. Review Project Plan Document	The project plan document (or a cited stand-alone document) that addresses: (a) Directed Planning Process Report, including DQOs, MQOs, and optimized Sampling and Analysis Plan (b) Revisions to documents in (a) and problems or deficiency reports (c) DQA Plan	• Identification of project documents • Clear understanding by the assessment team of project's DQOs and MQOs • Clear understanding of assumptions made during the planning process • DQOs (as established for assessment) if a clear description of the DQOs does not exist
2. Are the Samples Representative?	The project plan document (or a cited stand-alone document) that addresses: (a) The sampling portion of the Sampling and Analysis Plan (b) SOPs for sampling (c) Sample handing and preservation requirements of the analytical protocol specifications	• Documentation of all assumptions as potential limitations and, if possible, a description of their associated ramifications • Determination of whether the design resulted in a representative sampling of the population of interest • Determination of whether the sampling locations introduced bias • Determination of whether the sampling equipment used, as described in the sampling procedures, was capable of extracting a representative set of samples from the material of interest • Evaluation of the necessary deviations (documented), as well as those deviations resulting from misunderstanding or error, and a determination of their impact on the representativeness of the affected samples

DQA PROCESS	Input	Output for DQA Report
3. Are the Data Accurate?	The project plan documents (or a cited stand-alone document) which address: (a) The analysis portion of the Sampling and Analysis Plan (b) Analytical protocol specifications, including quality control requirements and MQOs (c) SOW (d) The selected analytical protocols and other SOPs (e) Ongoing evaluations of performance (f) Data Verification and Validation plans and reports	• Determination of whether the selected methods were appropriate for the intended applications • Identification of any potential sources of inaccuracy • Assessment of whether the sample analyses were implemented according to the analysis plan • Evaluation of the impact of any deviations from the analysis plan on the usability of the data set
4. Can a Decision be Made?	The project plan document (or a cited stand-alone document) that addresses: (a) The DQA plan, including the statistical tests to be used (b) The DQOs and the tolerable decision error rates	• Results of the statistical tests. If new tests were selected, the rationale for their selection and the reason for the inappropriateness of the statistical tests selected in the DQA plan • Graphical representations of the data set and parameter(s) of interest • Determination of whether the DQOs and tolerable decision error rates were met • Final determination of whether the data are suitable for decisionmaking, estimating, or answering questions within the levels of certainty specified during planning

9.6.1 Review of Project Documents

The first step of the DQA process is for the team to identify and become familiar with the DQOs of the project and the DQA plan. Like the planning process, the steps of the DQA process are iterative, but they are presented in this text in a step-wise fashion for discussion purposes. Members of the assessment team may focus on different portions of the project plan documents and different elements of the planning process. Some may do an in-depth review of the directed planning process during this step; others will perform this task during a later step. The assessment team should receive revisions to the project planning documents and should review deficiency reports associated with the project. The first two subsections below discuss the key project documents that should be reviewed, at a minimum.

9.6.1.1 The Project DQOs and MQOs

Since the usability of data is measured in terms of the project DQOs, the first step in the DQA process is to acquire a thorough understanding of the DQOs. If the DQA will be performed by more than one assessor, it is essential that the assessment team shares a common understanding of the project DQOs and tolerable decision error rates. The assessment team will refer to these DQOs continually as they make determinations about data usability. The results of the directed planning process should have been documented in the project plan documents. The project plan documents, at a minimum, should describe the DQOs and MQOs clearly and in enough detail that they are not subject to misinterpretation or debate at this last phase of the project.

If the DQOs and MQOs are not described properly in the project plan documents or do not appear to support the project decision, or if questions arise, it may be necessary to review other planning documents (such as memoranda) or to consult the project planning team or the core group (Section 2.4). If a clear description of the DQOs does not exist, the assessment team should record any clarifications the assessment team made to the DQO statement as part of the DQA report.

9.6.1.2 The DQA Plan

If the assessment team was not part of the directed planning process, the team should familiarize itself with the DQA plan and become clear on the procedures and criteria that are to be used for the DQA Process. If the assessment team was part of the planning process, but sufficient time has elapsed since the conclusion of planning, the assessment team should review the DQA plan. If the process is not clearly described in a DQA plan or does not appear to support the project decision, or if questions arise, it may be necessary to consult the project planning team or the core group. If necessary, the DQA plan should be revised. If it cannot be, any deviations from it should be recorded in the DQA report.

During DQA, it is important for the team, including the assessors and statistician, to be able to communicate accurately. Unfortunately, this communication can be complicated by the different meanings assigned to common words (e.g., samples, homogeneity). The assessment team should be alert to these differences during their deliberations. The assessment team will need to determine the usage intended by the planning team.

It is important to use a directed planning process to ensure that good communications exist from planning through data use. If the statistician and other experts are involved through the data life cycle and commonly understood terms are employed, chances for success are increased.

9.6.1.3 Summary of the DQA Review

The review of project documents should result in:

- An identification and understanding of project plan documents, including any changes made to them and any problems encountered with them;

- A clear understanding of the DQOs for the project. If a clear description of the DQOs does not exist, the assessment team should reach consensus on the DQOs prior to commencing the DQA and record the DQOs (as they were established for assessment) as part of the DQA report; and

- A clear understanding of the terminology, procedures, and criteria for the DQA process.

9.6.2 Sample Representativeness

MARLAP does not provide specific guidance on developing sampling designs or a sampling plan. The following discussion of sampling issues during a review of the DQA process is included for purposes of completeness.

"Sampling" is the process of obtaining a portion of a population (i.e., the material of interest as defined during the planning process) that can be used to characterize populations that are too large or complex to be evaluated in their entirety. The information gathered from the samples is used to make inferences whose validity reflects how closely the samples represent the properties and analyte concentrations of the population. "Representativeness" is the term employed for the degree to which samples properly reflect their parent populations. A "representative sample," as defined in ASTM D6044, is "a sample collected in such a manner that it reflects one or more characteristics of interest (as defined by the project objectives) of a population from which it was collected" (Figure 9.1). Samples collected in the field as a group and subsamples generated as a group in the laboratory (Appendix F) should reflect the population physically and chemically. A flaw in any portion of the sample collection or sample analysis design or their implementation can impact the representativeness of the data and the correctness of associated decisions. Representativeness is a complex issue related to analyte of interest, geographic and temporal units of concern, and project objectives.

The remainder of this subsection discusses the issues that should be considered in assessing the representativeness of the samples: the sampling plan (Section 9.6.2.1) and its implementation (Section 9.6.2.2). MARLAP recommends that all sampling design and statistical assumptions be identified clearly in project plan documents along with the rationale for their use.

9.6.2.1 Review of the Sampling Plan

The sampling plan and its ability to generate representative samples are assessed in terms of the project DQOs. The assessors review the project plan with a focus on the approach to sample collection, including sample preservation, shipping and subsampling in the field and laboratory, and sampling standard operating procedures (SOPs). Ideally the assessors would have been involved in the planning process and would be familiar with the DQOs and MQOs and the decisions made during the selection of the sampling and analysis design. If the assessors were part of the project planning team, this review to become familiar with the project plan will go

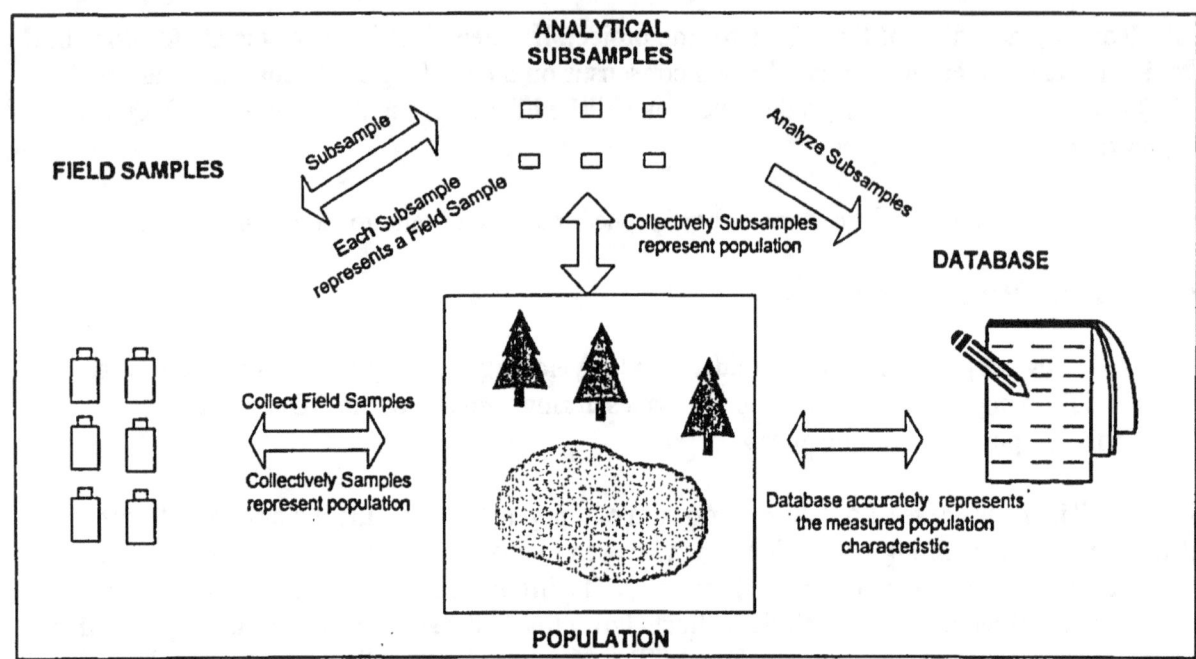

FIGURE 9.1 — Using physical samples to measure a characteristic of the population representatively.

quickly, and the team can focus on deviations from the plan that will introduce unanticipated imprecision or bias (Section 9.6.2.2).

APPROACH TO SAMPLE COLLECTION

Project plan documents (e.g., QAPP, SAP, Field Sampling Plan) should provide details about the approach to sample collection and the logic that was employed in its development. At this stage, the assessment team should evaluate whether the approach, as implemented, resulted in representative samples. For example, if the approach was probabilistic, the assessment team should determine if it was appropriate to assume that spatial or temporal correlation is not a factor, and if all portions of the population had an equal chance of being sampled. If an "authoritative" sample collection approach was employed (i.e., a person uses his knowledge to choose sample locations and times), the assessment team—perhaps in consultation with the appropriate experts (e.g., an engineer familiar with the waste generation process)—should determine if the chosen sampling conditions do or do not result in a "worst case" or "best case."

The assessment team should evaluate whether the chosen sampling locations resulted in a negative or positive bias, and whether the frequency and location of sample collection accounted for the population heterogeneity.

Optimizing the data collection activity (Section 2.5.4 and Appendix B3.8) involves a number of assumptions. These assumptions are generally employed to manage a logistical, budgetary, or other type of constraint, and are used instead of additional sampling or investigations. The assessment team needs to understand these assumptions in order to fulfill its responsibility to review and evaluate their continued validity based on the project's implementation. The assessment team should review the bases for the assumptions made by the planning team because they can result in biased samples and incorrect conclusions. For example, if samples are collected from the perimeter of a lagoon to characterize the contents of the lagoon, the planning team's assumption was that the waste at the lagoon perimeter has the same composition as that waste located in the less-accessible center of the lagoon. In this example, there should be information to support the assumption, such as historical data, indicating that the waste is relatively homogen-ous and well-mixed. Some assumptions will be stated clearly in project plan documents. Others may only come to light after a detailed review. The assessment team should review assumptions for their scientific soundness and potential impact on the representativeness of the samples.

Ideally, assumptions would be identified clearly in project plan documents, along with the rationale for their use. Unfortunately, this is uncommon, and in some cases, the planners may be unaware of some of the implied assumptions associated with a design choice. The assessment team should document any such assumptions in the DQA report as potential limitations and, if possible, describe their associated ramifications. The assessment team may also suggest additional investigations to verify the validity of assumptions which are questionable or key to the project.

SAMPLING SOPs

Standard operating procedures for sampling should be assessed for their appropriateness and scientific soundness. The assessment team should assess whether the sampling equipment and their use, as described in the sampling procedures, were capable of extracting a representative set of samples from the material of interest. The team also should assess whether the equipment's composition was compatible with the analyte of interest. At this stage, the assessment team assumes the sampling device was employed according to the appropriate SOP. Section 9.6.2.2 discusses implementation and deviations from the protocols.

In summary, the assessment team should investigate whether:

- The sampling device was compatible with the material being sampled and with the analytes of interest;

- The sampling device accommodated all particle sizes and did not discriminate against portions of the material being sampled;

- The sampling device avoided contamination or loss of sample components;

- The sampling device allowed access to all portions of the material of interest;

• The sample handling, preparation, and preservation procedures maintained sample integrity; and

• The field and laboratory subsampling procedures resulted in a subsample that accurately represents the contents of the original sample.

These findings should be detailed in the DQA report.

9.6.2.2 Sampling Plan Implementation

The products of the planning phase are integrated project plan documents that define how the planners intend the data collection process to be implemented. At this point in the DQA process, the assessment team determines whether sample collection was done according to the plan, reviews any noted deviations from the protocols, identifies any additional deviations, and evaluates the impact of these deviations on sample representativeness and the usability of the data. The success of this review will be a function of the documentation requirements specified during the planning process, and how thoroughly these requirements were met during sample collection.

The determination as to whether the plans were implemented as written typically will be based on a review of documentation generated during the implementation phase, through on-site assessments, and during verification, if sampling activities (e.g., sample preservation) were subjected to verification. In some instances, assessment team members may have firsthand knowledge from an audit that they performed, but in general the assessment team will have to rely upon documentation generated by others. The assessment team will review field notes, sample forms, chain-of-custody forms, verification reports, audit reports, deviation reports, corrective action documentation, QA reports, and reports to management. The assessment team also may choose to interview field personnel to clarify issues or to account for missing documentation.

Due to the uncontrolled environments from which most samples are collected, the assessment team expects to find some deviations even from the best-prepared plans. Those not documented in the project deficiency and deviation reports should be detailed in the DQA report. The assessment team should evaluate these necessary deviations, as well as those deviations resulting from misunderstanding or error, and determine their impact on representativeness of the affected samples. These findings also should be detailed in the DQA report.

In summary, the assessment team will develop findings and determinations regarding any deviations from the original plan, the rationale for the deviations, and if the deviations raise question of representativeness.

9.6.2.3 Data Considerations

Sample representativeness also can be evaluated in light of the resulting data. Favorable comparisons of the data to existing data sets (especially those data sets collected by different organizations and by different methods) offer encouraging evidence of representativeness, but not absolute confirmation of sample representativeness, since both data sets could suffer from the same bias and imprecision. The project plan documents should have referenced any credible and applicable existing data sets identified by the planning team. Comparisons to existing data sets may offer mutual support for the accuracy of each other, and when differences result they tend to raise questions about both data sets. Quite often, the DQA assessors are looking for confirmatory or conflicting information. How existing data sets are used during the DQA will be determined by how much confidence the assessors place in them. If they are very confident in the accuracy of existing data sets, then they may classify the new data as unusable if it differs from the existing data. If there is little confidence in the existing data set, then the assessors may just mention in the DQA report that the new data set was in agreement or not in agreement. However, if the planning team has determined that additional data were needed, they probably will not have sufficient confidence in the existing data set for purposes of decisionmaking.

Data comparability is an issue that could be addressed during validation to some degree, depending on the validation plan. However, at this point in the DQA, comparable data sets serve a different purpose. For example, the MDCs, concentration units, and the analytical methods may be the same and allow for data comparison in validation. However, the assessors during DQA would look for similarities and dissimilarities in reported concentrations for different areas of the populations, and whether any differences might be an indication of a bias or imprecision that makes the samples less representative. Temporal and spatial plots of the data also may be helpful in identifying portions of the sampled population that were over- or under-represented by the data collection activity.

The planning process and development of probabilistic sampling plans typically require assumptions regarding average concentrations and variances. If the actual average concentrations and variances are different than anticipated, it is important for the assessment team to evaluate the ramifications of these differences on sample representativeness. As reported values approach an action level, the greater the need for the sample collection activities to accurately represent the population characteristics of interest.

During the evaluation of sample representativeness, as discussed in the previous subsections, the assessment team has the advantage of hindsight, since they review the sample collection design in light of project outcomes and can determine if the sample collection design could have been optimized differently to better achieve project objectives. Findings regarding the representativeness of samples and how sampling can be optimized should be expeditiously passed to project managers if additional sampling will be performed.

In summary, results of the evaluation of the sample representativeness are:

- An identification of any assumptions that present limitations and, if possible, a description of their associated ramifications;

- A determination of whether the design resulted in a representative sampling of the population of interest;

- A determination of whether the specified sampling locations, or alternate locations as reported, introduced bias;

- A determination of whether the sampling equipment used, as described in the sampling procedures or as implemented, was capable of extracting a representative set of samples from the material of interest; and

- An evaluation of the necessary deviations from the plan, as well as those deviations resulting from misunderstanding or error, and a determination of their impact on the representativeness of the affected samples.

The product of this step is a set of findings regarding the impact of representativeness—or the lack thereof—that affects data usability. Findings and determinations regarding representative-ness will impact the usability of the resulting data to varying degrees. Some findings may be so significant (e.g., the wrong waste stream was sampled) that the samples can be determined to be non-representative and the associated data cannot be used; as a result, the DQA need not progress any further. Typically, findings will be subject to interpretation, and the impacts on representa-tiveness will have to be evaluated in light of other DQA findings to determine the usability of data.

9.6.3 Data Accuracy

The next step in the DQA process is the evaluation of the analysis process and accuracy of the resulting data. The term "accuracy" describes the closeness of the result of a measurement to the true value of the quantity being measured. The accuracy of results may be affected by both imprecision and bias in the measurement process, and by blunders and loss of statistical control (see Chapter 19, *Measurement Uncertainty*).

Since MARLAP uses "accuracy" only as a qualitative concept, in accordance with the *International Vocabulary of Basic and General Terms in Metrology* (ISO, 1993), the agreement between measured results and true values is evaluated quantitatively in terms of the "precision" and "bias" of the measurement process. "Precision" usually is expressed as a standard deviation, which measures the dispersion of results about their mean. "Bias" is a persistent deviation of results from the true value (see Section 6.5.5.7, "Bias Considerations").

During the directed planning process, the project planning team should have made an attempt to identify and control sources of imprecision and bias (Appendix B3.8). During DQA, the assessment team should evaluate the degree of precision and bias and determine its impact on data usability. Quality control samples are analyzed for the purpose of assessing precision and bias. Laboratory spiked samples and method blanks typically are used to assess bias, and duplicates are used to assess precision. Since a single measurement of a spike or blank principle cannot distinguish between imprecision and bias, a reliable estimate of bias requires a data set that includes many such measurements. Control charts of quality control (QC) data, such as field duplicates, matrix spikes, and laboratory control samples are graphical representations and primary tools for monitoring the control of sampling and analytical methods and identifying precision and bias trends (Chapter 18, *Laboratory Quality Control*).

Bias can be identified and controlled through the application of quantitative MQOs to QC samples, such as blanks, standard reference materials, performance testing samples, calibration check standards, and spikes samples. Blunders (e.g., a method being implemented incorrectly, such as reagents being added in the incorrect order) are usually identified and controlled by well-designed plans that specify quality assurance systems that detail needed training, use of appropriate SOPs, deficiency reporting systems, assessments, and quality improvement processes.

Bias in a data set may be produced by measurement errors that occur in steps of the measurement process that are not repeated. Imprecision may be produced by errors that occur in steps that are repeated many times. The distinction between bias and imprecision is complicated by the fact that some steps, such as instrument calibration and tracer preparation and standardization, are repeated at varying frequencies. For this reason, the same source of measurement error may produce an apparent bias in a small data set and apparent imprecision in a larger data set. During data assessment, an operational definition of bias is needed. This would normally be determined by the data assessment specialist(s) on the project planning team during the directed planning process. For example, a bias may exist if results for analytical spikes (i.e., laboratory control samples, matrix spike, matrix spike duplicate), calibration checks, and performance evaluation samples associated with the data set are mostly low or mostly high, if the results of method blank analyses tend to be positive or negative, or if audits uncover certain types of biased implementation of the SOPs. At times, the imprecision of small data sets can incorrectly indicate a bias, while at other times, the presence of bias may be masked by imprecision. For example, two or three samples may be all high or all low by chance, and may be a result of imprecision rather than bias. On the other hand, it is unlikely that ten samples would all be high or low, and such an occurrence would be indicative of bias. Statistical methods can be applied to imprecise data sets and used to determine if there are statistically significant differences between data sets or between a data set and an established value. If the true value or reference value (e.g., verified concentration for a standard reference material) is known, then statistics can be used to determine whether there is a bias.

Figure 9.2 employs targets to depict the impacts of imprecision and bias on measurement data. The true value is portrayed by the bulls-eye and is 100 units (e.g., ppm, dpm, Bq, pCi/g). Ideally, all measurements with the same true value would be centered on the target, and after analyzing a number of samples with the same true value, the reported data would be 100 units for each and every sample. This ideal condition of precise and unbiased data is pictured in Figure 9.2(a). If the analytical process is very precise but suffers from a bias, the situation could be as pictured in Figure 9.2(b) in which the data are very reproducible but express a significant 70 percent departure from the true value—a significant bias. The opposite situation is depicted in Figure 9.2(c), where the data are not precise and every sample yields a different concentration. However, as more samples are analyzed, the effects of imprecision tend to average out, and lacking any bias, the average measurement reflects the true concentration. Figure 9.2(d) depicts a situation where the analytical process suffers from both imprecision and bias. Even if innumerable samples with the same true value are collected and analyzed to control the imprecision, an incorrect average concentration still would be reported due to the bias.

Each target in Figure 9.2 has an associated frequency distribution curve. Frequency curves are made by plotting a concentration value versus the frequency of occurrence for that concentration. Statisticians employ frequency plots to display the precision of a sampling and analytical event, and to identify the type of distribution. The curves show that as precision decreases the curves flatten-out and there is a greater frequency of measurements that are distant from the average value (Figures 9.2c and d). More precise measurements result in sharper curves (Figures 9.2a and b), with the majority of measurements relatively closer to the average value. The greater the bias (Figures 9.2b and d), the further the average of the measurements is shifted from the true value. The smaller the bias (Figures 9.2a and c), the closer the average of the measurements is to the true value.

The remainder of this subsection focuses on the review of analytical plans (Section 9.6.3.1) and their implementation (Section 9.6.3.2) as a mechanism to assess the accuracy of analytical data and their suitability for supporting project decisions.

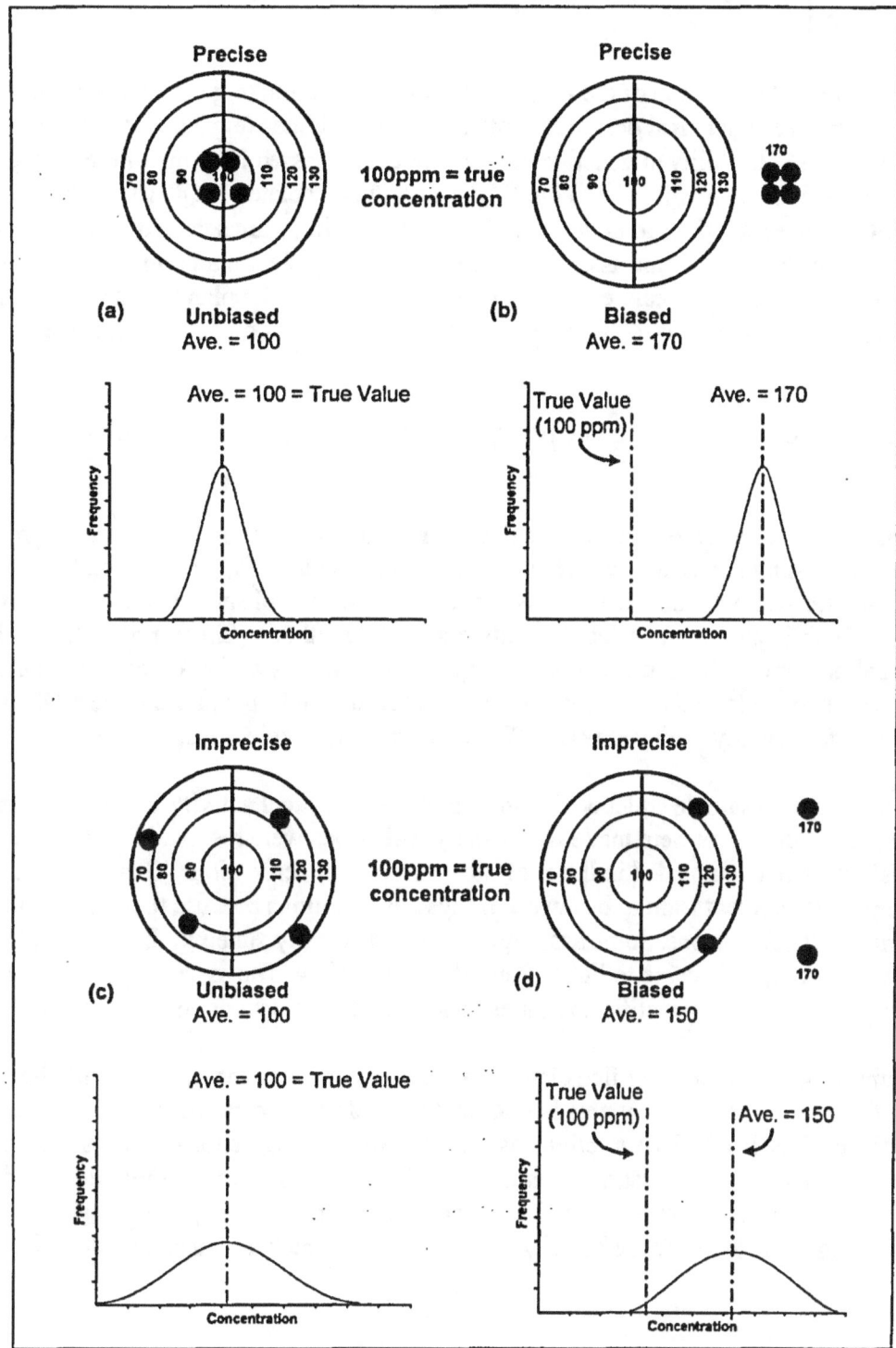

FIGURE 9.2 — Types of sampling and analytical errors.

9.6.3.1 Review of the Analytical Plan

The analytical plan is that portion of the project plan documentation (e.g., in QAPP or SAP) that addresses the optimized analytical design and other analytical issues (e.g., analytical protocol specifications, SOPs). Its ability to generate accurate data is assessed in terms of the project DQOs. The assessment team will refer to the DQOs and the associated MQOs as they review the analytical protocol specifications to understand how the planning team selected methods and developed the analytical plan. If the assessors were part of the project planning team, this review process will go quickly and the team can focus on deviations from the plan that will introduce unanticipated imprecision or bias. (The term "analytical plan" is not meant to indicate a separate document.)

REVIEW OF THE MQOs, ANALYTICAL PROTOCOL SPECIFICATIONS, AND OPTIMIZED ANALYTICAL DESIGN

The assessment team's review of the analytical plan first should focus on the analytical protocol specifications, including the MQOs, which were established by the project planning team (Chapter 3). The team should understand how the analytical protocol specifications were used to develop the SOW (Chapter 5) and select the radioanalytical methods (Chapter 6). If the project and contractual documentation are silent or inadequate on how they address these key issues, the assessment team may be forced to review the analytical results in terms of the project DQOs and determine if the data quality achieved was sufficient to meet the project's objectives.

As with the approach to sample collection, optimizing the analytical activity involved a number of assumptions. Assumptions were made when analytical issues were resolved during planning and the decisions were documented in the analytical protocol specifications (Chapter 3). It is important for the assessment team to be aware of these assumptions because they can result in biases and incorrect conclusions. Some assumptions will be clearly stated in the project plan documents. Others may only come to light after a detailed review. The assessment team should review assumptions for their scientific soundness and potential impact on the data results.

Ideally, assumptions would be identified clearly in project plan documents, along with the rationale for their use. Unfortunately, this is uncommon, and in some cases, the planners may be unaware of some of the implied assumptions associated with a design choice. The assessment team should document any such assumptions in the DQA report as potential limitations and, if possible, describe their associated ramifications. The assessment team may also suggest additional investigations to verify the validity of assumptions which are questionable or key to the project.

REVIEW OF THE ANALYTICAL PROTOCOLS

The analytical plan and the associated analytical protocols will be reviewed and assessed for their scientific soundness, applicability to the sample matrix and the ability to generate precise and unbiased data. The analytical protocols review should consider the entire analytical process, from sample preparation through dissolution and separations, counting, data reduction, and reporting. MARLAP, whose focus is on the analytical process, defines "analytical process" as including sample handling in the field (e.g., filtration, sample preservation) to ensure that all activities that could impact analyses would be considered. The assessment team should consider both sampling and analytical processes in assessing data quality—and such field activities as sample preservation—along with other issues that can affect representativeness (Section 9.6.2). The assessment team also should review the contract evaluation (under the performance-based approach) for the selection of the analytical protocols to assure that the documentation showed that the protocol could meet the analytical protocol specifications (which defines the MQOs).

Since the review of the analytical protocols will be performed with the advantage of hindsight gained from the data verification and data validation reports, the assessment team also should attempt to identify any flaws in the analytical protocols that may have resulted in noncompliance with MQOs. The identification of these flaws is essential if future analyses will be required.

REVIEW OF VERIFICATION AND VALIDATION PLANS

To understand how the verification and validations processes were implemented and the degree to which the assessors can rely upon their findings, the assessors should familiarize themselves with the verification and validation plans that were developed during the planning phase. A review of these plans will indicate the thoroughness of the evaluations and whether the issues deemed important to the assessors were evaluated.

9.6.3.2 Analytical Plan Implementation

After reviewing the analytical plan, the assessment team should assess whether sample analyses were implemented according to the analysis plan. Typically, the first two steps of the assessment phase—data verification and data validation—have laid most of the groundwork for this determination. However, the issue of whether the plan was implemented as designed needs to be reviewed one final time during the DQA process. This final review is needed since new and pertinent information may have been uncovered during the first steps of the DQA process.

The goal of this assessment of the analytical process with respect to the associated MQOs is to confirm that the selected method was appropriate for the intended application and to identify any potential sources of inaccuracy, such as:

- Laboratory subsampling procedures that resulted in the subsample that may not accurately represent the content of the original sample;

- Sample dissolution methods that may not have dissolved sample components quantitatively;

- Separation methods whose partitioning coefficients were not applicable to the sample matrix;

- Unanticipated self-absorption that biased test-source measurements;

- Non-selective detection systems that did not resolve interferences; or

- Data reduction routines that lacked needed resolution or appropriate interference corrections.

The success of the assessment of the analytical plan implementation will be a function of the documentation requirements specified during the planning process, and how thoroughly these requirements were met during sample analysis. In some instances, assessment team members may have firsthand knowledge from an audit that they performed, but in general the assessment team will have to rely upon documentation generated by others.

In addition to verification and validation reports, the assessment team will review pertinent documents such as: laboratory notebooks, instrument logs, quality control charts, internal sample-tracking documentation, audit reports, deviation reports, corrective action documentation, performance evaluation sample reports, QA reports, and reports to management provided for verification and validation. To clarify issues or to account for missing documentation, the assessment team may choose to interview laboratory personnel.

Verification and validation reports will be used to identify nonconformance, deviations, and problems that occurred during the implementation of the analytical plan. The challenge during DQA is to evaluate the impact of nonconformance, deviations, problems, and qualified data on the usability of the overall data set and the ability of the data set to support the decision.

Deviations from the plan will be encountered commonly and the assessment team will evaluate the impact of these deviations upon the accuracy of the analytical data. The deviations and the assessment team's related findings should be detailed in the data quality assessment report.

The prior verification and validation processes and the prior DQA steps involving the evaluation of sampling are all an attempt to define the quality of data by (1) discovering sources of bias, quantifying their impact, and correcting the reported data; and (2) identifying and quantifying data precision. The products of this step are a set of findings regarding the analytical process and their impact on data usability. Some findings may be so significant (e.g., the wrong analytical method was employed) that the associated data cannot be used, and as a result, the DQA need not progress any further. Typically, findings will be subject to interpretation and a final determination as to the impacts will have to wait until the data has been subjected to evaluations described in Section 9.6.4.

After reviewing the verification and validation reports, the outputs of the analytical data evaluation are:

- A determination of whether the selected analytical protocols and analytical performance specifications were appropriate for the intended application;

- An identification of any potential sources of inaccuracy; and

- A determination of whether sample analyses were implemented according to the analysis plan and the overall impact of any deviations on the usability of the data set.

9.6.4 Decisions and Tolerable Error Rates

A goal of DQA is to avoid making a decision based on inaccurate data generated by analytical protocols found to be out of control or on data generated from samples found to be nonrepresentative, and to avoid making decisions based on data of unknown quality. Preferably, a decision should be made with data of known quality (i.e., with data of known accuracy from samples of known representativeness) and within the degree of confidence specified during the planning phase.

This section focuses on the final determination by the assessment team, who uses the information taken from the previous assessment processes and statistics to make a final determination of whether the data are suitable for decision-making, estimating, or answering questions within the levels of certainty specified during planning.

9.6.4.1 Statistical Evaluation of Data

Statistics are used for the collection, presentation, analysis, and interpretation of data. The two major branches of statistics, "descriptive statistics" and "inferential statistics," are applicable to data collection activities. "Descriptive statistics" are those methods that describe populations of data. For example, descriptive statistics include the mean, mode, median, variance, and correlations between variables, tables, and graphs to describe a set of data. "Inferential statistics" use data taken from population samples to make estimates about the whole population ("inferential estimations") and to make decisions ("hypothesis testing"). Descriptive statistics is an important tool for managing and investigating data in order that their implications and significance to the project goals can be understood.

Sampling and inferential statistics have identical goals—to use samples to make inferences about a population of interest and to use sample data to make defensible decisions. This similarity is the reason why planning processes, such as those described in Chapter 2, couple sample collection activities with statistical techniques to maximize the representativeness of samples, the accuracy of data, and the certainty of decisions.

Due to the complexity of some population distributions (Attachment 19A) and the complex mathematics needed to treat these distributions and associated data, it is often best to consult with someone familiar with statistics to ensure that statistical issues have been addressed properly. However, it is critical for the non-statistician to realize that statistics has its limitations. The following statistical limitations should be considered when assessment teams and the project planning team are planning the assessment phase and making decisions:

- Statistics are used to measure precision and, when true or reference values are known, statistics can be applied to imprecise data to determine if a bias exists. Statistics do not address all types of sampling or measurement bias directly.

- If the characteristic of interest in a sample is more similar to that of samples adjacent to it than to samples that are further removed, the samples are deemed to be "correlated" and are not independent of each other (i.e., there is a serial correlation such that samples collected close in time or space have more similar concentrations than those samples further removed). Conventional parametric and non-parametric statistics require that samples be independent and are not applicable to populations that have significantly correlated concentrations.

The statistical tests typically are chosen during the directed planning process and are documented in the project plan documents (e.g., DQA plan, QAPP). However, there are occasions when the conditions encountered during the implementation phase are different than anticipated (e.g., data were collected without thorough planning, or data are being subjected to an unanticipated secondary data use). Under these latter conditions, the statistical tests will be chosen following data collection.

The statistical analysis of data consists of a number of steps. The following outline of these steps is typical of the analyses that a statistician would implement in support of a data quality assessment.

CALCULATE THE BASIC STATISTICAL PARAMETERS

Statistical "parameters" are fundamental quantities that are used to describe the central tendency or dispersion of the data being assessed. The mean, median, and mode are examples of statistical parameters that are used to describe the central tendency, while range, variance, standard deviation, coefficient of variation, and percentiles are statistical parameters used to describe the dispersion of the data. These basic parameters are used because they offer a means of understanding the data, facilitating communication and data evaluation, and generally are necessary for subsequent statistical tests.

GRAPHICAL REPRESENTATIONS

Graphical representations of the data are similar to basic statistical parameters in that they are a means of describing and evaluating data sets. Graphical representations of QC-sample results used to evaluate project-specific control limits and warning limits derived from the MQO criteria are discussed in Appendix C. Graphical representations of field data over space or time have the additional ability of offering insights, such as identifying temporal and spatial patterns, trends, and correlations. Graphical depictions are also an excellent means of communicating and archiving information.

REVIEW AND VERIFY TEST ASSUMPTIONS

Statistical tests are the mathematical structure that will be employed to evaluate the project's data in terms of the project decision, question, or parameter estimate. Statistical tests are not universally applicable, and their choice and suitability are based on certain assumptions. For example:

- Some tests are suitable for "normal" distributions, while others are designed for other types of distributions.

- Some tests assume that the data are random and independent of each other.

- Assumptions that underlie tests for "outliers" should be understood to ensure that hot spots or the high concentrations symptomatic of skewed distributions (e.g., lognormal) are not incorrectly censored.

- Assumptions are made regarding the types of population distributions whenever data are transformed before being subjected to a test.

- Assumptions of test robustness need to be reviewed in light of the analyte. For example, radiological data require statistical tests that can accommodate positive and negative numbers.

It is important that a knowledgeable person identify all assumptions that underlie the chosen statistical tests, and that the data are tested to ensure that the assumptions are met. If any of the assumptions made during planning proved to be not true, the assessment team should evaluate the appropriateness of the selected statistical tests. Any decision to change statistical tests should be documented in the DQA report.

APPLYING STATISTICAL TESTS

The chosen statistical tests will be a function of the data properties, statistical parameter of interest, and the specifics of the decision or question. For example, choice of the appropriate tests will vary according to whether the data are continuous or discrete; whether the tests will be single-tailed or double-tailed, whether a population is being compared to a standard or to a second population, or whether stratified sampling or simple random sampling was employed. Once the statistical tests are deemed appropriate, they should be applied to the data by an assessor who is familiar with statistics. The outputs from applying the statistical tests and comparisons to project DQOs are discussed in the following section. Appropriate statistical tests and guidance on their use are available from many sources, including EPA (2000b).

9.6.4.2 Evaluation of Decision Error Rates

The heterogeneity of the material being sampled and the imprecision of the sampling and analytical processes generate uncertainty in the reported data and in the associated decisions and answers. The project planning team, having acknowledging this decision uncertainty, will have chosen "tolerable decision errors rates" during the planning process, which balanced resource costs against the risk of making a wrong decision or arriving at a wrong answer. During this final step of DQA process, the assessment team will use the project's tolerable levels of decision error rates as a metric of success.

The DQA process typically corrects data for known biases and then subjects the data to the appropriate statistical tests to make a decision, answer a question, or supply an estimate of a parameter. The assessment team will compare statistical parameters—such as the sample mean and sample variance estimates employed during the planning process—to those that were actually obtained from sampling. If the distribution was different, if the mean is closer to the action level, or if the variance is greater or less than estimated, one or all of these factors could have an impact on the certainty of the decision. The assessment team also will review the results of the statistical tests in light of missing data, outliers, and rejected data. The results of the statistical tests are then evaluated in terms of the project's acceptable decision error rates. The assessment team determines whether a decision could or could not be made, or why the decision could not be made, within the project specified decision error rates.

In summary, outputs from this step are:

- Generated statistical parameters;

- Graphical representations of the data set and parameters of interest;

- If new tests were selected, the rationale for selection and the reason for the inappropriateness of the statistical tests selected in the DQA plan;

- Results of application of the statistical tests; and

- A final determination as to whether the data are suitable for decisionmaking, estimating, or answering questions within the levels of certainty specified during planning.

9.7   Data Quality Assessment Report

The DQA process concludes with the assessment team documenting the output of the statistical tests and the rationale for why a decision could or could not be made, or why the decision could not be made within the project specified decision error rates. The DQA report will document findings and recommendations and include or reference the supporting data and information. The DQA report will summarize the use of the data verification and data validation reports for data sets of concern, especially if rejected for usability in the project's decisionmaking. The report also will document the answers to the three DQA questions:

- Are the samples representative?
- Are the data accurate?
- Can a decision be made?

Although there is little available guidance on the format for a DQA report, the report should contain, at a minimum:

- An executive summary that briefly answers the three DQA questions and highlights major issues, recommendations, deviations, and needed corrective actions;

- A summary of the project DQOs used to assess data usability, as well as pertinent documentation such as the project plan document, contracts, and SOW;

- A listing of those people who performed the DQA;

- A summary description of the DQA process, as employed, with a discussion of any deviations from the DQA plan designed during the planning process (the DQA plan should be appended to the report);

- A summary of the data verification and data validation reports that highlights significant findings and a discussion of their impact on data usability (the data verification and data validation reports should be appended to the DQA report);

- A discussion of any missing documentation or information and the impact of their absence on the DQA process and the usability of the data;

- A thorough discussion of the three DQA questions addressing the details considered in Sections 9.6.2 through 9.6.4 (possible outputs to be incorporated in the report are listed at the conclusion of each these section);

- A discussion of deviations, sampling, analytical and data management problems, concerns, action items, and suggested corrective actions (the contents of this section should be highlighted in the executive summary if the project is ongoing and corrections or changes are needed to improve the quality and usability of future data); and

- A recommendation or decision on the usability of the data set for the project's decision-making.

Upon completion, the DQA report should be distributed to the appropriate personnel as specified in the DQA plan and archived along with supporting information for the period of time specified in the project plan document. Completion of the DQA report concludes the assessment phase and brings the data life cycle to closure.

9.8   Summary of Recommendations

- MARLAP recommends that the assessment phase of a project (verification, validation, and DQA processes) be designed during the directed planning process and documented in the respective plans as part of the project plan documents.

- MARLAP recommends that project objectives, implementation activities, and QA/QC data be well documented in project plans, reports, and records, since the success of the assessment phase is highly dependent upon the availability of such information.

- MARLAP recommends the involvement of the data assessment specialist(s) on the project planning team during the directed planning process.

- MARLAP recommends that the DQA process should be designed during the directed planning process and documented in a DQA plan.

- MARLAP recommends that all sampling design and statistical assumptions be clearly identified in project plan documents along with the rationale for their use.

9.9   References

9.9.1   Cited Sources

American Society for Testing and Materials (ASTM) D6044. *Guide for Representative Sampling and Management of Waste and Contaminated Media.* 1996.

American Society for Testing and Materials (ASTM) D6233. *Standard Guide for Data Assessment for Environmental Waste Management Activities.* 1998.

U.S. Environmental Protection Agency (EPA). 2000a. *Guidance for the Data Quality Objective Process* (EPA QA/G-4). EPA/600/R-96/055, Washington, DC. Available from www.epa.gov/quality/qa_docs.html.

U.S. Environmental Protection Agency (EPA). 2000b. *Guidance for Data Quality Assessment: Practical Methods for Data Analysis* (EPA QA/G-9). EPA/600/R-96/084, Washington, DC. Available from www.epa.gov/quality/qa_docs.html.

International Organization for Standardization (ISO). 1993. *International Vocabulary of Basic and General Terms in Metrology.* ISO, Geneva, Switzerland.

MARSSIM. 2000. *Multi-Agency Radiation Survey and Site Investigation Manual, Revision 1.* NUREG-1575 Rev 1, EPA 402-R-97-016 Rev1, DOE/EH-0624 Rev1. August. Available from www.epa.gov/radiation/marssim/.

U.S. Army Corps of Engineers (USACE). 1998. *Technical Project Planning (TPP) Process.* Engineer Manual EM-200-1-2.

U.S. Nuclear Regulatory Commission (NRC). 1998. *A Nonparametric Statistical Methodology for the Design and Analysis of Final Status Decommissioning Surveys.* NUREG 1505, Rev. 1.

9.9.2   Other Sources

American Society for Testing and Materials (ASTM). 1997. *Standards on Environmental Sampling,* 2nd Edition, PCN 03-418097-38. West Conshohocken, PA.

American Society for Testing and Materials (ASTM) D5956. *Standard Guide for Sampling Strategies for Heterogeneous Wastes.* 1996.

American Society for Testing and Materials (ASTM) D6051. *Guide for Composite Sampling and Field Subsampling for Environmental Waste Management Activities*. 1996.

American Society for Testing and Materials (ASTM) D6311. *Standard Guide for Generation of Environmental Data Related to Waste Management Activities: Selection and Optimization of Sampling Design*.1998.

American Society for Testing and Materials (ASTM) D6323. *Standard Guide for Laboratory Subsampling of Media Related to Waste Management Activities*. 1998.

U. S. Environmental Protection Agency (EPA). 2002. *Guidance for Quality Assurance Project Plans*. EPA QA/G-5. EPA/240/R-02/009. Office of Environmental Information, Washington, DC. Available at www.epa.gov/quality/qa_docs.html.

Taylor, J. K. 1990. *Quality Assurance of Chemical Measurements*. Lewis, Chelsea, Michigan.

APPENDIX A
DIRECTED PLANNING APPROACHES

A.1 Introduction

There are a number of approaches being used for directed planning of environmental operations. Some of these approaches were designed specifically for data collection activities; others are applications of more general planning philosophies. Many variations to these approaches have been made for specific applications. The following are some of the approaches being used:

- Data Quality Objectives (DQO);
- Observational Approach (OA);
- Streamlined Approach for Environmental Restoration (SAFER);
- Technical Project Planning (TPP);
- Expedited Site Characterization (ESC);
- Value Engineering;
- Systems Engineering;
- Total Quality Management (TQM); and
- Partnering.

Employing any of these approaches assures that sufficient planning is carried out to define a problem adequately, determine its importance, and develop an approach to solutions prior to spending resources.

This appendix discusses some elements that are common to direct planning processes (Section A.2) and provides in Sections A.3 through A.11 very brief descriptions of the planning approaches listed above. References are listed at the end of the appendix on each of the approaches to provide sources of more detailed information.

Several directed planning approaches have been implemented by the federal sector for environmental data collection activities. Project planners should be aware of agency requirements for planning. MARLAP does not endorse any one planning approach. Users of MARLAP are encouraged to consider all the available approaches and choose a directed planning process that is appropriate to their project and agency.

Contents

A.1 Introduction A-1
A.2 Elements Common to Directed Planning
     Approaches A-2
A.3 Data Quality Objectives Process A-2
A.4 Observational Approach A-3
A.5 Streamlined Approach for Environmental
     Restoration A-4
A.6 Technical Project Planning A-4
A.7 Expedited Site Characterization A-4
A.8 Value Engineering A-5
A.9 Systems Engineering A-6
A.10 Total Quality Management A-6
A.11 Partnering A-7
A.12 References A-7

A.2 Elements Common to Directed Planning Approaches

To achieve the benefits desired from directed planning, all of these approaches address the following essential elements:

1. *Defining the problem or need*: Identifying the problem(s) facing the stakeholder/customer that requires attention, or the concern that requires streamlining.

2. *Establishing the optimum result*: Defining the decision, response, product, or result that will address the problem or concern and satisfy the stakeholder/customer.

3. *Defining the strategy and determining the quality of the solution*: Laying out a decision rule or framework, roadmap, or wiring diagram to get from the problem or concern to the desired decision or product and defining the quality of the decision, response, product, or result that will be acceptable to the stakeholder/customer by establishing specific, quantitative, and qualitative performance measures (e.g., acceptable error in decisions, defects in product, false positive responses).

4. *Optimizing the design*: Determining what is the optimum, cost-effective way to reach the decision or create the product while satisfying the desired quality of the decision or product.

To most problem solvers, these four elements stem from the basic tenets of the scientific method, which *Webster's* defines as "principles and procedures for the systematic pursuit of knowledge involving the recognition and formulation of a problem, the collection of data through observation and experiment, and the formulation and testing of hypotheses."

Each approach requires that a team of customers, stakeholders, and decision makers defines the problem or concern; a team of technical staff or line operators have the specific knowledge and expertise to define and then provide the desired product; and both groups work together to understand each other's needs and requirements and to agree on the product to be produced. The approaches represent slightly different creative efforts in the problem-solving process. All are intended to facilitate the achievement of optimum results at the lowest cost, generally using team work and effective communication to succeed.

A.3 Data Quality Objectives Process

The Data Quality Objectives (DQO) process was created by the U. S. Environmental Protection Agency to promote effective communications between decisionmakers, technical staff, and stakeholders on defining and planning the remediation of environmental problems.

The DQO process consists of seven basic steps:

1. State the problem;
2. Identify the decision;
3. Identify inputs to the decision;
4. Define the study boundaries;
5. Develop a decision rule;
6. Specify limits on decision errors; and
7. Optimize the design.

Applying the DQO steps requires effective communication between the parties who have the problem and the parties who must provide the solution. Additional information about the DQO Process is provided in Appendix B.

A.4 Observational Approach

The Observational Approach (OA) emphasizes determining what to do next by evaluating existing information and iterating between collecting new data and taking further action. The name "observational approach" is derived from observing parameters during implementation. OA was developed by Karl Terzaghi (Peck, 1969) for geological applications. In mining operations, there may be substantial uncertainty in the location of valuable geological formations. Information on soil and mineral composition would help to identify such formations. Application of OA utilizes the sampling information on soil and mineral composition to direct the digging locations. OA should be encouraged in situations where uncertainty is large, the vision of what is expected or required is poor, and the cost of obtaining more certainty is very high.

The philosophy of OA when applied to waste site remediation is that remedial action can be initiated without fully characterizing the nature and extent of contamination. The approach provides a logical decision framework through which planning, design, and implementation of remedial actions can proceed with increased confidence. OA incorporates the concepts of data sufficiency, identification of reasonable deviations, preparation of contingency plans, observation of the systems for deviations, and implementation of the contingency plans. Determinations of performance measures and the quality of new data are done as the steps are implemented.

The iterative steps of site characterization, developing and refining a site conceptual model, and identifying uncertainties in the conceptual model are similar to traditional approaches. The concept of addressing uncertainties as reasonable deviations is unique to OA and offers a qualitative description of data sufficiency for proceeding with site remediation.

A.5 Streamlined Approach for Environmental Restoration

The Streamlined Approach for Environmental Restoration (SAFER) is an integration of the DQO process and OA developed by the U. S. Department of Energy (DOE). The planning and assessment steps of SAFER are the DQO process. The implementation steps of SAFER are the Observational Approach. The approach emphasizing team work between decisionmakers and technical staff reduces uncertainty with new data collection and manages remaining uncertainty with contingency plans. The labels in each SAFER step are slightly different from the DQO and OA steps, but the basic logic is the same. The SAFER planning steps are:

- Develop a conceptual model;
- Develop remedial objectives and general response actions;
- Identify priority problem(s);
- Identify reasonable deviations and possible contingencies;
- Pursue limited field studies to focus and expedite scoping;
- Develop the decision rule;
- Establish acceptable conditions and acceptable uncertainty for achieving objective; and
- Design the work plan.

A.6 Technical Project Planning

Technical Project Planning (TPP) (formerly Data Quality Design), developed by the U. S. Army Corps of Engineers, is intended for developing data collection programs and defining data quality objectives for hazardous, toxic, and radioactive waste sites (HTRW). This systematic process (USACE, 1998) entails a four-phase planning approach in which a planning team—comprised of decisionmakers, data users, and data providers—identifies the data needed to support specific project decisions and develops a data collection program to obtain those data. In Phase I, an overall site strategy and a detailed project strategy are identified. The data user's data needs, including the level of acceptable data quality, are defined in Phase II. Phase III entails activities to develop sampling and analysis options for the data needed. During phase IV, the TPP team finalizes a data collection program that best meets the decisionmakers' short- and long-term needs within all project and site constraints. The technical personnel complete Phase IV by preparing detailed project objectives and data quality objectives, finalizing the scope of work, and preparing a detailed cost estimate for the data collection program. The TPP process uses a multi-disciplinary team of decisionmakers, data users, and data implementors focused on site closeout.

A.7 Expedited Site Characterization

Expedited Site Characterization (ESC) was developed to support DOE's Office of Science and Technology's Characterization, Monitoring, and Sensor Technology (CMST) program

(Burton, 1993). The ESC process has been developed by American Society for Testing and Materials (ASTM) as a provisional standard for rapid field-based characterization of soil and groundwater (ASTM D585). The process is also known as QUICKSITE and "expedited site conversion." ESC is based on a core multi-disciplinary team of scientists participating throughout the processes of planning, field implementation, data integration, and report writing. ESC requires clearly defined objectives and data quality requirements that satisfy the needs of the ESC client, the regulatory authority, and the stakeholders. The technical team uses real-time field techniques, including sophisticated geophysical and environmental sampling methods and an on-site analytical laboratory, to collect environmental information. Onsite computer support allows the expert team to analyze data each day and decide where to focus data collection the next day. Within a framework of an approved dynamic work plan, ESC relies on the judgment of the technical team as the primary means for selecting the type and location of measurements and samples throughout the ESC process. The technical team uses on-site data reduction, integration and interpretation, and on-site decisionmaking to optimize the field investigations.

Traditional site investigations generally are based on a phased engineering approach that collects samples based on a pre-specified grid pattern and does not provide the framework for making changes in direction in the field. A dynamic work plan (Robatt, 1997; Robatt et al., 1998) relies—in part—on an adaptive sampling and analysis program. Rather than specify the sample analyses to be performed, the number of samples to be collected and the location of each sample, dynamic work plans specify the decisionmaking logic that will be used in the field to determine where the samples will be collected, when the sampling will stop, and what analyses will be performed. Adaptive sampling and analysis programs change or adapt based on the analytical results produced in the field (Johnson, 1993a, b; Robatt, 1998).

A.8   Value Engineering

Value methodology was developed by Lawrence D. Miles in the late 1940s. He used a function-based process ("functional analysis") to produce goods with greater production and operational efficiency. Value methodology has evolved and, depending on the specific application, is often referred to as "value engineering," "value analysis," "value planning," or "value management." In the mid-1960s value engineering was adopted by three federal organizations: the Navy Bureau of Shipyards and Docks, the U. S. Army Corp of Engineers, and the U. S. Bureau of Reclamation. In the 1990s, Public Law 104-106 (1996) and OMB Circulars A-131 (1993) and A-11 (1997) set out the requirements for the use of value engineering, as appropriate, to reduce nonessential procurement and program costs.

Value engineering is a systematic and organized decision-making process to eliminate, without impairing essential functions, anything that increases acquisition, operation, or support costs. The techniques used analyze the functions of the program, project, system, equipment, facilities, services, or supplies to determine "best value," or the best relationship between worth and cost.

The method generates, examines, and refines creative alternatives that would produce a product or a process that consistently performs the required basic function at the lowest life-cycle cost and is consistent with required performance, reliability, quality, and safety.

A standard job plan is used to guide the process. The six phases of the value engineering job plan are:

- Information;
- Speculation (or creative);
- Evaluation (or analysis);
- Evolution (or development);
- Presentation (or reporting); and
- Implementation (or execution).

Value engineering can be used alone or with other management tools, such as TQM and Integrated Product and Process Development (IPPD).

A.9 Systems Engineering

Systems engineering brings together a group of multi-disciplinary team members in a structured analysis of project needs, system requirements and specifications, and a least-cost strategy for obtaining the desired results. Systems engineering is a logical sequence of activities and decisions that transforms an operational need into a preferred system configuration and a description of system performance parameters. Problem and success criteria are defined through requirements analysis, functional analysis, and systems analysis and control. Alternative solutions, evaluation of alternatives, selection of the best life-cycle balanced solution, and the description of the solution through the design package are accomplished through synthesis and systems analysis and control.

The systems engineering process involves iterative application of a series of steps:

- Mission analysis or requirements understanding;
- Functional analysis and allocation;
- Requirements analysis;
- Synthesis; and
- System analysis and control.

A.10 Total Quality Management

Total Quality Management (TQM) is a customer-based management philosophy for continuously improving the quality of products (or how work is performed) in order to meet customer expectations of quality and to measure and produce results aligned with strategic objectives. TQM grew out of two systems developed by Walter Shewhart of Bell Laboratories in the 1920s. Statistical process control was used to measure variance in production systems and to monitor consistency and diagnose problems in work processes. The "Plan-Do-Check-Act" cycle applied a systematic approach to improving work processes. The work of Deming and others in Japan following World War II expanded the quality philosophy beyond production and inspection to all functions within an organization and defined quality as "fit for customer use."

TQM has been defined as "the application of quantitative methods and the knowledge of people to assess and improve (a) materials and services supplied to the organizations, (b) all significant processes within the organization, and (c) meeting the needs of the end-user, now and in the future" (Houston and Dockstader, 1997). The goal of TQM is to enhance effectiveness of providing services or products. This is achieved through an objective, disciplined approach to making changes in processes that affect performance. Process improvement focuses on preventing problems rather than fixing them after they occur. TQM involves everyone in an organization in controlling and continuously improving how work is done.

A.11 Partnering

Partnering is intended to bring together parties that ordinarily might have differing or competing interests to create a synergistic effect on an outcome each views as desirable. Partnering is a team building and relationship enhancing technique that seeks to identify and communicate the needs, expectations, and strengths of the participants. Partnering combines the talents of the participating organizations in order to develop actions that promote their common goals and objectives. In the synergistic environment of partnering, creative solutions to problems can be developed. Like TQM, partnering enfranchises all stakeholders (team members) in the decision process and holds them accountable for the end results. Each team member (customer, management, employee) agrees to share the risks and benefits associated with the enterprise. Like the other approaches, partnering places a premium on open and clear communication among stakeholders to define the problem and the solution, and to decide upon a course of action.

A.12 References and Other Sources

A.12.1 Data Quality Objectives

Guidance:

American Society for Testing and Materials (ASTM). D5792. *Standard Practice for Generation of Environmental Data Related to Waste Management Activities: Development of Data Quality Objectives*. West Conshohocken, PA.

U. S. Environmental Protection Agency (EPA). 1993. *Data Quality Objectives Process for Superfund.* EPA/540/G-93/071 (Interim Final Guidance). Office of Emergency and Remedial Response. OSWER Directive 9355.9-01. September.

U.S. Environmental Protection Agency (EPA). 2000. *Guidance for the Data Quality Objective Process* (EPA QA/G-4). EPA/600/R-96/055, Washington, DC. Available at www.epa.gov/quality/qa_docs.html.

Papers:

Blacker, S. M. 1993. "The Data Quality Objective Process—What It Is and Why It Was Created." *Proceedings of the Twentieth Annual National Energy and Environmental Quality Division Conference,* American Society for Quality Control.

Blacker, S. and D. Goodman. 1994a. "Risk-Based Decision Making An Integrated Approach for Efficient Site Cleanup." *Environmental Science & Technology,* 28:11, pp. 466A-470A.

Blacker, S. and D. Goodman. 1994b. "Risk-Based Decision Making Case Study: Application at a Superfund Cleanup." *Environmental Science & Technology,* 28:11, pp. 471A-477A.

Blacker, S. M. and P. A. Harrington. 1994. "Use of Process Knowledge and Sampling and Analysis in Characterizing FFC Act Waste — Applying the Data Quality Objective (DQO) Process to Find Solutions." *Proceedings of the Twenty First Annual National Energy and Environmental Quality Division Conference,* American Society for Quality Control.

Blacker, S. M. and J. Maney. 1993. "The System DQO Planning Process." *Environmental Testing and Analysis.* July/August.

Blacker, S. M., J. D. Goodman and J. M. Clark. 1994. "Applying DQOs to the Hanford Tank-Waste Remediation." *Environmental Testing and Analysis,* 3:4, p. 38.

Blacker, S., D. Neptune, B. Fairless and R. Ryti. 1990. "Applying Total Quality Principles to Superfund Planning." *Proceedings of the 17th Annual National Energy Division Conference,* American Society for Quality Control.

Carter, M. and D. Bottrell. 1994. "Report on the Status of Implementing Site-Specific Environmental Data Collection Project Planning at the Department of Energy's (DOE) Office of Environmental Restoration and Waste Management (EM)." *Proceedings of the Waste Management '94 Conference.* Vol 2, pp. 1379-1383.

Goodman, D. and S. Blacker. 1997. "Site Cleanup: An integrated Approach for Project Optimization to Minimize Cost and Control Risk." In: *The Encyclopedia of Environmental Remediation*. New York: John Wiley & Sons.

Michael, D. I. 1992. "Planning Ahead to Get the Quality of RI Data Needed for Remedy Selection: Applying the DQO Process to Superfund Remedial Investigations." *Proceedings of the Air and Waste Management Association 85ᵗʰ Annual Meeting*.

Michael, D. I. and E. A. Brown. 1992. "Planning Tools that Enhance Remedial Decision Making." *Proceedings of the Nineteenth Annual Energy and Environmental Quality Division Conference*, American Society for Quality Control.

Neptune, M. D. and S. M. Blacker. 1990. "Applying Total Quality Principles to Superfund Planning: Part I: Upfront Planning in Superfund." *Proceedings of the 17ᵗʰ Annual National Energy Division Conference*, American Society for Quality Control.

Neptune, D., E. P. Brantly, M. J. Messner and D. I. Michael. 1990. "Quantitative Decision-Making in Superfund: A Data Quality Objectives Case Study." *Hazardous Material Control*, 3, pp. 18-27.

Ryti, R. T. and D. Neptune. 1991. "Planning Issues for Superfund Site Remediation." *Hazardous Materials Control*, 4, pp. 47-53.

A.12.2 Observational Approach

Papers:

Brown, S. M. 1990. "Application of the Observational Method to Groundwater Remediation." *Proceedings of HAZMAT'90*, Atlantic City, NJ.

Ferguson, R. D., G. L. Valet, and F. J. Hood. 1992. *Application of the Observational Approach, Weldon Springs Case Study*.

Mark, D. L. et al. 1989. "Application of the Observational Method to an Operable Unit Feasibility Study - A Case Study." *Proceedings of Superfund'89*, Hazardous Material Control Research Institute, Silver Spring, MD, pp. 436-442.

Myers, R. S. and Gianti, S. J. 1989. "The Observational Approach for Site Remediation at Federal Facilities." *Proceedings of Superfund'89*, Hazardous Material Control Research Institute, Silver Spring, MD.

Peck, R. B. 1969. "Ninth Rankine Lecture, Advantages and Limitations of the Observational Method in Applied Soil Mechanics." *Geotechnique*, 19, No. 2, pp.171-187.

Smyth, J. D. and R. D. Quinn. 1991. "The Observational Approach in Environmental Restoration." *Proceedings of the ASCE National Conference of Environmental Engineering*, Reno, NV.

Smyth, J. D., J. P. Amaya and M. S. Peffers. 1992. "DOE Developments: Observational Approach Implementation at DOE Facilities." *Federal Facilities Environmental Journal*, Autumn, pp. 345-355.

Smyth, J. D., J. P. Kolman, and M. S. Peffers. 1992. "Observational Approach Implementation Guidance: Year-End Report." Pacific Northwest Laboratory Report PNL-7999.

A.12.3 Streamlined Approach for Environmental Restoration (Safer)

Guidance:

U. S. Department of Energy (DOE). 1993. *Remedial Investigation/Feasibility Study (RI/FS) Process, Elements and Techniques Guidance, Module 7 Streamlined Approach for Environmental Restoration*, Office of Environmental Guidance, RCRA/CERCLA Division and Office of Program Support, Regulatory Compliance Division Report DOE/EH-94007658.

Papers:

Bottrell, D. 1993. "DOE's Development and Application of Planning to Meet Environmental Restoration and Waste Management Data Needs." *Proceedings of the Twentieth Annual National Energy & Environmental Quality Division Conference*, American Society for Quality Control.

Dailey, R., D. Lillian and D. Smith. 1992. "Streamlined Approach for Environmental Restoration (SAFER): An Overview." Proceedings of the 1992 Waste Management and Environmental Sciences Conference.

Gianti, S., R. Dailey, K. Hull and J. Smyth. 1993. "The Streamlined Approach For Environmental Restoration." *Proceedings of Waste Management '93*, 1, pp. 585-587.

Smyth, J. D. and J. P. Amaya. 1994. *Streamlined Approach for Environmental Restoration (SAFER): Development, Implementation and Lessons Learned*. Pacific Northwest Laboratory Report PNL-9421/UC-402, Richland, WA.

A.12.4 Technical Project Planning

Guidance:

U. S. Army Corps of Engineers (USACE). 1995. *Technical Project Planning Guidance for Hazardous, Toxic and Radioactive Waste (HTRW) Data Quality Design.* Engineer Manual EM-200-1-2 (superceded by EM-200-1-2, 1998).

U. S. Army Corps of Engineers (USACE). 1998. *Technical Project Planning Process.* Engineer Manual EM-200-1-2.

A.12.5 Expedited Site Characterization

Guidance:

American Society for Testing and Materials (ASTM) D585. *Standard Provisional Guide for Expedited Site Characterization of Hazardous Waste Contaminated Sites.* West Conshohocken, PA.

Papers:

Bottrell, D. 1993. "DOE's Development and Application of Planning Processes to Meet Environmental Restoration and Waste Management Data Needs." *Proceedings of the Twentieth Annual National Energy & Environmental Quality Division Conference,* American Society for Quality Control.

Burton, J. C., et al. 1993. "Expedited Site Characterization: A Rapid Cost-Effective Process for Preremedial Site Characterization." *Proceeding of Superfund XIV,* Vol. II, Hazardous Materials Research and Control Institute, Greenbelt, MD, pp. 809-826.

Burton, J. C. 1994. "Expedited Site Characterization for Remedial Investigations at Federal Facilities." *Proceedings Federal Environmental Restoration III and Waste Minimization II Conference,* Vol. II, pp. 1407-1415.

Johnson, R. 1993a "Adaptive Sampling Program Support for Expedited Site Characterization." *ER'93 Environmental Remediation Conference Proceedings.*

Johnson, R. 1993b. "Adaptive Sampling Program Support for the Unlined Chromic Acid Pit, Chemical Waste Landfill, Sandia National Laboratory, Albuquerque, New Mexico." ANL-EAD/TM-2.

Robatt, A. 1997. "A Guideline for Dynamic Work Plans and Field Analytics: The Keys to Cost Effective Site Cleanup." Tufts University Center for Field Analytical Studies and Technology and U.S. EPA, Region 1, Hazardous Waste Division.

Robatt, A. 1998. "A Dynamic Site Investigation: Adaptive Sampling and Analysis Program for Operable Unit 1 at Hanscom Air Force Base, Bedford, Massachusetts." Tufts University Center for Field Analytical Studies and Technology and U.S. EPA, Region 1, Office of Site Remediation and Restoration, Boston, MA.

Robbat, A., S. Smarason, and Y. Gankin. 1998. "Dynamic Work Plans and Field Analytics, The Key to Cost-Effective Hazardous Waste Site Investigations," *Field Analytical Chemistry and Technology* 2:5, pp. 253-65.

Starke, T. P., C. Purdy, H. Belencan, D. Ferguson and J. C. Burton. 1995. "Expedited Site Characterization at the Pantex Plant." *Proceedings of the ER'95 Conference*.

A.12.6 Value Engineering

Guidance:

The February 1996 Amendment to the Office of Federal Procurement Policy Act (41 U.S.C. 401 et. seq.) (Public Law 104-106, Sec 4306 amended this.)

Federal Acquisitions Regulations. FAR, Part 48, Value Engineering.

Federal Acquisitions Regulations. FAR, Part 52.248-1,-2,-3, Value Engineering Solicitation Provisions and Contract Clauses.

National Defense Authorization Act for Fiscal Year 1996. PL 104-106, Law Requiring Value Engineering in Executive Agencies. February 10, 1996.

Office of Management and Budget (OMB). 1993. *OMB Circular A-131, Value Engineering.*

Office of Management and Budget (OMB). 1997. *OMB Circular A-11, Preparation and Submission of Budget Estimates.*

U. S. Army. Value Engineering. *Army Regulation AR 5-4*, Chapter 4 (Reference only).

U. S. Army Corps of Engineers (USACE). *Engineer Regulation.* ER 5-1-11.

U. S. Department of Energy. 1997. *Value Management.* Good Practice Guide (GPG-FM-011).

U. S. Department of the Interior (DOI). 1995. *Departmental Manual, Management Systems and Procedures*, Part 369, Value Engineering, Chapter 1, General Criteria and Policy. May 18, 1995.

Books:

Fallon, C. 1990. *Value Analysis*. The Miles Value Foundation, 2nd Edition.

Kauffman, J. J. 1985. *Value Engineering for the Practitioner*. North Carolina State University, Raleigh, NC.

Miles, L. D. 1989. *Techniques of Value Analysis and Engineering*. McGraw-Hill Book Company, New York, NY.

Mudge, A. E. 1989. *Value Engineering, A Systematic Approach*. J. Pohl Associates.

Parker, D. 199x. *Value Engineering Theory*. The Miles Value Foundation.

Papers:

Al-yousefi, A. 1996. "Total Value Management (TVM): A VE-TQM Integration," *Proceedings of the 1996 SAVE Conference*, Society of American Value Engineers.

Blumstein, G. 1996. "FAST Diagramming: A Technique to Facilitate Design Alternatives," *Proceedings of the 1996 SAVE Conference*, Society of American Value Engineers.

Maynor, D. 1996. *Value Engineering for Radiation Hazards Remediation at the Fernald OU1, Ohio*. U.S. DOE, Ohio Field Office.

Morrel, C. 1996. *Value Engineering for Radiation Hazards Remediation at Fernald OU4, Ohio*. U.S. DOE Reclamation Technical Service Center.

Wixson, J. R. 1987. "Improving Product Development with Value Analysis/Value Engineering: A Total Management Tool," *Proceedings of the Society of American Value Engineers*, 22, pp.51-66.

A.12.7 Systems Engineering

Guidance:

Electronic Industries Alliance (EIA). 1994. *Systems Engineering*. Standard EIA/IS-632.

Electronic Industries Alliance (EIA). 1997. *Upgrade IS-632, Process for Engineering a System.* EIA/SP-3537 Part 1: Process Characteristics and EIA/SP-4028 Part 2: Implementation Guidance.

International Electrical and Electronics Engineers (IEEE). 1994. *Standard for Application and Management of the Systems Engineering Process.* P1220.

U. S. Department of Defense (DOD). 1992. *Systems Engineering.* MIL-STD-499B.

U. S. Department of Energy (DOE). 1996. *Project Execution and Engineering Management Planning.* Good Practice Guide GPG-FM-010.

Books:

Boardman, J. 1990. *Systems Engineering: An Introduction.* New York: Prentice Hall.

Chestnut, H. 1967. *System Engineering Methods.* New York: John Wiley & Sons.

Churchman, C. W. 1968. *The Systems Approach.* New York: Dell Publishing Co., Inc.

Eisner, H. 1998. *Computer-Aided Systems Engineering (CASE).* Englewood Cliffs: Prentice Hall.

Goode, H. H. 1957. *Systems Engineering: An Introduction to the Design of Large-Scale Systems.* New York: McGraw-Hill.

Machol, R. E. 1965. *Systems Engineering Handbook.* New York: McGraw-Hill.

Smith, D. B. 1974. *Systems Engineering and Management.* Reading, MA: Addison-Wesley Publ. Co.

Wymore, A. W. 1976. *Systems Engineering Methodology for Interdisciplinary Teams.* New York: John Wiley & Sons.

Papers:

Bensoussan, A. 1982. "Analysis and Optimization of Systems." *Proceedings of the Fifth International Conference on Analysis and Optimization of Systems,* Versailles, France. December 14-17.

David, H.T. and S. Yoo. 1993. "Where Next? Adaptive Measurement Site Selection for Area Remediation." In: *Environmental Statistics, Assessment and Forecasting* (Richard Cathern, Ed.). Lewis Publishers, MI.

Ljunggren M. and J Sundberg. 1996. "A Systems Engineering Approach to National Solid Waste Management -- Case Study, Sweden." *Proceedings of the 12th International Conference on Solid Waste Management*. November 17-20.

Pacific Northwest Laboratory. 1995. *A Systems Engineering Analysis to Examine the Economic Impact for Treatment of Tritiated Water in the Hanford K-Basin*. Report No. PNL-SA-24970. Richland, WA.

A.12.8 Total Quality Management

Guidance:

U. S. Department of the Army. 1992. *The Leadership for Total Army Quality Concept Plan.*

U. S. Department of Energy (DOE). 1993. *Total Quality Management Implementation Guidelines*. DOE/HR-0066.

U. S. Office of Personnel Management (OPM), Federal Quality Institute. 1990. *Federal Total Quality Management Handbook, How to Get Started, Booklet 1: Implementing Total Quality Management*, U. S. Government Printing Office.

Books:

Berk, J. and S. Berk. 1993. *Total Quality Management: Implementing Continuous Improvement*. Sterling Publishing Co. Inc., New York, NY.

Carr, D. K. and I. D. Littman. 1993. *Excellence in Government*. Coopers and Lybrand, Arlington, VA.

Dobyns, L. and C. Crawford-Mason. 1994. *Thinking about Quality: Progress, Wisdom and the Deming Philosophy*. New York: Times Books.

Harrington, H. J. 1991. *Business Process Improvement*. New York: McGraw-Hill.

Koehler, J. W. and J. M. Pankowski. 1996. *Quality Government. Designing, Developing and Implementing TQM*. Delray Beach, FL: St. Lucie Press.

Rao, A, L.P. Carr, I. Dambolena, R.J. Kopp, J. Martin, F. Rafii, and P.F. Schlesinger. 1996. *Total Quality Management: A Cross Functional Perspective*. New York: John Wiley & Sons.

Walton, M. 1990. *Deming Management at Work*. New York: Putnam.

Papers:

Blacker, S. 1990. "Applying Total Quality Concepts to Environmental Data Operations." *Proceedings of the Eighth International Conference*, International Society for Quality Control.

Breisch, R.E. 1996. "Are You Listening?" *Quality Progress*, pp. 59-62.

Houston, A. and Dockstader, S. L. 1997. *Total Quality Leadership: A Primer*. Department of the Navy, Total Quality Leadership Office Publication Number 97-02.

Kidder, P. J. and B. Ryan. 1996. "How the Deming Philosophy Transformed the Department of the Navy." *National Productivity Review* 15:3.

A.12.9 Partnering

Guidance:

U. S. Department of the Army. 1993. Engineering and Design Quality Management, Appendix B Partnering. ER-1110-1-12.

Books:

Hrebniak, L. 1994. *We Force in Management: How to Build and Sustain Cooperation*. New York: Free Press.

Maurer, R. 1992. *Caught in the Middle: A Leadership Guide for Partnership in the Workplace*. Portland, OR: Productivity Press.

Poirier, C. C. 1994. *Business Partnering for Continuous Improvement: How to Forge Enduring Alliances Among Employees, Suppliers, and Customers*. New York: Berrett-Koehler.

Papers:

Brown, T. L. 1993. "Is there Power in Partnering?" *Industry Week*, 242:9, p. 13.

Covey, S. R. 1993. "Win-Win Partnerships." *Executive Excellence*, 10:11, pp. 6-7.

Chem-Nuclear Systems, Inc. (CNSI). 1996. *Community Partnering Plan: Pennsylvania Low-Level Radioactive Waste Disposal Facility*. S80-PL-021, Revision 0. Commonwealth of Pennsylvania, Department of Environmental Protection, Bureau of Radiation.

Mosley, D. and C. C. Moore. 1994. "TQM and Partnering: An Assessment of Two Major Change Strategies." *PMNETwork*, 18:9, pp. 22-26.

Sanders, S. R. and M. M. Moore. 1992. "Perceptions on Partnering in the Public Sector." *Project Management Journal*, 23:4, pp. 13-19.

Simmons, J. 1989. "Partnering Pulls Everything Together." *Journal for Quality & Participation*, 12, pp. 12-16.

U. S. Army Corps of Engineers (USACE). 1996. "U.S. Corps of Engineers Adopts Partnering." National Academy of Public Administration Foundation, Washington, DC.

... ...

... ...
... ...

APPENDIX B
THE DATA QUALITY OBJECTIVES PROCESS

B.1   Introduction

This appendix provides information about the basic framework of the DQO process (ASTM 5792; EPA, 2000; NRC, 1998; MARSSIM, 2000). The DQO planning process empowers both data users and data suppliers to take control and resolve issues in a stepwise fashion. It brings together at the right time all key players from the data user and data supplier constituencies and enables each participant to play a constructive role in clearly defining:

- The problem that requires resolution;
- What type, quantity, and quality of data the decisionmaker needs to resolve that problem;
- Why the decisionmaker needs that type and quality of data;
- How much risk of making a wrong decision is acceptable; and
- How the decisionmaker will use the data to make a defensible decision.

The DQO process provides a logic for setting well-defined, achievable objectives and developing a cost-effective, technically sound sampling and analysis design. It balances the data user's tolerance for uncertainty with the available resources for obtaining data. The number of visible and successful applications of the DQO process has proven its value to the environmental community. The DQO process is adaptable depending on the complexity of the project and the input from the decisionmakers. Some users have combined DQO planning with remedy selection for restoration projects (e.g., DOE's Streamlined Approach for Environmental Restoration—see Section A.5 in Appendix A). Other users have integrated their project scoping meetings with the DQO process. Much of the information that is developed during the DQO process is useful for developing the project plan documents (Chapter 4) and implementing the data validation process (Chapter 8) and the data quality assessment (DQA) process (Chapter 9).

Since its inception, the term "data quality objectives" has been adopted by many organizations, and the definition has been adapted and modified (see box on next page). Throughout this document, MARLAP uses EPA's (2000) definition of DQOs: "Qualitative and quantitative statements derived from the DQO process that clarify study objectives, define the appropriate type of data, and specify the tolerable levels of potential decision errors that will be used as the basis for establishing the quality and quantity of data needed to support decisions."

Contents

B.1 Introduction B-1
B.2 Overview of the DQO Process B-2
B.3 The Seven Steps of the DQO Process B-3
B.4 References B-24
Attachment B1: Decision Error Rates and the Gray
    Region for Decisions About Mean
    Concentrations B-26
Attachment B2: Decision Error Rates and the Gray
    Region for Detection Decisions B-36

Definitions of Data Quality Objectives

(1) Statements on the level of uncertainty that a decisionmaker is willing to accept in the results derived from environmental data (ASTM 5283; EPA, 1986).

(2) Qualitative and quantitative statements derived from the DQO process that clarify study objectives, define the appropriate type of data, and specify the tolerable levels of potential decision errors that will be used as the basis for establishing the quality and quantity of data needed to support decisions (EPA, 2000).

(3) Qualitative and quantitative statements derived from the DQO process describing the decision rules and the uncertainties of the decision(s) within the context of the problem(s) (ASTM D5792).

(4) Qualitative and quantitative statements that specify the quality of the data required to support decisions for any process requiring radiochemical analysis (radioassay) (ANSI N42.23).

B.2 Overview of the DQO Process

The DQO process (Figure B.1) consists of seven steps (EPA, 2000). In general, the first four steps require the project planning team to define the problem and qualitatively determine required data quality. The next three steps establish quantitative performance measures for the decision and the data. The final step of the process involves developing the data collection design based on the DQOs, which is dependent on a clear understanding of the first six steps.

Although the DQO process is described as a sequence of steps, it is inherently iterative. The output from each step influences the choices that will be made in subsequent steps. For instance, a decision rule cannot be created without first knowing the problem and desired decision. Similarly, optimization of the sampling and analysis design generally cannot occur unless it is clear what is being optimized —the results of the preceding steps. Often the outputs of one step will trigger the need to rethink or address issues that were not evaluated thoroughly in prior steps. These iterations lead to a more focused sampling and analysis design for resolving the defined problem. The

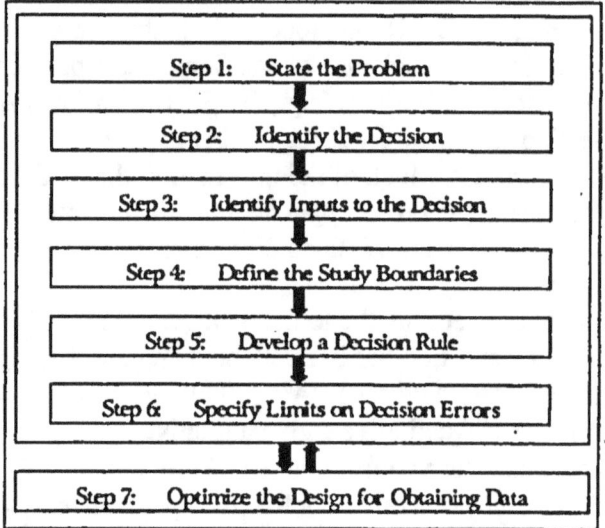

FIGURE B.1 — Seven steps of the DQO process

first six steps should be completed before the sampling and analysis design is developed, and every step should be completed before data collection begins. The DQO process is considered complete with the approval of an optimal design for sampling and analysis to support a decision or when available historical data are sufficient to support a decision.

In practice, project planning teams often do a cursory job on the first four steps, wanting to get into technical design issues immediately. Without carefully defining the problem and the desired result, the project planning team may develop a design that is technically sound but answers the wrong question, or answers the questions only after the collection of significant quantities of unnecessary data. Time spent on the first four steps is well spent. Extra effort must be given to assure that Steps 1 to 4 are adequately addressed.

When applying the DQO process, or any planning approach, it is important to document the outputs of each step to assure that all participants understand and approve the interim products, and that they have a clear record of their progress. It is sometimes useful to circulate an approval copy with signature page to ensure agreement of the stakeholders.

B.3 The Seven Steps of the DQO Process

Each step of the DQO process will be discussed in the following sections. Not all items will be applicable to every project. The project planning team should apply the concepts that are appropriate to the problem.

B.3.1 DQO Process Step 1: State the Problem

The first step is to define the problem clearly. The members of the project planning team present their concerns, identify regulatory issues and threshold levels, and review the site history. The project planning team should develop a concise description of the problem. Some elements to include in the description might be the study objectives, regulatory context, groups who have an interest in the study, funding and other resources available, previous study results, and any obvious sampling design constraints. The more facts, perceptions and concerns of the key stakeholders—including important social, economic, or political issues—that are identified during this step, the better the chances are that the issues driving the decisions and actions will be identified.

The primary decisionmaker should be identified. The resources and relevant deadlines to address the problem are also defined at this time. If possible, a "project conceptual model" should be developed. This will help structure and package the diverse facts into an understandable picture of what the various issues are and how those issues can be focused into a specific problem. The expected outputs of Step 1 are:

- A conceptual model that packages all the existing information into an understandable picture of the problem;

- A list of the project planning team members and identification of the decisionmaker;

- A concise description of the problem; and

- A summary of available resources and relevant deadlines for the study.

B.3.2 DQO Process Step 2: Identify the Decision

During Step 2 of the DQO process, the project planning team defines what decision must be made or what question the project will attempt to resolve. The decision (or question) could be simple, like whether a particular discharge is or is not in compliance, or the decision could be complex, such as determining if observed adverse health is being caused by a nonpoint source discharge. Linking the problem and the decision focuses the project planning team on seeking only that information essential for decisionmaking, saving valuable time and money.

The result may be a comprehensive decision for a straightforward problem, or a sequence of decisions for a complex problem. For complex problems with multiple concerns, these concerns should be ranked in order of importance. Often a complex concern is associated with a series of decisions that need to be made. Once these decisions have been identified, they should be sequenced in a logical order so the answer to one decision provides input in answering the next decision. It may be helpful to develop a logic-flow diagram (decision framework), arraying each element of the issue in its proper sequence along with its associated decision that requires an answer.

The term "action level" is used in this document to denote the numerical value that will cause the decisionmaker to choose one of the alternative actions. The action level may be a derived concentration guideline level, background level, release criteria, regulatory decision limit, etc. The action level is often associated with the type of media, analyte, and concentration limit. Some action levels, such as release criteria for license termination, are expressed in terms of dose or risk. The release criteria typically are based on the total effective dose equivalent (TEDE), the committed effective dose equivalent (CEDE), risk of cancer incidence (morbidity), or risk of cancer death (mortality), and generally cannot be measured directly. A radionuclide-specific predicted concentration or surface area concentration of specific nuclides that can result in a dose (TEDE or CEDE) or specific risk equal to the release criterion is called the "derived concentration guideline level" (DCGL). A direct comparison can be made between the project's analytical measurements and the DCGL (MARSSIM, 2000).

The project planning team should define the possible actions that may be taken to solve the problem. Consideration should be given to the option of taking no action. A decision statement can then be developed by combining the decisions and the alternative actions. The decision rule and the related hypothesis test will be more fully developed in the DQO process at Steps 5 and 6.

By defining the problem and its associated decision clearly, the project planning team has also begun to define the inputs and boundaries (DQO process Steps 3 and 4). At the end of Step 2, the project planning team has:

- Identified the principal decisions or questions;

- Defined alternative actions that could be taken to solve the problem based on possible answers to the principal decisions and questions;

- Combined the principal decisions and questions and the alternative actions into decision statements that expresses a choice among alternative actions; and

- Organized multiple decisions.

B.3.3 DQO Process Step 3: Identify Inputs to the Decision

During Step 3, the project planning team makes a formal list of the specific information required for decisionmaking. The project planning team should determine what information is needed and how it can be acquired. The project planning team should specify if new measurements are required for the listed data requirements. The data required are based on outcomes of discussion during the previous two steps. The project planning team should define the basis for setting the action level. Depending on the level of detail of the discussion during the previous steps, then efforts associated with Step 3 may be primarily to capture that information. If the first two steps have not defined the inputs with enough specificity, then those inputs should be defined here. However, before going further, the output should be reviewed to assure that the problem, the decision steps and the input are compatible in complete agreement.

An important activity during Step 3 is to determine if the existing data or information, when compared with the desired information, has significant gaps. If no gaps exist, then the existing data or information may be sufficient to resolve the problem and make the decision. (Although there may be no gaps in the data, the data may not have enough statistical power to resolve the action level. See Step 6 for more discussion.) In order to optimize the use of resources, the project planning team should maximize the use of historical information. If new data are required, then this step establishes what new data (inputs) are needed. The specific environmental variable or characteristic to be measured should be identified. The DQO process clearly links sampling and analysis efforts to an action and a decision. This linkage allows the project planning team to determine when enough data have been collected.

If the project planning team determines that collection of additional data is needed, the analytical laboratory acquisition strategy options should be considered at this stage. Identifying suitable contracting options should be based on the scope, schedule, and budget of the project, and the capability and availability of laboratory resources during the life of the project, and other technical considerations of the project. If an ongoing contract with a laboratory is in place, it is advisable to involve them with the radioanalytical specialists as early as possible.

The project planning team should ensure that there are analytical protocols available to provide acceptable measurements. If analytical methods do not exist, the project planning team will need to consider the resources needed to develop a new method, reconsider the approach for providing input data, or perhaps reformulate the decision statement.

The expected outputs of Step 3 are:

- A list of information needed for decisionmaking;
- Determination of whether data exist and are sufficient to resolve the problem;
- Determination of what new data, if any, are required;
- Definition of the characteristics that define the population and domain of interest;
- Definition of the basis for the action level;
- Confirmation that appropriate analytical protocols exist to provide the necessary data; and
- A review of the planning output to assure the problem, decision, and inputs are fully linked.

B.3.4 DQO Process Step 4: Define the Study Boundaries

In Step 4, the project planning team specifies the spatial and temporal boundaries covered by the decision statement. The spatial boundaries define the physical aspects to be studied in terms of geographic area, media, and any appropriate subpopulations (e.g., an entire plant, entire river basin, one discharge, metropolitan air, emissions from a power plant). When appropriate, divide the population into strata that have relatively homogeneous characteristics. The temporal boundaries describe the time frame the study data will represent (e.g., possible exposure to local residents over a 30-year period) and when samples should be taken (e.g., instantaneous samples, hourly samples, annual average based on monthly samples, samples after rain events). Changing conditions that could impact the success of sampling and analysis and interpretation need to be considered. These factors include weather, temperature, humidity, or amount of sunlight and wind.

The scale of the decision is also defined during this step. The selected scale should be the smallest, most appropriate subset of the population for which decisions will be made based on the spatial or temporal boundaries. During Step 4, the project planning team also should identify practical constraints on sampling and analysis that could interfere with full implementation of the data collection design. These include time, personnel, equipment, and seasonal or meteorological conditions when sampling is not possible or may bias the data.

In practice, the study boundaries are discussed when the project planning team and decision-maker agree on the problem and its associated decision. For instance, a land area that may be contaminated or a collection of waste containers would be identified as part of the problem and decision definition in Steps 1 and 2. The boundaries also would be considered when determining inputs to the decision in Step 3. If the study boundaries had not been addressed before Step 4 or if new issues were raised during Step 4, then Steps 1, 2, and 3 should be revisited to determine how Step 4 results are now influencing the three previous steps.

The outputs of Step 4 are:

- A detailed description of the spatial and temporal boundaries of the problem; and
- Any practical constraints that may interfere with the sampling and analysis activities.

B.3.5 Outputs of DQO Process Steps 1 through 4 Lead Into Steps 5 through 7

At this stage in the DQO process, the project planning team has defined with a substantial degree of detail the problem, its associated decision, and the inputs and boundaries for addressing that problem. The project planning team knows whether it needs new data to fill specific gaps and what that data should be. The remaining three steps are highly technical and lead to the selection of the sampling and analysis design. Even when new data are not required (i.e., a data collection design is not needed), the project planning team should continue with Steps 5 and 6 of the DQO process. By establishing the formal decision rule and the quantitative estimates of tolerable decision error rates, the project planning team is assured that consensus has been reached on the actions to be taken and information to establish criteria for the DQA process.

It is important to emphasize that every effort must be made to assure that Steps 1 through 4 are adequately addressed. If the necessary time is taken in addressing the first four steps carefully and assuring consensus among the project planning team, then the three remaining steps are less difficult.

B.3.6 DQO Process Step 5: Develop a Decision Rule

In Step 5, the project planning team determines the appropriate statistical parameter that characterizes the population, specifies the action level, and integrates previous DQO process outputs into a single "if ..., then ..." statement (called a "decision rule") that describes a logical basis for choosing among alternative actions.

The four main elements to the decision rule are:

A. THE PARAMETER OF INTEREST. A descriptive measure (e.g., mean, median, or proportion) that specifies the characteristic or attribute that the decisionmaker would like to know and that the data will estimate. The characteristics that define the population and domain of interest was established in Step 3.

B. THE SCALE OF DECISIONMAKING. The smallest, most appropriate subset for which decisions will be made. The scale of decisionmaking was defined in Step 4.

C. THE ACTION LEVEL. A threshold value of the parameter of interest that provides the criterion for choosing among alternatives. Action levels may be based on regulatory standards or they may be derived from project- and analyte-specific criteria such as dose or risk analysis. The basis for the action level was determined in Step 3.

D. THE ALTERNATIVE ACTIONS. The actions the decisionmaker would take, depending on the "true value" of the parameter of interest. The alternative actions were determined in Step 2.

The decision rule is a logical, sequential set of steps to be taken to resolve the problem. For example, "If one or more conditions exists then take action 1, otherwise take action 2."

The outputs of Step 5 are:

- The action level;
- The statistical parameter of interest; and
- An "if ..., then ..." statement that defines the conditions that would cause the decisionmaker to choose among alternative courses of action.

PROCEDURE FOR DEVELOPING A DECISION RULE

The outcome of a decision rule is a result: often to take action or not to take action. The decision rule is an "If..., then..." statement that defines the conditions that would cause the decisionmaker to choose an action. The decision rule establishes the exact criteria for making that choice. There are four main elements to a decision rule:

A. The *parameter of interest*. For example, the mean or median of the concentration of an analyte.
B. The *area over which the measurements are taken*. For example, in MARSSIM, a survey unit.
C. The *action level*. For example, in MARSSIM, the action level is called the DCGL.
D. *Alternative actions*. For example, if the mean is greater than the action level, then corrective action must be taken, otherwise the survey unit may be released.

A decision rule is action oriented, so a decision rule has the general form:

> If the value of parameter A, over the area B, is greater than C, then take action D, otherwise take action D*.

For example, if:

(A)   the true *mean concentration of* ^{238}U in the
(B)   *surface soil* of the survey unit is greater than

(C)   *30 pCi/g*, then
(D)   *remove the soil* from the site; otherwise,
(D*)   *leave the soil* in place.

The decisionmaker and planning team should be comfortable with the decision rule regarding the criteria for taking action before any measurements are taken. The input to a decision rule is the result of measurements. A decision will be made, and action taken, based upon those results.

There is uncertainty with every scientific measurement taken. Sampling uncertainty is due to the natural spatial and temporal variation in contaminant concentrations across a site. Measurement uncertainty is the variability in a combination of factors that arise during sample analysis. Because there is uncertainty in measurement results, the decision based on them could be incorrect. Controlling decision error is the subject of Step 6 of the DQO process.

B.3.7 DQO Process Step 6: Specify the Limits on Decision Errors

In this step, the project planning team assesses the potential consequences of making a wrong decision and establishes a tolerable level for making a decision error. The project planning team defines the types of decision errors (Type I and II) and the tolerable limits on the decision error rates. In general, a Type I error is deciding against the default assumption (the null hypothesis) when it is actually true; a Type II error is not deciding against the null hypothesis when it is actually false (see Attachment B1 and Appendix C for detailed discussions). The limits imposed on the probability of making decision errors will be used to establish measurement performance criteria for the data collection design.

Traditionally, the principles of statistical hypothesis testing have been used to determine tolerable levels of decision error rates. Other approaches applying decision theory have been applied (Bottrell et al., 1996a, b). Based on an understanding of the possible consequences of making a wrong decision in taking alternative actions, the project planning team chooses the null hypotheses and judges what decision error rates are tolerable for making a Type I or Type II decision error.

The project planning team also specifies a range of possible values where the consequences of decision errors are relatively minor (the gray region). Specifying a gray region is necessary because variability in the population and imprecision in the measurement system combine to produce variability in the data such that the decision may be "too close to call" when the true value is very near the action level. The width of the gray region establishes the distance from the action level where it is most important that the project planning team control Type II errors. For additional information on the gray region, hypothesis testing, and decision errors, see EPA (2000) and NRC (1998).

The tolerable decision error rates are used to establish performance goals for the data collection design. Overall variability in the result can be attributed to several sources, including sample location, collection, and handling; laboratory handling and analysis; and data handling and analysis. In many environmental cases, sampling is a much larger source of uncertainty than laboratory analyses. The goal is to develop a sampling and analysis design that reduces the chance of making a wrong decision. The greater certainty demanded by the decisionmakers, the more comprehensive and expensive the data collection process is likely to be. In this step, the project planning team has to come to an agreement on how to determine acceptable analytical uncertainty and how good the overall data results are required to be. The team has to reach a consensus on the trade off between the cost of more information and the increased certainty in the resulting decision.

Often the project planning team does not feel comfortable with the concepts and terminology of hypothesis testing (Type I and Type II errors, gray region, critical region, tolerable decision error rates). As a result, the project planning team may have difficulty with (or want to skip) this step of the directed planning process. If these steps are skipped or insufficiently addressed, it is more likely that the data will not be of the quality needed for the project. Attachment B1 gives additional guidance on these concepts. MARLAP recommends that for each radionuclide of concern, an action level, gray region, and limits on decision error rates be established during a directed planning process. A stepwise procedure for accomplishing this is given at the end of this section.

Figure B.2 summarizes the outputs of the decisions made by the project planning team in a decision performance goal diagram (EPA, 2000). The horizontal axis represents the (unknown) true value of the parameter being estimated. The vertical axis represents the decisionmaker's desired probability of concluding that the parameter exceeds an action limit. The "gray region" (bounded on one side by the action level) defines an area where the consequences of decision error are relatively minor (in other words, it defines how big a divergence from the action level we wish to distinguish). The gray region is related to the desired precision of the measurements. The height of the indicated straight lines to the right and left of the gray region depict the decisionmaker's tolerance for Type I and Type II errors.

For purposes of this example, the default assumption (null hypothesis) was established as "the measured concentration exceeds the action level" (Figure B.2a). A Type I error consists in making a decision *not* to take action (e.g., remediate) when that action was in fact required (e.g., analyte concentrations are really above an action level). The desired limit on the probability of making a Type I error is set at 5 percent if the true concentration is between 100 and 150 and at 1 percent if the true concentration exceeds 150. A Type II error is understood as taking an action when in fact that action is not required (e.g., analyte concentrations are really below the action level). The desired limit on the probability of making a Type II error is set at 5 percent if the true concentrations is less than 25 and 10 percent if the true concentrations is between 25 and 75.

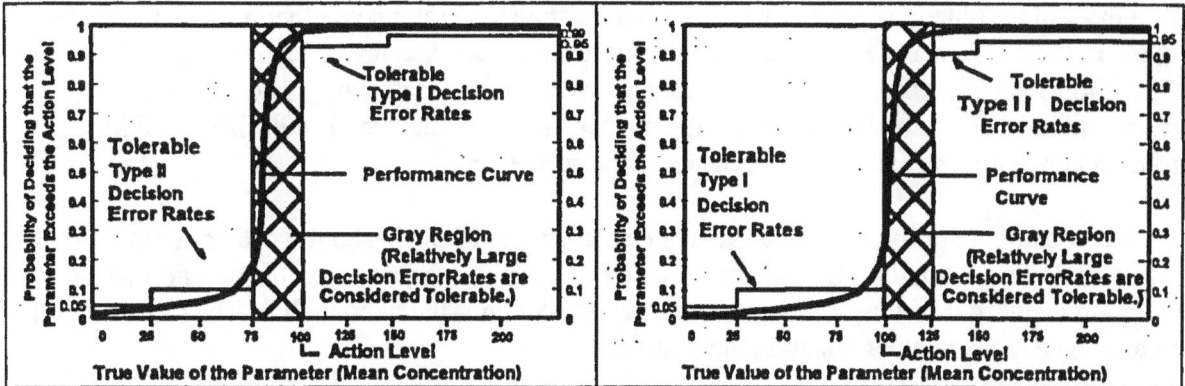

Figure B.2(a) — Decision performance goal diagram null hypothesis: the parameter exceeds the action level.

Figure B.2(b) — Decision performance goal diagram null hypothesis: the parameter is less than the action level.

In Figure B.2(b), the default assumption (null hypothesis) was established as "the measured concentration is less than the action level." The Type I error is understood as taking an action when in fact that action is *not* required (e.g., analyte concentrations are really below the action level). The desired limit on the probability of making a Type I error is set at 5 percent if the true concentration is less than 25, and at 10 percent if the true concentration is between 25 and 100. The Type II error is understood as making a decision not to take action to solve an environmental problem (e.g., to remediate) when that action was in fact required (e.g., analyte concentrations are really above an action level). The desired limit on the probability of making a Type II error is set at 10 percent if the true concentrations is between 125 and 150 and at 5 percent if the true concentrations is over 150.

The output of Step 6 is:

• The project planning team's quantitative measure of tolerable decision error rates based on consideration of project resources.

PROCEDURE FOR SPECIFYING LIMITS ON DECISION ERRORS—AN EXAMPLE

Decisionmakers are interested in knowing the true state of some parameter for which action may be proposed. In Step 5 of the DQO process, the parameter, the action level, and the alternative actions were specified in a decision rule. But, decisionmakers cannot positively know the true state because there will always be the potential for uncertainty in estimating the parameter from data. There will be sampling uncertainty, due to spatial and temporal variability in concentrations across the site and from one sample to the next. There will also be analytical measurement uncertainty due to the variability in the measurement process itself. Since it is impossible to eliminate uncertainty, basing decisions on measurement data opens the possibility of making a decision error. Recognizing that decision errors are possible because of uncertainty is the first step in controlling them.

As an example problem, suppose that a decision must be made about whether or not a particular survey unit at a site meets established criteria for residual radioactivity concentrations. Table B.1(a) shows the two possible decision errors that might occur in deciding whether or not a survey unit has been remediated sufficiently so that it may be released. The decision will be based on concentration measurements taken in the survey unit.

As another example problem, suppose that a decision must be made about whether or not a sample contains a particular radionuclide. Table B.1(b) shows the two possible decision errors that might occur in deciding whether or not a sample contains the radionuclide. The decision will be based on a measurement taken on the sample.

TABLE B.1 — Possible decision errors

(a) For survey unit release	
Decision	**True State**
Deciding a survey unit meets the release criterion	when it actually does not
Deciding a survey unit does not meet the release criterion ...	when it actually does
(b) For radionuclide detection	
Decision	**True State**
Deciding a sample contains the radionuclide	when it actually does not
Deciding a sample does not contain the radionuclide	when it actually does

The probability of making a decision error can be controlled by the use of statistical hypothesis testing. In statistical hypothesis testing, data are used to select between a chosen baseline condition (null hypothesis) and an alternative condition. The test can then be used to decide if there is sufficient evidence to indicate that the baseline condition is unlikely and that the alternative condition is more consistent with the data. Actions appropriate to the alternative conditions would then be appropriate. Otherwise, the default baseline condition remains in place as the basis for decisions and actions. The burden of proof is placed on rejecting the baseline condition. The structure of statistical hypothesis testing maintains the baseline condition as being true until significant evidence is presented to indicate that the baseline condition is not true.

The selection of the baseline condition is important to the outcome of the decision process. The same set of sample data from a survey unit might lead to different decisions depending on what is chosen as the baseline condition.

In deciding if a sample analyzed for a particular radionuclide actually contains that radionuclide, the two possibilities for the baseline condition are:

1) The sample contains the radionuclide, or
2) The sample does not contain the radionuclide.

In this instance, suppose Condition 2, the sample does not contain the radionuclide, is taken as the baseline.[1] The measurement result must be high in order to dismiss the assumption that the sample does not contain the radionuclide. If the measurement is high enough, it is no longer credible that the sample does not contain the radionuclide. Therefore it will be decided that the sample does contain the radionuclide. The framework of statistical hypothesis testing allows one to quantify what is meant by "high enough" and "no longer credible." The measurement value that is considered just "high enough" that the baseline is "no longer credible" is called the "critical value." The baseline condition is called the "null hypothesis," usually denoted H_0. The alternate condition is called the alternative hypothesis, usually denoted H_1 or H_A.

Note that if a poor measurement is made—for example, if the sample containing a concentration near the minimum detectable concentration (MDC) is not counted as long as specified in the standard operating procedures—it will be less likely that a result that is clearly above the variability in the measurement of a blank sample will be obtained. Thus, it will be less likely that a sample with a concentration of the radionuclide near the MDC will be detected with greater than the 95 percent probability that is usually specified in MDC calculations. This is another consequence of the structure of statistical hypothesis testing that maintains the baseline condition until convincing evidence is found to the contrary. Poor or insufficient data often will result in the null hypothesis being retained even when it is not true.

In choosing the baseline condition, it is usually prudent to consider which condition will cause the least harm if it is the one that is acted upon, even if it is not true. This is because the baseline will continue to be assumed true unless the data are clearly in conflict with it.

In deciding if a survey unit meets the release criteria for a particular radionuclide, the two possibilities for the baseline condition are:

1) The survey unit does not meet the release criteria, or
2) The survey unit meets the release criteria.

Condition 1 is usually taken as the baseline. This means that the measurement result must be low in order to dismiss the assumption that the survey unit does not meet the release criteria. If the measurement is low enough, it is no longer credible. Therefore it will be decided that the survey unit does meet the release criteria. Again, the framework of statistical hypothesis testing allows one to quantify what is meant by "low enough" and "no longer credible." The null hypothesis, H_0, is that the survey unit does not meet the release criteria; the alternative hypothesis, H_A, is the survey unit does meet the release criteria. By phrasing the null hypothesis this way, the benefit of performing a good survey is that it will be more likely that a survey unit that *should* be released

---

[1] Condition 1 could only be used if it were phrased in reference to a particular concentration, e.g. the sample contains the radionuclide in concentration in excess of x pCi/g. Condition 2 implies a concentration of zero.

will be released. On the other hand, a poor survey will generally result in retaining the assumption that the release criterion has not been met even if it has. This arrangement provides the proper incentive for good survey work.

The term "Type I error" is assigned to the decision error made by concluding the null hypothesis is not true, when it actually is true. The term "Type II error" is assigned to the decision error made by concluding the null hypothesis is true, when it actually is not true. The possibility of a decision error can never be totally eliminated, but it can be controlled.

When the decision is to be based on comparing the average of a number of measurements from samples taken over some specified area, sampling uncertainty can be reduced by collecting a larger number of samples. Measurement uncertainty can be reduced by analyzing individual samples several times or using more precise laboratory methods. Which uncertainty is more effective to control depends on their relative magnitude. For much environmental work, controlling the sampling uncertainty error by increasing the number of field samples is usually more effective than controlling measurement uncertainty by repeated radiochemical analyses.

One thing is certain, however, that reducing decision errors requires the expenditure of more resources. Drastically controlling decision error probabilities to extremely small values may be unnecessary for making a reasonable decision. If the consequences of a decision error are minor, a reasonable decision might be made based on relatively crude data. On the other hand, if the consequences of a decision error are severe, sampling and measurement uncertainty should be controlled as much as reasonably possible. How much is enough? It is up to the decisionmaker and the planning team to decide how much control is enough. They must specify tolerable limits on the probabilities for decision errors. If necessary, efforts to reduce sampling and measurement uncertainty to meet these specified limits can then be investigated.

Throughout the remainder of this example, the decision to be made is going to be based on comparing the average of a number of measurements from samples taken over a specific area to a pre-determined limit. The goal of the decisionmaker and planning team is to design a sampling plan that controls the chance of making a decision error to a tolerable level. The strategy outlined below can be used to specify limits on decision errors:

I. Determine the potential range of the parameter of interest.

II. Choose the null hypothesis and identify the Type I and Type II decision errors.

III. Specify a range of concentrations where the consequences of decision errors are relatively minor.

IV. Assign tolerable decision error rates outside of the range specified in III.

I. DETERMINE POTENTIAL RANGE OF THE PARAMETER OF INTEREST

Establish the range of average concentrations likely to be encountered in the survey unit. One must have some idea of the concentration range in order to specify the type of analysis to be done and the sensitivity it must have. It is also the starting point for deciding what differences in concentration are important to detect.

In the example shown in Figure B.3, the project planning team considers a range of feasible concentrations for the radionuclide to be between 0–50 pCi/g. This is based on prior experience of the site, scoping, characterization, and remediation-control survey data.

II. CHOOSE THE NULL HYPOTHESIS AND IDENTIFY DECISION ERRORS

The decision rule states that the action level will be 30 pCi/g for the radionuclide. The project planning team states the null hypothesis as—

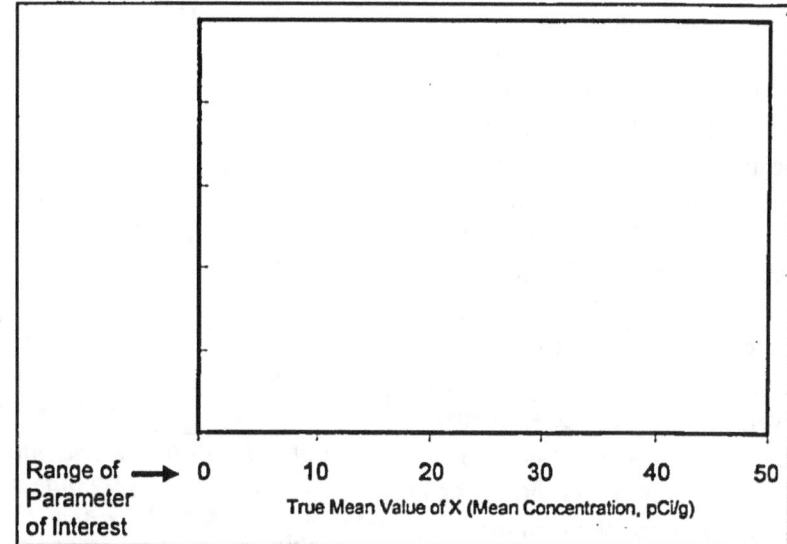

FIGURE B.3 — Plot is made showing the range of the parameter of interest on the x-axis

H_0: *The survey unit concentration exceeds the action level.*

The corresponding decision errors are defined as in Table B.2.

TABLE B.2 — Example of possible decision errors with null hypothesis that the average concentration in a survey unit is above the action level

Decision	True State	Consequences	Probability
Deciding a survey unit is below the action level...	...when it actually is above the action level (H_0).	Type I error	α
Deciding a survey unit is above the action level...	...when it actually is below the action level (H_A).	Type II error	β

Now that a null hypothesis has been chosen, the meaning of a Type I and a Type II decision error is also defined. In Figure B.4, a line is added showing the action level. A Type I error occurs when the null hypothesis is incorrectly rejected. This means that it is decided that a survey unit with a true mean concentration above the action level may be released. This is the only kind of decision error that can occur if the true concentration is at or above the action level. A Type II error occurs when the null hypothesis is *not* rejected when it is *false*. This means that it is decided that a survey unit with a true mean concentration below the action level may not be released. This is the only kind of decision error that can occur if the true concentration is below the action level. The type of decision error possible at a given value of the true concentration is shown, and a y-axis for displaying control limits on making decision errors,

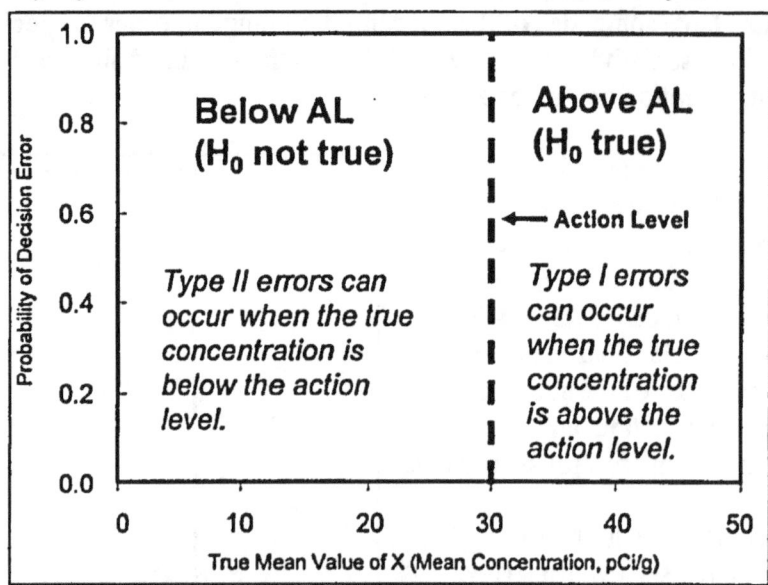

FIGURE B.4 — A line showing the action level, the type of decision error possible at a given value of the true concentration, and a y-axis showing the acceptable limits on making a decision error have been added to Figure B.3

once they have been specified by the project planning team, are also shown in Figure B.4.

III. SPECIFY A RANGE OF CONCENTRATIONS WHERE THE CONSEQUENCES OF DECISION ERRORS ARE RELATIVELY MINOR

The gray region, or region of uncertainty, indicates an area where the consequences of a Type II decision error are relatively minor. It may not be reasonable to attempt to control decision errors within the gray area. The resources expended to distinguish small differences in concentration could well exceed the costs associated with making the decision error.

In this example, the question is whether it would really make a major difference in the action taken if the concentration is called 30 pCi/g when the true value is 26 or even 22 pCi/g. If not, the gray region might extend from 20 to 30 pCi/g . This is shown in Figure B.5.

The width of the gray region reflects the decisionmaker's concern for Type II decision errors near the action level. The decisionmaker should establish the gray region by balancing the resources needed to "make a close call" versus the consequences of making a Type II decision error. The cost of collecting data sufficient to distinguish small differences in concentration could exceed the cost of making a decision error. This is especially true if the consequences of the error are judged to be minor.

There is one instance where the consequences of a Type II decision error might be considered major. That is when expensive remediation actions could be required that are not necessary to protect public health. It could be argued that this is always the case when the true concentration is less than the action level. On the other hand, it can be also be argued that remediation of concentrations near, even though not above the action level, will still carry some benefit. To resolve the issue,

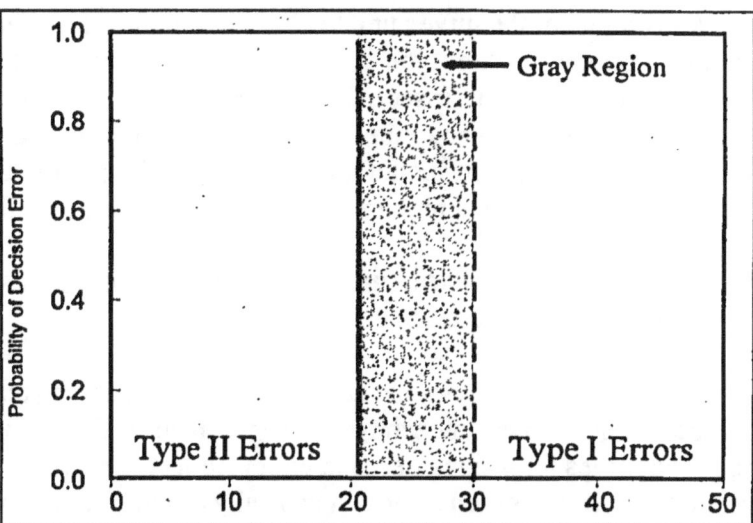

FIGURE B.5 — The gray region is a specified range of values of the true concentration where the consequences of a decision error are considered to be relatively minor

however, the project planning team knows that not all values of the average concentration below the action level are equally likely to exist in the survey unit. Usually, there is some knowledge, if only approximate, of what the average value of the concentration in the survey unit is. This information can be used to set the width of the gray region. If the planning team is fairly confident that the concentration is less than 20 pCi/g but probably more than 10 pCi/g, they would be concerned about making Type II errors when the true concentration is between 10 and 20 pCi/g. However, they will be much less concerned about making Type II errors when the true

concentration is between 20 and 30 pCi/g. This is simply because they do not believe that the true concentration is likely to be in that range. Figure B.6 shows three possible ways that the project planning team might decide to set the gray region. In "A" the project planning team believes the true concentration remaining in the survey unit is about 15 pCi/g, in "B" they believe it to be about 20 pCi/g, and in "C" about 25 pCi/g. In each case, they are less concerned about a decision error involving a true concentration greater than what is estimated to actually remain. They have used

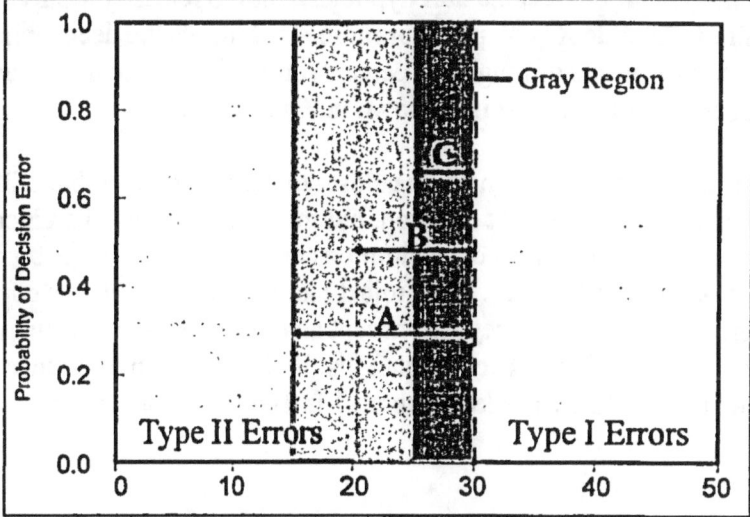

FIGURE B.6 — Three possible ways of setting the gray region. In (A) the project planning team believes the true concentration remaining in the survey unit is about 15 pCi/g, in (B) about 20 pCi/g and in (C) about 25 pCi/g

their knowledge of the survey unit to choose the range of concentration where it is appropriate to expend resources to control the Type II decision error rate. The action level, where further remediation would be necessary, defines the upper bound of the gray region where the probability of a Type I error should be limited. The lower bound of the gray region defines the concentration below which remediation should not be necessary. Therefore, it defines where the probability of a Type II error that would require such an action should be limited.[2]

IV. ASSIGN TOLERABLE PROBABILITY VALUES FOR THE OCCURRENCE OF DECISION ERRORS OUTSIDE OF THE RANGE SPECIFIED IN III

As part of the DQO process, the decisionmaker and planning team must work together to identify possible consequences for each type of decision error. Based on this evaluation, desired limits on the probabilities for making decision errors are set over specific concentration ranges. The risk associated with a decision error will generally be more severe as the value of the concentration moves further from the gray region. The tolerance for Type I errors will decrease as the concentration increases. Conversely, the tolerance for Type II errors will decrease as the concentration deceases.

In the example, the decisionmaker has identified 20–30 pCi/g as the area where the consequences of a Type II decision error would be relatively minor. This is the gray region. The tolerable limits on Type I decision errors should be smallest for cases where the decisionmaker has the greatest concern for making an incorrect decision. This will generally be at relatively high values of the true concentration, well above the action level. Suppose, in the example, that the decisionmaker is determined to be nearly 99 percent sure that the correct decision is made, namely, *not* to reject the null hypothesis, *not* to release the survey unit, if the true concentration of radionuclide X is 40 pCi/g or more. That means the decisionmaker is only willing to accept a Type I error rate of roughly 1 percent, or making an incorrect decision 1 out of 100 times at this concentration level. This is shown in Figure B.7(a).

If the true concentration of X is closer to the action level, but still above it, the decisionmaker wants to make the right decision, but the consequences of an incorrect decision are not considered as severe at concentrations between 30 and 40 pCi/g as they are when the concentration is over 40 pCi/g. The project planning team wants the correct action to be taken at least 90 percent of the time. They will accept an error rate not worse than about 10 percent. They will only accept a data collection plan that limits the potential to incorrectly decide not to take action when it is actually needed to about 1 in 10 times. This is shown in Figure B.7(b).

---

[2] Had the null hypothesis been chosen differently, the ranges of true concentration where Type I and Type II errors occur would have been reversed.

The decisionmaker and project planning team are also concerned about wasting resources by cleaning up sites that do not represent any substantial risk. Limits of tolerable probability are set low for extreme Type II errors, i.e. failing to release a survey unit when the true concentration is far below the gray region and the action level. They want to limit the chances of deciding to take action when it really is not needed to about 1 in 20 if the true concentration is less than 10 pCi/g. This is shown in Figure B.7(c).

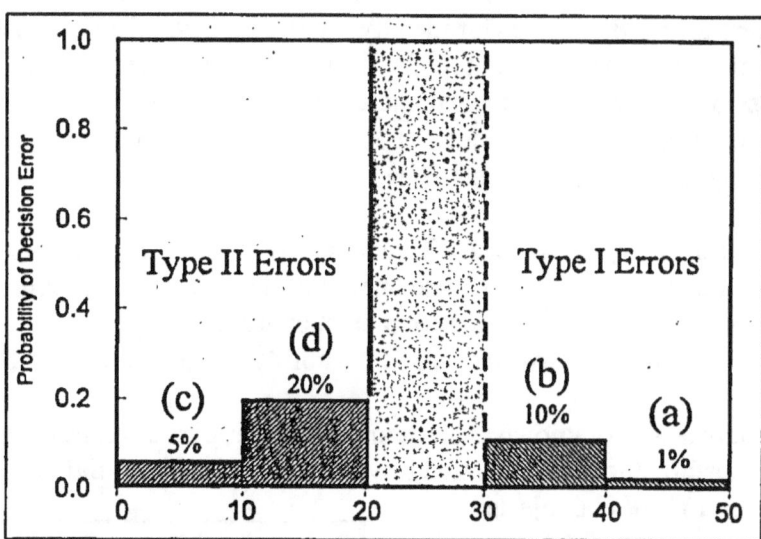

FIGURE B.7 — Example decision performance goal diagram

They are more willing to accept higher decision error rates for concentrations nearer to the gray region. After all, there is some residual risk that will be avoided even though the concentration is below the action level. A Type II error probability limit of 20 percent in the 10-20 pCi/g range is agreed upon. They consider this to be an appropriate transition between a range of concentrations where Type II errors are of great concern (<10 pCi/g) to a range where Type II errors are of little concern. The latter is, by definition, the gray region, which is 20-30 pCi/g in this case . The chance of taking action when it is not needed within the range 10-20 pCi/g is set at roughly 1 in 5. This is shown in Figure B.7(d).

Once the limits on both types of decision error rates have been specified, the information can be displayed on a decision performance goal diagram, as shown in Figure B.7, or made into a decision error limits table, as shown in Table B.3. Both are valuable tools for visualizing and evaluating proposed limits for decision errors.

TABLE B.3 — Example decision error limits table

True Concentration	Correct Decision	Tolerable Probability of Making a Decision Error
0 – 10 pCi/g	Does not exceed	5%
10 – 20 pCi/g	Does not exceed	20%
20 – 30 pCi/g	Does not exceed	gray region: decision error probabilities not controlled
30 – 40 pCi/g	Does exceed	10%
40 – 50 pCi/g	Does exceed	1%

There are no fixed rules for identifying at what level the decisionmaker and project planning team should be willing to tolerate the probability of decision errors. As a guideline, as the possible true values of the parameter of interest move closer to the action level, the tolerance for decision errors usually increases. As the severity of the consequences of a decision error increases, the tolerance decreases.

The ultimate goal of the DQO process is to identify the most resource-effective study design that provides the type, quantity, and quality of data needed to support defensible decisionmaking. The decisionmaker and planning team must evaluate design options and select the one that provides the best balance between cost and the ability to meet the stated DQOs.

A statistical tool known as an estimated power curve can be extremely useful when investigating the performance of alternative survey designs. The probability that the null hypothesis *is* rejected when it *should* be rejected is called the statistical power of a hypothesis test. It is equal to one minus the probability of a Type II error $(1-\beta)$. In the example, the null hypothesis is false whenever the true concentration is less than the action level. Figure B.8 shows the power diagram constructed from Figure B.7 by replacing the desired limits on Type II error probabilities, β, with the power, $1-\beta$. The desired limits on Type I error probabilities, α, are carried over without modification, as is the gray region. Drawing a smooth decreasing function through the desired limits results in the

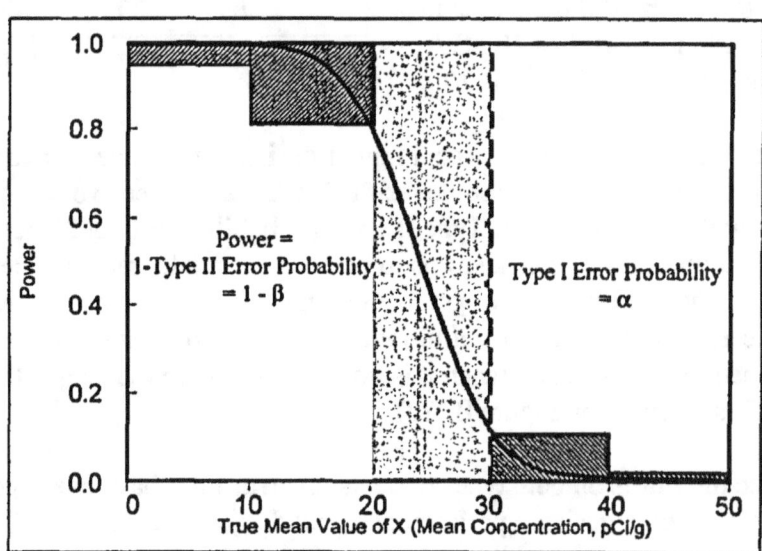

FIGURE B.8 — A power curve constructed from the decision performance goal diagram in Figure B.7

desired power curve. A decision performance goal diagram with an estimated power curve can help the project planning team visually identify information about a proposed study design.

Statisticians can determine the number of measurements needed for a proposed survey design from four values identified on the decision performance goal diagram:

(1) The tolerable limit for the probability of making Type I decision errors, α, at the action level AL).

(2) The tolerable limit for the probability of making Type II decision errors, β, along the

lower bound of the gray region (LBGR).

(3) The width of the gray region, $\Delta = AL - LBGR$, where the consequences of Type II decision errors are relatively minor.

(4) The statistical expression for the total expected variability of the measurement data in the survey unit, σ.

The actual power curve for the statistical hypothesis test can be calculated using these values, and can be compared to the desired limits on the probability of decision errors.

The estimated number of measurements required for a proposed survey design depends heavily on the expected variability of the measurement data in the survey unit, σ. This may not always be easy to estimate from the information available. However, the impact of varying this parameter on the study design is fairly easy to determine during the planning process. Examining a range of reasonable values for σ may not result in great differences in survey design. If so, then a crude estimate for σ is sufficient. If not, the estimate for σ may need to be refined, perhaps by a pilot study of 20 to 30 samples. If the change in the number of samples (due to refining the estimate of σ) is also about 20 to 30 in a single survey unit, it may be better to simply use a conservative estimate of σ that leads to the larger number of samples rather than conduct a pilot study to obtain a more accurate estimate of σ. On the other hand, if several or many similar survey units will be subject to the same design, a pilot study may be worthwhile.

The example in Figure B.9 shows that the probability of making a decision error for any value of the true concentration can be determined at any point on the power curve. At 25 pCi/g, the probability of a Type II error is roughly 45–50 percent. At 35 pCi/g, the probability of a Type I error is roughly 3 percent.

The larger the number of samples required to meet the stated DQOs, the greater the costs of sampling and analysis for a proposed plan. Specifying a narrow gray region and/or very small limits on decision error probabilities indicate a high level of certainty is needed and a larger number of samples will be required.

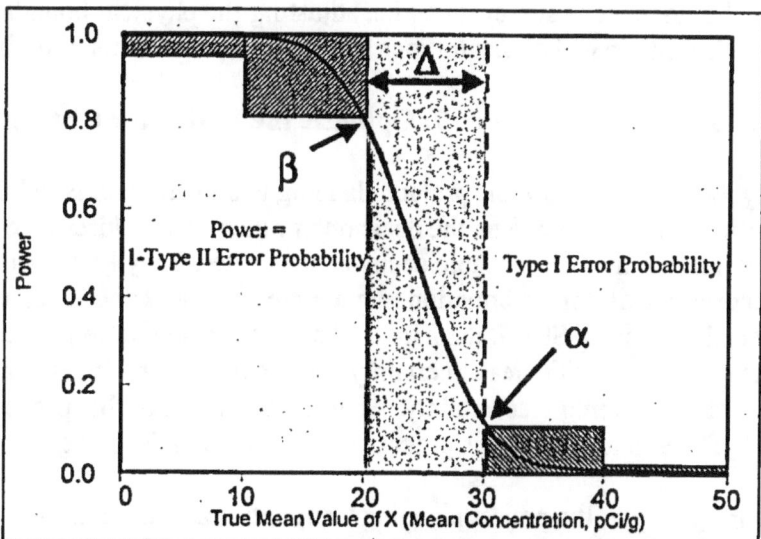

FIGURE B.9 — Example power curve showing the key parameters used to determine the appropriate number of samples to take in the survey unit

Specifying a wide gray region and/or larger limits on decision error probabilities indicates a lower level of certainty is required. A smaller number of samples will be necessary. The required level of certainty should be consistent with the consequences of making decision errors balanced against the cost in numbers of samples to achieve that level of certainty.

If a proposed survey design fails to meet the DQOs within constraints, the decisionmaker and planning team may need to consider:

- ADJUSTING THE ACCEPTABLE DECISION ERROR RATES. For example, the decisionmaker may be unsure what probabilities of decision error are acceptable. Beginning with extremely stringent decision error limits with low risk of making a decision error may require an extremely large number of samples at a prohibitive cost. After reconsidering the potential consequences of each type of decision error, the decisionmaker and planning team may be able to relax the tolerable rates.

- ADJUST THE WIDTH OF THE GRAY REGION. Generally, an efficient design will result when the relative shift, Δ/σ, lies between the values of 1 and 3. A narrow gray region usually means that the proposed survey design will require a large number of samples to meet the specified DQOs. By increasing the number of samples, the chances of making a Type II decision error is reduced, but the potential costs have increased. The wider the gray region, the less stringent the DQOs. Fewer samples will be required, costs will be reduced but the chances of making a Type II decision error have increased. The relative shift, Δ/σ, depends on the width of the gray region, Δ, and also on the estimated data variability, σ. Better estimates of either or both may lead to a more efficient survey design. In some cases it may be advantageous to try to reduce σ by using more precise measurement methods or by forming more spatially homogeneous survey units, i.e. adjusting the physical boundaries of the survey units so that the anticipated concentrations are more homogeneous with them.

B.3.8 DQO Process Step 7: Optimize the Design for Obtaining Data

By the start of Step 7, the project planning team has established their priority of concerns, the definition of the problem, the decision or outcome to address the posed problem, the inputs and boundaries, and the tolerable decision error rates. They have also agreed on decision rules that incorporate all this information into a logic statement about what action to take in response to the decision. During Step 7, the hard decisions are made between the planning team's desire to have measurements with greater certainty and the reality of the associated resource needs (time, cost, etc.) for obtaining that certainty. Another viewpoint of this process is illustrated in Attachment B1. The application of this process to MDC calculations is given in Attachment B2.

During Step 7, the project planning team optimizes the sampling and analytical design and establishes the measurement quality objectives (MQOs) so the resulting data will meet all the established constraints in the most resource-effective manner. The goal is to determine the most efficient design (combination of sample type, sample number and analytical procedures) to meet all the constraints established in the previous steps. Once the technical specialists and the rest of the project planning team come to agreement about the sampling and analysis design, the operational details and theoretical assumptions of the selected design should be documented.

If a proposed design cannot be developed to meet the limits on decision error rates within budget or other constraints, then the project planning team will have to consider relaxing the error tolerance, adjusting the width of the gray region, redefining the scale of decision, or committing more funding. There is always a trade off among quality, cost, and time. The project planning team will need to develop a consensus on how to balance resources and data quality. If the proposed design requires analysis using analytical protocols not readily available, the project planning team must consider the resources (time and cost) required to develop and validate a method, generate method detection limits relevant to media of concern, and develop appropriate QA/QC procedures and criteria (Chapter 6, *Selection and Application of an Analytical Method*).

If the project entails a preliminary investigation of a site or material for which little is known, the planners may choose to employ MQOs and requirements that typically are achieved by the selected sampling and analytical procedures. At this early point in the project, the lack of detailed knowledge of the site or material may postpone the need for the extra cost of more expensive sampling and analytical procedures and large numbers of samples, until more site or material knowledge is acquired. The less-demanding MQOs, however, should be adequate to further define the site or material. For situations when the measured values are distant from an action level the MQO-compliant data could also be sufficient to support the project decision.

The planning of data collection activities typically is undertaken to determine if a characteristic of an area or item does or does not exist above an action level. Since the area of interest (population) is usually too large to be submitted to analyses, in its entirety, these data collection activities generally include sampling. If sampling is done correctly, the field sample or set of field samples will represent the characteristics of interest and, if analyzed properly, the information gleaned from the samples can be used to make decisions about the larger area. However, if errors occur during implementation of the project, the samples and associated data may not accurately reflect the material from which the samples were collected and incorrect decisions could be made.

The planning team attempts to anticipate, quantify, and minimize the uncertainty in decisions resulting from imprecision, bias, and blunders—in other words, attempts to manage uncertainty by managing its sources. The effort expended in managing uncertainty is project dependent and depends upon what constitutes an acceptable level of decision uncertainty and the proximity of the data to a decision point. For example, Figure B.10(a) presents a situation where the data have significant variability. Yet the variability of the data does not materially add to the uncertainty of the decision since the measurements are so far removed from the action level. More resources could be expended to control the variability. However, the additional expenditure would be unnecessary, since they would not alter the decision or measurably increase confidence in the decision.

In contrast, Figure B.10(b) depicts data with relatively little variability, yet this level of variability is significant since the measured data are adjacent to the action level, which results in increased uncertainty in the decision. Depending upon the consequences of an incorrect decision, it may be advisable to expend more resources with the intention of increasing confidence in the decision.

The outputs of Step 7 are:

* The most resource-effective design for sampling and analysis that will obtain the specific amount and quality of data needed to resolve the problem within the defined constraints; and

* Detailed plans and criteria for data assessment.

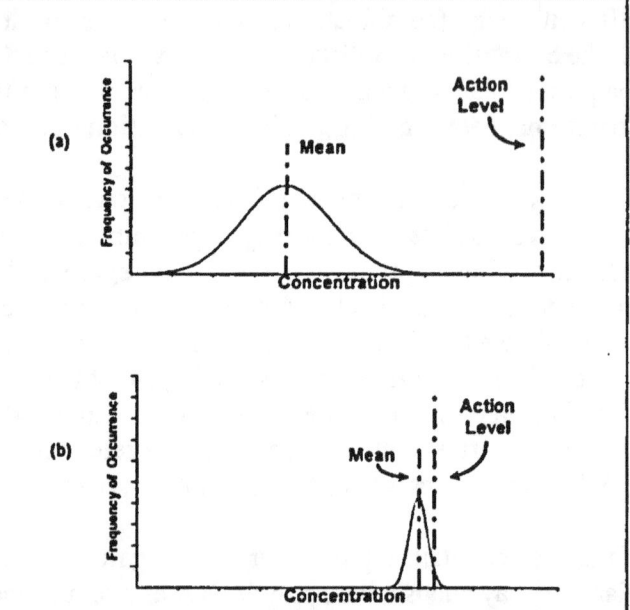

Figure B.10 — How proximity to the action level determines what is an acceptable level of uncertainty

B.4 References

American National Standards Institute (ANSI) N42.23. *American National Standard Measurement and Associated Instrument Quality Assurance for Radioassay Laboratories.* 2003.

American Society for Testing and Materials (ASTM) D5283. *Standard Practice for Generation of Environmental Data Related to Waste Management Activities: Quality Assurance and Quality Control Planning and Implementation.* 1992.

American Society for Testing and Materials (ASTM) D5792. *Standard Practice for Generation of Environmental Data Related to Waste Management Activities: Development of Data Quality Objectives,* 1995.

American Society for Testing and Materials (ASTM) D6051. *Standard Guide for Composite Sampling and Field Subsampling for Environmental Waste Management Activities.* 1996.

Bottrell, D., S. Blacker, and D. Goodman. 1996a. "Application of Decision Theory Methods to the Data Quality Objectives Process." *In,* Proceedings of the Computing in Environmental Resource Management Conference, Air and Waste Management Association.

Bottrell, D., N. Wentworth, S. Blacker, and D. Goodman. 1996b. "Improvements to Specifying Limits on Decision Errors in the Data Quality Objectives Process." *In*, Proceedings of the Computing in Environmental Resource Management Conference, Air and Waste Management Association.

U.S. Environmental Protection Agency (EPA). 1986. *Development of Data Quality Objectives, Description of Stages I and II.* Washington, DC.

U.S. Environmental Protection Agency (EPA). 2000. *Guidance for the Data Quality Objective Process* (EPA QA/G-4). EPA/600/R-96/055, Washington, DC. Available at www.epa.gov/quality1/qa_docs.html.

MARSSIM. 2000. *Multi-Agency Radiation Survey and Site Investigation Manual, Revision 1.* NUREG-1575 Rev 1, EPA 402-R-97-016 Rev1, DOE/EH-0624 Rev1. August. Available at www.epa.gov/radiation/marssim/.

U.S. Nuclear Regulatory Commission (NRC). 1998. *A Nonparametric Statistical Methodology for the Design and Analysis of Final Status Decommissioning Surveys.* NUREG-1505, Rev. 1.

ATTACHMENT B1
Decision Error Rates and the Gray Region for Decisions About Mean Concentrations

B1.1   Introduction

This attachment presents additional information on decision error rates and the gray region. The project planning team will need to specify a range of possible values where the consequences of decision errors are relatively minor—the "gray region." Specifying a gray region is necessary because variability in the population and imprecision in the measurement system combine to produce variability in the data such that the decision may be "too close to call" when the true value is very near the action level. The gray region establishes the minimum separation from the action level, where it is most important that the project planning team control Type II errors.

B1.2   The Region of Interest

The first step in constructing the gray region is setting the range of concentrations that is a region of interest (a range of possible values). This normally means defining the lowest and highest average concentrations at which the contaminant is expected to exist. Usually there is an action level (such as the derived concentration guideline level, DCGL, a regulatory limit) that should not be exceeded. If the project planning team wants a method to measure sample concen-

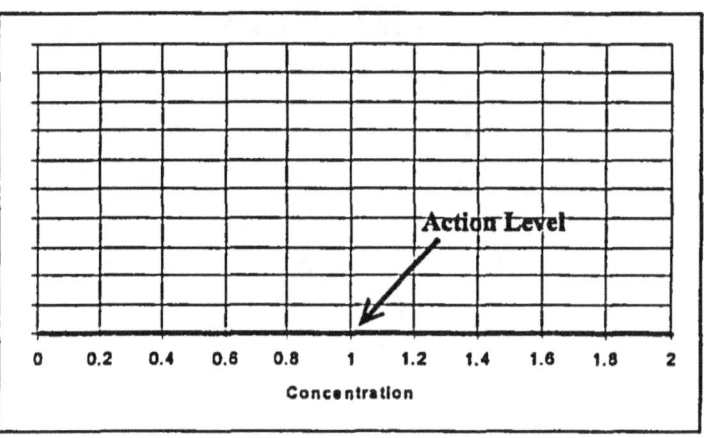

FIGURE B1.1 — The action level is 1.0

trations around this level, they would not select one that worked at concentrations at 10 to 100 times the action level, nor would they select one that worked from zero to half the action level. They would want a method that worked well around the action level—perhaps from 0.1 to 10 times the action level, or from one-half to two times the action level. For the purpose of the example in this attachment, the action level is 1.0 and the project planning team selected a region of interest that is zero to twice the action level (0–2), as shown on the x-axis in Figure B1.1.

B1.3   Measurement Uncertainty at the Action Level

The action level marks the concentration level that the project planning team must be able to distinguish. The project planning team wants to be able to tell if the measured concentration is above or below the action level. Does this mean that the project planning team needs to be able to distinguish 0.9999 times the action level from 1.0001 times the action level? Sometimes, but not usually. This is fortunate, because current measurement techniques are probably not good enough to distinguish that small a difference in concentrations.

How close to the action level can the project planning team plan to measure? This example assumes that the standard uncertainty (1 sigma, σ) of the measured concentration is 10 percent of the action level. With that kind of measurement "precision," can the project planning team tell the difference between a sample with 0.9 times the action level from one right at the action level? Not always. Figure B1.2 shows the distribution of the concentration that is measured (assuming a normal distribution). This means that about 16 percent of

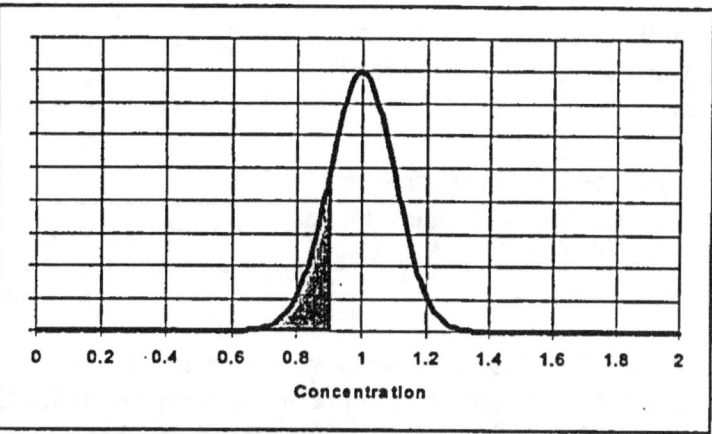

FIGURE B1.2 — The true mean concentration is 1.0. The standard uncertainty of the distribution of measured concentrations is 0.1.

the time, the measured concentration (in the shaded area) will appear to be 0.9 times the action level or less, even though the true concentration is exactly equal to the action level.

Similarly, about 16 percent of the time, the measured concentration will appear to be at or above the action level (as shown in the shaded area in Figure B1.3), even though the true concentration is only 0.9 times the action level.

The problem is, when there is only the measurement result to go by, the project planning team cannot tell the difference with confidence. If the measured concentration is 0.9, it is more likely that the true concentration is 0.9 than it is 1.0, but there remains a chance that it is really 1.0. The moral of the story is that measurement variability causes some ambiguity about what the true concentration is. This translates into some uncertainty in the decisionmaking process. This uncertainty can be controlled with careful planning, but it can never be eliminated. On the other hand, the ambiguity caused by measurement variability really only affects the ability to distinguish between concentrations that are "close together." In our example, 0.9 and 1.0 are "close together" not because 0.1 is a small difference, but because there is a great degree of overlap between the curves shown in Figures B1.2 and B1.3. The peaks of the two curves are separated by 0.1, but each curve spreads out over a value several times this amount on both sides. The most common statistical measure of the amount of this spread is the standard deviation. The standard deviation in this case is 0.1, the same as the amount of separation between the peaks. If the peaks were separated by 0.3, i.e. 3 standard deviations, there would be far less overlap, and far less ambiguity. There would be very little uncertainty in deciding which curve a single measurement belonged to, and consequently whether the mean was 0.7 or 1.0.

From this discussion, at least two very important conclusions can be drawn:

(1) True mean concentrations that are "very close together" are not easily distinguished by a single measurement.

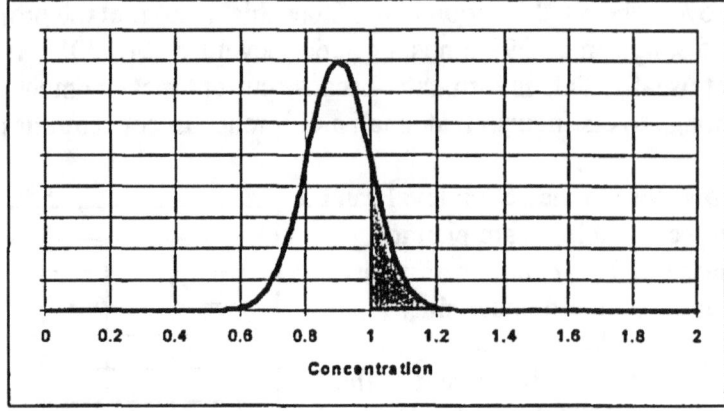

FIGURE B1.3 — The true mean concentration is 0.9. The standard uncertainty of the distribution of measured concentrations is 0.1.

(2) A useful way for determining what is meant by "very close together" is by measuring the separation in concentration in standard deviation units. Concentrations that are one or fewer standard deviations apart are close together, whereas concentrations that are three or more standard deviations apart are well separated.

From conclusion (1), it is immediately apparent that no matter how small the measurement variability is, there must be some separation between the concentration values to be distinguished. It is not possible to determine whether or not the concentration is on one side or the other of "a bright line" (e.g. above or below the action level). Instead, one must be content to pick two concentrations separated by a finite amount and attempt to tell them apart. These two concentrations define what is known as the gray region, because one cannot be certain about deciding whether concentrations that lie between the two boundaries are above or below the action level. To illustrate this with the example, if the measured concentration is 0.95—exactly in the middle of the gray region between the two concentrations to be distinguished— it is equally likely that the true concentration is 0.9 as it is 1.0 (Figure B1.4).

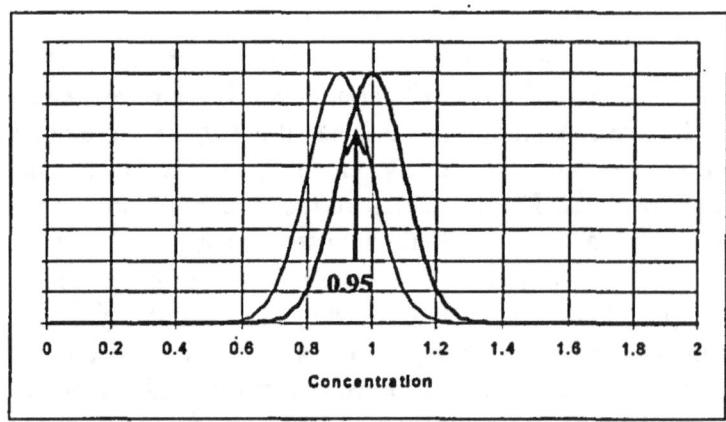

FIGURE B1.4 — If 0.95 is measured, is the true mean concentration 1.0 (right) or 0.9 (left)? The standard uncertainty of the distribution of measured concentrations is 0.1.

To formalize this process of distinguishing whether the true concentration is above our upper bound or below our lower bound, two hypotheses will be defined and a statistical hypothesis test will be used to decide between the two.

B1.4   The Null Hypothesis

How does the project planning team decide whether the true concentration is above or below the gray region? By formulating hypotheses. Suppose it has been decided that it is important to distinguish whether the true mean concentration is above 1.0 or below 0.9. These concentrations then correspond to the "upper bound of the gray region" (UBGR) and to the "lower bound of the gray region" (LBGR), respectively.

The project planning team starts by asking which mistake is worse: (1) deciding the true concentration is less than the action level when it is actually above, or (2) deciding the true concentration is above the action level when it is actually below?

Mistake (1) may result in an increased risk to human health in the general population following site release, while mistake (2) may result in increased occupational risks or a waste of resources that might have been used to reduce risks elsewhere.

The way to avoid the "worse mistake" is to assume the worse case is true, i.e., make the worse case the baseline or null hypothesis. For example, to avoid mistake (1), deciding the true concentration is less than the action level when it is actually above, the null hypothesis should be that the true concentration is above the action level. Only when the data provide convincing evidence to the contrary will it be decided that the true concentration is less than the action level. Borderline cases will default to retaining (not rejecting) the null hypothesis.

Note that while the null hypothesis must be, in fact, either true or false, the data cannot prove that it is true or false with absolute certainty. When the probability of obtaining the given data is sufficiently low under the conditions specified by the null hypothesis, it is evidence to decide that the null hypothesis should be rejected. On the other hand, if the null hypothesis is not rejected, it is not the same as proving that the null hypothesis is true. It only means that there was not enough evidence, based on the probability of observing the data obtained, to decide to reject it.

Notice that in Figure B.2 (Section B.3.7 on page B-11), the risk that is elevated in the gray region is that of making a Type II error. That is, in the gray region, the Type II error rate exceeds the tolerable limit set at the boundary of the gray region. The Type I error rate remains fixed. (It is fixed at exactly the value used to determine the critical value for the statistical test.) A Type II error is incorrectly accepting (failing to reject) the null hypothesis when it is false. So another way to think about choosing the null hypothesis is to decide which mistake is less tolerable, and framing the null hypothesis so that kind of mistake corresponds to a Type I error (i.e., incorrectly rejecting the null hypothesis when it is actually true).

Another pragmatic consideration is that the project planning team really does not want to make a mistake that is likely to remain undiscovered or will be difficult or expensive to correct if it is discovered. If the project planning team decides the true concentration is less than the action level, the team is not likely to look at the data again. That would mean that the mistake would probably not be discovered until much later (e.g. during a confirmatory survey), if at all. On the other hand, if the project planning team decides that the true concentration is over the action level when it really is not, the project planning team will discover the mistake while they are trying to figure out how to take action (i.e., to remediate). This is a pragmatic reason to set the null hypothesis so as to assume the true concentration exceeds the action level. This null hypothesis will not be rejected unless the project planning team is certain that the true concentration is below the action level. This way of choosing the null hypothesis will not work when the action level is so low compared with the expected data variability that no reasonable values of Type II error rates can be achieved. This can occur, for example, when the action level is close to (or even equal to) zero. In that case, if the action level is chosen to be the UBGR, the lower bound might have to be negative. It is impossible to demonstrate that the true concentration is less than some negative value, because negative concentrations are not possible. In such cases, there may be no alternative but to choose as the null hypothesis that the action level is met. Then a concentration that is unacceptably higher than the action level is chosen for the UBGR.

CASE 1: ASSUME THE TRUE CONCENTRATION IS OVER 1.0

If a true concentration of 1.0 or more is over a regulatory limit, such as a DCGL, the project planning team will not want to make mistake (1) above. So they generally will choose as the null hypothesis that the true concentration exceeds the action level of 1.0. How sure does the project planning team need to be? To be 95 percent sure, they would have to stay with their assumption that the true concentration is over 1.0 unless the measured concentration is 0.84 or less (Figure B1.5). The project planning team knows that they will only observe a concentration less than 0.84 about 5 percent of the time when the true concentration is really 1.0. That is, the measurement has to be less than 0.84 to be 95 percent sure the true concentration is less than 1.0. This is an example of a *decision rule* being used to decide between two alternative hypotheses. If a concentration of less than 0.84 is observed, one can decide that the true concentration is less than 1.0—

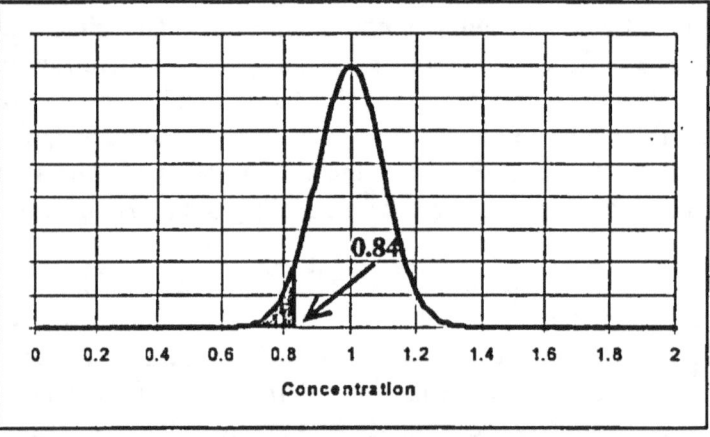

FIGURE B1.5 — When the true mean concentration is 1.0, and the standard uncertainty of the distribution of measured concentrations is 0.1, a measured concentration of 0.84 or less will be observed only about 5 percent of the time i.e., the null hypothesis is rejected. Otherwise, if a concentration over 0.84 is observed, there is not enough evidence to reject the null hypothesis, and one retains the assumption that the true concentration is over 1.0.

But what if the true concentration is 0.9 or less? Under the null hypothesis, how often will the project planning team say that the true concentration is over 1.0 when it is really only 0.84? As seen in Figure B1.6, there is only a 50-50 chance of making the right decision when the true concentration really is 0.84. That is the price of being sure the action level is not exceeded. The Type II error rate, when the true concentration is 0.9, is over 50 percent.

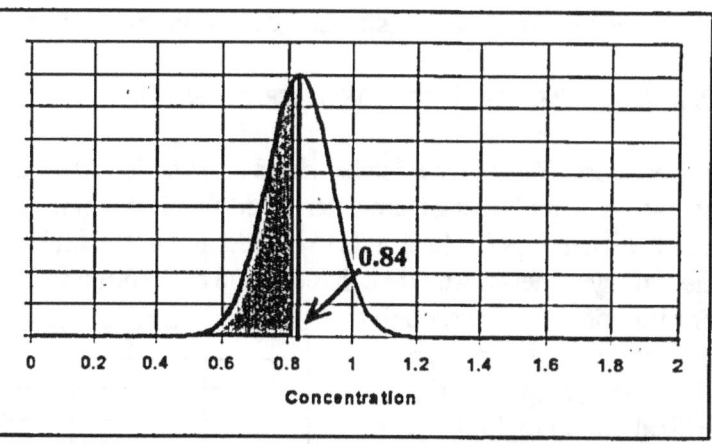

FIGURE B1.6 — When the true mean concentration is 0.84, and the standard uncertainty of the distribution of measured concentrations is 0.1, a measured concentration of 0.84 or less will be observed only about half the time

How low does the true concentration have to be in order to have a pretty good chance of deciding that the true concentration is below the limit? To be 95 percent sure, the true concentration needs to be twice as far below the action level as the decision point (i.e., critical value), namely at about 0.68. That is, the project planning team will need a concentration of 0.68 or less to be 95 percent sure that they will be able to decide the true concentration is less than 1.0 (see the unshaded portion in Figure B1.7). The "critical value" (or decision point) is the measured value that divides the measurement results into two different sets: (1) those values that will cause the null hypothesis to be rejected and (2) those values that will leave the null hypothesis as the default. In other words, it is only when the true concentration is 0.68 or less that the project planning team can be pretty sure that they will decide the true concentration is less than 1.0. Notice that the project planning team could change the decision rule. For example, they could decide that if the measured concentration is less than 0.9, they will reject the null hypothesis. Examining Figures B1.2

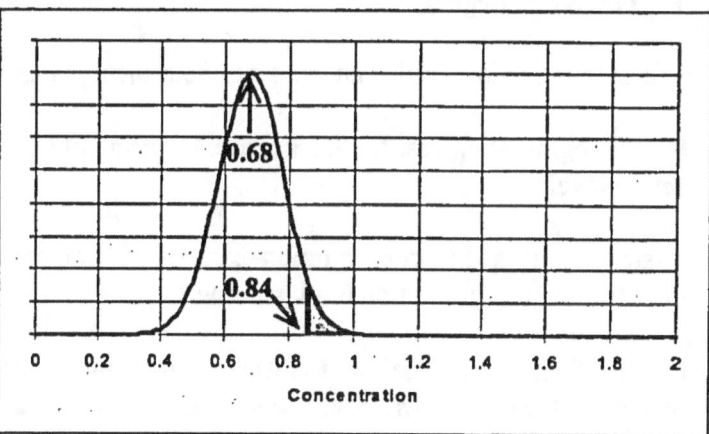

FIGURE B1.7 — When the true mean concentration is 0.68 and the standard uncertainty of the distribution of measured concentrations is 0.1, a measured concentration over 0.84 will be observed only about 5 percent of the time

and B1.3 once again, the Type I error rate will be about 16 percent instead of 5 percent. However, the Type II error rate will decrease from 50 percent to 16 percent. Fortunately, by moving the decision point—called the "critical value"—the error rates can be adjusted. However, reducing one error rate necessarily increases the other. The only way to decrease both decision error rates is to reduce the uncertainty (standard deviation) of the distribution of measured concentrations.

CASE 2: ASSUME THE TRUE CONCENTRATION IS 0.9

As stated previously, the mistake that is most serious determines the null hypothesis. Suppose that the project planning team determined that it is worse to decide that the true concentration is over 1.0 when it is 0.9 (than it is to decide it is 0.9 when it is 1.0). Then, the default assumption (the null hypothesis) would be that the true concentration is less than 0.9, unless the measured

concentration is large enough to convince the planning team otherwise. Using a decision rule (critical value) of 1.06, the planning team can decide the true concentration is over 1.0 with only a 5 percent chance that it is actually 0.9 or less (Figure B1.8). The team will have to have a true concentration of 1.22 or more to be 95 percent sure that they will be able to decide the true concentration is over 1.0.

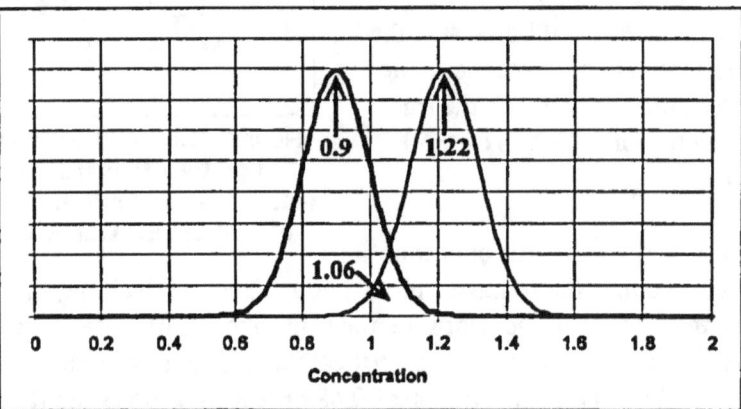

Figure B1.8 — The true mean concentration is 0.9 (left) and 1.22 (right). The standard uncertainty of the distribution of measured concentrations is 0.1.

B1.5 The Gray Region

In the previous sections of this attachment, the project planning team:

- Set the region of interest for the measured concentrations between zero and about twice the action level;

- Assumed that the true concentration exceeds 1.0, unless they measure "significantly" below that, the default assumption (null hypothesis);

- Defined "significantly below" to mean a concentration that would be observed less than 5 percent of the time, when the true concentration is actually 1.0. To describe their uncertainty, the project planning team used the normal distribution, with a relative standard deviation of 10 percent at the action level, as a model;

- Developed an operational decision rule: If the measured concentration is less than 0.84, then decide the true concentration is less than 1.0. Otherwise, decide there is not enough reason to change the default assumption (null hypothesis); and

- Found using this operational decision rule that they were pretty sure (95 percent) of deciding that the true concentration is less than 1.0 only when the true concentration is actually 0.68 or less.

If the true concentration is between 0.68 and 1.0, all the project planning team really can say is that the probability of correctly deciding that the true concentration is less than 1.0 will be between 5 percent (when the true concentration is just under 1.0) and 95 percent (when the true concentration is 0.68). In other words, when the true concentration is in the range of 0.68 to 1.0, the probability of incorrectly deciding that the true concentration is not less than 1.0 (i.e., the probability of making a Type II error) will be between 5 percent (when the true concentration is 0.68) and 95 percent (when the true concentration is just under 1.0). This range of concentrations, 0.68 to 1.0, is the "gray region."

When the null hypothesis is that the true concentration exceeds the action level (1.0), the gray region is bounded from above by the action level. This is where α (the desired limit on the Type I error rate) is set. It is bounded from below at the concentration where β (the desired limit on the Type II error rate) is set. There is some flexibility in setting the LBGR. If the project planning team specifies a concentration, they can calculate the probability β. If they specify β, they can calculate the value of the true concentration that will be correctly detected as being below 1.0 with probability $1-\beta$.

Often it will make sense to set the LBGR at a concentration at, or slightly above, the project planning team's best estimate of the true concentration based on all of the information that is available to them. Then the width of the gray region will truly represent the minimum separation in concentration that it is important to detect, namely, that between the action level and what actually is believed to be there.

In our example, the project planning team found that they needed the true concentration to be 0.68 or less to be at least 95 percent sure that they will correctly decide (by observing a measured value of 0.84 or less) that the true concentration is less than 1.0. If the project planning team is not satisfied with that, the team can find that a true concentration of 0.71 will be correctly detected 90 percent of the time (also by observing a measured value of 0.84 or less). The critical value, or decision point, is determined by α, not β.

If the project planning team decides to raise the LBGR (i.e., narrow the gray region) the Type II error rate at the LBGR goes up. If they lower the LBGR (i.e., widen the gray region) the Type II error rate at the LBGR goes down. Nothing substantive is really happening. The project planning team is merely specifying the ability to detect that the null hypothesis is false (i.e., reject the null hypothesis because it is not true) at a particular concentration below the action level called the

LBGR.

If the project planning team wants to make a substantive change, they need to change the probability that an error is made. That is, they need to change the uncertainty (standard deviation) of the measurements. Suppose the relative standard deviation of the measurements at the action level is 5 percent instead of 10 percent. Then the value of the true concentration that will be correctly detected to be below the action level (by observing a measured value of 0.92 or less) 95 percent of the time, is 0.84. Cutting the standard deviation of the measurement in half has cut the (absolute) width of the gray region in half, but left the width of the gray region in standard deviations unchanged. Previously, with σ = 10 percent, the width of the gray region was 1.0 –

0.68 = 0.32 = 3.2 (0.10) = 3.2σ. As Figure B1.9 illustrates, with σ = 5 percent, the width of the gray region is 1.0 – 0.84 = 0.16 = 3.2 (0.05) = 3.2σ.

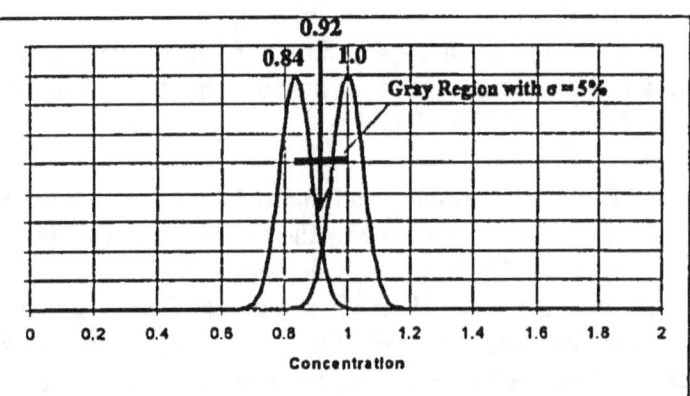

What is important is the width of the gray region in standard deviations; not the width of the gray region in concentration. In order to achieve the same specified Type II error rate at the LBGR, the action level and the LBGR must be separated by the same number of standard deviations. The width of the gray region (action level minus LBGR) will be denoted

FIGURE B1.9 — The true mean concentration is 0.84 (left) and 1.0 (right). The standard uncertainty of the distribution of measured concentrations is 0.05. The relative shift is 3.2.

by delta (Δ), the "shift." Δ/σ is how many standard deviations wide the gray region is. Δ/σ is called the "relative shift."

If the gray region is less than one standard deviation wide, the Type II error rate may be high at the LBGR. The only way to improve the situation would be to decrease the standard deviation (i.e., increase the relative shift, Δ/σ). This can be done by employing a more precise measurement method or by averaging several measurements. When the width of the gray region is larger than about three standard deviations (i.e., Δ/σ exceeds 3), it may be possible to use a simpler, less expensive measurement method or take fewer samples. Unnecessary effort should not be expended to achieve values of Δ/σ greater than 3.

B1.6 Summary

The mistake that is "worse" defines the null hypothesis and also defines a "Type I" error. The probability of a Type I error happening is called the "Type I error rate," and is denoted by alpha (α). Under the original null hypothesis (Case 1: Assume the true concentration is over 1.0), a

Type I error would be deciding that the concentration was less than 1.0 when it really was not. In general, a Type I error is deciding against the null hypothesis when it is actually true. (A Type I error is also called a "false positive." This can be confusing when the null hypothesis appears to be a "positive" statement. Therefore, MARLAP uses the neutral terminology.)

The "less serious" mistake is called a Type II error, and the probability of it happening is the "Type II error rate," denoted by beta (β). Under the original null hypothesis that the concentration was 1.0 or more, a Type II error would be deciding that the concentration was more than 1.0 when it really was not. In general, a Type II error is not deciding against the null hypothesis when it is actually false.

In both Case 1 and Case 2, the probability of both Type I errors and Type II errors were set to 5 percent. The probabilities were calculated at multiples of the standard deviation, assuming a normal distribution. The data may not always be well described by a normal distribution, so a different probability distribution may be used. However, the probability of a Type I error is always calculated as the probability that the decisionmaker will reject the null hypothesis when it is actually true. This is simple enough, as long as there is a clear boundary for the parameter of interest.

The parameter of interest in both Case 1 and Case 2 was the true concentration. The true concentration had a limit of 1.0. Therefore, all the project planning team had to do was calculate the probability that they would get a measured concentration that would cause them to decide that the true concentration was less than 1.0, even though it was equal to 1.0. In the example, the project planning team actually started with the probability (5 percent) and worked out the critical value. The "critical value" (or decision point) is the measured value that divides the measurement results into two different sets: (1) those values that will cause the null hypothesis to be rejected and (2) those values that will leave the null hypothesis as the default.

The Type I and Type II error rates, α and β, often are both set at 5 percent. This is only by tradition. Neither error rate needs to be set at 5 percent, nor do they have to be equal. The way the project planning team should set the value is by examining the consequences of making a Type I or a Type II error. What consequences will happen as a result of making each type of error? This is a little different than the criterion that was used to define the null hypothesis. It may be that in some circumstances, a Type II error is riskier than a Type I error. In that case, consider making α bigger than β.

ATTACHMENT B2
Decision Error Rates and the Gray Region for Detection Decisions

B2.1   Introduction

This section is provided to present some additional discussion on the subject of applying the Data Quality Objectives (DQO) process to the problem of measurement detection capability. In particular, "not detected" does not mean zero radioactivity concentration. To understand this, one needs to examine the concept of "minimum detectable concentration" (MDC). This involves the DQO process and limiting decision error rates.

B2.2   The DQO Process Applied to the Detection Limit Problem

STEP 1. PROBLEM STATEMENT
To determine if the material that is being measured contains radioactivity.

STEP 2. IDENTIFY THE DECISION
Decide if the material contains radioactivity at a level that requires action.

STEP 3. IDENTIFY INPUTS TO THE DECISION
What level of radioactivity in the material is important to detect?

STEP 4. DEFINE THE STUDY BOUNDARIES
How much material is to be measured, what instrumentation/analysis is available, how much time and resources are available for the measurements.

STEP 5. DEVELOP A DECISION RULE
This is an "if...then" rule that specifies the parameter of interest to be measured, and an action level against which it is compared in order to choose between alternative actions. At this stage, it is assumed that the true value of the parameter can be measured exactly without uncertainty. Such a decision rule in this case might be "If the true concentration in the sample is greater than zero, appropriate action will be taken. Otherwise, no action is required."

STEP 6. SPECIFY LIMITS ON DECISION ERROR RATES
Develop an operational rule so that when the measurement is made, a decision on the appropriate action to take can be made. This rule takes into account that there is uncertainty in any measurement, and therefore there is the possibility of making decision errors. When the material is processed and inserted into an instrument, the measurement is made and the instrument output is a result that is a number. The decision rule involves taking that numerical result and comparing it to a pre-determined number called the critical value. If the

result is greater than the critical value, the decision is made to treat the material as containing radioactivity above the action level, and then taking the appropriate action. The critical value will vary depending on the limits on decision errors rates that are specified.

The material either contains radioactivity or it does not. Unfortunately, it is impossible to determine absolutely whether the material does or does not contain radioactivity. Decisions can only be based on the result of measurements. There are four possibilities:

- The material *does not* contain radioactivity, and the measurement results in a value below the critical value and so it is *decided* that it *does not* contain radioactivity.

- The material *does* contain radioactivity, and the measurement results in a value above the critical value and so it is *decided* that it *does* contain radioactivity.

- The material *does not* contain radioactivity, and the measurement results in a value above the critical value and so it is *decided* that it *does* contain radioactivity. This would be a decision error.

- The material *does* contain radioactivity, and the measurement results in a value below the critical value and so it is *decided* that it *does not* contain radioactivity. This also would be a decision error.

Note that one never knows if a decision error is made, one only knows the result of the measurement. Measurements are not perfect, people make mistakes, and decision errors are unavoidable. However, recognizing that decision errors exist does allow their severity to be controlled. Several steps are necessary in order to create the framework for controlling decision error rates. These are described in the following sections.

B2.3   Establish the Concentration Range of Interest

Step three of the DQO process determined a level of radioactivity concentration in the material that is important to detect. This is also sometimes called an *action level* (such as the DCGL, a regulatory limit) that should not be exceeded. It is also important to define a region of interest ranging from the lowest to the highest average concentrations at which the contaminant is expected to exist. If the project planning team wants a method to measure sample concentrations around the action level, they would not select one that only worked at concentrations at 10 to 100 times the action level, nor would they select one that only worked from zero to half the action level. They would want a method that worked well around the action level—perhaps from 0.1 to 10 times the action level, or from one-half to two times the action level. For the purpose of the example in this attachment, the action level is 1.0 and the project planning team selected a region of interest that is zero to twice the action level (0–2), as shown on the x-axis of Figure B2.1. The first thing to notice is that Figure B2.1 ranges from – 1 to 2 and not 0 to 2. Why is this?

If a blank sample is placed in the instrument, the "true concentration" is zero. The instrument will produce a reading that is a number, and not necessarily the same reading each time. This is shown in Figure B2.2(a). Usually, the instrument output must be converted to a concentration value using a calibration factor. For simplicity, this example will assume that the

Figure B2.1 — Region of interest for the concentration around the action level of 1.0

calibration factor is 100, and in the remaining figures the measurement results will be shown directly in concentration. The zero point of concentration is at the average instrument reading when "nothing" (a blank) is being measured. In Figure B2.2(a) this is 100. The distribution of

many measurements of nothing will look like Figure B2.2(b). This is obtained from Figure B2.2(a) by subtracting the average blank reading (100) and dividing by the calibration factor (also 100). The spread in these measurement results is characterized by the standard deviation of this distribution. In Figure B2.2(b), the standard deviation is 0.2. For the problem to be actually addressed, the standard deviation may be larger or smaller than this, but it will not be zero. There is always some variability in measurements, and this will always cause some uncertainty about whether or not the decisions based on these measurements are correct.

Consider a possible decision rule: Decide that there is radioactivity in the sample if the measurement result is greater than zero. (This means that the "critical value" is zero.)

Figure B2.2 shows that if the critical value for the decision is made equal to zero, a decision that there is radioactivity in the

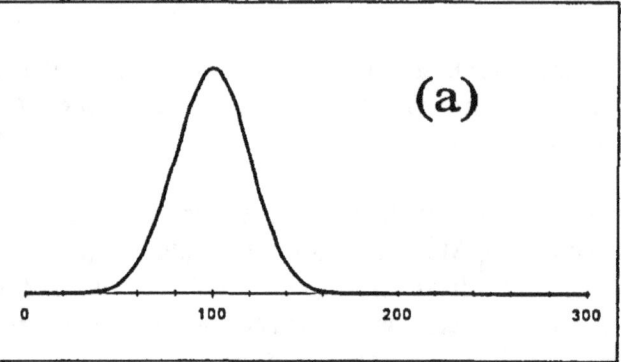

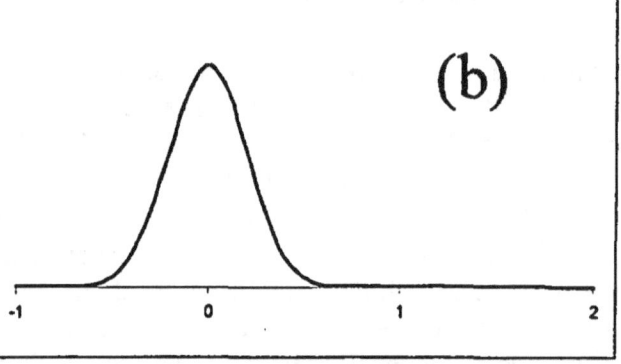

FIGURE B2.2 (a) — The distribution of blank (background) readings. (b) The true concentration is 0.0. The standard deviation of the distribution of measured concentrations is 0.2.

sample will be made about half the time, even when there is nothing to measure. Also, notice that unless the instrument reading is negative, it is not possible to decide that there is no radioactivity in the sample. There is nothing contradictory about this. The zero point on the x-axis was chosen simply to be the average measurement of "nothing." About half the time a measurement of nothing will be larger, and about half the time it will be smaller. This does not imply anything about concentrations being negative. It is about the variability of measurement readings, not the true concentration.

Decisionmakers might not be too happy about a decision rule that will lead them to the wrong conclusion half of the time. How can this be improved? Notice that if the critical value is made larger, the wrong conclusion (that there is radio-activity when there is none) will be made less often. If the critical value for the example is 0.2, it will be decided that there is radioactivity in the sample when the measurement result is greater than 0.2. From the example in Figure B2.3, this will be estimated to happen about 16 percent of the time.

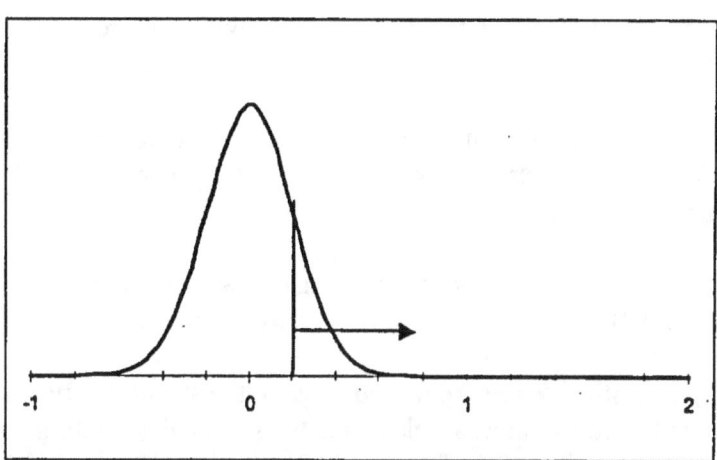

FIGURE B2.3 — The true concentration is 0.0, and the standard deviation of the distribution of measured concentrations is 0.2. A critical value of 0.2 would be exceeded about 16 percent of the time.

By making the critical value larger and larger, the probability can be reduced practically to zero of deciding that there is radioactivity when there is not. This apparently happy solution comes at a price. To see that, just consider the opposite situation. Suppose, instead of "nothing," there is a concentration of 0.2 in the sample (in this example, units are irrelevant). If a sample with this concentration is measured often, the distribution of results might look like Figure B2.4.

Notice that with a critical value of 0.2, a decision that there is radioactivity in this sample will only be made about half the time. Even if the critical value were zero,

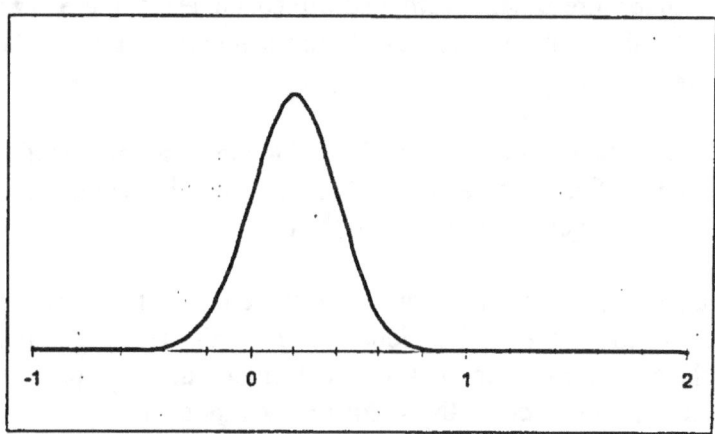

Figure B2.4 — The true concentration is 0.2 and the standard deviation of the distribution of measured concentrations is 0.2

a decision that there is radioactivity in the sample would only be made about 84 percent of the time.

As shown above, there are two types of decision errors that can be made: that there is radioactivity when there is not, or that there is no radioactivity when there is. What the figures show is that by making the critical value for the decision rule bigger, one can reduce the chances of making first kind of decision error, but doing so will increase the chance of making the second kind of decision error. Making the critical value for the decision rule smaller will reduce chances of the second kind of decision error, but will increase the chance of the first kind of decision error.

This example used a measurement variability (standard deviation) of 0.2. What if the variability is larger or smaller? By looking at the figures, one can conclude that *no matter what the variability actually is*:

(1) If a critical value of zero is used, one will conclude that there is radioactivity in a sample that actually contains nothing about half the time.

(2) If a critical value is selected equal to the standard deviation, one will conclude that there is radioactivity in a sample that actually contains nothing about 16 percent of the time. (A slight modification of the figures would show that if the critical value equals two times the standard deviation, one will conclude that there is radioactivity in a sample that actually contains nothing about 2.5 percent of the time.)

(3) If a critical value of zero is used, one will conclude that there is *no* radioactivity in a sample that actually contains a concentration that is numerically equal to the standard deviation about 16 percent of the time. (A slight modification of the figures would show that if the critical value were equal to zero, one will conclude that there is *no* radioactivity in a sample that actually contains a concentration equal to twice the standard deviation about 2.5 percent of the time.)

(4) If a critical value is selected equal to the standard deviation, one will conclude that there is *no* radioactivity in a sample that actually contains a concentration numerically equal to the standard deviation about half the time.

The key is to notice that it is not the numerical value of the variability alone nor the numerical value of the concentration alone, that determines the probability of a decision error. It is the ratio of the concentration to the standard deviation that is important. In essence, the standard deviation determines the *scale* of the x-axis (concentration axis) for this problem. Background determines the zero point of the concentration axis.

The MDC for a measurement process is the concentration that the sample must contain so that the probability of a Type II decision error is limited to 5 percent. (Other values may be chosen, but 5 percent is most commonly used in this context.)

This means that if a sample containing a concentration equal to the MDC is measured, about 95 percent of the time the measurement result will lead to the decision that the sample contains radioactivity. This is shown in Figure B2.5.

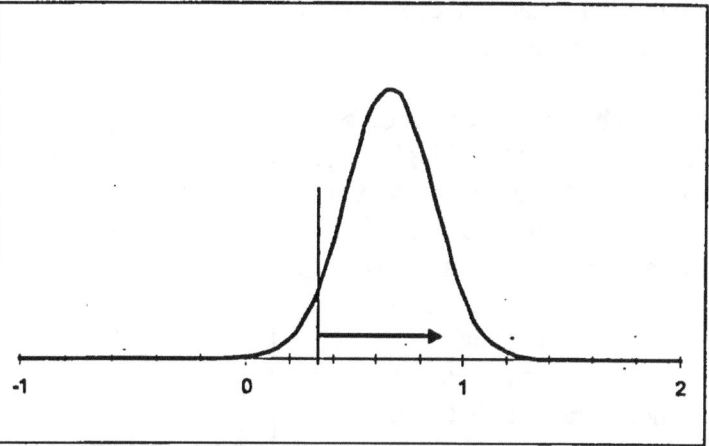

FIGURE B2.5 — The true value of the concentration is 0.66 and the standard deviation of the distribution of measured concentrations is 0.2. A critical value of 0.33 will be exceeded about 95 percent of the time.

However, if a sample containing a blank is measured, the probability that the measurement result will lead to the decision that the sample contains radioactivity will be only about 5 percent. This is shown in Figure B2.6.

For this example, Figure B2.7 summarizes the relationship among the distribution of measurements on a blank, the critical value, the MDC, and the action level.

The critical value used to limit the decision error of concluding that there is radioactivity in a sample that actually contains a blank to 5 percent, is about 1.5 or 2 times the

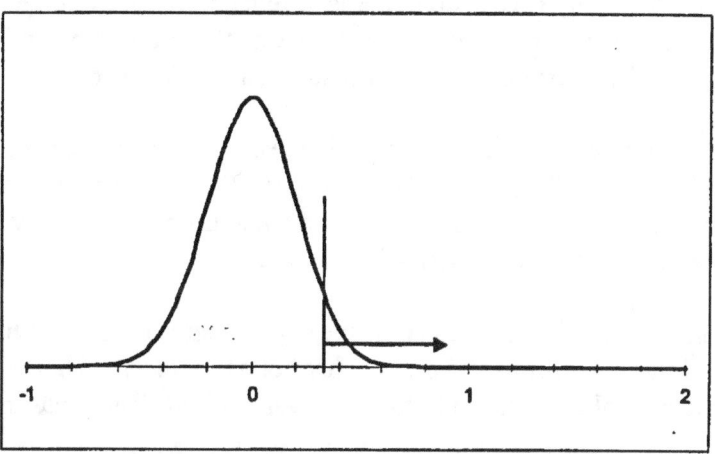

FIGURE B2.6 — The true value of the measured concentration is 0.0 and the standard deviation of the measured concentrations is 0.2. A critical value of 0.33 would be exceeded about 5 percent of the time.

measurement variability when measuring a blank. Limiting the decision error of concluding that there is no radioactivity in a sample that actually contains a concentration equal to the MDC, results in an MDC that is usually about twice the critical value. Consequently, the MDC is usually about 3 or 4 times the measurement variability when measuring a blank.

B2.4 Estimate the Measurement Variability when Measuring a Blank

The measurement variability when measuring a blank is thus a key parameter for planning. The best way to get a handle on this is by making many measurements of a blank sample and computing the standard deviation of the measurement results.

What can be concluded about the ability to measure "nothing" (i.e., no analyte)? Radioactivity present at a concentration less than the MDC may be detected, but less than 95 percent of the time. If the "true concentration" is at half the MDC (right at the critical value), the presence of radioactivity will be detected about half the time, and about half the time it will not. Concentrations lower than the critical value will be detected less

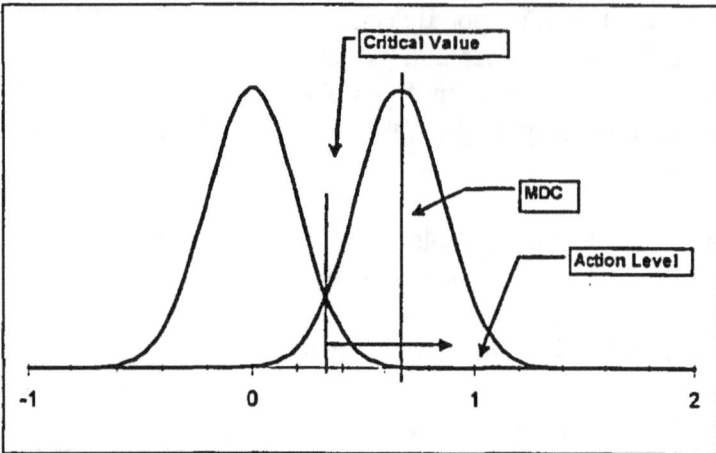

Figure B2.7 — The standard deviation of the normally distributed measured concentrations is 0.2. The critical value is 0.33, the MDC is 0.66 and the action level is 1.0.

often. The only way to do better is to reduce the measurement variability. This usually can only be done by either taking more measurements or by using an instrument or measurement process that has less variability when measuring a blank sample.

So what does it mean if a sample is measured, and a decision was made that there was no radioactivity? (This is another way of saying that no radioactivity was detected.) By itself, such a statement means nothing, and has no value unless one knows *the level of radioactivity that could be detected if it were there*—i.e. the MDC.

Similarly, a criterion for action specifying that no radioactivity be detected in a sample must be qualified by information on how hard one must look. That is, the MDC must be specified, which in turn implies a certain limit on the variability of the measurement procedure.

In either case, one can never measure zero. One can only decide from a measurement, with a prescribed limit on the probability of being wrong, that if enough radioactivity were there, it would be found. If it is not found, it does not mean it is not present; it only means that whatever might be there is unlikely to be more than the MDC.

In conclusion, an action level must be determined, and the MDC must be below it. Only then can radioactivity concentrations of concern can be detected with any degree of certainty. Conversely, specifying a measurement process implies an action level (level of concern) that is at or above the MDC.

APPENDIX C
MEASUREMENT QUALITY OBJECTIVES
FOR METHOD UNCERTAINTY AND
DETECTION AND QUANTIFICATION CAPABILITY

C.1 Introduction

This appendix expands on issues related to measurement quality objectives (MQOs) for several method performance characteristics which are introduced in Chapter 3, *Key Analytical Planning Issues and Developing Analytical Protocol Specifications*. Specifically, this appendix provides the rationale and guidance for establishing project-specific MQOs for the following method performance characteristics: method uncertainty, detection capability and quantification capability. In addition, it provides guidance in the development of these MQOs for use in the method selection process and guidance in the evaluation of laboratory data based on the MQOs. Section C.2 is a brief overview of statistical hypothesis testing as it is commonly used in a directed planning process, such as the Data Quality Objectives (DQO) Process (EPA, 2000). More information on this subject is provided in Chapter 2, *Project Planning Process* and Appendix B, *The Data Quality Objectives Process*. Section C.3 derives MARLAP's recommended criteria for establishing project-specific MQOs for method uncertainty, detection capability, and quantification capability. These criteria for method selection will meet the requirements of a statistically based decision-making process. Section C.4 derives MARLAP's recommended criteria for evaluation of the results of quality control analyses by project managers and data reviewers (see also Chapter 8, *Radiochemical Data Verification and Validation*).

It is assumed that the reader is familiar with the concepts of measurement uncertainty, detection capability, and quantification capability, and with terms such as "standard uncertainty," "minimum detectable concentration," and "minimum quantifiable concentration," which are introduced in Chapter 1, *Introduction to MARLAP*, and discussed in more detail in Chapter 20, *Detection and Quantification Capabilities*. MARLAP also uses the term "method uncertainty" to refer to the predicted uncertainty of the result that would be measured if the method were applied to a hypothetical laboratory sample with a specified analyte concentration. The method uncertainty is a characteristic of the analytical method and the measurement process.

C.2 Hypothesis Testing

Within the framework of a directed planning process, one considers an "action level," which is the contaminant concentration in either a population (e.g., a survey unit) or an individual

Contents

C.1 Introduction C-1
C.2 Hypothesis Testing C-1
C.3 Development of MQOs for Analytical Protocol Selection C-3
C.4 The Role of the MQO for Method Uncertainty in Data Evaluation C-9
C.5 References C-17

item (e.g., a laboratory sample) that should not be exceeded. Statistical hypothesis testing is used to decide whether the actual contaminant concentration, denoted by X, is greater than the action level, denoted by AL. For more information on this topic, see EPA (2000), MARSSIM (2000), NRC (1998), or Appendix B of this manual.

In hypothesis testing, one formulates two hypotheses about the value of X, and evaluates the measurement data to choose which hypothesis to accept and which to reject.[1] The two hypotheses are called the *null hypothesis* H_0 and the *alternative hypothesis* H_1. They are mutually exclusive and together describe all possible values of X under consideration. The null hypothesis is presumed true unless the data provide evidence to the contrary. Thus the choice of the null hypothesis determines the burden of proof in the test.

Most often, if the action level is not zero, one assumes it has been exceeded unless the measurement results provide evidence to the contrary. In this case, the null hypothesis is H_0: $X \geq$ AL and the alternative hypothesis is H_1: $X <$ AL. If one instead chooses to assume the action level has not been exceeded unless there is evidence to the contrary, then the null hypothesis is H_0: $X \leq$ AL and the alternative hypothesis is H_1: $X >$ AL. The latter approach is the only reasonable one if AL = 0, because it is virtually impossible to obtain statistical evidence that an analyte concentration is exactly zero.

For purposes of illustration, only the two forms of the null hypothesis described above will be considered. However, when AL > 0, it is also possible to select a null hypothesis that states that X does not exceed a specified value less than the action level (NRC, 1998). Although this third scenario is not explicitly addressed below, the guidance provided here can be adapted for it with few changes.

In any hypothesis test, there are two possible types of decision errors. A *Type I* error occurs if the null hypothesis is rejected when it is, in fact, true. A *Type II* error occurs if the null hypothesis is not rejected when it is false.[2] Since there is always measurement uncertainty, one cannot eliminate the possibility of decision errors. So instead, one specifies the maximum Type I decision error rate α that is allowable when the null hypothesis is true. This maximum usually occurs when $X =$ AL. The most commonly used value of α is 0.05, or 5 %. One also chooses another concentration, denoted here by DL (the "discrimination limit"), that one wishes to be able to distinguish reliably from the action level. One specifies the maximum Type II decision error rate

---

[1] In hypothesis testing, to "accept" the null hypothesis only means not to reject it, and for this reason many statisticians avoid the word "accept" in this context. A decision not to reject the null hypothesis does not imply the null hypothesis has been shown to be true.

[2] The terms "false positive" and "false negative" are synonyms for "Type I error" and "Type II error," respectively. However, MARLAP deliberately avoids these terms here, because they may be confusing when the null hypothesis is an apparently "positive" statement, such as $X \geq$ AL.

β that is allowable when $X = $ DL, or, alternatively, the "power" $1 - \beta$ of the statistical test when $X = $ DL. The *gray region* is then defined as the interval between the two concentrations AL and DL.

The gray region is a set of concentrations close to the action level, where one is willing to tolerate a Type II decision error rate that is higher than β. For concentrations above the upper bound of the gray region or below the lower bound, the decision error rate is no greater than the specified value (either a or β as appropriate). Ideally, the gray region should be narrow, but in practice, its width is determined by balancing the costs involved, including the cost of measurements and the estimated cost of a Type II error, possibly using prior information about the project and the parameter being measured.

If H_0 is $X \geq$ AL (presumed contaminated), then the upper bound of the gray region is AL and the lower bound is DL. If H_0 is $X \leq$ AL (presumed uncontaminated), then the lower bound of the gray region is AL and the upper bound is DL. Since no assumption is made here about which form of the null hypothesis is being used, the lower and upper bounds of the gray region will be denoted by LBGR and UBGR, respectively, and not by AL and DL. The width of the gray region (UBGR $-$ LBGR) is denoted by Δ and called the *shift* or the required *minimum detectable difference* in concentration (EPA, 2000; MARSSIM, 2000; NRC, 1998). See Appendix B, *The Data Quality Objectives Process*, for graphical illustrations of these concepts.

Chapter 3 of MARLAP recommends that for each radionuclide of concern, an action level, gray region, and limits on decision error rates be established during a directed planning process. Section C.3 presents guidance on the development of MQOs for the selection and development of analytical protocols. Two possible scenarios are considered. In the first scenario, the parameter of interest is the mean analyte concentration for a sampled population. The question to be answered is whether the population mean is above or below the action level. In the second scenario a decision is to be made about individual items or specimens, and not about population parameters. This is the typical scenario in bioassay, for example. Some projects may involve both scenarios. For example, project planners may want to know whether the mean analyte concentration in a survey unit is above an action level, but they may also be concerned about individual samples with high analyte concentrations.

C.3 Development of MQOs for Analytical Protocol Selection

This section derives MARLAP's recommendations for establishing MQOs for the analytical protocol selection and development process. Guidance is provided for establishing project-specific MQOs for method uncertainty, detection capability, and quantification capability. Once selected, these MQOs are used in the initial, ongoing, and final evaluations of the protocols. MARLAP considers two scenarios and develops MQOs for each.

SCENARIO I: A Decision Is to Be Made about the Mean of a Sampled Population

In this scenario the total variance of the data, σ^2, is the sum of two components

$$\sigma^2 = \sigma_M^2 + \sigma_S^2$$

where σ_M^2 is the average analytical method variance (M = "method" or "measurement") and σ_S^2 is the variance of the contaminant concentration in the sampled population (S = "sampling"). The sampling standard deviation σ_S may be affected by the spatial and temporal distribution of the analyte, the extent of the survey unit, the physical sample sizes, and the sample collection procedures. The analytical standard deviation σ_M is affected by laboratory sample preparation, subsampling, and analysis procedures. The value of σ_M may be estimated by the *combined standard uncertainty* of a measured value for a sample whose concentration equals the hypothesized population mean concentration (see Chapter 19, *Measurement Uncertainty*).

The ratio Δ / σ, called the "relative shift," determines the number of samples required to achieve the desired decision error rates α and β. The target value for this ratio should be between 1 and 3, as explained in MARSSIM (2000) and NRC (1998). Ideally, to keep the required number of samples low, one prefers that $\Delta / \sigma \approx 3$. The cost in number of samples rises rapidly as the ratio Δ / σ falls below 1, but there is little benefit from increasing the ratio much above 3.

Generally, it is easier to control σ_M than σ_S. If σ_S is known (approximately), a target value for σ_M can be determined. For example, if $\sigma_S < \Delta / 3$, then a value of σ_M no greater than $\sqrt{\Delta^2 / 9 - \sigma_S^2}$ ensures that $\sigma \leq \Delta / 3$, as desired. If $\sigma_S > \Delta / 3$, the requirement that the total σ be less than $\Delta / 3$ cannot be met regardless of σ_M. In the latter case, it is sufficient to make σ_M negligible in comparison to σ_S. Generally, σ_M can be considered negligible if it is no greater than about $\sigma_S / 3$.

Often one needs a method for choosing σ_M in the absence of specific information about σ_S. In this situation, MARLAP recommends the requirement $\sigma_M \leq \Delta / 10$ by default. The recommendation is justified below.

Since it is desirable to have $\sigma \leq \Delta / 3$, this condition is adopted as a primary requirement. Assume for the moment that σ_S is large. Then σ_M should be made negligible by comparison. As mentioned above, σ_M can be considered negligible if it is no greater than $\sigma_S / 3$. When this condition is met, further reduction of σ_M has little effect on σ and therefore is usually not cost-effective. So, the inequality $\sigma_M \leq \sigma_S / 3$ is adopted as a second requirement.

Algebraic manipulation of the equation $\sigma^2 = \sigma_M^2 + \sigma_S^2$ and the required inequality $\sigma_M \leq \sigma_S / 3$ gives

$$\sigma_M \le \frac{\sigma}{\sqrt{10}}$$

The inequalities $\sigma \le \Delta / 3$ and $\sigma_M \le \sigma / \sqrt{10}$ together imply the requirement

$$\sigma_M \le \frac{\Delta}{3\sqrt{10}}$$

or approximately

$$\sigma_M \le \frac{\Delta}{10}$$

The required upper bound for the standard deviation σ_M will be denoted by σ_{MR}. MARLAP recommends the equation

$$\sigma_{MR} = \frac{\Delta}{10}$$

by default as a requirement in Scenario I when σ_S is unknown. This upper bound was derived from the assumption that σ_S was large, but it also ensures that the primary requirement $\sigma \le \Delta / 3$ will be met if σ_S is small. When the analytical standard deviation σ_M is less than σ_{MR}, the primary requirement will be met unless the sampling variance, σ_S^2, is so large that σ_M^2 is negligible by comparison, in which case little benefit can be obtained from further reduction of σ_M.

The recommended value of σ_{MR} is based on the assumption that any known bias in the measurement process has been corrected and that any remaining bias is much smaller than the shift, Δ, when a concentration near the gray region is measured. (See Chapter 6, which describes a procedure for testing for bias in the measurement process.)

Achieving an analytical standard deviation σ_M less than the recommended limit, $\Delta / 10$, may be difficult in some situations, particularly when the shift, Δ, is only a fraction of UBGR. When the recommended requirement for σ_M is too costly to meet, project planners may allow σ_{MR} to be larger, especially if σ_S is believed to be small or if it is not costly to analyze the additional samples required because of the larger overall data variance ($\sigma_M^2 + \sigma_S^2$). In this case, project planners may choose σ_{MR} to be as large as $\Delta / 3$ or any calculated value that allows the data quality objectives to be met at an acceptable cost.

The true standard deviation, σ_M, is a theoretical quantity and is never known exactly, but the laboratory may estimate its value using the methods described in Chapter 19, and Section 19.5.13 in particular. The laboratory's estimate of σ_M will be denoted here by u_M and called the "method uncertainty." The method uncertainty, when estimated by uncertainty propagation, is the predicted value of the combined standard uncertainty ("one-sigma" uncertainty) of the analytical result for a laboratory sample whose concentration equals UBGR. Note that the term "method uncertainty" and the symbol u_M actually apply not only to the method but to the entire measurement process.

In theory, the value σ_{MR} is intended to be an upper bound for the true standard deviation of the measurement process, σ_M, which is unknown. In practice, σ_{MR} is actually used as an upper bound for the method uncertainty, u_M, which may be calculated. Therefore, the value of σ_{MR} will be called the "required method uncertainty" and denoted by u_{MR}. As noted in Chapter 3, MARLAP recommends that project planners specify an MQO for the method uncertainty, expressed in terms of u_{MR}, for each analyte and matrix.

The MQO for method uncertainty is expressed above in terms of the required standard deviation of the measurement process for a laboratory sample whose analyte concentration is at or above UBGR. In principle the same MQO may be expressed as a requirement that the minimum quantifiable concentration (MQC) be less than or equal to UBGR. Chapter 20 defines the MQC as the analyte concentration at which the relative standard deviation of the measured value (i.e., the relative method uncertainty) is $1 / k_Q$, where k_Q is some specified positive value. The value of k_Q in this case should be specified as $k_Q = $ UBGR $/ u_{MR}$. In fact, if the lower bound of the gray region is zero, then one obtains $k_Q = 10$, which is the value most commonly used to define the MQC in other contexts. In practice the requirement for method uncertainty should only be expressed in terms of the MQC when $k_Q = 10$, since to define the MQC with any other value of k_Q may lead to confusion.

EXAMPLE C.1 Suppose the action level is 1 Bq/g and the lower bound of the gray region is 0.6 Bq/g. If decisions are to be made about survey units based on samples, then the required method uncertainty at 1 Bq/g is

$$u_{MR} = \frac{\Delta}{10} = \frac{1\,\text{Bq/g} - 0.6\,\text{Bq/g}}{10} = 0.04\,\text{Bq/g}$$

If this uncertainty cannot be achieved, then an uncertainty as large as $\Delta / 3 = 0.13$ Bq/g may be allowed if σ_S is small or if more samples are taken per survey unit.

EXAMPLE C.2 Again suppose the action level is 1 Bq/g, but this time assume the lower bound of the gray region is 0 Bq/g. In this case the required method uncertainty at 1 Bq/g is

$$u_{MR} = \frac{\Delta}{10} = \frac{1\,\text{Bq/g} - 0\,\text{Bq/g}}{10} = 0.1\,\text{Bq/g}$$

A common practice in the past has been to select an analytical method based on the *minimum detectable concentration* (MDC), which is defined in Chapter 20, *Detection and Quantification Capabilities*. For example, the Multi-Agency Radiation Survey and Site Investigation Manual (MARSSIM, 2000) says:

> During survey design, it is generally considered good practice to select a measurement system with an MDC between 10-50% of the DCGL [action level].

Such guidance implicitly recognizes that for cases when the decision to be made concerns the mean of a population that is represented by multiple laboratory samples, criteria based on the MDC may not be sufficient and a somewhat more stringent requirement is needed. It is interesting to note that the requirement that the MDC (about 3 times σ_M) be 10 % to 50 % of the action level is tantamount to requiring that σ_M be 0.03 to 0.17 times the action level—in other words, the relative standard deviation should be approximately 10 % at the action level. Thus, the requirement is more naturally expressed in terms of the MQC.

SCENARIO II: Decisions Are to Be Made about Individual Items

In this scenario, the total variance of the data equals the analytical variance, σ_M^2, and the data distribution in most instances should be approximately normal. The decision in this case may be made by comparing the measured concentration, x, plus or minus a multiple of its combined standard uncertainty to the action level. The combined standard uncertainty, $u_c(x)$, is assumed to be an estimate of the true standard deviation of the measurement process as applied to the item being measured; so, the multiplier of $u_c(x)$ equals $z_{1-\alpha}$, the $(1-\alpha)$-quantile of the standard normal distribution (see Appendix G, *Statistical Tables*).

Alternatively, if AL = 0, so that any detectable amount of analyte is of concern, the decision may involve comparing x to the critical value of the concentration, x_C, as defined in Chapter 20, *Detection and Quantification Capabilities*.

Case II-1: Suppose the null hypothesis is $X \geq$ AL, so that the action level is the upper bound of the gray region. Given the analytical variance σ_M^2, only a measured result that is less than about $UBGR - z_{1-\alpha}\sigma_M$ will be judged to be clearly less than the action level. Then the desired power of the test $1 - \beta$ is achieved at the lower bound of the gray region only if $LBGR \leq UBGR - z_{1-\alpha}\sigma_M - z_{1-\beta}\sigma_M$. Algebraic manipulation transforms this requirement to

$$\sigma_M \leq \frac{UBGR - LBGR}{z_{1-\alpha} + z_{1-\beta}} = \frac{\Delta}{z_{1-\alpha} + z_{1-\beta}}$$

Case II-2: Suppose the null hypothesis is $X \leq$ AL, so that the action level is the lower bound of the gray region. In this case, only a measured result that is greater than about $LBGR + z_{1-\alpha}\sigma_M$

will be judged to be clearly greater than the action level. The desired power of the test $1 - \beta$ is achieved at the upper bound of the gray region only if UBGR $\geq$ LBGR $+ z_{1-\alpha}\sigma_M + z_{1-\beta}\sigma_M$. Algebraic manipulation transforms this requirement to

$$\sigma_M \leq \frac{\text{UBGR} - \text{LBGR}}{z_{1-\alpha} + z_{1-\beta}} = \frac{\Delta}{z_{1-\alpha} + z_{1-\beta}}$$

So, in either case, the requirement remains that:

$$\sigma_M \leq \frac{\Delta}{z_{1-\alpha} + z_{1-\beta}}$$

Therefore, MARLAP recommends the use of the equation

$$u_{MR} = \sigma_{MR} = \frac{\Delta}{z_{1-\alpha} + z_{1-\beta}}$$

as an MQO for method uncertainty when decisions are to be made about individual items (i.e., laboratory samples) and not about population parameters.

If both α and β are at least 0.05, one may use the value $u_{MR} = 0.3\Delta$.

The recommended value of u_{MR} is based on the assumption that any known bias in the measurement process has been corrected and that any remaining bias is small relative to the method uncertainty.

If LBGR $= 0$, then $\Delta =$ UBGR and $\sigma_{MR} = \Delta / (z_{1-\alpha} + z_{1-\beta})$ implies

$$\sigma_M \leq \frac{\text{UBGR}}{z_{1-\alpha} + z_{1-\beta}}$$

This requirement is essentially equivalent to requiring that the MDC not exceed UBGR. Thus, when LBGR $= 0$, the MQO may be expressed in terms of the detection capability of the analytical method.

Note that when AL $=$ LBGR $= 0$, the MQO for detection capability may be derived directly in terms of the MDC, since the MDC is defined as the analyte concentration at which the probability of detection is $1 - \beta$ when the detection criterion is such that the probability of false detection in a sample with zero analyte concentration is at most α.

> **EXAMPLE C.3** Suppose the action level is 1 Bq/L, the lower bound of the gray region is 0.5 Bq/L, $\alpha = 0.05$, and $\beta = 0.10$. If decisions are to be made about individual items, then the required method uncertainty at 1 Bq/L is
>
> $$u_{MR} = \frac{\Delta}{z_{1-\alpha} + z_{1-\beta}} = \frac{1\,\text{Bq/L} - 0.5\,\text{Bq/L}}{z_{0.95} + z_{0.90}} = \frac{0.5\,\text{Bq/L}}{1.645 + 1.282} = 0.17\,\text{Bq/L}.$$

C.4 The Role of the MQO for Method Uncertainty in Data Evaluation

This section provides guidance and equations for determining warning and control limits for QC sample results based on the project-specific MQO for method uncertainty. In the MARLAP Process as described in Chapter 1, these warning and control limits are used in the ongoing evaluation of protocol performance (see Chapter 7, *Evaluating Methods and Laboratories*) and in the evaluation of the laboratory data (see Chapter 8, *Radiochemical Data Verification and Validation*).

C.4.1 Uncertainty Requirements at Various Concentrations

When project planners follow MARLAP's recommendations for establishing MQOs for method uncertainty for method selection and development, the maximum allowable standard deviation, σ_{MR}, at the upper bound of the gray region is specified. During subsequent data evaluation, the standard deviation at any concentration less than UBGR should be at most σ_{MR}, and the relative standard deviation at any concentration greater than UBGR should be at most σ_{MR}/UBGR, which will be denoted here by φ_{MR}. Note that, since the true standard deviation can never be known exactly, in practice the requirement is expressed in terms of the required method uncertainty, u_{MR}, to which the combined standard uncertainty of each result may be compared.

> **EXAMPLE C.4** Consider the preceding example, in which AL = UBGR = 1 Bq/L, LBGR = 0.5 Bq/L, and $u_{MR} = 0.17$ Bq/L. In this case the combined standard uncertainty for any measured result, x, should be at most 0.17 Bq/L if $x < 1$ Bq/L, and the relative combined standard uncertainty should be at most 0.17 / 1, or 17 %, if $x > 1$ Bq/L.

In Scenario I, where decisions are made about the mean of a population based on multiple physical samples (e.g., from a survey unit), if the default value $u_{MR} = \Delta / 10$ is assumed for the required method uncertainty, then the required bound for the analytical standard deviation as a function of concentration is as shown in Figure C.1. The figure shows that the bound, u_{Req}, is constant at all concentrations, x, below UBGR, and u_{Req} increases with x when x is above UBGR. So, $u_{Req} = u_{MR}$ when $x < $ UBGR and $u_{Req} = x \cdot u_{MR}$ /UBGR when $x > $ UBGR.

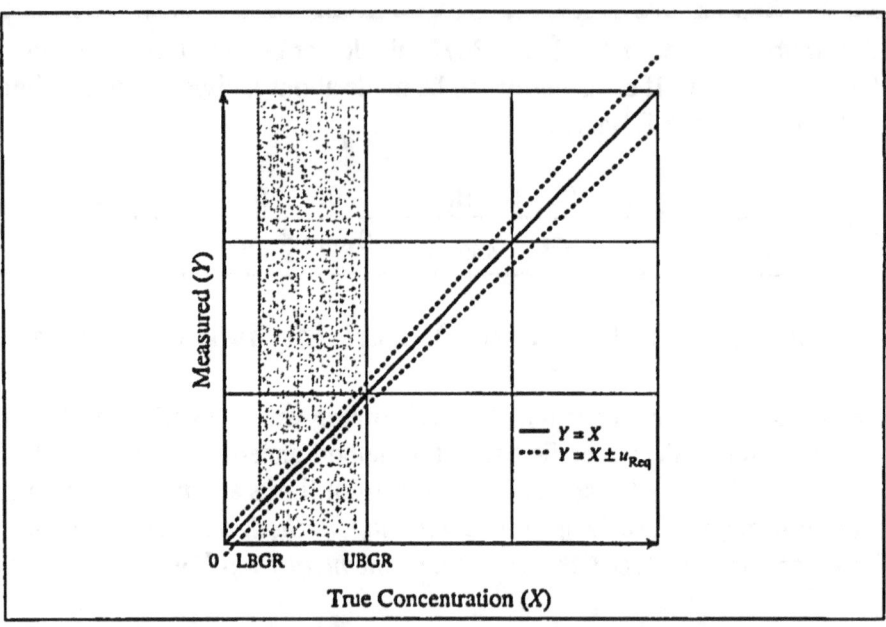

FIGURE C.1 — Required analytical standard deviation (u_{Req})

These requirements can be relaxed somewhat for samples with very high analyte concentrations as long as the project's requirements for decision uncertainty are met. However, MARLAP does not provide specific guidance to address this issue for Scenario I.

In Scenario II, where decisions are made about individual physical samples, it is possible to widen the required bounds for the standard deviation at any concentration outside the gray region. For example, suppose UBGR = AL, LBGR is set at some concentration below UBGR, and the decision error probabilities α and β are specified. Then the project planners require the probability of a Type I error not to exceed α when the true concentration is at or above UBGR, and they require the probability of a Type II error not to exceed β when the true concentration is at or below LBGR. The decision rule is based on the combined standard uncertainty of the measurement result: any sample whose measured concentration, x, exceeds AL minus $z_{1-\alpha}$ times the combined standard uncertainty, $u_c(x)$, is assumed to exceed the action level. So, assuming $u_c(x)$ is an adequate estimate of the analytical standard deviation, the planners' objectives are met if

$$u_c(x) \leq \begin{cases} \dfrac{\text{UBGR} - x}{z_{1-\alpha} + z_{1-\beta}}, & \text{if } x \leq \text{LBGR} \\[2ex] \dfrac{x - \text{LBGR}}{z_{1-\alpha} + z_{1-\beta}}, & \text{if } x \geq \text{UBGR} \\[2ex] \dfrac{\Delta}{z_{1-\alpha} + z_{1-\beta}}, & \text{if LBGR} \leq x \leq \text{UBGR} \end{cases}$$

EXAMPLE C.5 Consider the earlier example in which AL = UBGR = 1.0 Bq/L, LBGR = 0.5 Bq/L, $a = 0.05$, $\beta = 0.10$, and $u_{MR} = 0.17$ Bq/L. The less restrictive uncertainty requirement can be expressed as

$$
u_c(x) \leq \begin{cases} \dfrac{1.0\,\text{Bq/L} - x}{2.927}, & \text{if } x \leq 0.5\,\text{Bq/L} \\[2ex] \dfrac{x - 0.5\,\text{Bq/L}}{2.927}, & \text{if } x \geq 1.0\,\text{Bq/L} \\[2ex] 0.17, & \text{if } 0.5\,\text{Bq/L} \leq x \leq 1.0\,\text{Bq/L} \end{cases}
$$

So, if $x = 0$, the requirement is $u_c(x) \leq (1\ \text{Bq/L}) / 2.927 = 0.34$ Bq/L, and, if $x = 2$ Bq/L, the requirement is $u_c(x) \leq (2\ \text{Bq/L} - 0.5\ \text{Bq/L}) / 2.927 = 0.51$ Bq/L, which is approximately 26 % in relative terms.

C.4.2 Acceptance Criteria for Quality Control Samples

The next issue to be addressed is how to set warning and control limits for quality control (QC) sample results. These limits will be used by project data assessors to determine whether the laboratory appears to be meeting MQOs. Presumably the lab has stricter internal QC requirements (see Chapter 18, *Laboratory Quality Control*).

The development of acceptance criteria for QC samples will be illustrated with an example. Assume UBGR = 5 Bq/g (soil) and LBGR = 1.5 Bq/g. The width of the gray region is $\Delta = 5 - 1.5 = 3.5$ Bq/g. Project planners, following MARLAP's guidance, choose the required method uncertainty at 5 Bq/g (UBGR) to be

$$
u_{MR} = \frac{\Delta}{10} = 0.35\,\text{Bq/g}
$$

or 7 %. So, the maximum standard uncertainty at analyte concentrations less than 5 Bq/g should be $u_{MR} = 0.35$ Bq/g, and the maximum *relative* standard uncertainty at concentrations greater than 5 Bq/g should be $\varphi_{MR} = 0.07$, or 7 %.

Although it is possible to relax these uncertainty criteria for samples with very high analyte concentrations, MARLAP recommends that the original criteria be used to develop acceptance limits for the results of QC sample analyses.

C.4.2.1 Laboratory Control Samples

It is assumed that the concentration of a laboratory control sample (LCS) is high enough that the relative uncertainty limit $\varphi_{MR} = 0.07$ is appropriate. The *percent deviation* for the LCS analysis is defined as

$$\%D = \frac{SSR - SA}{SA} \times 100\ \%$$

where
  SSR   is the measured result (spiked sample result) and
  SA   is the spike activity (or concentration) added.

It is assumed that the uncertainty of SA is negligible; so, the maximum allowable relative standard deviation of $\%D$ is the same as that of the measured result itself, or $\varphi_{MR} \times 100\ \%$. Then the 2-sigma warning limits for $\%D$ are $\pm 2\varphi_{MR} \times 100\ \%$ and the 3-sigma control limits are $\pm 3\varphi_{MR} \times 100\ \%$. (In situations where φ_{MR} is very small, the uncertainty of SA should not be ignored.)

The requirements for LCSs are summarized below.

Laboratory Control Samples

Statistic: $\%D = \dfrac{SSR - SA}{SA} \times 100\ \%$

Warning limits: $\pm 2\varphi_{MR} \times 100\ \%$
Control limits: $\pm 3\varphi_{MR} \times 100\ \%$

EXAMPLE C.6

(UBGR = 5 Bq/g, $u_{MR} = 0.35$ Bq/g, $\varphi_{MR} = 0.07$.)

Suppose an LCS is prepared with a concentration of SA = 10 Bq/g and the result of the analysis is 11.61 Bq/g with a combined standard uncertainty of 0.75 Bq/g. Then

$$\%D = \frac{11.61\ \text{Bq/g} - 10\ \text{Bq/g}}{10\ \text{Bq/g}} \times 100\ \% = 16.1\ \%$$

The warning limits in this case are

$$\pm 2\varphi_{MR} \times 100\% = \pm 14\%$$

and the control limits are

$$\pm 3\varphi_{MR} \times 100\% = \pm 21\%$$

So, the calculated value of %D is above the upper warning limit but below the control limit.

C.4.2.2 Duplicate Analyses

Acceptance criteria for duplicate analysis results depend on the sample concentration, which is estimated by the average $\bar{x}$ of the two measured results x_1 and x_2.

$$\bar{x} = \frac{x_1 + x_2}{2}$$

When $\bar{x} <$ UBGR, the warning limit for the absolute difference $|x_1 - x_2|$ is

$$2u_{MR}\sqrt{2} \approx 2.83\, u_{MR}$$

and the control limit is

$$3u_{MR}\sqrt{2} \approx 4.24\, u_{MR}$$

Only upper limits are used, because the absolute value $|x_1 - x_2|$ is being tested.

When $\bar{x} \geq$ UBGR, the acceptance criteria may be expressed in terms of the *relative percent difference* (RPD), which is defined as

$$\text{RPD} = \frac{|x_1 - x_2|}{\bar{x}} \times 100\%$$

The warning limit for RPD is

$$2\varphi_{MR}\sqrt{2} \times 100\% \approx 2.83\, \varphi_{MR} \times 100\%$$

and the control limit is

$$3\varphi_{MR}\sqrt{2} \times 100\% \approx 4.24\, \varphi_{MR} \times 100\%$$

The requirements for duplicate analyses are summarized below.

Duplicate Analyses

If $\bar{x} <$ UBGR:
    Statistic: $|x_1 - x_2|$
    Warning limit: $2.83\, u_{MR}$
    Control limit: $4.24\, u_{MR}$

If $\bar{x} \geq$ UBGR:

    Statistic: $RPD = \dfrac{|x_1 - x_2|}{\bar{x}} \times 100\ \%$

    Warning limit: $2.83\, \varphi_{MR} \times 100\ \%$
    Control limit: $4.24\, \varphi_{MR} \times 100\ \%$

EXAMPLE C.7

(UBGR = 5 Bq/g, $u_{MR} = 0.35$ Bq/g, $\varphi_{MR} = 0.07$)

Suppose duplicate analyses are performed on a laboratory sample and the results of the two measurements are

    $x_1 = 9.0$ Bq/g with combined standard uncertainty $u_c(x_1) = 2.0$ Bq/g
    $x_2 = 13.2$ Bq/g with combined standard uncertainty $u_c(x_2) = 2.1$ Bq/g

The duplicate results are evaluated as follows.

$$\bar{x} = \frac{9.0\,\text{Bq/g} + 13.2\,\text{Bq/g}}{2} = 11.1\ \text{Bq/g}$$

Since $\bar{x} \geq 5$ Bq/g, the acceptance criteria are expressed in terms of RPD.

$$RPD = \frac{|9.0\,\text{Bq/g} - 13.2\,\text{Bq/g}|}{11.1\,\text{Bq/g}} \times 100\ \% = 37.84\ \%$$

The warning and control limits for RPD are

    Warning limit = $2.83 \times 0.07 \times 100\ \% = 19.81\ \%$
    Control limit = $4.24 \times 0.07 \times 100\ \% = 29.68\ \%$

In this case, the value of RPD is above the control limit. (Also note that the relative standard uncertainties are larger than the 7 % required for concentrations above 5 Bq/g.)

C.4.2.3 Method Blanks

Case 1. If an aliquant of blank material is analyzed, or if a nominal aliquant size is used in the data reduction, the measured blank result is an activity concentration. The target value is zero, but the measured value may be either positive or negative. So, the 2-sigma warning limits are $\pm 2u_{MR}$ and the 3-sigma control limits are $\pm 3u_{MR}$.

Case 2. If no blank material is involved (only reagents, tracers, etc., are used), the measured result may be a total activity, not a concentration. In this case the method uncertainty limit u_{MR} should be multiplied by the nominal or typical aliquant size, m_S. Then the 2-sigma warning limits are $\pm 2u_{MR}m_S$ and the 3-sigma control limits are $\pm 3u_{MR}m_S$.

The requirements for method blanks are summarized below.

Method Blanks

Concentration:
    Statistic: Measured concentration
    Warning limits: $\pm 2u_{MR}$
    Control limits: $\pm 3u_{MR}$

Total Activity:
    Statistic: Measured total activity
    Warning limits: $\pm 2u_{MR}m_S$
    Control limits: $\pm 3u_{MR}m_S$

EXAMPLE C.8

(UBGR = 5 Bq/g, u_{MR} = 0.35 Bq/g, φ_{MR} = 0.07)

Suppose a method blank is analyzed and the result of the measurement is

x = 0.00020 Bq with combined standard uncertainty $u_c(x)$ = 0.00010 Bq

Assuming the nominal aliquant mass is 1.0 g, or m_S = 0.001 g, the result is evaluated by comparing x to the warning and control limits:

$$\pm 2u_{MR}m_S = \pm 0.00070 \text{ Bq}$$
$$\pm 3u_{MR}m_S = \pm 0.00105 \text{ Bq}$$

In this case x is within the warning limits.

C.4.2.4 Matrix Spikes

The acceptance criteria for matrix spikes are more complicated than those described above for laboratory control samples because of pre-existing activity in the unspiked sample, which must be measured and subtracted from the activity measured after spiking. The *percent deviation* for a matrix spike is defined as

$$\%D = \frac{SSR - SR - SA}{SA} \times 100 \%$$

where

SSR   is the spiked sample result
SR   is the unspiked sample result
SA   is the spike concentration added (total activity divided by aliquant size).

However, warning and control limits for %D depend on the measured values; so, %D is not a good statistic to use for matrix spikes. A better statistic is the "Z score":

$$Z = \frac{SSR - SR - SA}{\varphi_{MR}\sqrt{SSR^2 + \max(SR, UBGR)^2}}$$

where "max(x, y)" denotes the maximum of x and y. Then warning and control limits for Z are set at ± 2 and ± 3, respectively. (It is assumed again that the uncertainty of SA is negligible.) The requirements for matrix spikes are summarized below.

Matrix Spikes

Statistic: $Z = \dfrac{SSR - SR - SA}{\varphi_{MR}\sqrt{SSR^2 + \max(SR, UBGR)^2}}$

Warning limits: ± 2
Control limits: ± 3

EXAMPLE C.9

(UBGR = 5 Bq/g, u_{MR} = 0.35 Bq/g, φ_{MR} = 0.07)

Suppose a matrix spike is analyzed. The result of the original (unspiked) analysis is

---

$SR = 3.5$ Bq/g with combined standard uncertainty $u_c(SR) = 0.29$ Bq/g

the spike concentration added is

$SA = 10.1$ Bq/g with combined standard uncertainty $u_c(SA) = 0.31$ Bq/g

and the result of the analysis of the spiked sample is

$SSR = 11.2$ Bq/g with combined standard uncertainty $u_c(SSR) = 0.55$ Bq/g

Since SR is less than UBGR (5), max(SR, UBGR) = UBGR = 5. So,

$$Z = \frac{SSR - SR - SA}{\varphi_{MR}\sqrt{SSR^2 + UBGR^2}} = \frac{11.2 \text{ Bq/g} - 3.5 \text{ Bq/g} - 10.1 \text{ Bq/g}}{0.07\sqrt{(11.2 \text{ Bq/g})^2 + (5 \text{ Bq/g})^2}} = -2.80$$

So, Z is less than the lower warning limit (–2) but slightly greater than the lower control limit (–3).

C.5   References

U.S. Environmental Protection Agency (EPA). 2000. *Guidance for the Data Quality Objectives (DQO) Process*. EPA QA/G-4. EPA/600/R-96/055. Office of Environmental Information, Washington, DC. Available at www.epa.gov/quality/qa_docs.html.

MARSSIM. 2000. *Multi-agency Radiation Survey and Site Investigation Manual* (MARSSIM) Rev. 1. NUREG-1575, Nuclear Regulatory Commission, Washington, DC. EPA 402-R-97-016, Environmental Protection Agency, Washington, DC. Available from www.epa.gov/radiation/marssim/.

Nuclear Regulatory Commission (NRC). 1998. *A Nonparametric Statistical Methodology for the Design and Analysis of Final Status Decommissioning Surveys*. NUREG-1505. NRC, Washington, DC.

APPENDIX D
CONTENT OF PROJECT PLAN DOCUMENTS

D.1 Introduction

Project plan documents were discussed in Chapter 4, *Project Plan Documents*. This appendix will discuss appropriate content of plan documents. The content of project plan documents, regardless of the document title or format, will include similar information, including the project description and objectives, identification of those involved in the project activities and their responsibilities and authorities, enumeration of the quality control (QC) procedures to be followed, reference to specific standard operating procedures (SOPs) that will be followed for all aspects of the projects, and Health and Safety protocols.

The discussion of project plan document content in this appendix will rely on EPA's guidance on elements for a QA project plan (QAPP). MARLAP selected EPA's QAPP as a model for content of a project plan document since it is closely associated with the data quality objective (DQO) planning process and because other plan documents lack widely accepted guidance regarding content. MARLAP hopes that presentation of a project plan document in one of the most commonly used plan formats will facilitate plan writing by those less familiar with the task, provide a framework for reviewing plan documents, and aid in tracking projects.

The discussion of plan content in Sections D2 to D5 follows the outline developed by EPA requirements (EPA, 2001) and guidance (EPA, 2002) for QAPPs for environmental data operations. The QAPP elements are presented in four major sections (Table D.1) that are referred to as "groups":

- Project Management;
- Measurement/Data Acquisition;
- Assessment/Oversight; and
- Data Validation and Usability.

There are many formats that can be used to present the project plan elements. MARLAP does not recommend any particular plan format over another. The project planning team should focus on the appropriate content of plan documents needed to address the necessary quality assurance (QA), QC, and other technical activities that must be implemented to ensure that the results of the work performed will satisfy the stated performance criteria. Table D.2 provides a crosswalk between the table of

Contents

D.1 Introduction	D-1
D.2 Group A: Project Management	D-5
D.3 Group B: Measurement/Data Acquisition	D-14
D.4 Group C: Assessment/Oversight	D-26
D.5 Group D: Data Validation and Usability	D-28
D.6 References	D-31

contents of two example project plan documents—a QAPP and a work plan—and EPA's (2002) project plan document elements.

TABLE D.1—QAPP groups and elements [a,b]

GROUP	ID	ELEMENT	APPENDIX SECTION	MARLAP CHAPTER
A Project Management	A1	Title and Approval Sheet	D2.1	NA
	A2	Table of Contents	D2.2	NA
	A3	Distribution List	D2.3	NA
	A4	Project/Task Organization	D2.4	2
	A5	Problem Definition/Background	D2.5	2
	A6	Project/Task Description	D2.6	2
	A7	Quality Objectives and Criteria for Measurement Data	D2.7	2, 3
	A8	Special Training Requirements/Certifications	D2.8	7
	A9	Documentation and Record	D2.9	7, 16
B Measurement/Data Acquisition	B1	Sampling Process Design	D3.1	NA
	B2	Sample Methods Requirements	D3.2	NA
	B3	Sample Handling and Custody Requirements	D3.3	11
	B4	Analytical Methods Requirements	D3.4	6
	B5	QC Requirements	D3.5	18
	B6	Instrument/Equipment Testing, Inspection and Maintenance Requirements	D3.6	15
	B7	Instrument Calibrations and Frequency	D3.7	18
	B8	Inspection/Acceptance Requirements for Supplies and Consumables	D3.8	NA
	B9	Data Acquisition Requirements (Non-direct Measurements)	D3.9	2
	B10	Data Management	D3.10	16
C Assessment/Oversight	C1	Assessments and Response Actions	D4.1	7
	C2	Reports to Management	D4.2	9
D Data Validation and Usability	D1	Verification and Validation Requirements	D5.1	8
	D2	Verification and Validation Methods	D5.2	8
	D3	Reconciliation with Data Quality Objectives	D5.3	9

(a) Based on EPA, 2002.

(b) MARLAP recommends a graded approach to project plan documents. All elements may not be applicable, especially for a small project. See Section 4.3, "A Graded Approach to Project Plan Documents" and Section 4.5.3, "Plan Content for Small Projects."

This appendix also will discuss how the project plan document is linked to the outputs of the project planning process. Directed project planning is discussed in Chapter 2, *Project Planning Process*. The discussion of project plan documents in this appendix will use the DQO process

(EPA, 2000a) as a model for directed planning (see Appendix B, *The Data Quality Objectives Process*). References will be made in this appendix to the steps of the DQO process, where appropriate, to illustrate the linkage between the direct planning process and plan documents.

TABLE D.2—Comparison of project plan contents
I. Example QAPP[a] using EPA guidance[b] and EPA QAPP elements[b]

QA PROJECT PLAN FOR RADIOLOGICAL MONITORING TABLE OF CONTENTS	EPA G-5 QA PROJECT PLAN ELEMENTS
Title Page Approval Sheet Distribution List	A1 Title and Approval Sheet A3 Distribution List
1.0 Table of Contents	A2 Table of Contents
2.0 Project Description 2.1 Site History 2.2 Project Objectives and Requirements 2.3 DQOs	A5 Problem Definition/Background A6 Project/Task Description
3.0 Project Organization and Responsibility	A4 Project/Task Organization
4.0 QA Objectives for Measurement Data (Precision, Accuracy, Representativeness, Comparability, Completeness)	A7 Quality Objectives and Criteria for Measurement Data
5.0 Sampling Procedures, including QC [Cited Field Sampling and Analysis Plan]	B1 Sampling Process Designs B2 Sampling Methods Requirements
6.0 Sample Custody 6.1 Sample 6.2 Sample Identification 6.3 COC Procedures	B3 Sample Handling and Custody Requirements
7.0 Calibration Procedures and Frequency (Field and Laboratory)	B7 Instrument Calibration and Frequency
8.0 Analytical Procedures 8.1 Background 8.2 Specific Analytical Procedures 8.3 Test Methods 8.4 Control of Testing 8.5 Limits of Detection	B4 Analytical Methods Requirements B6 Instrument/Equipment Testing, Inspection, and Maintenance Requirements
9.0 Data Reduction, Validation and Reporting and Record	B10 Data Management D1 Data review, Validation, and Verification Requirements A9 Documentation and Records
10.0 Internal QC Checks	B5 Quality Control Requirements
11.0 Performance and Systems Audits 11.1 Systems Audits 11.2 Surveillance 11.3 Performance Audits 11.4 Resolution of Discrepancies 11.5 Review of Contractor Procedures	C1 Assessment and Response Actions
12.0 Preventive Maintenance	B6 Instrument/Equipment Testing, Inspection, and Maintenance Requirements
13.0 Specific Routine Procedures to Assess Data Precision, Accuracy, Completeness	D3 Reconciliation with DQOs

QA PROJECT PLAN FOR RADIOLOGICAL MONITORING TABLE OF CONTENTS	EPA G-5 QA PROJECT PLAN ELEMENTS
14.0 Corrective Action	
15.0 QA Report to Management	C2 Response to Management
16.0 References	
	A8 Special Training Requirements/Certification
	B8 Inspection/Acceptance Requirements for Supplies and Consumables
	B9 Data Acquisition Requirement for Non-direct Measurements
	D2 Verification and Validation Methods

II. Example work plan[c] and EPA QA/G-5 QAPP elements[b]

Work Plan Table of Contents	EPA QAPP Elements
Cover Letter	A3 Distribution List
Title Page (including Document Number, Prepared by/Prepared for Identification)	A1 Title and Approval Sheet
Approvals	A1 Title and Approval Sheet
Table of Contents	A2 Table of Contents
1 Introduction/Background	
Site and Regulatory Background	A5 Problem Definition/Background
Project Scope and Purpose	A6 Project/Task Description
Project Organization and Management	A4 Project/Task Organization
Data Quality Objectives and Approach	A7 Quality Objectives and Criteria for Measurement Data
Environmental Setting	A5 Problem Definition/Background
Sampling Site Selection, Locations and Identification	B1 Sampling Process Design
2 Sampling and Analysis Plan	
Objective	B1 Sampling Process Design
QA Objectives for Field Measurements, Laboratory Measurements (including Calibration Procedures and Frequency)	A7 Quality Objectives and Criteria for Measurement Data B7 Instrument Calibrations and Frequency
Sample Collection Procedures	B2 Sample Methods Requirements
Sample Identification, Handling and Transport	B3 Sample Handling and Custody Requirements
Sample Analysis	B4 Analytical Methods Requirements
Sample Tracking and Records	B10 Data Management
Data Reduction, Validation and Reporting	D1 Data Review, Verification, and Validation Requirements D2 Verification and Validation Methods
Internal QC Checks	B5 QC Requirements
3 QA Project Plan	
QA Training and Awareness	

Work Plan Table of Contents	EPA QAPP Elements
Performance and Systems Audits	C1 Assessments and Response Actions
Preventive Maintenance	B6 Instrument/Equipment Testing, Inspection, and Maintenance Requirements
Quality Improvement	B6 Instrument/Equipment Testing, Inspection, and Maintenance Requirements
QA Reports to Management	C2 Reports to Management
Purchase Items and Service Control	B8 Inspection/Acceptance Requirements for Supplies and Consumables
4 Data and Records Management Plan Objectives Data Management Document Control Records Management System Administrative Records	A9 Documentation and Record B10 Data Management
5 Data Interpretation Plan Approach for Data Evaluation Data Interpretation and Comparisons	D3 Reconciliation with DQOs
6 Risk Analysis Plan	——
7 Health and Safety Plan	——
	B9 Data Acquisition Requirements (Non-direct Measurements)
	A8 Special Training Requirements/Certifications

(a) Plan elements adapted from DOE, 1997.
(b) EPA, 2002
(c) Plan elements adapted from DOE, 1996.

It should be noted that although the project plan documents will address both sampling and analysis, MARLAP does not provide guidance on sampling design issues or sample collection. Discussion in D3.1, "Sample Process Design," and D3.2, "Sample Methods Requirements," are provided for completeness and consistency.

D.2 Group A: Project Management

This group consists of nine elements that address project management issues such as organization of the plan itself, management systems, and a description of project goals, participants and activities. These elements ensure that the project goals are clearly stated, the approach to be used is understood, and the project planning decisions are documented.

D.2.1 Project Management (A1): Title and Approval Sheet

The project title sheet should:

• Clearly identify the project in an unambiguous manner;

- Include references to organizational identifiers such as project numbers (when appropriate);

- Clearly label and distinguish between draft and approved versions;

- Include the date of issuance of drafts or final approved version;

- Include revision or version numbers;

- Indicate if the document represents only a portion of the QAPP (e.g., Volume 1 of 4 Volumes);

- Include names of the organization(s) preparing the plan document and, if different, for whom the plan was prepared; and

- Identify clearly on the title page if the document is a controlled copy and subjected to no-copying requirements. If so, indicate the document control number.

QAPPs should be reviewed on an established schedule. QAPPs should be kept current and revised when necessary. Documented approval, as an amendment to the QAPP, should be obtained for modifications to the QAPP.

The approval sheet documents that the QAPP has been reviewed and approved prior to implementation. The approval sheet should consist of the name, title, organization, signature and signature date for:

- The project manager or other person with overall responsibility for the project;

- The QA manager or other person with overall responsibility for the quality of the project outputs;

- The project managers or QA managers for all organizations (e.g., sampling organization, laboratories, data validators) implementing project activities; and

- The representative of any oversight or regulatory organization.

The project manager or other person with overall responsibility for the project should require an approved QA program, management plan, or quality manual that supports all technical operations, including data collection and assessment activities.

D.2.2 Project Management (A2): Table of Contents

The table of contents should:

- List all sections and subsections of the document, references, glossaries, acronyms and abbreviations, appendices (including sections and subsections) and the associated page numbers;

- List all attachments and the associated page numbers;

- List all tables and associated page numbers;

- List all figures and diagrams and associated page numbers; and

- List titles of other volumes, if the QAPP consists of more than one volume.

A document control format is useful in maintaining reference to the latest version of the planned document, especially when only portions of a document have been copied and are being used to implement or discuss project activities.

D.2.3 Project Management (A3): Distribution List

The distribution list should identify all individuals, along with their titles and organizations, who will receive copies and revisions of the approved QAPP and subsequent revisions. Listed individuals should include, at a minimum, all managers and QA personnel responsible for the implementation and quality of the data collection activities. The project planning team or the core group (Section 2.4) should be included on the document distribution list.

D.2.4 Project Management (A4): Project/Task Organization

This QAPP element should:

- Identify the individuals and/or organizations participating in the project, as well as contact information (address, telephone number, fax number, e-mail). The stakeholders, data users, decision makers, and technical planning team members, and the person or organization that will be responsible for project implementation, are identified during the directed planning process (Appendix B, *The DQO Process*, Steps 1 and 7).

- Discuss the roles and responsibilities of the individuals and/or organizations that participate in the data collection, including the roles and responsibilities of the data users, decision makers, and QA manager.

- Include an organizational chart clearly showing the relationship, lines of authority and communication, and mechanisms for information exchange among all project participants.

Complex projects may require more than one organizational chart to properly describe the relationships among participants. At times, to clearly detail an organizations responsibilities and communications, a general inter-organizational chart with primary contacts, responsibilities, and communications may need to be accompanied by secondary charts that describe intra-organizational contacts, responsibilities, and lines of communication.

One of the keys to successful projects is communication. The QAPP should identify the point of contact for resolving field and laboratory problems. The QAPP may also summarize the points of contact for dissemination of data to managers, users and the public.

D.2.5 Project Management (A5): Problem Definition/Background

The "Problem Definition/Background" element (A5) and the subsequent elements "Project/Task Description" (A6) and "Quality Objectives and Criteria" (A7) constitute the project description. Separating the project description into three elements focuses and encourages the plan authors to address all key issues (identification of problem to be solved, description of site history, description of tasks and the quality objectives and data-acceptance criteria), some of which can be overlooked if a larger, less-focused section is written. Table D.3 provides bulleted components for these three elements. This section and sections D.2.6 and D.2.7 provide a more detailed discussion of these elements.

TABLE D.3—Content of the three elements that constitute the project description

Problem Definition/Background (A5)	Project/Task Description (A6)	Objectives and Criteria (A7)
• Serves as an Introduction • Identifies the "problem to be solved" or the "question to be answered" • Identifies the regulatory, legal or "informational needs" drivers • Presents the historical perspective	• Describes measurements • Identifies regulatory standards and action levels • Identifies special personnel, procedural and equipment requirements • Summarizes assessment tools • Details schedule and milestones • Identifies record and report requirements	**Quality Objectives** • Problem definition/Site history • Data inputs • Population boundaries • Tolerable decision error rates **Criteria for Measurement Data** • Measurement quality objectives (MQOs; such as the measurement uncertainty at some concentration; the detection capability; the quantification capability; the range; the specificity; and the ruggedness of the method)

The Problem Definition/Background element provides the implementation team with an understanding of the pertinent context of the project. This section does not discuss the details of project activities, which are described in a subsequent project management element. Much of the information needed for this element was collected and discussed during Step 1 of the DQO process (Appendix B3.1). The decision statement was developed during Step 2 of the DQO process.

The "Problem Definition/Background" element should:

* Introduce the project;

* Identify the "problem to be solved" or the "question to be answered" upon successful completion of the project—the decision rule (Appendix B3.6);

* Discuss the assumptions, limitations, and scope of the project;

* Identify the regulatory, legal, or "informational needs" drivers that are the underlying reasons for the project; and

* Describe the context of the project so that it can be put into a historical perspective. This section may include a description and maps of a facility or site, its location, its use, site topography, geology and hydrogeology, past data collection activities, historical data including analytes and concentrations, past and present regulatory status, past releases, seriousness and potential risk of any release, site maps, and utilities.

If the data collection activity is in support of a technology evaluation, it should also discuss the purpose of the demonstrations, how the technology works, its operating conditions, any required utilities, its effluents and waste by-products and residues, past and expected efficiencies, and multi-media mass-balances by analyte and matrix.

D.2.6 Project Management (A6): Project/Task Description

This element of the QAPP provides a discussion of the project and underlying tasks for the implementation teams. It should provide a description of the work to be performed to resolve the problem or answer the question, including the following information:

* A description of the measurements and the associated QA/QC procedures that are to be made during the course of the project. DQO Step 3 describes existing and needed data inputs, while Step 7 yields the optimized sampling and analytical designs as well as quality criteria.
  o Identification of the analytes of interest.

- A summary (preferably a table) of samples type (e.g., grab, spatial or temporal composite), number of samples, analyte or analyte class (e.g., ^{99}Tc, transuranic, gamma emitters) and analytical protocol specifications or method.

- A discussion of applicable regulatory standards or action levels to which measurements will be compared. Identify any applicable regulatory standard (e.g., gross alpha drinking water maximum contamination limit), or applicable or relevant and appropriate requirements (ARARs) that will be used as a metric or action level during decision-making. The DQO Step 6 details action levels and tolerable decision errors that will be the basis for decisions.

- Identify any special requirements required to implement project tasks.
  - Identify any special training (e.g., hazardous waste site health and safety training (29 CFR 1910.120), radiation safety training).
  - Identify any special protective clothing and sampling equipment.
  - Identify any boundary conditions (e.g., only sample after a rainfall of more than 1 inch).
  - Specify any special document format, chain-of-custody, or archival procedures.
  - Identify any special sample handling (e.g., freezing of tissue samples), instrumentation, or non-routine analytical protocols that are required to achieve specified performance criteria (e.g., very low detection limits) (see also Chapter 3, *Key Analytical Planning Issues and Developing Analytical Protocol Specifications*).

- Summarize the assessment tools that will be employed to determine whether measurement data complied with performance criteria and are suitable to support decision-making. Include a schedule of the assessment events. Assessment tools include performance evaluations, program technical reviews, surveillance, technical and systems audits, and verification and validation. Briefly outline:
  - A first tier of reviews (e.g., when field or lab personnel check each other's notes or calculations).
  - Reviews of the work, notes and calculations of subordinates by the supervisor (e.g., review and sign all notebook entries).
  - The percentage of data subject to review by internal QA staff.
  - Data verification and validation to be performed by an independent party and the guidelines or plan to be used.
  - Assessment of project activities to be conducted by personnel independent of project activities (e.g., performance evaluation samples, surveillance, audits).
  - Assessment of how results of the project will be reconciled with the project DQOs ("data quality assessment").

- Supply a schedule that includes start and completion dates for tasks and a list of completion dates for important milestones. Dates can be calendric, or as number of days following approval of the QAPP, or number of days following commencement of field operations. DQO Steps 1 and 4 identify deadlines and other constraints that can impact scheduling.

- Identify the records and reports that will be required. This should be a brief but complete listing of necessary reports and records (e.g., field and lab notebooks, sample logbooks, spectra, sample tracking records, laboratory information system print-outs, QA reports, corrective action reports).

- Identify whether the original documents are required or if photocopies are sufficient. More detailed information will be presented in "Documentation and Records" (A9) and "Data Management" (B10).

D.2.7 Project Management (A7): Quality Objectives and Criteria for Measurement Data

This element addresses two closely related but different issues, quality objectives for the project and criteria used to evaluate the quality of measurement data. The element summarizes outputs from all steps of the DQO process. A fundamental principle underlying plan documents is that requirements for the data quality must be specified by the project planning team and documented. By clearly stating the intended use of the data and specifying qualitative and quantitative criteria for system performance, a critical link between the needs of the project planning team and the performance requirements to be placed on the laboratory data is established. (See Chapter 3 for a discussion of MQOs.)

D.2.7.1 Project's Quality Objectives

The project's quality objectives or data quality objectives (DQOs) are qualitative and quantitative statements that:

- Clarify the intended use of the data (e.g., data will be used to determine if lagoon sediment contains ^{232}Th at concentrations greater than or equal to the action level);

- Define the type and quantity of data per matrix needed to support the decision (e.g., ^{232}Th concentrations in 300 composite sediments samples each composite consisting of 10 samples randomly collected from a 100 m^2 sampling grid adjacent to the point of discharge);

- Identify the conditions under which the data should be collected (e.g., sediment samples collected from the top 6 cm of sediment within a 100 m radius of the point of discharge into lagoon #1, following de-watering of the lagoon and prior to sediment removal); and

- Specify tolerable limits on the probability of making a decision error due to uncertainty in the data and any associated action levels (e.g., 95 percent confidence that the true concentration is actually below the action level).

Authors of project plan documents are often encouraged to condense the DQO outputs in a summary statement. This approach can have value as long as critical information is not lost in the summary process and the original information is cited and available for all project participants. The following is an example of a DQO summary statement:

"The purpose of this project is to determine, to within a lateral distance of 10 m, the extent of ^{232}Th in soil along a pipeline at concentrations at or above 1,145 mBq/g, with a Type I error rate less than or equal to 5 percent; and to define within 1 m the vertical extent of measured ^{232}Th concentrations greater than 7,400 mBq/g."

D.2.7.2 Specifying Measurement Quality Objectives

Measurement quality objectives (MQOs) or measurements performance criteria are essential to the success of a project since they establish the necessary quality of the data. The quality of data can vary as a result of the occurrence and magnitude of three different types of errors (Taylor, 1990).

- BLUNDERS—mistakes that occur on occasion and produce erroneous results (e.g., mis-labeling or transcription errors);

- SYSTEMATIC ERRORS—mistakes that are always the same sign and magnitude and produce bias (i.e., they are constant no matter how many measurements are made); and

- RANDOM ERRORS—mistakes that vary in sign and magnitude and are unpredictable on an individual basis (i.e., random differences between repetitive readings) but will average out if enough measurements are taken.

The frequent occurrence of these types of errors is the reason why data quality is subject to question, why there is uncertainty when using data to make decisions and why measurement performance criteria are necessary.

During the DQO process, project DQOs are used to establish the MQOs. An MQO is a statement of a performance objective or requirement for a particular method performance characteristic. Examples of method performance characteristics include the measurement uncertainty at some concentration; the detection capability; the quantification capability; the range; the specificity; and the ruggedness of the method. MQOs for the project should be identified and described within this element of the QAPP. MARLAP provides guidance for developing MQOs for select method performance characteristics in Chapter 3 (*Key Analytical Planning Issues and Developing Analytical Protocol Specifications*) and Appendix C (*MQOs for Method Uncertainty and Detection and Quantification Capability*).

D.2.7.3 Relation between the Project DQOs, MQOs, and QC Requirements

The ultimate goal of all data collection operations is the collection of appropriately accurate data. Appropriately accurate data are data for which errors caused by imprecision and bias are controlled such that it is suitable for use in the context outlined by the DQOs (i.e., the overall error is less than that specified in the acceptable decision error). During the optimization of design in the planning process, DQO-specified decision error rates are translated into MQOs with the intention of monitoring, detecting, quantifying and controlling imprecision and analytical bias. During optimization, precautions are also incorporated into the design with the intention of preventing blunders and types of non-measurable bias not susceptible to measurement by QC samples.

The MQOs provide acceptance or rejection criteria for the quality control samples whose types and frequency are discussed in the Quality Control Requirements element (B5) (Appendix C). QC samples and the project's associated MQOs are key—but not the sole mechanisms—for monitoring the achievement of DQOs.

In summary, translating acceptable decision error rates into a design that will produce data of appropriate precision and bias is often a complex undertaking. The team must consider the synergistic and antagonistic interactions of the different options for managing errors and uncertainty. Accurate data require not only control of imprecision, but must also control the various forms of bias.

D.2.8 Project Management (A8): Special Training Requirements/Certification

All project personnel should be qualified and experienced in their assigned task(s). The purpose of this element is to add additional information regarding special training requirements and how they will be managed during implementation of the project. This element should:

- Identify and describe any mandated or specialized training or certifications that are required;
- Indicate if training records or certificates are included in the QAPP as attachments;
- Explain how training will be implemented and certifications obtained; and
- Identify how training documentation and certification records will be maintained.

D.2.9 Project Management (A9): Documentation and Record

This element of the QAPP will identify which records are critical to the project, from data generation in the field to final use. It should include what information needs to be contained in these records and reports, the formats of the records and reports, and a brief description of document control procedures. The following are suggested records and content:

- SAMPLE COLLECTION RECORDS should include sampling procedures, the names of the persons conducting the activity, sample number, sample collection points, maps and diagrams, equipment/protocol used, climatic conditions, and unusual observations. Bound field notebooks, pre-printed forms, or computerized notebooks can serve as the recording media. Bound field notebooks are generally used to record raw data and make references to prescribed procedures, changes in planned activities and implementation of corrective actions. Preferably, notebooks will contain pre-numbered pages with date and signature lines and entries will be made in ink. Field QC issues such as field, trip, and equipment rinsate blanks, co-located samples, field-spiked samples, and sample preservation should be documented. Telephone logbooks and air bill records should be maintained.

- SAMPLE TRACKING RECORDS document the progression of samples as they travel from the original sampling location to the laboratory and finally to their disposal or archival. These records should contain sample identification, the project name, signatures of the sample collector, the laboratory custodian and other custodians, and the date and time of receipt. The records should document any sample anomalies. If chain-of-custody (COC) is required for the project, the procedures and requirements should be outlined (Chapter 11, *Sample Receipt, Inspection, and Tracking*).

- ANALYTICAL QC issues that should be documented include standard traceability, and frequency and results of QC samples, such as, method and instrument blanks, spiked samples, replicates, calibration check standards and detection limit studies.

- ANALYTICAL RECORDS should include standard operating procedures for sample receipt, preparation, analysis and report generation. Data report formats and the level of supporting information is determined by data use and data assessment needs.

- PROJECT ASSESSMENT RECORDS should include audit check lists and reports, performance evaluation (PE) sample results, data verification and validation reports, corrective action reports. The project may want to maintain copies of the laboratory proposal package, pre-award documentation, initial precision and bias test of the analytical protocol and any corrective action reports.

The QAPP should indicate who is responsible for creating, tracking, and maintaining these records and when records can be discarded, as well as any special requirements for computer, microfiche, and paper records.

D.3 Group B: Measurement/Data Acquisition

The Measurement/Data Acquisition group consists of 10 elements that address the actual data collection activities related to sampling, sample handling, sample analysis and the generation of data reports. Although these issues may have been previously considered by project management elements, the project management section of the QAPP dealt with the overall perspective. The measurement/data section contains the details covering design and implementation to ensure that appropriate protocols are employed and documented. This section also addresses quality control activities that will be performed during each phase of data collection from sampling to data reporting.

D.3.1   Measurement/Data Acquisition (B1): Sampling Process Design

This element of the QAPP describes the finalized sampling design that will be used to collect samples in support of project objectives. The design should describe the matrices to be sampled, where the samples will be taken, the number of samples to be taken, and the sampling frequency. A map of the sampling locations should be included to provide unequivocal sample location determination and documentation.

If a separate sampling and analysis plan or a field sampling and analysis plan has been developed, it can be included by citation or as an appendix. This element will not address the details of standard operating procedures for sample collection, which will be covered in subsequent elements. This element will describe the sampling design and the underlying logic, so that implementation teams can understand the rationale behind and better implement the sampling effort. Understanding the rationale for the decisions will help if plans have to be modified due to conditions in the field. DQO Step 7 establishes the rationale for and the details of the sampling design.

This element should restate the outputs of the planning process and any other considerations and assumptions that impacted the design of the sampling plan, such as:

- The number of samples, including QC samples, sample locations and schedule, and rationale for the number and location of samples;

- A brief discussion of how the sampling design will facilitate the achievement of project objectives;

- A discussion of the population boundaries (temporal and spatial) and any accessibility limitations;

- A description of how the sampling design accommodates potential problems caused by the physical properties of the material being sampled (e.g., large particle size), the characteristic of concern (e.g., potential losses due to the volatility of tritium) or heterogeneity;

- A discussion of the overarching approach to sampling design (e.g., worse case or best case sampling versus average value) and assumptions made in selecting this approach (e.g., an assumption that the darkened soil adjacent to the leaking tank would present a worse case estimate of soil contamination);

- A listing of guidance and references that were relied upon when designing the sampling plan;

- Identification of the characteristics of interest (e.g., ^{99}Tc activity), associated statistical parameters (e.g., mean, standard deviations, 99th percentile), and acceptable false error rates (e.g., false negative rate of less than 5%);

- Identification of relevant action level and how data will be compared to the action level (Appendix B3.2);

- A discussion of the anticipated range of the characteristic of interest and assumed temporal and spatial variations (heterogeneity), anticipated variance, anticipated sources and magnitude of error (e.g., heterogeneity of material being sampled, sampling imprecision, analytical imprecision), anticipated mean values and distribution of measurements and the basis (e.g., historical data, similar processes or sites) for any associated assumptions;

- If any level of bias is assumed, what is the assumed magnitude and the basis of the assumption (e.g., historical data, typical analytical bias for matrix type);

- It is usually assumed that the magnitude of measurements made at individual sampling locations are independent of each other (e.g., no correlation of concentration with location). Geostatistical approaches may be more appropriate if measurements are significantly correlated with locations (e.g., serial-correlation, auto-correlation) since serial-correlation can bias estimates of variance and invalidate traditional probabilistic techniques such as hypothesis testing; and

- A discussion of the rationale for choosing non-routine sampling protocols and why these non-routine protocols are expected to produce acceptable precision and bias.

D.3.2 Measurement/Data Acquisition (B2): Sampling Methods Requirements

This element of the QAPP describes the detailed sampling procedures that will be employed during the project. The preliminary details of sampling methods to be employed were established during Step 7 of the DQO process. The selected sampling procedures should be appropriate to (1) ensure that a representative sample is collected, (2) avoid the introduction of contamination during collection, and (3) properly preserve the sample to meet project objectives. Written SOPs should be included as attachments to the QAPP. This element and the appendices or other documents that it references should in total contain all the project specific details needed to successfully implement the sampling effort as planned. If documents to be cited in the QAPP are not readily available to all project participants, they must be incorporated as appendices. All sampling personnel should sign that they have read the sampling procedures and the health and safety procedures.

Correct sampling procedures and equipment used in conjunction with a correct sampling design should result in a collection of samples that in total will represent the population of interest. A detailed discussion of sampling procedures, equipment and design are beyond the scope of MARLAP. In general, the selected procedures must be designed to ensure that the equipment is used properly and that the collected samples represent the individual sampling unit from which samples are collected. The sampling equipment should be chemically and physically compatible with the analyte of concern as well as the sample matrix. The sampling design should facilitate access to individual sampling units, result in an appropriate mass/volume of sample such that it meets or exceeds minimum analytical sample sizes, accommodates short-range heterogeneity (*i.e.*, does not preclude large particle sizes or lose small particles) and reduce or prevent loss of volatile components, if appropriate.

This element of the QAPP should:

- Identify the sampling methods to be used for each matrix, including the method number if a standardized method. If methods are to be implemented differently than specified by the standard method or if the standard method offers alternatives for implementation, the differences and alternatives should be specified;

- Identify the performance requirements of the sampling method. If the sampling method of choice is unlikely to be able to achieve the level of performance demanded by the project DQO, the project planning team should be notified;

- Identify the required field QC samples (e.g., trip blank, co-located duplicate);

- Identify any sample equipment preparation (e.g., sharpening of cutting edges, degreasing and cleaning) or site preparation (e.g., removal of overburden, establishing dust-free work space for filtering) for each method;

- Identify and preferably generate a list of equipment and supplies needed. For example, the sampling devices, decontamination equipment, sampling containers, consumables (e.g., paper towels), chain-of-custody seals and forms, shipping materials (e.g., bubble-pack, tape), safety equipment and paper work (e.g., pens, field books);

- Identify and detail logistical procedures for deployment, sample shipment and demobilization. If a mobile lab will be used, explain its role and the procedures for sample flow to the mobile lab and data flow to the data-user;

- Identify, preferably in a tabular form, sample container types, sizes, preservatives, and holding times;

- Identify procedures that address and correct problems encountered in the field (variances and nonconformance to the established sampling procedures);

- Identify for each sampling method, decontamination procedures and the procedures for disposing of contaminated equipment and used-decontamination chemicals and waters;

- Identify the disposal procedures for waste residuals generated during the sampling process (e.g., purged well waters, drilling dregs) for each method; and

- Identify oversight procedures (e.g., audits, supervisor review) that ensure that sampling procedures are implemented properly. The person responsible for implementing corrective actions should be identified.

D.3.3 Measurement/Data Acquisition (B3): Sample Handling and Custody Requirements

This element of the QAPP details how sample integrity will be maintained and how the sample history and its custody will be documented ensuring that (1) samples are collected, transferred, stored, and analyzed by authorized personnel, (2) the physical, chemical and legal integrity of samples is maintained, and (3) an accurate written record of the history of custody is maintained. DQO Step 1 describes the regulatory situation which can be used to identify the appropriate level of sample tracking. The QAPP should state whether COC is required. Sample handling, tracking and COC requirements are discussed in detail in Chapter 11, *Sample Receipt, Inspection, and Tracking.*

In the QAPP, the following elements should be documented:

- INTEGRITY OF SAMPLE CONTAINERS: Describe records to be maintained on the integrity of sample container and shipping container seals upon receipt. Describe records to be maintained if specially prepared or pre-cleaned containers are required.

- SECURITY: If wells are being sampled, whether the wellheads were locked or unlocked should be noted. Security of remote sampling sites or automatic samplers not maintained in locked cages should be discussed.

- SAMPLE IDENTIFICATION: The assignment of sample numbers and the labeling of sample containers is explained. If samples are to be assigned coded sample identifications (IDs) to preclude the possibility of bias during analysis, the sample code is one of the few items that will not be included in the QAPP, since the lab will receive a copy. The code and sample ID assignment process will have to be described in a separate document, which is made available to the field team and the data validators. An example of a sample label should be included in the QAPP.

- TRACKING OR CUSTODY IN THE FIELD: Procedures for sample tracking or custody while in the field and during sample shipment should be described. When COC is required, a copy of the COC form and directions for completion should be included. A list of all materials needed for tracking or custody procedures should be provided (e.g., bound notebooks, shipping containers, shipping labels, tape, custody seals, COC forms).

- SAMPLE PRESERVATION: Sample preservation procedures, if desired, should be clearly described. Preservation of radiological samples is discussed in Chapter 10, *Field and Sampling Issues that Affect Laboratory Measurements*.

- TRACKING OR CUSTODY IN THE LABORATORY: A decision must be made as to whether the laboratory in general is considered a secure area such that further security is not required once the sample is officially received by the laboratory or whether internal tracking or custody procedures will be required as the samples are handled by different personnel within the lab. The laboratory's sample receipt SOP, laboratory security procedures, and—if needed—internal tracking or custody procedures should be described.

- SPECIAL REQUIREMENT: Any special requirements, such as shipping of flammable or toxic samples, or requirements for verification of sample preservation upon sample receipt by the laboratory should be clearly described.

- ARCHIVAL: Document the rationale for the request to archive samples, extracts, and digestates. Describe how samples, extracts, and digestates will be archived. Identify how long samples, extracts, digestates, reports, and supporting documentation must be maintained.

D.3.4 Measurement/Data Acquisition (B4): Analytical Methods Requirements

This element of the QAPP should identify the analytical protocol specifications (APSs) including the MQOs that were employed by the laboratory to select the analytical protocols. (See Chapter 3 for guidance on developing APSs.) This element integrates decisions from three DQO steps: Step 3 which identified the analyte of interest and needed inputs to the decision, Step 6 that identifies the allowable uncertainty, and Step 7 that identifies the optimized analytical design. Input from all three steps drive the choice of analytical protocols. The discussion of the selected analytical protocols should address: subsampling, sample preparation, sample clean-up, radiochemical separations, the measurement system, confirmatory analyses and pertinent data calculation and reporting issues. A tabular summary of the analytical protocol by matrix type can facilitate reference for both the plan document development team and the laboratory analytical team.

This element of the QAPP should clearly describe the expected sample matrices (e.g., groundwater with no sediments, soils with no rocks larger than 2 cm in diameter) and what should be done or who should be contacted if sample matrices are different than expected. Subsampling is a key link in the analytical process which is often overlooked during planning leaving important decisions to laboratory staff, this element should specify appropriate subsampling procedures.

This QAPP element should:

- Identify the laboratories supplying analytical support. If more than one laboratory will be used, detail the analyses supplied by each laboratory;

- Identify analyses to be performed in the field using portable equipment or by a mobile lab;

- Identify the sample preparation techniques. Non-routine preparatory protocols, such as novel radiochemical separations, should be described in detail and documented in an SOP including pertinent literature citations and the results of validations studies and other performance data, when they exist;

- Identify the analytical protocols to be used. The protocol documentation should describe all necessary steps including the necessary reagents, apparatus and equipment, standards preparation, calibration, sample introduction, data calculation, quality control, interferences, and waste disposal;

- If the selected analytical protocols have not been demonstrated for the intended application, the QAPP should include information about the intended procedure, how it will be validated, and what criteria must be met before it is accepted for the project's application (Chapter 6, *Selection and Application of an Analytical Method*);

- If potential analytical protocols were not identified during the project planning process and existing analytical protocols can not meet the MQOs, an analytical protocol will have to be developed and validated (Section 6.6, "Method Validation"). If this issue was not identified by the project planning team, the project planning team must be contacted because the original project objectives and the associated MQOs may have to be revisited and changed (Appendix B);

- If both high concentration and low concentration samples are expected, discuss how the two sample types will be identified and handled in a manner that will prevent cross-contamination or other analytical problems;

- Discuss reporting requirements (e.g., suitable data acquisition and print-outs or electronic data archival that will capture all necessary information), the proper units (dry weight versus wet weight), the method to be employed to report the final result and its uncertainty, and reporting package format requirements; and

• Identify oversight procedures (e.g., QC samples, audits, supervisor review) for ensuring that analytical procedures are implemented properly and procedures for correcting problems encountered in the laboratory. The person responsible for implementing corrective actions in the lab should be identified.

The project plan document should be a dynamic document, used and updated over the life of the project as information becomes available or changes. For example, under a performance based approach, the analytical protocols requirements in the project plan documents should initially reflect the Analytical Protocol Specifications established by the project planning team and issued in the statement of work (or task order). When the analytical laboratory has been selected (Appendix E, *Contracting Laboratory Services*) the project plan document should be updated to reflect the identification of the selected laboratory and the analytical protocols, that is, the actual analytical protocols to be used should be included by citation or inclusion of the SOPs as appendices.

D.3.5 Measurement/Data Acquisition (B5): Quality Control Requirements

This element of the QAPP should include enough detail that the use and evaluation of QC sample results and corrective actions will be performed as planned and support project activities. The QC acceptance limits and the required corrective actions for nonconformances should be described. DQO Step 7 identified the optimized analytical design and the desired MQOs which will help determine the QC acceptance criteria. Refer to Chapter 18, *Laboratory Quality Control*, for a detailed discussion of radioassay QC and quality indicators. A discussion of QC requirements in the QAPP should include the following information:

• A list of all QC sample types by matrix;

• The frequency of QC sample collection or analysis, preferably a tabular listing;

• A list of QC sample acceptance criteria or warning limits and control limits;

• Procedures for documenting QC sample results;

• Equations and calculations used to evaluate QC sample results and to determine measurement performance acceptability;

• Actions to be taken if QC samples fail to meet the acceptance criteria; and

• Identification of the appropriate responsible person to whom QC reports should be sent.

Acceptance criteria for QC samples should be based on the project MQOs, in particular the MQO for measurement uncertainty at some concentration. Appendix C provides guidance on developing acceptance criteria for QC samples based on the project's MQO for the method's measurement uncertainty at some concentration, typically the action level.

D.3.6 Measurement/Data Acquisition (B6): Instrument/Equipment Testing, Inspection, and Maintenance Requirements

The QAPP should include a discussion of testing, inspection and maintenance requirements that will be followed to ensure that equipment and instrumentation will be in working order during implementation of project activities. An instrument or testing equipment will be deemed to be in working order if it is maintained according to protocol and it has been inspected and tested and meets acceptance criteria.

This element of the QAPP should:

- Discuss the maintenance policy for all essential instrumentation and equipment, what it involves, its frequency, whether it is performed by internal staff or if it is a contracted service, and whether an inventory of spare parts is maintained;

- Describe the inspection protocols for instrumentation and equipment. This ranges from the routine inspections (i.e, gases, nebulizers, syringes and tubing) prior to instrument or equipment use and more detailed inspections employed while troubleshooting an instrument or equipment problem. Mandatory inspection hold points, beyond which work may not proceed, should be identified; and

- Address the frequency and details of equipment and instrument testing. This may involve the weighing of volumes to test automatic diluters or pipets, the use of a standard weight prior to weighing sample aliquots to the use of standards to test sophisticated instrumentation. If standards (e.g., National Institute of Standards and Technology [NIST] standard reference material [SRM]) are used during testing, the type, source and uncertainty of standard should be identified.

There is not always a clear distinction between the testing component of this element and the previous element addressing the use of QC samples to determine whether an instrument is within control. In any case, it is important to describe in either of these elements of the QAPP, all procedures that are deemed important to determining whether an instrument/equipment is in working order and within control.

D.3.7 Measurement/Data Acquisition (B7): Instrument Calibration and Frequency

This element of the QAPP details the calibration procedures including standards, frequencies, evaluation, corrective action measures and documentation. Summary tables may be used to complement the more detailed discussions in the text. The following issues should be addressed in this element:

- Identify all tools, gauges, sampling devices, instruments, and test equipment that require calibration to maintain acceptable performance;

- Describe the calibration procedures in enough detail in this element or by citation to readily available references so that the calibration can be performed as intended;

- Identify reference equipment (e.g., NIST thermometers) and standards, their sources, and how they are traceable to national standards. Where national standards are not available, describe the procedures used to document the acceptability of the calibration standard used;

- Identify the frequency of calibration and any conditions (e.g., failed continuing calibration standard, power failure) that may be cause for unscheduled calibration;

- Identify the procedure and the acceptance criteria (i.e., in control) to be used to evaluate the calibration data;

- Identify the corrective actions to be taken if the calibration is not in control. When calibration is out of control, describe the evaluations to be made to determine the validity and acceptability of measurements performed since the last calibration; and

- Identify how calibration data will be documented, archived and traceable to the correct instrument/equipment.

See Chapter 15, *Quantification of Radionuclides*, for a discussion of radiochemical instrument calibration.

D.3.8 Measurement/Data Acquisition (B8): Inspection/Acceptance Requirements for Supplies and Consumables

This element of the QAPP deals with inspecting and accepting all supplies and consumables that may directly or indirectly affect the quality of the data. For some projects, this information may be provided by citation to a chemical safety and hygiene plan. The contents of this element should contain enough supportive information that the project and the data will be sufficient to undergo solicited and unsolicited reviews. The following detail should be included in this element, so the inspection process can be accurately implemented:

- Identify and document all supplies and consumables (e.g., acids, solvents, preservatives, containers, reagents, standards) that have the potential of directly or indirectly impacting the quality of the data collection activity;

- Identify the significant criteria that should be used when choosing supplies and consumables (e.g., grade, purity, activity, concentration, certification);

- Describe the inspection and acceptance procedures that will be used for supplies or consumables, including who is responsible for inspection, the timing of inspections and the acceptance and rejection criteria. This description should be complete enough to allow replication of the inspection process. Standards for receiving radiological packages are provided in 10 CFR 20 Section 20.1906 "Procedures for Receiving and Opening Packages" or an Agreement State equivalent;

- Describe the procedures for checking the accuracy of newly purchased standards, other than SRMs, by comparison to other standards purchased from other sources;

- Identify any special handling and storage (e.g., refrigerated, in the dark, separate from high concentration standards, lead shielding) conditions that must be maintained;

- Describe the method of labeling, dating and tracking supplies and consumables and the disposal method for when their useful life has expired; and

- Describe the procedures and indicate by job function who is responsible for documenting the inspection process and the status of inventories.

D.3.9 Measurement/Data Acquisition (B9): Data Acquisition Requirements for Non-Direct Measurement Data

This element of the QAPP addresses the use of existing data. Non-direct measurement data is defined as existing data that is independent of the data generated by the current project's sampling and analytical activities. Non-direct data may be of the same type (e.g., mBq/g of ^{232}Th in soil) that will complement the data being collected during the project. Other non-direct data may be of a different type such as weather information from the National Weather Service, or geological and hydrogeological data from the U.S. Geological Survey.

To achieve project objectives it is important that the data obtained from non-direct sources be subjected to scrutiny prior to acceptance and use. Use of existing data is discussed during Step 1 and 3 of the DQO process. If existing data of the same type is to be used to achieve project objectives, it has to be evaluated in terms of its ability to comply with MQOs established in DQO Step 7. The limitations on the use of non-direct measurements should be established by the project planning team.

This element should:

- Identify the type and source of all non-direct data that will be needed to achieve the project objectives;

- State whether the same quality criteria and QC sample criteria will be applied to the non-direct measurement data. If the same criteria cannot be applied, then identify criteria that will be acceptable for the non-direct data but at the same time won't bias or significantly add to the uncertainty of decisions for the project;

- Identify whether the data will support qualitative decisions (e.g., rain occurred on the third day of sampling) or if the data will be used quantitatively (e.g., used to calculate a mean concentration that will be compared to an action level);

- Identify whether enough information exists to evaluate the quality of the non-direct data (e.g., spike and collocated sample data, minimum detectable concentrations, reported measurement uncertainties); and

- If the non-direct data are to be combined with project-collected data, identify the criteria that will be used to determine if the non-direct data are comparable (e.g., sampled the same population, same protocol).

D.3.10 Measurement/Data Acquisition (B10): Data Management

This element of the QAPP should present an overview of the data management process from the receipt of raw data to data storage. The overview should address all interim steps, such as, data transformations, transmittals, calculations, verifications, validations and data quality assessments. The procedures should address how internal checks for errors are made. Laboratories should follow accepted data management practices (EPA, 1995). Applicable SOPs should be included as attachments to the QAPP. (See Chapter 16, *Data Acquisition, Reduction, and Reporting for Nuclear Counting Instrumentation*, for a discussion of radiochemical data generation and reduction.)

The discussion of data management should address the following issues:

- DATA RECORDING: The process of the initial data recording steps (e.g., field notebooks, instrument printouts, electronic data storage of alpha and gamma spectra) should be described. Examples of unique forms or procedures should be described. Describe the procedures to be used to record final results (e.g., negative counts) and the uncertainty.

- CONVERSIONS AND TRANSFORMATIONS: All data conversions (e.g., dry weight to wet weight), transformations (conversion to logs to facilitate data analysis) and calculation of statistical parameters (e.g., uncertainties) should be described, including equations and the rationale for the conversions, transformations and calculations. Computer manipulation of data should be specified (e.g., software package, macros).

- DATA TRANSMITTALS: Data transmittals occur when data are sent to another location or person or when it is converted to another format (incorporated into a spreadsheet) or media (printed reports keyed into a computer database). All transmittals and associated QA/QC steps taken to minimize transcription errors should be described in enough detail to ensure their proper implementation.

- DATA REDUCTIONS: Identify and explain the reasons for data reductions. Data reduction is the process of changing the number of data items by arithmetic or statistical calculations, standard curves, or concentration factors. A laboratory information management system may use a dilution factor or concentration factor to change raw data. These changes often are irreversible and in the process the original data are lost.

- DATA VERIFICATION, VALIDATION AND ASSESSMENTS: Since these assessment issues are discussed in a subsequent element of the QAPP (D2), only an overview should be provided identify the timing and frequency of these assessments.

- DATA TRACKING, STORAGE AND RETRIEVAL: Describe the system for tracking and compiling data as samples are being analyzed, how data are stored, and the mechanism for retrieving data (e.g., from archived back-up tapes or disks).

- SECURITY: Describe procedures for data and computer security.

D.4 Group C: Assessment/Oversight

The elements of this group are intended to assess progress during the project, facilitate corrective actions in a timely manner (Section D.4.1), and provide reports to management (Section D.4.2). It should be stressed that early detection of problems and weaknesses—before project commencement or soon thereafter—and initiation of corrective actions are important for a project's success. The focus of the elements in this group is the implementation of the project as defined in the QAPP. This group is different from the subsequent group, data validation and usability, which will assesses project data after the data collection activity is complete.

D.4.1 Assessment/Oversight (C1): Assessment and Response Actions

The QAPP authors have a range of assessment choices that can be employed to evaluate on-going project activities, which include surveillance, peer review, systems reviews, technical systems audits (of field and laboratory operations), and performance evaluations. A detailed discussion of laboratory evaluation is presented in Chapter 7, *Evaluating Methods and Laboratories*. It is important to schedule assessments in a timely manner. An assessment has less value if its findings become available after completion of the activity. The goal is to uncover problems and weaknesses before project commencement or soon thereafter and initiate corrective actions so the project is a success.

This element of the QAPP should:

- Identify all assessments by type, frequency and schedule;
- Identify the personnel who will implement the assessments;
- Identify the criteria, documents, and plans upon which assessments will base their review;
- Describe the format of assessment reports;
- Identify the time frame for providing the corrective action plan; and
- Identify who is responsible for approving corrective actions and ensuring that they are implemented.

D.4.2 Assessment/Oversight (C2): Reports to Management

Reports to management are a mechanism for focusing management's attention on project quality and on project issues that may require the management's level of authority. To be effective reports to management and management's review and response must be timely. The benefit of these status reports is the opportunity to alert management of data quality problems, propose viable solutions and procure additional resources.

At the end of the project, a final project report which includes the documentation of the DQA findings should be prepared (Chapter 9, *Data Quality Assessment*). It may also be beneficial for future planning efforts for the project planning team to provide a summary of the "lesson learned" during the project, such as key issues not addressed during planning and discovered in implementation or assessment, specialist expertise needed on the planning team, experience with implementing performance-based analytical protocol selection.

This element of the QAPP should address the following issues:

- Identify the various project reports that will be sent to management;

- Identify non-project reports that may discuss issues pertinent to the project (e.g., backlog reports);

- Identify QA reports that provide documentary evidence of quality (e.g., results of independent performance testing, routine QC monitoring of system performance);

- Identify the content of "reports to management" (e.g., project status, deviations from the QAPP and approved amendments, results of assessments, problems, suggested corrective actions, status on past corrective actions);

- Identify the frequency and schedule for reports to management;

- Identify the organization or personnel who are responsible for authoring reports; and

- Identify the management personnel who will receive and act upon the assessment reports.

D.5 Group D: Data Validation and Usability

This group of elements ensures that individual data elements conform to the project specific criteria. This section of the QAPP discusses data verification, data validation and data quality assessment (DQA), three processes employed to accept, reject or qualify data in an objective and consistent manner. Although there is good agreement as to the range of issues that the three elements, in total, should address, within the environmental community there are significant differences as to how verification, validation and DQA are defined. The discussion of this group of elements will use the definitions which are defined Chapter 8, *Radiochemical Data Verification and Validation.*

D.5.1 Data Validation and Usability (D1): Verification and Validation Requirements

This element of the QAPP addresses requirements for both data verification and data validation. The purpose of this element is to clearly state the criteria for deciding the degree to which each data item and the data set as a whole has met the quality specifications described in the "Measurement/Data Acquisition" section of the QAPP. The strength of the conclusions that can be drawn from the data is directly related to compliance with and deviations from the sampling and analytical design. The requirements can be presented in tabular or narrative form.

Verification procedures and criteria should be established prior to the data evaluation. Requirements for data verification include the following criteria:

- Criteria for determining if specified protocols were employed (e.g., compliance with essential procedural steps);

- Criteria for determining if methods were in control (e.g., QC acceptance criteria);

- Criteria for determining if a data report is complete (e.g., list of critical components that constitute the report);

- Criteria for determining if the analysis was performed according to the QAPP and the SOW;

- Criteria and codes used to qualify data; and

- Criteria for summarizing and reporting the results of verification.

A discussion of verification can be found in Chapter 8.

Data validation should be performed by an organization independent of the group that generated the data to provide an unbiased evaluation. Validation procedures and criteria should be established prior to the data evaluation. Requirements for data validation include the following:

- An approved list of well-defined MQOs and the action level(s) relevant to the project DQOs;

- Criteria for assigning qualifiers based on the approved list of MQOs;

- Criteria for identifying situations when the data validator's best professional judgement can be employed and when a strict protocol must be followed; and

- Criteria for summarizing and reporting the results of validation.

A discussion of validation can be found in Chapter 8.

D.5.2 Data Validation and Usability (D2): Verification and Validation Methods

D.5.2.1 Data Verification

Data verification or compliance with the SOW is concerned with: complete, consistent, compliant and comparable data. Since the data verification report documents whether laboratory conditions and operations were compliant with the SOW, the report is often used to determine payment for laboratory services. Chapter 5, *Obtaining Laboratory Services,* discusses the need to prepare a SOW for all radioanalytical laboratory work regardless of whether the work is contracted out or performed in-house.

This element of the QAPP should address the following issues to ensure that data verification will focus on the correct issues:

- Identify the documents (e.g., other QAPP sections, SOW, contracts, standard methods) that describe the deliverables and criteria that will be used to evaluate compliance;

- Identify the performance indicators that will be evaluated (e.g., yield, matrix spikes, replicates). See Chapter 18, *Laboratory Quality Control*, for a discussion of radiochemistry performance indicators;

- Identify the criteria that will be used to determine "in-control" and "not-in-control" conditions;

- Identify who will perform data verification;

- Describe the contents of the verification report (e.g., a summary of the verification process as applied; required project activities not performed or not on schedule or not according to required frequency; procedures that were performed but did not meet acceptance criteria; affected samples; exceptions); and

- Identify who will receive verification reports and the mechanism for its archival.

D.5.2.2 Data Validation

Chapter 8, *Radiochemical Data Verification and Validation*, discusses radiochemical data validation in detail. MARLAP recommends that a data validation plan document be included as an appendix to the QAPP. The data validation report will serve as the major input to the process that evaluates the reliability of measurement data.

This element of the QAPP should address the following issues:

- Describe the deliverables, measurement performance criteria and acceptance criteria that will be used to evaluate data validity;

- Identify who will perform data validation;

- Describe the contents of the validation report (e.g., a summary of the validation process as applied; summary of exceptional circumstances; list of validated samples, summary of validated results); and

- Identify who will receive validation reports and the mechanism for its archival.

D.5.3 Data Validation and Usability (D3): Reconciliation with Data Quality Objectives

This element of the QAPP describes how project data will be evaluated to determine its usability in decision-making. This evaluation is referred to as the "data quality assessment." DQA is the process that scientifically and statistically evaluates project-wide knowledge in terms of the project objectives to assess the usability of data. DQA should be ongoing and integrated into the project data collection activities. On project diagrams and data life cycles, it is often shown as the last phase of the data collection activity. However, like any assessment process, DQA should be considered throughout the data collection activity to ensure usable data. EPA guidance (EPA, 2000b) provides a detailed discussion of that part of the DQA process that addresses statistical manipulation of the data. In addition to statistical considerations, the DQA process integrates and considers information from the validation report, assessment reports, the field, the conceptual model and historical data to arrive at its conclusions regarding data usability. DQA is discussed in Chapter 9, *Data Quality Assessment*.

The DQA considers the impact of a myriad of data collection activities in addition to measurement activities. This element of the QAPP should direct those performing the DQA to:

- Review the QAPP and DQOs;
- Review the validation report;
- Review reports to management;
- Review identified field, sampling, sample handling, analytical and data management problems associated with project activities;
- Review all corrective actions; and
- Review all assessment reports and findings (e.g., surveillances, audits, performance evaluations, peer reviews, management and technical system reviews).

In addition to the above, this element of the QAPP should address the following issues:

- Identify who will perform the DQA;
- Identify what issues will be addressed by the DQA;
- Identify any statistical tests that will be used to evaluate the data (e.g., tests for normality);
- Describe how MQOs will be used to determine the usability of measurement data (i.e., did the measurement uncertainty in the data significantly affect confidence in the decision?);
- Describe how the representativeness of the data will be evaluated (e.g., review the sampling strategy, the suitability of sampling devices, subsampling procedures, assessment findings);
- Describe how the potential impact of non-measurable factors will be considered;
- Identify what will be included in the DQA report; and
- Identify who will receive the report and the mechanism for its archival.

D.6 References

U.S. Department of Energy (DOE). 1996. *Project Plan for the Background Soils Project for the Paducah Gaseous Diffusion Plant, Paducah, Kentucky*. Report DOE/OR/07-1414&D2. May.

U.S. Department of Energy (DOE). 1997. *Quality Assurance Project Plan for Radiological Monitoring at the U.S. DOE Paducah Gaseous Diffusion Plant, Paducah, Kentucky.* February.

U.S. Environmental Protection Agency (EPA). 1995. Good Automated Laboratory Practices. Report 2185, EPA, Washington, DC.

U.S. Environmental Protection Agency (EPA). 2000a. *Guidance for the Data Quality Objective Process* (EPA QA/G-4). EPA/600/R-96/055, Washington, DC. Available from www.epa.gov/quality/qa_docs.html.

U.S. Environmental Protection Agency (EPA). 2000b. *Guidance for Data Quality Assessment: Practical Methods for Data Analysis.* EPA QA/G-9. EPA/600/R-96/084. Office of Environmental Information, Washington, DC. Available at www.epa.gov/quality/qa_docs.html.

U.S. Environmental Protection Agency (EPA). 2001. *EPA Requirements for Quality Assurance Project Plans.* EPA QA/R-5. EPA/240/B-01/003. Office of Environmental Information, Washington, DC. Available at www.epa.gov/quality/qa_docs.html.

U. S. Environmental Protection Agency (EPA). 2002. *Guidance for Quality Assurance Project Plans.* EPA QA/G-5. EPA/240/R-02/009. Office of Environmental Information, Washington, DC. Available at www.epa.gov/quality/qa_docs.html.

Taylor, J. K. 1990. *Quality Assurance of Chemical Measurements.* Lewis, Chelsea, Michigan.

APPENDIX E
CONTRACTING LABORATORY SERVICES

E.1 Introduction

This appendix provides general guidance on federal contracting and contracting terminology as used for negotiated procurements. Federal agencies, and laboratories doing business with them, must follow applicable provisions of the *Federal Acquisition Regulations* (FAR) and agency-specific supplements. The examples provided in this appendix are based primarily on procedures followed by the U.S. Geological Survey (USGS).

This appendix addresses selecting a laboratory to establish services that supplement an agency's in-house activities through the contracting of additional outside support. This appendix offers a number of principles that may be used when selecting a service provider, establishing a contractual agreement, and later working with a contract laboratory. These principles may also be applied to contractors that are located outside of the United States. In such cases, legal counsel will need to review and advise an agency concerning pertinent issues related to international contracts.

This appendix also covers laboratory audits that are part of a final selection process and other activities that take place until the contract is concluded. Chapter 5 (*Obtaining Laboratory Services*) supports this appendix with a general description on how to obtain laboratory services. Chapter 7 (*Evaluating Methods and Laboratories*) complements this appendix by considering information related to laboratory evaluations that are conducted throughout the term of a project—whether or not this work is specifically covered by a contract.

Obtaining support for laboratory analyses is already a practice that is familiar to a number of federal and state agencies. The following discussion will apply:

- *Agency*: A federal or state government office or department, (or potentially any other public or private institution) that offers a solicitation or other mechanism to obtain outside services;

- *Proposer*: A person, firm, or commercial facility that submits a proposal related to providing services; and

- *Contractor*: A person or firm that is awarded the contract and is engaged in providing analytical services.

Contents

E.1 Introduction E-1
E.3 Request for Proposals—The Solicitation ... E-7
E.4 Proposal Requirements E-10
E.5 Proposal Evaluation and Scoring Procedures E-21
E.6 The Award E-31
E.7 For the Duration of the Contract E-31
E.8 Contract Completion E-36
E.9 References E-36

Furthermore, the size and complexity of some agency projects will clearly exceed the extent of the information presented here. In its present form, this appendix serves to touch on many of the issues and considerations that are common to all projects, be they large or small.

MARLAP draws attention to another dimension of the overall contracting process by considering how the data quality objectives (DQOs) and measurement quality objectives (MQOs) are incorporated into every stage of a project—as described earlier in greater detail (Chapters 2, *Project Planning Process*, and 3, *Key Analytical Planning Issues and Developing Analytical Protocol Specifications*). In this regard, an agency's project managers and staff are given an opportunity to consider options with some foresight and to examine the larger picture, which concerns planning short- or long-term projects that utilize a contractor's services. As services are acquired, and later as work is performed, the specific concepts and goals outlined by the DQOs and MQOs will be revisited. This becomes an iterative process that offers the possibility to further define objectives as work is conducted. Whenever the DQOs or MQOs are changed, the contract should be modified to reflect the new specifications. Employing the MQOs and tracking the contractor's progress provides a means by which project managers and contract-laboratory technical staff can return and review the project at any point during the contract period. This allows for repeated evaluations to further optimize a project's goals and, if anticipated in the contract's language, perhaps even provides for the option to revise or redirect the way performance-based work is conducted.

The Office of Federal Procurement Policy (OFPP, 1997) has developed a Performance-Based Service Contracting review checklist to be used as a guide in developing a performance-based solicitation. The checklist contains minimum required elements that should be present for a contract to be considered performance-based. Performance-Based Service Contracting focuses on three elements: a performance work statement; a quality assurance project plan (QAPP); and appropriate incentives, if applicable. The performance work statement defines the requirements in terms of the objective and measurable outputs. The performance work statement should answer five basic questions: what, when, where, how many, and how well. The work statement should structure and clearly define the requirements, performance standards, acceptable quality levels, methods of surveillance, incentives if applicable and evaluation criteria. A market survey should be conducted so that the marketplace and other stakeholders are provided the opportunity to comment on draft performance requirements and standards, the proposed QA project plan, and performance incentives, if applicable.

A number of benefits arise from establishing a formal working relationship between an agency and a contractor. For example:

- A contract is a legal document that clearly defines activities and expectations for the benefit of both parties engaged in the contractual relationship.

- The process of drafting language to cover legal considerations may well include contributions from legal staff. Legal guidance may be obtained as needed at any time during the planning stages or later when a contract is in place. However, the core of a contractor's proposal, and eventually the contract itself, provide the foundation of technical work that is required to complete a project or attain an ongoing program goal. *In this regard, aside from legal issues that are an integral part of every contract, this appendix's principal focus is on the laboratory process or technical work-related content of the contract.*

- The statement of work (SOW) first appears as part of the agency's request for proposal (RFP) and later is essentially incorporated into the proposal by the proposer when responding to the RFP. When work is underway, the SOW becomes a working document that both the agency and contractor refer to during the life of the contract.

- Legal challenges concerning project results (i.e., laboratory data) may arise during the contract period. The language in a contract should offer sufficient detail to provide the means to circumvent potential or anticipated problems. For example, attention to deliveries of samples to the laboratory on weekends and holidays or data reporting requirements that are designed to support the proper presentation of data in a legal proceeding are important aspects of many federal- and state-funded contracts.

Overall, this appendix incorporates a sequence that includes both a planning and a selection process. Figure E-1 illustrates a series of general steps from planning before a contract is even in place to the ultimate termination of the contract. An agency first determines a need as part of planning, and along the way advertises this need to solicit proposals from outside service providers who operate analytical laboratory facilities. Planning future work, advertising for, and later selecting services from proposals submitted to an agency takes time—perhaps six or more months pass before a laboratory is selected, a contract is in place, and analytical work begins. The total working duration of a contract, for example, might cover services for a brief time (weeks or months) and in other cases, many contracts may run for a preset one-year period or for a more extended period of three to five years with optional renewal periods during that time.

The MARLAP user will find that planning employs a thought process much like that used to prepare an RFP. In general, one starts with questions that define a project's needs. Further, by developing Analytical Protocol Specifications (APSs) which include specific MQOs, one enters an iterative process such that—at various times—data quality is checked in relation to work performed both in-house and by the outside service provider. Overall, planning results in the development of a project plan document (e.g., QAPP). During planning, a project manager and the agency staff can consider both routine and special analytical services that may be required to provide data of definable quality. The SOW serves to integrate all technical and quality aspects of the project, and to define how specific quality-assurance and quality-control activities are implemented during the time course of a contract. Also, at an early stage in planning, the agency may choose to assemble a team to serve as the Technical Evaluation Committee (TEC; Section

E.5.1). The main role of the TEC is in selecting the contract laboratory by reviewing proposals and by auditing laboratory facilities. The TEC is discussed later in this appendix, however, the key issue here concerns the benefit to establishing this committee early on, even to the point of including TEC members in the initial planning activities. The result is a better informed evaluation committee and a team of individuals that can help make adjustments when the directed planning process warrants an iterative evaluation of the way work is performed under the contract. Overall, planning initiates the process that characterizes the nature of the contracting process to follow.

E.2 Procurement of Services

Recognizing that the procurement process differs from agency to agency, the following guidance provides a general overview to highlight considerations that may already be part of—or be incorporated into—the current practice. First, the request for specific analytical services can be viewed as a key product of both the agency's mission and the directed planning process. As agency staff ask questions, list key considerations to address during the work, and in turn define objectives, they

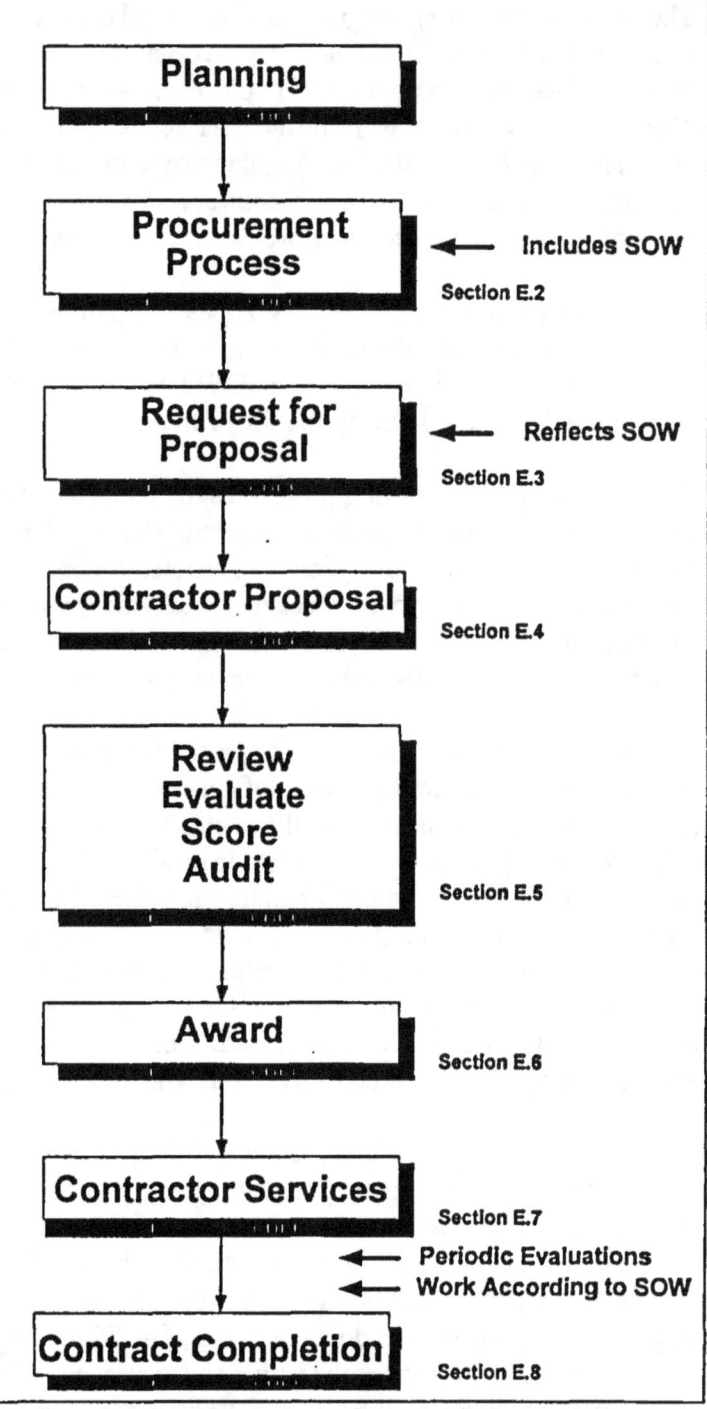

FIGURE E.1 — General sequence initiating and later conducting work with a contract laboratory

also eliminate unnecessary options to help focus on the most suitable contracting options that satisfy the APSs. Thereafter, the scope of the work, schedule, manpower constraints, availability of in-house engineering resources, and other technical considerations all enter into estimating and defining a need for project support. This approach refines the objectives and establishes needs that may be advertised in a solicitation for outside services. The resulting work or project plan should clearly articulate what is typically known but not limited to the following:

- Site conditions;
- Analytes of interest;
- Matrices of concern;
- How samples are to be collected and handled;
- Custody requirements;
- Data needs and APSs, including the MQOs;
- Stipulated analytical methods, if required;
- Applicable regulations; and
- Data reporting.

All of this defines the scope of work, such that the agency can initiate a formal request for proposals or arrange for an analysis request as part of a less formal procurement.

E.2.1 Request for Approval of Proposed Procurement Action

If required within an agency, a request is processed using forms and related paperwork to document information typically including, but not limited to, the following:

- Identification of product or service to be procured;
- Title of program or project;
- Description of product or service;
- Relationship of product or service to overall program or project;
- Funding year, projected contract life, amounts, etc.;
- Name and phone number of Project Officer(s);
- Signature of Project Officer and date
- Name and phone number of Contracting Officer; and
- Signature of Contracting Officer and date.

An agency may also be required to collect or track information for an RFP with regard to:

- New procurements: type of contract, grant, agreement, proposal, etc. Continuing procurements: pre-negotiated options, modifications, justification for noncompetitive procurement, etc.
- Source information: small business or other set aside, minority business, women-owned business, etc.

In addition to the information listed above, agency-specific forms used to initiate a procurement request may also provide a place to indicate agency approval with names, signature lines, and date spaces for completion by officials in the office responsible for procurement and contracts. An agency administrator or director above the level of the office of procurement may also sign this form indicating agency approval.

E.2.2 Types of Procurement Mechanisms

Table E.1 lists many of the procurement options available to the project manager. Each option offers a solution to a specific need. For example, a purchase order is typically appropriate for tasks with a somewhat limited scope and thus is perhaps most useful when samples are to be processed on a one-time basis. In some cases where only one or a limited number of vendors can fulfill the needs of the project, e.g., low-level tritium analysis by helium ingrowth within a specified time period, a sole source solicitation is commonly used.

TABLE E.1— Examples of procurement options to obtain materials or services

Procurement Mechanism	Example of Specific Use or Application
Purchase order	In-house process handled through purchasing staff; limited to small needs without a formal request or used in conjunction with a solicitation (competitive process) and a limited amount of funding; commonly used to purchase equipment and supplies, but may be used for processing samples.
Sole source solicitation	In specific instances, a single or a limited number of service providers are able to offer specific services.
Request for Quotation (RFQ)	Formal, main process for establishing contracts—generally addresses a major, long-term need for contractor support; this is a competitive process based mainly on cost.
Request for Proposal (RFP)	Formal, main process for establishing contracts—generally addresses a major, long-term need for contractor support; this is a competitive process based mainly on technical capability.
Modification to an existing contract or delivery order	This approach meets a need that is consistent with the type of contract that is in place, e.g., agency amends contract to add a method for sample processing that is similar to work already covered.
Basic Ordering Agreement (BOA)	Work is arranged with a pre-approved laboratory as described in Section E.2.2.

The process leading to a formal contract provides a more comprehensive view of nearly every aspect of the work that an agency expects from a contractor. The formal process includes three types of procurement: request for quotation (RFQ), request for proposal (RFP), and the basic ordering agreement (BOA). The RFQ solicits bidders to provide a quotation for laboratory services that have been detailed in the solicitation. The specifications may include the technical, administrative, and contractual requirements for a project. For the RFQ, the contract typically is awarded to the lowest bidder that can fulfill the contract specifications without regard to the quality of the service. What appears to be a good price may not entail the use of the best or most appropriate method or technology. There may be significant advantages in seeking to acquire high-technology services as a primary focus in advance of, or along with, concerns pertaining to price.

For an RFP, there is considerably more work for the agency and the laboratory. The laboratory must submit a formal proposal addressing all key elements of the solicitation that include how, why, what, when, where and by whom the services are to be performed. The TEC or Contracting Officer must review all proposals, rank them according to a scoring system and finally assess the cost effectiveness of the proposals before making the final award.

The BOA provides a process that serves to pre-approved service providers. This includes a preliminary advertisement for a particular type of work, such as radioanalytical services. The agency then selects and approves a number of candidates that respond to the advertisement. With this approach, the agency assembles a potential list of approved laboratories that are contacted as needed to support specific needs. The agency may choose to simply write a task order (defining a specific scope of work) with a specific pre-approved laboratory, or the agency may initiate a competitive bidding process for the task order between several or all members on the list of pre-approved laboratories. Once chosen, the laboratory may be guided by a combined statement of work or task order that is issued by the agency.

Mechanisms that permit an agency to obtain analyses for a limited number of samples—without an established contractual relationship with a specific contractor—may simply be necessitated by the small number of samples, time constraints where specific analyses are not part of an existing contract, limitations related to funding, or other consideration. The formal business and legal requirements of a long-term relationship warrant a stronger contractual foundation for work conducted in a timely fashion, on larger numbers of samples, and over specified periods of time. The contracts described above, with the exception of a BOA, are considered "requirement" contracts and requires the group initiating the solicitation to use only the contracted laboratory, without exception, for the contract period to perform the sample analyses.

E.3 Request for Proposals—The Solicitation

To appreciate the full extent of a competitive process leading to a formal working relationship—between an agency and a contractor—*the primary example used hereafter is the solicitation and selection process that starts with the issuance of a* RFP, as shown in Figure E-1.

Federal announcements of RFPs can be found on the *Federal Business Opportunities* (FBO) web site (http://vsearch1.eps.gov/servlet/SearchServlet); many agencies also announce their own RFPs on their own web sites. FBO primarily provides a synopsis or brief description of the type of work the agency is interested in purchasing. States and local governments also solicit proposals and announce the availability of work in USABID (a compilation of solicitations from hundreds of city, county, and state agencies). Internet sites that offer access to the FBO and USABID listings can be located through electronic searches using web browser software. Once a site is located, the information can be viewed through public access or commercial Internet-based services. In other cases, a state or federal agency may maintain a mailing list with names and addresses for potentially interested parties. This might include contractors that previously supported the Agency or others who have volunteered information for the mailing list.

Once the RFP, state advertisement, or other form of solicitation is publicized, interested parties can contact the appropriate Agency to obtain all the specific information relevant to completing a candidate laboratory's contract proposal. For the present discussion, this information is contained in the text of the RFP document. The RFP may be accompanied by a cover letter stating an invitation to applicants and general information related to the content of a proposal and specific indication for the types of sections or sub-sections the proposal will contain. For example, a proposal divided into three sections technical proposal, representations and certifications, and price proposal allows the agency to separate pricing from technical information. In this way, the agency considers each candidate first on technical merits before the price of services enters the selection process.

The agency's RFP is designed to provide a complete description of the proposed work. For example, a RFP should inform all candidate laboratories (i.e., proposers) of the estimated number of samples that are anticipated for processing under the contract. The description of work in the RFP as described in the SOW serves to indicate the types of radionuclide analyses required for the stated sample types and the number of samples to undergo similar or different processing protocols. The estimate also has a bearing on cost and other specific project details as described in the SOW. Additional information provided with the RFP serves to instruct the proposer regarding other technical requirements (APSs), the required number of copies of each section of the proposal, proposal deadline, address where proposals are to be sent, and other general concerns or specifications relevant to the solicitation.

The cover letter may indicate how each proposer will be notified if its proposal is dropped from the competitive range of candidates during the selection process. The letter may also include precautionary notes concerning whom to contact or not contact at the agency regarding the potential contract during the competitive process. Finally, if particular sources are encouraged to apply (e.g., minority or small business), this information will be mentioned in the agency's invitation to apply.

E.3.1 Market Research

The Office of Federal Procurement Policy (OFPP, 1997) recommends that the marketplace and other stakeholders be provided the opportunity to comment on draft performance requirements and standards. This practice allows for feedback from those people working in the technical community so that their comments may be incorporated into the final RFP and potential proposers can develop intelligent proposals.

E.3.2 Period of Contract

The time and resources involved in writing and awarding a major contract generally make it impractical and cost ineffective to award contracts for less than one or more years. While contracts running for shorter terms are sometimes established, single or multiple year terms are commonly used to provide the necessary services for some federal or state programs. Monitoring programs are likely to go long periods of time with renewals or RFPs that continue the work into the future. Elsewhere, relatively large projects conducting radiation survey and site investigations may require a contract process that, for the most part, estimates the time services will be needed to finish work through to the completion of a final status survey. In this case, the contract may specify any length of time, but also include the option to renew the contract for a period of time to bring the project to a close. The relationship between the length of a contract and the type of project can be part of the structured planning process that seeks to anticipate every facet of a project from start to finish.

Multi-year contracts typically are initiated with an award for the first year or two followed by an additional number of one-year options. In this way, a five-year contract is awarded for one year (or two) with four (or three) one-year option periods to complete the contract's full term. The government must exercise its option for each period beyond the base term. Problems that arise during any year may result in an agency review of the MQOs or an examination of the current working relationship that may result in the agency's decision to not extend the contract into the next option year.

E.3.3 Subcontracts

For continuity or for quality assurance (QA), the contract may require one laboratory to handle the entire analytical work load. However, subcontracting work with the support of an additional laboratory facility may arise if the project plan calls for a large number of samples requiring quick turnaround times and specific methodologies that are not part of the primary laboratory's support services. A proposer may choose to list a number of subcontractors in the proposal. The listing may or may not include other laboratories with whom the proposer has an existing or prior working relationship. The choice of subcontracting firms may be limited during the proposal process. There may be many qualified service providers to meet specific project needs. However, once work is under way, using a limited number of laboratories that qualify for this secondary role helps maintain greater control of quality and thus the consistency of data coming from more than a single laboratory alone. Furthermore, the contractor may prefer working with a specific subcontractor, but this arrangement is subject to agency approval.

The use of multiple service providers adds complexity to the agency's tasks of auditing, evaluating, and tracking services. The prime contractor and their subcontractor(s) are held to the same terms and conditions of the contract. The prime contractor is responsible for the performance of its subcontract laboratories. In some instances, certain legal considerations related to chain of custody, data quality and reporting, or other concern may limit an agency's options and thus restrict the number of laboratories that are part of any one contract.

However, the decision to use an approved subcontractor, or which subcontractor to use, is strictly up to the prime. Under federal contracting regulations, agencies may not direct a laboratory to use a particular subcontractor, nor may the federal customer deal directly with the subcontractor (without going through the prime contractor). This may create a more convoluted or complex relationship among the customer, prime contractor, and subcontractors, which in turn may affect turnaround times and quality. Consequently, a laboratory's quality manual and proposal should address its approach for ensuring timely and accurate communications between the customer, prime contractor, and subcontractor(s).

E.4 Proposal Requirements

The agency's RFP will state requirements that each proposer is to cover in its proposal. The proposal document itself becomes first the object of evaluation and is a reflection of how the contract and the SOW are structured. Whether one works with a formal contract or a simpler analysis request, the agency and contractor need to agree to all factors concerning the specific analytical work. Where written agreements are established, the language should be specific to avoid disputes. Clear communication and complete documentation are critical to a project's success. For example, the agency's staff asks questions of itself during the planning process to create and later advertise a clearly stated need in the RFP. The contractor then composes a proposal that documents relevant details concerning their laboratory's administrative and technical personnel, training programs, instrumentation, previous project experience, etc. Overall, the proposer should make an effort to address every element presented in the RFP. The proposer should be as clear and complete as possible to ensure a fair and proper evaluation during the agency's selection process.

The planning process will reveal numerous factors related to technical requirements necessary to tailor a contract to specific project needs. The following sections may be reviewed by agency staff (radiochemist or TEC) during planning to determine if additional needs are required beyond those listed in this manual. agency personnel should consider carefully the need to include every necessary detail to make a concise RFP. The proposer can read the same sections to anticipate the types of issues that are likely to appear in an RFP and that may be addressed in a proposal.

E.4.1 RFP and Contract Information

There are two basic areas an agency can consider when assembling information to include in an RFP. The proposer is expected to respond with information for each area in its proposal. The first area includes a listing of *General Laboratory Requirements and Activities*. The second area, *Technical Components to Laboratory Functions*, complements the first, but typically includes more detailed information.

1) General laboratory requirements
   - Personnel;
   - Facilities;
   - Meeting contract data quality requirements;
   - Schedule;
   - Quality manual;
   - Data deliverables including electronic format;
   - Licenses and certifications; and
   - Experience: previous and current contracts; quality of performance.

2) Technical components to laboratory functions
   - Standard operating procedures;
   - Instrumentation
   - Training
   - Performance evaluation programs; and
   - Quality system.

The laboratory requirements and technical components indicated above are addressed in this appendix. Beyond this, there are additional elements that may be required to appear with detailed descriptions in an RFP and later in a formal proposal. One significant portion of the RFP, and a key element appearing later in the contract itself, is the SOW. This is the third area a proposer is to address, and information in a SOW may vary depending on the nature of the work.

The agency will provide specifications in the RFP regarding the work the contractor will perform. This initiates an interaction between a proposer and the agency and further leads to two distinct areas of contractor-agency activity. The first concerns development and submitting of proposals stating how the laboratory work will be conducted to meet specific agency needs. The second concerns agency evaluations of the laboratory's work according to contract specifications (Section E.5) and the SOW. Once the contract is awarded, a contractor is bound to perform the work as proposed.

Specific sections of each contract cover exactly what is expected of the contractor and its analytical facilities to fulfill the terms and conditions of the contract. The SOW describes the required tasks and deliverables, and presents technical details regarding how tasks are to be executed. A well written SOW provides technical information and guidance that directs the contractor to a practice that is technically qualified, meets all relevant regulatory requirements, and appropriately coordinates all work activities. A sample checklist for key information that may be in a SOW is presented in Table E.2. Note that not all topics in the list are appropriate for each project, and in some cases, only a subset is required. The list may also be considered in relation to less formal working relationships (e.g., purchase order), as well as tasks covered in formal contracts.

TABLE E.2 — SOW checklists for the agency and proposer

SAMPLE HISTORY
    ____ General background on the problem
    ____ Site conditions
    ____ Regulatory background
    ____ Sample origin
    ____ Analytes and interferences (chemical forms and estimated concentration range)
    ____ Safety issues
    ____ Data use
        ____ Regulatory compliance
        ____ Litigation

ANALYSIS RELATED
    ____ Number of samples
    ____ Matrix
    ____ Container type and volume
    ____ Receiving and storage requirements
    ____ Special handling considerations
    ____ Custody requirements
    ____ Preservation requirements, if any
    ____ Analytes of interest (specific isotopes or nuclide)
    ____ Measurement Quality Objectives
    ____ Proposed method (if appropriate) and method validation documentation
    ____ Regulatory reporting time requirement (if applicable)
    ____ Analysis time requirements (time issues related to half-lives)
    ____ QC requirements (frequency, type, and acceptance criteria)
    ____ Waste disposal issues during processing
    ____ Licenses and accreditation

OVERSIGHT
    ____ Quality manual
    ____ Required Performance Evaluation Program participation
    ____ Criteria for (blind) QC
    ____ Site visit/data assessment
    ____ Audit (if any)

REPORTING REQUIREMENTS
    ____ Report results as gross, isotopic....
    ____ Reporting units
    ____ Reporting basis (dry weight, ...)
    ____ How to report measurement uncertainties
    ____ Reporting Minimum Detectable Concentration and Minimum Quantifiable Concentration
    ____ Report contents desired and information for electronic data transfer
    ____ Turn-around time requirements
    ____ Electronic deliverables
    ____ Data report format and outline

TABLE E.2 — SOW checklists for the agency and proposer

```
NOTIFICATION
____    Exceeding predetermined Maximum Concentration Levels - when applicable
____    Batch QC failures or other issues
____    Failure to meet analysis or turnaround times
____    Violations related to radioactive material license
____    Change of primary staff associated with contract work

SCHEDULE
____    Expected date of delivery
____    Method of delivery of samples
____    Determine schedule (on batch basis)
____    Method to report and resolve anomalies and nonconformance in data to the client
____    Return of samples and disposition of waste

CONTACT
____    Name, address, phone number of responsible parties
```

E.4.2 Personnel

The education, working knowledge, and experience of the individuals that supervise operations, conduct analyses, operate laboratory instruments, process data, and create the deliverables is of key importance to the operation of a laboratory. The agency is essentially asking who is sufficiently qualified to meet the proposed project's needs. (The answer to this question may come from an agency's guidance or other specific requirements generated by the structured planning process.) The laboratory staff that will perform the analyses should be employed, trained, and qualified prior to the award of the contract.

In response to the RFP, the proposer should include a listing of staff members capable of managing, receiving, logging, preparing, and processing samples; providing reports in the format specified by the project; preparing data packages with documentation to support the results; maintaining the chain of custody; and other key work activities. The laboratory should list the administrative personnel and appoint a technical person to be a point of contact for the proposed work. This person should fully understand the project's requirements and be reasonably available to respond to every project need. A proposal should include the educational background and a brief resume for all key personnel. The level of training for each technician should be included.

Tables E.3 and E.4 are examples that briefly summarize the suggested minimum experience, education, and training for the listed positions. Note, some agency-specific requirements may exceed the suggested qualifications and this issue should be explored further during the planning process. The goal is to provide basic guidance with examples that the MARLAP user can employ as a starting point during planning. Once specific requirements are established, this information will appear in the RFP. Table E.3 provides a listing for the types of laboratory technical supervisory personnel that are likely to manage every aspect of a laboratory's work. Each position title is given a brief description of responsibilities, along with the minimum level of education and experience. Table E.4 presents descriptions for staff members that may be considered optional personnel or, in some cases, represent necessary support that is provided by personnel with other position titles. Table E.5 indicates the minimum education and experience for laboratory technical staff members. In some cases, specific training may add to or be substituted for the listed education or experience requirement. Training may come in a number of forms, such as instrument-specific classes offered by a manufacturer, to operational or safety programs given by outside trainers or the laboratory's own staff.

TABLE E.3 — Laboratory technical supervisory personnel listed by position title and examples for suggested minimum qualifications

All personnel are responsible to perform their work to meet all terms and conditions of the contract

Technical Supervisory Personnel		
Position Title and Responsibilities	Education	Experience
Radiochemical Laboratory Supervisor, Director, or Manager. Responsible for all technical efforts of the radiochemical laboratory.	Minimum of bachelor's degree in any scientific/engineering discipline, with training in radiochemistry, radiation detection instrumentation, statistics, and QA.	Minimum of three years of radioanalytical laboratory experience, including at least one year in a supervisory position. Training in laboratory safety, including radiation safety.
Quality Assurance Officer Responsible for overseeing the quality assurance aspects of the data and reporting directly to upper management.	Minimum of bachelor's degree in any scientific/engineering discipline, with training in physics, chemistry, and statistics.	Minimum of three years of laboratory experience, including at least one year of applied experience with QA principles and practices in an analytical laboratory or commensurate training in QA principles.

TABLE E.4 — Laboratory technical personnel listed by position title and examples for suggested minimum qualifications and examples of optional staff members

Optional Technical Personnel		
Position Title and Responsibilities	Education	Experience
Systems Manager Responsible for the management and quality control of all computing systems; generating, updating, and quality control for deliverables.	Minimum of bachelor's degree with intermediate courses in programming, information management, database management systems, or systems requirements analysis.	Minimum of three years experience in data or systems management of programming, including one year experience with the software being utilized for data management and generation of deliverables.
Programmer Analyst Responsible for the installation, operation, and maintenance of software and programs, generating, updating, and quality of controlling analytical databases and automated deliverables.	Minimum of bachelor's degree with intermediate courses in programming, information management, information systems, or systems requirements analysis.	Minimum of two years experience in systems or applications programming, including one year experience with the software being utilized for data management and generation of deliverables.

TABLE E.5 — Laboratory technical personnel listed by position title and examples for suggested minimum qualifications

All personnel are responsible to perform their work to meet all terms and conditions of the contract.

Technical Staff		
Position Title	**Education**	**Experience**
Gamma Spectrometrist	Minimum of bachelor's degree in chemistry or any physical scientific/engineering discipline. Training courses in gamma spectrometry.	Minimum two years experience in spectrometric data interpretation. Formal training or one year experience with spectral analysis software used to analyze data.
Alpha Spectrometrist	Minimum of bachelor's degree in chemistry or any physical scientific/engineering discipline. Training courses in alpha spectrometry.	Formal training or one year experience with spectral analysis software used to analyze data.
Radiochemist	Minimum of bachelor's degree in chemistry or any physical scientific/engineering discipline. In lieu of the educational requirement, two years of additional, equivalent radioanalytical experience may be substituted.	Minimum of two years experience with chemistry laboratory procedures, with at least one year of radiochemistry in conjunction with the educational qualifications, including (for example): 1) Operation and maintenance of radioactivity counting equipment; 2) Alpha/gamma spectrometric data interpretation; 3) Radiochemistry analytical procedures; and 4) Sample preparation for radioactivity analysis.
Counting Room Technician	Minimum of bachelor's degree in chemistry or any scientific/engineering discipline.	Minimum of one year experience in a radioanalytical laboratory.
Laboratory Technician	Minimum of high school diploma and a college level course in general chemistry or equivalent—or college degree in another scientific discipline (e.g., biology, geology, etc.)	Minimum of one year experience in a radioanalytical laboratory.

E.4.3 Instrumentation

A proposer's laboratory must have in place and in good working order the types and required number of instruments necessary to perform the work advertised by the agency. Specific factors are noted in the RFP, such as: an estimate for the number of samples, length of the contract, and expected turnaround times which influence the types of equipment needed to support the contract.

Analytical work can be viewed as a function of current technology. Changes may occur from time to time, especially in relation to scientific advancements in equipment, software, etc. Instrumentation represents the mechanical interface between prepared samples and the data generated in the laboratory. The capacity to process larger and larger numbers of samples while sustaining the desired level of analytical sensitivity and accuracy is ultimately a function of the laboratory's equipment, and the knowledge and experience of the individuals who operate and maintain the instruments. Additional support for the laboratory's on-line activities or the state of readiness to maintain a constant or an elevated peak work load comes in the form of back-up instruments that are available at all times. Information concerning service contracts that provide repairs or replacement when equipment fails to perform is important to meeting contract obligations. Demonstrating that this support will be in place for the duration of the contract is a key element for the proposer to clearly describe in a proposal.

E.4.3.1   Type, Number, and Age of Laboratory Instruments

A description of the types of instruments at a laboratory is an important component of the proposal. The number of each type of instrument available for the proposed work should be indicated in the proposal. This includes various counters, detectors, or other systems used for radioanalytical work. A complete description for each instrument might include the age or acquisition date. This information may be accompanied by a brief description indicating the level of service an instrument provides at its present location.

E.4.3.2   Service Contract

The types and numbers of service contracts may vary depending on the service provider. Newly purchased instruments will be covered by a manufacturer's warranty. Other equipment used beyond the initial warranty period may either be supported by extensions to the manufacturer's warranties or by other commercial services that cover individual instrument or many instruments under a site-wide service contract. Whatever type of support is in place, the contractor will need to state how having or not having such service contracts affects the laboratory's ability to meet the terms of the contract and the potential impact related to the SOW.

E.4.4 Narrative to Approach

A proposal can "speak" to the agency's evaluation team by providing a logical and clearly written narrative of how the proposer will attend to every detail listed in the RFP. This approach conveys key information in a readable format to relate a proposer's understanding, experience, and working knowledge of the anticipated work. In this way, the text also illustrates how various components of the proposal work together to contribute to a unified view of the laboratory functions given the proposed work load as described in the RFP and as detailed in the SOW. The next four sections provide examples of proposal topics for which the proposer may apply a narrative format to address how the laboratory is qualified to do the proposed work.

E.4.4.1   Analytical Methods or Protocols

The proposer should list all proposed methods they plan to use. The proposal should also furnish all required method validation documentation to gain approval for use. When addressing use of methods, the proposer can describe how a method exhibits the best performance and also offer specific solutions to meet the agency's needs.

E.4.4.2 Meeting Contract Measurement Quality Objectives

The agency's planning process started with a review of questions and issues concerned with generating specific project APSs/MQOs. Stating how a proposer intends to meet the APSs/ MQOs data quality requirements adds an important section to the proposal. This allows the competing laboratories to demonstrate that they understand the requirements of the contract and their individual approaches to fulfilling these requirements. Further evidence in support of the proposer's preparations to meet or exceed the agency's data quality needs is generally covered in a contract laboratory's quality manual (Section E.4.5).

E.4.4.3 Data Package

The proposer responds to the RFP by stating how data will be processed under the contract. A narrative describing the use of personnel, equipment, and facilities illustrates every step in obtaining, recording, storing, formatting, documenting and reporting sample information and analytical results. The specific information related to all these activities and the required information as specified by the SOW is gathered into a data package. For example, a standard data package includes a case narrative, the results (in the format specified by the agency), a contractor data review checklist, any nonconformance memos resulting from the work, agency and contractor-internal chains of custody, sample and quality control (QC) sample data (this includes a results listing, calculation file, data file list, and the counting data) and continuing calibration data, and standard and tracer source-trace information, when applicable. At the inception of a project, initial calibration data are provided for detectors used for the work. If a detector is recalibrated, or a new detector is placed in service, initial calibration data are provided whenever those changes apply to the analyses in question.

Specific data from the data package may be further formatted in reports, including electronic formats, as the required deliverables which the contractor will send to the agency. The delivery of this information is also specified according to a set schedule.

E.4.4.4 Schedule

The RFP will provide information that allows the proposer to design a schedule that is tailored to the agency's need. For example, samples that are part of routine monitoring will arrive at the laboratory and the appropriate schedule reflects a cycle of activity from sample preparation to delivering a data package to the agency. This type of schedule is repeatedly applied to each set of samples. Other projects, surveys, or studies may follow a time line of events from start to completion, with distinct sets of samples and unique needs that arise at specific points in time. The proposer will initially outline a schedule that may utilize some cycling of activities at various stages of the work, but overall the nature of the work may change from stage to stage. The schedule in this case will reflect how the contractor expects to meet certain unique milestones on specific calendar dates.

Some projects will have certain requirements to process samples according to a graded processing schedule. The SOW should provide the requirements for the radiological holding time and sample processing turnaround time. Radiological holding time refers to the time required to process the sample—the time differential from the sample receipt date to the final sample matrix counting date. The sample processing turnaround time normally means the time differential from the receipt of the sample at the laboratory (receipt date) to the reporting of the analytical results to the agency (analytical report date). As such, the turnaround time includes the radiological holding time, the time to generate the analytical results, and the time to report the results to the agency.

Typically, three general time-related categories are stated: routine, expedited, and rush. Routine processing is normally a 30-day turnaround time, whereas expedited processing may have a turnaround time greater than five days but less than 30 days. Rush sample processing may have a radiological holding time of less than five days. For short-lived nuclides, the RFP should state the required radiological holding time, wherein the quantification of the analyte in the sample must be complete within a certain time period. The reporting of such results may be the standard 30-day turnaround time requirement. The agency should be reasonable and technically correct in developing the required radiological holding and turnaround times.

The RFP should specify a schedule of liquidated or compensatory damages that should be imposed when the laboratory is noncompliant relative to technical requirements, radiological holding times, or turnaround times.

E.4.4.5   Sample Storage and Disposal

The RFP should specify the length of time the contractor must store samples after results are reported. In addition, it should state who is economically and physically responsible for the disposal of the samples. The laboratory should describe how the samples will be stored for the specified length of time and how it plans to dispose of the samples in accordance with local, state and federal regulations.

E.4.5 Quality Manual

Only those radiochemistry laboratories that adhere to well-defined quality procedures—pertaining to data validation, internal and external laboratory analytical checks, instrument precision and accuracy, personnel training, and setting routine laboratory guidelines—can insure the highest quality of scientifically valid and defensible data. In routine practice, a laboratory prepares a written description of its quality manual that addresses, at a minimum, the following items:

- Organization and management;
- Quality system establishment, audits, essential quality controls and evaluation and data verification;
- Personnel (qualifications and resumes);
- Physical facilities (accommodations and environment);
- Equipment and reference materials;
- Measurement traceability and calibration;
- Test methods and standard operating procedures (methods);
- Sample handling, sample acceptance policy and sample receipt;
- Records;
- Subcontracting analytical samples;
- Outside support services and supplies; and
- Complaints.

The quality manual may be a separately prepared document that may incorporate or reference already available and approved laboratory standard operating procedures (SOPs). This manual provides sufficient detail to demonstrate that the contractor's measurements and data are appropriate to meet the MQOs and satisfy the terms and conditions of the contract. The manual should clearly state the objective of the SOP, how the SOP will be executed, and which performance standards will be used to evaluate the data. Work-related requirements based on quality assurance are also an integral part of the SOW.

When a proposal is submitted for review, the contracting laboratory generally sends along a current copy of its quality manual. Additional details pertaining to the content of a quality manual can be found in NELAC (2002), ANSI/ASQC E-4, EPA (1993, 2001, 2002), ISO/IEC 17025, and MARSSIM (2000).

E.4.6 Licenses and Accreditations

All laboratories must have appropriate licenses from the U.S. Nuclear Regulatory Commission (NRC) or other jurisdictions (Agreement State, host nation, etc.) to receive, possess, use, transfer, or dispose of radioactive materials (i.e., those licensable as indicated in 10 CFR 30.70, Schedule A—Exempt Concentrations). A license number and current copy of a laboratory's licenses are typically requested with paperwork that one submits to obtain radionuclide materials—for example, when ordering and arranging to use laboratory standards. Overall, a laboratory's license permits work with certain radionuclides and limits to the quantity of each radionuclide at the laboratory. A proposer's license should allow for new work with the types and anticipated amounts of radionuclides as specified in an RFP. Part of the licensing requirement ensures that the laboratory maintains a functioning radiation safety program and properly trains its personnel in the use and disposal of radioactive materials. For more complete information on license requirements, refer to either the NRC, the appropriate state office, or 10 CFR 30.

The laboratory may need to be certified for radioassays by the state in which the lab resides. The RFP should request a copy of the current standing certification(s) to be submitted with the proposal. If the agency expects a laboratory to process samples from numerous states across the United States, then additional certifications for other states may or will be required. To request that a proposer arrange for certification in multiple states prior to submitting a proposal may be viewed as placing an unfair burden on a candidate laboratory who as yet to learn if it will be awarded a contract. Additional fees, for each state certification, potentially add to a proposer's cost to simply present a proposal. In such cases, an agency may indicate that additional certification(s)—above that already held for the laboratory's state of residence—may be required once the contract is awarded and just prior to initiating the work.

E.4.7 Experience

The contractor, viewed as a single entity made of all its staff members, may have an extensive work history as is exemplified through the number and types of projects and contracts that were previously or are currently supported by its laboratory services. This experience is potentially an important testimonial to the kind of work the contractor is presently able to handle with a high degree of competence. The agency's evaluation team will review this information relative to the need(s) stated in the RFP. The more applicable the track record, the stronger a case the proposer has when competing for the award.

E.4.7.1 Previous or Current Contracts

In direct relation to the preceding section, the proposer's staff should respond directly to the RFP when asked to provide a list of contracts previously awarded and those they are presently fulfilling. Of primary importance, the list should contain contracts that are similar to the one under consideration (i.e., similar work load and technical requirements), with the following information:

- Name of the company or agency awarding the contract;
- Address;
- Phone number;
- Name of contact person; and
- Scope of contract.

E.4.7.2 Quality of Performance

The agency's TEC (Section E.5.1) is likely to check a laboratory's results for its participation in a proficiency program which is sponsored by one of several federal agencies. For example, the

U.S. Department of Energy (DOE), and National Institute of Standards and Technology (NIST) offer proficiency programs. Records for the laboratory's results may be reviewed to cover a number of years. This review indicates quality and consistency in relation to the types of samples that the federal agency sends to each laboratory. Thus, at designated times during each year, a laboratory will receive, process, and later report findings for proficiency program samples. This routine is also required for certification by an agency, such as the U.S. Environmental Protection Agency (EPA) for drinking water analysis. In this case, to obtain or maintain a certification, the laboratory must pass (i.e., successfully analyze) on the basis of a specific number of the total samples.

E.5   Proposal Evaluation and Scoring Procedures

The initial stages of the evaluation process separate technical considerations from cost. Cost will enter the selection process later on. The agency's TEC will consider all proposals and then make a first cut (Table E.6 and Section E.5.3 below), whereby some proposals are eliminated based on the screening process. This selection from among the candidates is based on predetermined criteria that are related to the original MQOs and how a proposer's laboratory is technically able to support the contract. A lab that is obviously unequipped to perform work according to the SOW is certain to be dropped early in the selection process. In some cases, the stated ability to meet the analysis request should be verified by the agency, through pre-award audits and proficiency testing, as described below. Letters notifying unsuccessful bidders may be sent at this time. For information concerning a proposer's response to this letter, see Section E.5.7.

E.5.1   Evaluation Committee

The agency personnel initially involved in establishing a new contract and starting the selection process include the Contracting Officer (administrative, nontechnical) and Contracting Officer's Representative (technical staff person advising the Contracting Officer). Once all proposals are accepted by the agency, a team of technical staff members score the technical portion of the proposal. The team is lead by a chairperson who oversees the activities of this TEC. It is recommended that all members of the TEC have a technical background relevant to the subject matter of the contract.

One approach to evaluation includes sending copies of all proposals to each member of the committee for individual scoring (Table E.6). The agency, after an appropriate length of time, may conduct a meeting or conference call to discuss the scores and reach a unified decision. Using this approach, each proposal is given a numerical score and these are listed in descending order. A "break-point" in the scores is chosen. All candidates above this point are accepted for a continuation of the selection process. Those below the break point may be notified at this point in time. Note that evaluations performed by some agencies may follow variations on this scoring and decision process.

The TEC must have a complete technical understanding of the subject matter related to the proposed work and the contract that is awarded at the end of the selection process. These individuals are also responsible for responding to any challenge to the agency's decision to award the contract. Their answers to such challenges are based on technical merit in relation to the proposed work (Section E.5.7).

E.5.2 Ground Rules — Questions

The agency's solicitation should clearly state if and when questions from an individual proposer will be allowed during the selection process. Information furnished in the agency's response is simultaneously sent to all competing laboratories.

E.5.3 Scoring/Evaluating Scheme

The agency should prepare an RFP that includes information concerning scoring of proposals or weights for areas of evaluation. This helps a proposer to understand the relative importance of specific sections in a proposal and how a proposal will be scored. In this case, the method of evaluation and the scoring of specific topic areas is outlined in the solicitation. If this information is not listed in the solicitation and because evaluation formats differ agency to agency, proposers may wish to contact the agency for additional agency-specific details concerning this process.

An agency may indicate the relative weight an evaluation area holds with regard to the proposed work for two principle reasons. First, the request is focused to meet a need for a specific type of work for a given study, project, or program. This initially allows a proposer to concentrate on areas of greatest importance. Second, if the contractor submits a proposal that lacks sufficient information to demonstrate support in a specific area, the agency can then indicate how the proposal does not fulfill the need as stated in the request.

Listed below is an example of some factors and weights that an agency might establish before an RFP is distributed:

Description	Weight
Factor I ....Technical Merit	25
Factor II ...Proposer's Past Performance	25
Factor III ...Understanding of the Requirements	15
Factor IV ..Adequacy and Suitability of Laboratory Equipment and Resources	15
Factor V ...Academic Qualifications and Experience of Personnel	10
Factor VI ..Proposer's Related Experience	10

The format presented above assigns relative weights for each factor—with greater weight given to more important elements of the proposal. Technical merit (Factor I) includes technical approach, method validation, and the ability to meet the MQOs, etc. Factor II includes how well the proposer performed in previous projects or related studies. A proposer's understanding (Factor III) is demonstrated by the laboratory's programs, commitments as well as certifications, licenses, etc., to ensure the requirements of the RFQ will be met. Adequacy and suitability (Factor IV) is generally an indication that the laboratory is presently situated to accept samples and conduct the work as proposed. Factor V focuses on topics covered previously in Section E.4.2 while the proposer's experience (Factor VI) is considered in Section E.4.7.

An agency may use a Technical Evaluation Sheet—in conjunction with the Proposal Evaluation Plan as outlined in the next section (Table E.6)—to list the total weight for each factor and to provide a space for the evaluator's assigned rating. The evaluation sheet also provides areas to record the RFP number, identity of the proposer, and spaces for total score, remarks, and evaluator's signature. The scoring and evaluation scheme is based on additional, more detailed, considerations which are briefly discussed in the next three sections (E.5.3.1 to E.5.3.3)

E.5.3.1   Review of Technical Proposal and Quality Manual

Each bidding-contractor laboratory will be asked to submit a technical proposal and a copy of its quality manual. This document is intended to address all of the technical and general laboratory requirements. The proposal and quality manual are reviewed by members of the TEC who are both familiar with the proposed project and are clearly knowledgeable in the field of radiochemistry.

Table E.6 is an example of a proposal evaluation plan (based on information from the U.S. Geological Survey). This type of evaluation can be applied to proposals as they are considered by the TEC.

TABLE E.6 — Example of a proposal evaluation plan

Proposal Evaluation
Objective: To ensure impartial, equitable, and comprehensive evaluation of proposals from contractors desiring to accomplish the work as outlined in the Request for Proposals and to assure selection of the contractor whose proposal, as submitted, offers optimum satisfaction of the government's objective with the best composite blend of performance, schedules, and cost.
Basic Philosophy: To obtain the best possible technical effort which satisfies all the requirements of the procurement at the lowest overall cost to the government.
Evaluation Procedures
1. Distribute proposals and evaluation instructions to Evaluation Committee.
2. Evaluation of proposals individually by each TEC member. Numerical values are recorded with a concise narrative justification for each rating.

3. The entire committee by group discussion prepares a consensus score for each proposal. Unanimity is attempted, but if not achieved, the Chairperson shall decide the score to be given.

4. A Contract Evaluation Sheet listing the individual score of each TEC member for each proposal and the consensus score for the proposal is prepared by the Chairperson. The proposals are then ranked in descending order.

5. The Chairperson next prepares an Evaluation Report which includes a Contract Evaluation Sheet, the rating sheets of each evaluator, a narrative discussion of the strong and weak points of each proposal, and a list of questions which must be clarified at negotiation. This summary shall be forwarded to the Contracting Officer.

6. If required, technical clarification sessions are held with acceptable proposers.

7. Analysis and evaluation of the cost proposal will be made by the Contracting Officer for all proposals deemed technically acceptable. The Chairperson of the TEC will perform a quantitative and qualitative analysis on the cost proposals or those firms with whom cost negotiations will be conducted.

Evaluation Criteria

The criteria to be used in the evaluation of this proposal are selected before the RFP is issued. In accordance with the established agency policy, TEC members prepare an average or consensus score for each proposal on the basis of these criteria and only on these criteria.

A guideline for your numerical rating and rating sheets with assigned weights for each criteria are outlined next under Technical Evaluation Guidelines for Numerical Rating.

Technical Evaluation Guidelines for Numerical Rating

1. Each item of the evaluation criteria will be based on a rating of 0 to 10 points. Therefore, each evaluator will score each item using the following guidelines:

    a. *Above normal*: 9 to 10 points (a quote element which has a high probability of exceeding the expressed RFP requirements).

    b. *Normal*: 6 to 8 points (a quote element which, in all probability, will meet the minimum requirements established in the RFP and Scope of Work).

    c. *Below normal*: 3 to 5 points (a quote element which may fail to meet the stated minimum requirements, but which is of such a nature that it has correction potential).

    d. *Unacceptable*: 0 to 2 points (a quote element which cannot be expected to met the stated minimum requirements and is of such a nature that drastic revision is necessary for correction).

2. Points will be awarded to each element based on the evaluation of the quote in terms of the questions asked.

3. The evaluator shall make no determination on his or her own as to the relative importance of various items of the criteria. The evaluator must apply a 0 to 10 point concept to each item without regard to his or her own opinion concerning one item being of greater significance than another. Each item is given a predetermined weight factor in the Evaluation Plan when the RFP is issued and these weight factors must be used in the evaluation.

E.5.3.2   Review of Laboratory Accreditation

A copy of the current accreditation(s) should be submitted with the proposal. The agency should confirm the laboratory's accreditation by contacting the federal or state agency that provided the accreditation. In some cases, a public listing or code number is provided. Confirming that a specific code number belongs to a given laboratory will require contacting the agency that issued the code.

E.5.3.3   Review of Experience

The laboratory should furnish references in relation to its past or present work (Section E.4.7.1). To the extent possible, this should be done with regard to contracts or projects similar in composition and size to the proposed project. One or more members of the TEC are responsible for developing a list of pertinent questions and then contacting each reference listed by the proposer. The answers obtained from each reference are recorded for use later in the evaluation process. In some cases, the laboratory's previous performance for the same agency should be given special consideration.

E.5.4   Pre-Award Proficiency Samples

Some agencies may elect to send proficiency or performance testing (PT) samples to the laboratories that meet a certain scoring criteria in order to demonstrate the laboratory's analytical capability. The composition and number of samples should be determined by the nature of the proposed project. The PT sample matrix should be composed of well-characterized materials. It is recommended that site-specific PT matrix samples or method validation reference material (MVRM; Chapter 6, *Section and Application of an Analytical Method*) be used when available. The matrix of which the PT sample is composed must be well characterized and known to the agency staff who supply the sample to the candidate laboratory. For example, if an agency is concerned with drinking water samples, then the agency's laboratory may use its own source of tap water as a base for making PT samples. This water, with or without additives, may be supplied for this purpose.

Each competing lab should receive an identical set of PT samples. The RFP should specify who will bear the cost of analyzing these samples, as well as the scoring scheme, (e.g., pass/fail) or a sliding scale. Any lab failing to submit results should be automatically disqualified. The results should be evaluated and each lab given a score. This allows the agency to narrow the selection further—after which only two or three candidate laboratories are considered.

At this point, two additional selection phases remain. A visit to each candidate's facilities comes next (Section E.5.5) and thereafter, once all technical considerations are reviewed, the cost of the contractor's service is examined last (Section E.5.6).

E.5.5 Pre-Award Audit

A pre-award audit, which may be an initial audit, is often performed to provide assurance that a selected laboratory is capable of performing the required analyses in accordance with the SOW. In other words, *is the laboratory's representation (proposal) realistic when compared to the actual facilities?* To answer this question, auditors will be looking to see that a candidate laboratory appears to have all the required elements to meet the proposed contract's needs. In some cases, it may be appropriate to conduct both a pre-award audit, followed by an evaluation after the work begins (see Section E.6.7 for information on ongoing laboratory evaluations).

The two or three labs with the highest combined scores (for technical proposals and proficiency samples) may be given an on-site audit.

The pre-award audit is a key evaluating factor that is employed before the evaluation committee makes a final selection. Many federal agencies, including DOE, EPA, and USGS, have developed forms for this purpose. Some of the key items to observe during an audit include:

- Sample Security – Will the integrity of samples be maintained for chain of custody? If possible, examine the facility's current or past chain-of-custody practice.

- Methods – Are copies of SOP's available to every analyst? In some cases, one may check equations used to identify and quantitate the radionuclides of interest. Additional concerns include the potential for interferences, total propagated uncertainty, decision levels, and minimum detectable concentrations.

- Method Validation Documentation – Verify the method validation documentation provided in the response to the RFP. Have there been any QA/QC issues related to the methods? Are the identified staff (provided in the RFP) qualified to perform the methods?

- Adherence to SOPs – This may include looking to see that sample preparation, chemical analysis, and radiometric procedures are performed according to the appropriate SOP.

- Internal QC – Check the files and records.

- External QC/PT samples – Check files and records pertaining to third-party programs.

- Training – Check training logs. Examine analysts' credentials, qualifications, and proficiency examination results.

- Instrumentation – Check logs. Are instruments well maintained, is there much down time, are types and numbers listed in technical proposal correct? Look for QC chart documentation.

- Instrumentation – Calibration records. Do past and current calibration records indicate that the laboratory's instruments are capable of providing data consistent with project needs? Look at instrumentation characteristics, including resolution, detection efficiency, typical detection limits, etc. Are materials that are traceable to a national standards organization (such as NIST in the United States) used for detector calibration and chemical yield determinations?

- Personnel – Talk with and observe analysts. Verbal interaction with laboratory staff during an audit helps auditors to locate the information and likewise provide evidence for the knowledge and understanding of persons who conduct work in the candidate laboratory.

- Log-In – Is this area well-organized to reduce the possibility of sample mix-ups?

- Tracking – Is there a system of tracking samples through the lab?

Information about each laboratory may be gathered in various ways. One option available to the agency is to provide each candidate laboratory with a list of questions or an outline for information that will be collected during the audit (Table E.7). The agency's initial contact with the laboratory can include a packet with information about the audit and questions that the laboratory must address prior to the agency's on-site visit. For example, from the checklist presented in Table E.7, one can see the laboratory will be asked about equipment. In advance of the audit, laboratory personnel can create a listing of all equipment or instruments that will be used to support the contract. Table E.7 also indicates information to be recorded by the auditors during the visit. The audit record includes the agency's on-site observations, along with the laboratory's prepared responses.

TABLE E.7— Sample checklist for information recorded during a pre-award laboratory audit

```
Laboratory:
Date:
Auditors:
    1.
    2.
A. Review packet that was sent to laboratory for completion:
    1.  Laboratory Supervisor
    2.  Laboratory Director
    3.  Current Staff
    4.  Is the laboratory responsible for all analyses? If not, what other laboratory(s) is (are) responsible?
    5.  Agency responsible for [drinking water] program in the state.
    6.  Does the laboratory perform analyses of environmental samples around nuclear power facilities, or  from
        hospitals, colleges, universities, or other radionuclide users?
    7.  Agency responsible for sample collections in item 6.

B. Laboratory Facilities:
    1.  Check all items in the laboratory packet.
```

TABLE E.7— Sample checklist for information recorded during a pre-award laboratory audit

2. Comments
3. Is there a Hot Laboratory or a designated area for samples from a nuclear power facility that would represent a nuclear accident or incident? Is this documented in the SOP or QA Manual?

C. Laboratory Equipment and Supplies:
    1. Check all items on the laboratory packet. Includes analytical balances, pH meters, etc.
    2. Comments
    3. Radiation counting instruments:
        a. Thin window gas-flow proportional counters
        b. Windowless gas-flow proportional counters
        c. Liquid scintillation counter
        d. Alpha scintillation counter
        e. Radon gas-counting system
        f. Alpha spectrometer
        g. Gamma spectrometer systems:
            1. Ge (HPGe) detectors
            2. NaI detectors
            3. Multichannel analyzer(s)

D. Analytical Methodology:
    1. Check all items on the laboratory packet.
    2. Comments

E. Sample Collection, Handling, and Preservation:
    1. Check all items on the laboratory packet.
    2. Comments

F. Quality Assurance Section:
    1. Examine laboratory SOP
        a. Comments

    2. Examine laboratory's quality manual
        a. Comments

    3. Performance Evaluation Studies (Blind)
        a. Comments and results

    4. Maintenance records on counting instruments and analytical balances.
        a. Comments and results

    5. Calibration data
        a. Gamma Spectrometer system
            1. Calibration source
            2. Sufficient energy range
            3. Calibration frequency
            4. Control charts
                a. Full Peak Efficiency
                b. Resolution
                c. Background

TABLE E.7— Sample checklist for information recorded during a pre-award laboratory audit

    b. Alpha/Beta counters
        1. Calibration source
        2. Calibration frequency
        3. Control charts
            a. Alpha
            b. Beta
            c. Background

    c. Radon counters
        1. Calibration source
        2. Frequency of radon cell background checks

    d. Liquid Scintillation Analyzer
        1. Calibration sources
        2. Calibration frequency
        3. Control charts
            a. H-3
            b. C-14
            c. Background
            d. Quench

6. Absorption and Efficiency curves:
    a. Alpha absorption curve
    b. Beta absorption curve
    c. Ra-226 efficiency determination
    d. Ra-228 efficiency determination
    e. Sr-89, Sr-90, and Y-90 efficiency determinations
    f. Uranium efficiency determination

7. Laboratory QC Samples
    a. Spikes
    b. Replicates/duplicates
    c. Blanks
    d. Cross check samples
    e. Frequency of analysis
    f. Contingency actions if control samples are out of specification
    g. Frequency of analysis

E. Records and Data Reporting
    1. Typical data package
    2. Electronic data deliverable format
    3. Final data report

H. Software Verification and Validation
    1. Instrumentation and Equipment Control and Calibrations
    2. Analytical Procedure Calculations/Data Reduction
    3. Record Keeping/Laboratory/Laboratory Information Management System/Sample Tracking
    4. Quality Assurance Related — QC sample program/instrument QC

E.5.6 Comparison of Prices

To this point, the selection process focuses on technical issues related to conducting work under the proposed contract. Keeping this separate from cost considerations simplifies the process and helps to sustain reviewer objectivity. Once the scoring of labs is final, the price of analyses may be reviewed and compared. Prices are now considered along with inspection results. This part of the process is best performed by technical personnel, including members of the TEC who work in either a laboratory or the field setting, and who possess the knowledge to recognize a price that is reasonable for a given type of analysis. Various scenarios may apply where prices differ:

- Candidates are dropped generally if their proposed prices are extreme.

- Laboratories that score well—aside from their prices that may still be on the high side—are given an opportunity to rebid with a best and final cost. This lets laboratories know they have entered the final stage of the selection process.

A final ranking is based on the technical evaluation, including the proficiency examination and audit if conducted, and the best-and-final prices submitted by each laboratory.

While there is no way to determine how evaluations may be conducted in the future, some extra consideration may be given to proposals that offer greater technical capabilities (i.e., those that house state-of-the-art or high-tech analytical services) as opposed to fulfilling the minimum requirements of the RFP.

E.5.7 Debriefing of Unsuccessful Vendors

At an appropriate time in the selection process, all unsuccessful bidders are sent a letter outlining the reasons that they were not awarded the contract. As noted previously, the RFP should be very explicit in illustrating what a proposal should contain and which areas carry more or less weight with regard to the agency's evaluation. If so, the agency is able to provide a written response to specifically identify areas of the proposal where the contractor lacks the appropriate services or is apparently unable to present a sufficiently strong case documenting an ability to do the work. Also, as stated previously, the proposer must present as clear a case as possible and write into the proposal all relevant information. A simple deletion of key information will put a capable proposer out of the running in spite of the experience, support, and services they are able to render an agency.

If a contractor wishes an individual debriefing, the agency can arrange to have the TEC meet with the contractor's representatives. This meeting allows for an informal exchange to further explore issues to the satisfaction of the proposer. This exchange may offer the agency an opportunity to restate and further clarify the expected minimum qualifications that are required of the proposer.

A more formal approach contesting the agency's decision follows after a protest is lodged by the contractor. In this case, the agency's TEC and the contractor's representatives are accompanied by legal council for both sides.

E.6 The Award

The selection process ends when the agency personnel designate which contractor will receive the award. Several steps follow in advance of formally presenting the award. This essentially includes in-house processing, a review by the agency's legal department, and a final review by the contract staff. These activities verify that the entire selection process was followed properly and that the contract's paperwork is correct. The agency's contracts office then signs the proper documents and the paperwork is sent to the contractor. The contract becomes effective as of the date when the government's contracting officer signs.

E.7 For the Duration of the Contract

After the award is made, the agency enters into a working relationship with the contract laboratory and work begins. Over the period of the contract, the agency will send samples, receive deliverables, and periodically check the laboratory's performance. The work according to the SOW and the activities associated with performance checks and laboratory evaluations are topics covered beginning with the next section. Furthermore, as data are delivered to the agency, invoices will be sent by the contractor to the agency. The agency will process the invoices in steps: that receipt of data is initially confirmed, the results are appropriate (i.e., valid), and finally that the invoice is paid. This activity may occur routinely as invoices arrive—weekly, monthly, or at some other time interval throughout the course of a contract.

Keep in mind that the structured planning process is iterative in nature and may come into play at any point during a contract period. For example, federal or state laboratories engaging contract-support services may be involved in routine monitoring of numerous sampling sites. For sets of samples that are repeatedly taken from a common location over the course of years, only the discovery of unique results or change in performance-based methods may instigate an iteration and a review of the MQOs. For other types of projects, such as a location undergoing a MARSSIM-site survey, the project plan may change as preliminary survey work enters a period of discovery—e.g., during a scoping or characterization survey (MARSSIM, 2000). Even during a final status survey, discovery of some previously unknown source of radioactive contamination may force one to restate not only the problem, but to reconsider every step in the planning process. Modification of a contract may be necessary to address these circumstances.

E.7.1 Managing a Contract

Communication is key to the successful management and execution of the contract. Problems, schedule, delays, potential overruns, etc., can only be resolved quickly if communications between the laboratory and agency are conducted promptly.

A key element in managing a contract is the timely verification (assessment) of the data packages provided by the laboratory. Early identification of problems allows for corrective actions to improve laboratory performance and, if necessary, the cessation of laboratory analyses until solutions can be instituted to prevent the production of large amounts of data which are unusable. Note that some sample matrices and processing methods can be problematic for even the best laboratories. Thus the contract manager must be able to discern between failures due to legitimate reasons and poor laboratory performance.

E.7.2 Responsibility of the Contractor

First and foremost, the responsibility of the laboratory is to meet the performance criteria of the contract. If the SOW is appropriately written, this provides guidance necessary to ensure the data produced will meet the project planning goals and be of definable quality. Similarly, the laboratory must communicate anticipated or unforeseen problems as soon as possible. Again, this could easily occur with complex, unusual, or problematic sample matrices. Communication is vital to make sure that matrix interferences are recognized as early as possible, and that subsequent analyses are planned accordingly.

The laboratory's managers must plan the analysis—that is, have supplies, facilities, staff, and instruments available as needed—and schedule the analysis to meet the agency's due date. In the latter case, a brief buffer period might be included for unanticipated problems and delays, thus allowing the laboratory the opportunity to take appropriate corrective action on problems encountered during an analysis.

E.7.3 Responsibility of the Agency

During the period of the contract, the agency is responsible for employing external quality assurance oversight. Thus the performance of the laboratory should be monitored continually to insure the agency is receiving compliant results. Just because a laboratory produces acceptable results at the beginning of its performance on a contract does not necessarily mean that it will continue to do so throughout the entire contract period. For example, the quality of the data can degenerate at times when an unusually heavy workload is encountered by an environmental laboratory. One way to monitor this performance is to review the results of internal and external quality assurance programs. This may in part take the form of site visits (including onsite audits), inclusion of QC samples, evaluation of performance in Performance Evaluations or intercomparison programs, desk audits, and data assessments.

E.7.4 Anomalies and Nonconformance

The contractor must document and report all deviations from the method and unexpected observations that may be of significance to the data user. Such deviations should be documented in the narrative section of the data package produced by the contract laboratory. Each narrative should be monitored closely to assure that the laboratory is documenting departures from contract requirements or acceptable practice. The agency's reviewer should assure that the reason(s) given for the departures are clearly explained and are credible. The repeated reporting of the same deviation may be an indication of internal laboratory problems.

E.7.5 Laboratory Assessment

As work under a contract progresses over time, there are two principle means to assess a laboratory's performance: by having the laboratory process quality control samples (Section E.7.5.1 and E.7.5.2), and by agency personnel visiting the laboratory to conduct on-site evaluations (Section E.7.5.3).

E.7.5.1 Performance Testing and Quality Control Samples

A laboratory's performance is checked in one of several ways, including the use of agency PT samples, the laboratory's QC samples, laboratory participation in a performance evaluation program, agency certification program, and through agency audits, which may include an on-site visit.

There are several approaches to determining that an analysis is accurate and that the data reflect a true result. One check on each analysis comes from the laboratory's own QC measures. The contractor will routinely run standards, prepared spiked samples, and blanks, along with the samples submitted by the agency. Calibrations are also performed and a laboratory technician is expected to record information to document instrument performance.

Another avenue for monitoring performance comes with measures taken by the agency, including the incorporation of a number of double-blind PT samples, with each batch of samples sent to the contract laboratory. The preparation of double-blind PT samples for matrices other than water is difficult. A sample designated as a *blind PT sample* is one that the contractor knows is submitted by the agency for performance testing purposes. A *double-blind sample* is presented to the laboratory as if it were just another sample with no indication that this is for performance testing purposes. In the former case, the samples may be labeled in such a manner that the laboratory recognizes these as PT samples. In the latter case, unless the agency takes steps to use very similar containers and labeling as that for the field samples, the laboratory may recognize the double-blind PT samples for what they are. This in effect compromises the use of a double-blind sample. In each case, the agency knows the level or amount of each radionuclide in the blind sample.

When the analysis for a set of samples is complete and data are sent to the agency, the agency in turn checks the results for the PT samples and then performs data validation. In the case of characterization studies, one may continue to check results for PT samples, but data validation packages may not be required. If the double-blind results are not within reasonable limits, the agency will need to examine how these specific data may indicate a problem. In the meantime, work on subsequent sample sets cannot go forward until the problem is resolved. Some or all samples in the questionable batch may need to be reanalyzed depending on the findings for the PT samples. This is a case where storage of samples by the laboratory—e.g., from three to six months after analyses are performed—allows the agency to back track and designate specific samples for further or repeated analyses. The one exception to going back and doing additional analyses arises for samples containing radionuclides with short half lives. This type of sample requires a more immediate assessment to allow for repeated analyses, if needed.

Where data validation is required, the agency will routinely look at results for the PT samples that are added to the sample sets collected in the field. An additional QA measure includes a routine examination—for example, on a monthly or quarterly basis—of the laboratory's results for their own internal QC samples. This includes laboratory samples prepared as spikes, duplicates, and blanks that are also run along with the agency samples.

The agency can also schedule times to monitor a contractor laboratory's participation in a performance evaluation program—for example, those supported by the DOE, EPA, NIST, or NRC. Each laboratory, including the agency's own facilities, are expected to participate in such programs. The agency will also check to see if a laboratory's accreditation (if required) is current and this is something that should be maintained along with participation in a federally sponsored performance evaluation program. In general, the states accredit laboratories within their jurisdiction.

E.7.5.2   Laboratory Performance Evaluation Programs

Participating in an interlaboratory testing program (such as the PE programs mentioned in Section E.7.5.1) is the best way for a laboratory to demonstrate or an agency to evaluate a laboratory's measurement quality in comparison to other laboratories or to performance acceptance criteria. Furthermore, because MARLAP promotes consistency among radiochemistry laboratories, it is scientifically, programmatically, and economically advantageous to embrace the concept of a common basis for radioanalytical measurements—a measurement quality system that is ultimately traced to a national standards organization. ANSI N42.23 defines a system in which the quality and traceability of service laboratory measurements can be demonstrated through reference (and monitoring) laboratories. The service (in this case the contracted) laboratory should analyze traceable reference performance testing materials to examine the bias and precision of an analytical methodology or an analyst. Traceable reference material, a sample of known analyte concentration, is prepared from standard reference material obtained from a national standards body (NIST in the United States) or derived reference material supplied by a traceable radioactive source manufacturer (compliance with ANSI N42.22 for source manufac-turers in the United States). Demonstration of measurement performance and traceability shall be conducted at an appropriate frequency.

E.7.5.3   Laboratory Evaluations Performed During the Contract Period

An audit before awarding a contract emphasizes an examination of availability of instruments, facilities, and the potential to handle the anticipated volume of work. This also includes recognizing that the proper personnel are in place to support the contract. After the award, a laboratory evaluation will place additional weight on how instruments and personnel are functioning on a daily basis. Thus, logbooks, charts, or other documentation that are produced as the work progresses are now examined. This type of evaluation during the contract period uses an approach that differs from the pre-award audit (Section E.5.5). The format and documentation for an on-site audit may differ from agency to agency. An agency may wish to examine the EPA forms (EPA, 1997) and either adopt these or modify them to accommodate radionuclide work that includes sample matrices other than water or additional nuclides not presently listed.

There are two types of evaluations or audits that can be performed during the life of a contract. The first involves agency personnel that visit the contractor's facilities. The second approach includes activities conducted by agency personnel without visiting the laboratory.

In the former case, agency personnel examine documentation at the laboratory, including each instrument's logbook which is used to record background values, or to ensure that QC charts are current. During this type of evaluation, the agency and contractor personnel have an opportunity to communicate face-to-face, which is a benefit to both parties when clarification or additional detail is needed. For example, this audit's goal essentially is to check the capability of the laboratory to perform the ongoing work according to the contract work. In this case, an auditor may request to see one or more data packages, and then follow the information described in each package—including such items as sample tracking and documentation concerning sample preparation and analysis—to verify that the laboratory is now accomplishing the work as described by the SOW and in conformance with the quality manual.

In the latter case, one conducts what might be called a *desk audit*, where agency personnel review the contract and examine records or documentation that have come in as part of the project's deliverables. For the most part, the agency should constantly be monitoring activities under the contract, and in this sense, a desk audit is a daily activity without a formal process being applied at any specific point in time. However, depending on the agency's practice, if on-site visits are not made, then a desk audit becomes the only means to track activities under the contract. One approach to a desk audit is thus a periodic review—for example, every 6 or 12 months—of QC records to track the laboratory's performance over that period of time. This allows the agency to determine if there are deviations, shifts, or other trends that appear over time.

Each evaluation presents an additional opportunity to monitor various laboratory parameters, such as turnaround time. This is most important in cases when samples contain radionuclides having short half lives. During an on-site evaluation, the agency is able to determine if additional emphasis is required to tighten the time frame between sample receipt and analysis. The personal interaction between agency and laboratory permits a constructive dialog and facilitates an understanding of the possible means to increase or maintain the efficiency when processing and analyzing samples at the contractor's facility.

E.8   Contract Completion

There are several general areas of concern at the close of a contract that may be addressed differently depending on the agency or nature of the project under a given contract. For example, agency personnel who monitor contracts will review invoices to be certain that work is complete and that the corresponding results are considered acceptable. Once such monitoring activity provides the proper verification that the work is complete, then the agency's financial office processes all related bills and makes final payment for the work.

The laboratory should send in final deliverables, including routine submissions of raw data or records, as is the practice under the contract. Also, when applicable, agency-owned equipment shared with the laboratory during the contract period will be returned. The disposition of samples still in storage at the contractor's facility and additional records or other raw data must be decided and specified. The agency may wish to receive all or part of these items—otherwise, disposal of sample materials and documents held by the contractor must be arranged.

In some cases, work under the contract may create conditions where more time is necessary to process samples that remain or to process additional work that arises during the latter part of the contract period. Depending on the agency, funding, nature of the project, or other factor, the contract may be extended for a period of time, which may vary from weeks to months. Otherwise, once the contract comes to a close, the work ceases.

E.9   References

American National Standards Institute (ANSI). N42.22. *American National Standard—Traceability of Radioactive Sources to the National Institute of Standards and Technology (NIST) and Associated Instrument Quality Control.* 1995.

American National Standards Institute (ANSI). N42.23. *American National Standard—Measurement and Associated Instrument Quality Assurance for Radioassay Laboratories.* 2003.

American Society for Quality Control (ANSI/ASQC) E-4. *Specifications and Guidelines for Quality Systems for Environmental Data Collection and Environmental Technology Programs*. 1995. American Society for Quality Control, Milwaukee, Wisconsin.

U.S. Environmental Protection Agency (EPA). 1993. *Quality Assurance for Superfund Environmental Data Collection Activities*. Publication 9200.2-16FS, EPA, Office of Solid Waste and Emergency Response, Washington, DC.

U.S. Environmental Protection Agency (EPA). 1997. *Manual for the Certification of Laboratories Analyzing Drinking Water*. EPA 815-B-97-001. Office of Ground Water and Drinking Water, Washington, DC. Available at: www.epa.gov/safewater/certlab/labfront.html.

U.S. Environmental Protection Agency (EPA). 2001. *EPA Requirements for Quality Assurance Project Plans* (EPA QA/R-5). EPA/240/B-01/003. Office of Environmental Information, Washington, DC. Available at: www.epa.gov/quality/qa_docs.html.

U. S. Environmental Protection Agency (EPA). 2002. *Guidance for Quality Assurance Project Plans*. EPA QA/G-5. EPA/240/R-02/009. Office of Environmental Information, Washington, DC. Available at www.epa.gov/quality/qa_docs.html.

International Standards Organization/International Electrotechnical Commission (ISO/IEC) 17025. *General Requirements for the Competence of Testing and Calibration Laboratories*, International Organization for Standardization, Geneva, Switzerland. December 1999, 26 pp.

MARSSIM. 2000. *Multi-Agency Radiation Survey and Site Investigation Manual, Revision 1*. NUREG-1575 Rev 1, EPA 402-R-97-016 Rev1, DOE/EH-0624 Rev1. August. Available from www.epa.gov/radiation/marssim/.

National Environmental Laboratory Accreditation Conference (NELAC). 2002. *Quality Systems* Appendix D, *Essential Quality Control Requirements*. Revision 17. July 12. Available at: www.epa.gov/ttn/nelac/2002standards.html.

Office of Federal Procurement Policy (OFPP). 1997. *Performance-Based Service Contracting (PBSC) Solicitation/Contract/Task Order Review Checklist*. August 8. Available at: www.arnet.gov/Library/OFPP/PolicyDocs/pbscckls.html.

U.S. Code of Federal Regulations, Title 10, Part 30, *Rules of General Applicability to Domestic Licensing of Byproduct Material*.

GLOSSARY

absorption (10.3.2): The uptake of particles of a gas or liquid by a solid or liquid, or uptake of particles of a liquid by a solid, and retention of the material throughout the external and internal structure of the uptaking material. Compare with *adsorption*.

abundance (16.2.2): See *emission probability per decay event*.

accreditation (4.5.3, Table 4.2): A process by which an agency or organization evaluates and recognizes a program of study or an institution as meeting certain predetermined qualifications or standards through activities which may include performance testing, written examinations or facility audits. *Accreditation* may be performed by an independent organization, or a federal, state, or local authority. *Accreditation* is acknowledged by the accrediting organizations issuing of permits, licences, or certificates.

accuracy (1.4.8): The closeness of a measured result to the true value of the quantity being measured. Various recognized authorities have given the word *accuracy* different technical definitions, expressed in terms of *bias* and *imprecision*. MARLAP avoids all of these technical definitions and uses the term "accuracy" in its common, ordinary sense, which is consistent with its definition in ISO (1993a).

acquisition strategy options (2.5, Table 2.1): Alternative ways to collect needed data.

action level (1.4.9): The term *action level* is used in this document to denote the value of a quantity that will cause the decisionmaker to choose one of the alternative actions. The *action level* may be a *derived concentration guideline level (DCGL)*, background level, release criteria, *regulatory decision limit*, etc. The *action level* is often associated with the type of media, *analyte* and concentration limit. Some *action levels*, such as the release criteria for license termination, are expressed in terms of dose or risk. See *total effective dose equivalent (TEDE)* and *committed effective dose equivalent (CEDE)*.

activity, chemical (a) (10.3.5): (1) A thermodynamic quantity used in place of molal concentration in equilibrium expressions for reactions of real (nonideal) solutions. Activity indicates the actual behavior of *ions* in solution as a result of their interactions with the *solvent* and with each other. *Ions* deviate from ideal behavior as their concentration in solution increases and are not as effective in their chemical and physical behavior as their molar concentration would indicate. Thus, their effective concentration, a, is less than their stoichiometric concentration, c. (2) A measure of the effective molal concentration, c, in moles/Kg, of an *ion* under real (nonideal) solution conditions.

( ) Indicates the section in which the term is first used in MARLAP.
Italicized words or phrases have their own definitions in this glossary.

activity, of radionuclides (A) **(2.5.4.1):** Mean rate of nuclear decay occurring in a given quantity of material.

activity coefficient (γ) **(14.6.1):** (1) A fractional number that represents the extent that *ions* deviate from ideal behavior in solution (see *activity, chemical*). The *activity coefficient* multiplied times the molal concentration of *ions* in solution equals the chemical *activity*: $a = \gamma \cdot c$, where γ ≤ 1; thus, the *activity coefficient* is a correction factor applied to molal concentrations. At infinite dilution where behavior is ideal, $\gamma = 1.0$, but it decreases as the concentration of *ions* increases. (2) The ratio of effective (apparent) concentration of an *ion* in solution to the stoichiometric concentration, $\gamma = a/c$.

adsorption **(6.5.1.1):** Uptake of particles of a gas, liquid, or solid onto the surface of another substance, usually a solid. Compare with *absorption*.

adsorption chromatography **(14.7.1):** A chromatographic method that partitions (separates) components of a mixture through their different adsorption characteristics on a stationary solid phase and their different solubilities in a mobile liquid phase.

affinity chromatography **(14.7.1):** A chromatographic method that partitions (separates) proteins and nucleic acids in a mobile phase based on highly selective, very specific complementary bonds with antibody groups (*ligands*) that are chemically bonded to an inert solid matrix acting as the *stationary phase*.

aliquant **(3.3.1.2):** A representative portion of a homogeneous *sample* removed for the purpose of analysis or other chemical treatment. The quantity removed is not an evenly divisible part of the whole sample. An "aliquot" (a term not used in MARLAP) by contrast, is an evenly divisible part of the whole.

alternate analyte **(2.5):** *Analyte* whose concentration, because of an established relationship (e.g., secular equilibrium) can be used to quantitatively determine the concentration of a *target analyte*. An *alternate analyte* may be selected for analysis in place of a *target analyte* because of ease of analysis, lower analytical costs, better methodologies available, etc. (see *alternate radionuclide*).

alternate radionuclide **(3.3.4):** An "easy-to-measure" *radionuclide* that is used to estimate the amount of a radionuclide that is more difficult or costly to measure. Known or expected relationships between the radionuclide and its alternate can be used to establish a factor for amount of the hard-to-measure radionuclide (see *alternate analyte*).

( ) Indicates the section in which the term is first used in MARLAP.
Italicized words or phrases have their own definitions in this glossary.

alternative hypothesis (H_1 or H_A) **(2.5, Table 2.1):** One of two mutually exclusive statements tested in a statistical hypothesis test (compare with *null hypothesis*). The *null hypothesis* is presumed to be true unless the test provides sufficient evidence to the contrary, in which case the *null hypothesis* is rejected and the *alternative hypothesis* is accepted.

analyte **(1.4.7):** The component (e.g., a *radionuclide* or chemical compound) for which a *sample* is analyzed.

analysis **(3.3.1):** Analysis refers to the identification or quantification process for determining a *radionuclide* in a *radionuclide*/matrix combination. Examples of analyses are the measurement of ^{3}H in water, ^{90}Sr in milk, ^{239}Pu in soil, etc.

analytical data requirements **(1.1):** Measurement performance criteria used to select and decide how the laboratory analyses will be conducted and used for the initial, ongoing, and final evaluation of the laboratory's performance and the laboratory data. The project-specific *analytical data requirements* establish measurement performance criteria and decisions on how the laboratory analyses will be conducted (e.g., method selection, etc.) in a *performance-based approach* to data quality.

analytical method **(1.4.6):** A major component of an analytical protocol that normally includes written procedures for sample digestion, chemical separation (if required), and counting (*analyte* quantification through radioactive decay emission or atom-counting measurement techniques. Also called *laboratory method*.

analytical performance measure **(2.3.3):** A qualitative or quantitative aspect of the analysis, initially defined based on the *analyte*, its desired detection level and the sample matrix. See also *measurement quality objectives*.

analytical plan **(9.6.3):** The portion of the *project plan documents* that addresses the optimized analytical design and other analytical issues (e.g., *analytical protocol specifications, standard operating procedures*).

analytical process **(1.3):** The *analytical process* is a general term used by MARLAP to refer to a compilation of actions starting from the time a *sample* is collected and ending with the reporting of data. These are the actions that must be accomplished once a sample is collected in order to produce analytical data. These actions typically include field sample preparation and preservation, sample receipt and inspection, laboratory sample preparation, *sample dissolution, chemical separations*, preparation of samples for instrument measurements, instrument measurements, data reduction, data reporting, and the *quality control* of the process.

( ) Indicates the section in which the term is first used in MARLAP.
Italicized words or phrases have their own definitions in this glossary.

analytical protocol **(1.4.3):** A compilation of specific procedures/methods that are performed in succession for a particular analytical process. With a performance-based approach, there may be a number of appropriate analytical protocols for a particular analytical process. The *analytical protocol* is generally more inclusive of the activities that make up the analytical process than is the *analytical method*. See also *analytical process*.

analytical protocol specification **(*APS*) (1.4.10):** The output of a *directed planning process* that contains the project's analytical data needs and requirements in an organized, concise form. The level of specificity in the APSs should be limited to those requirements that are considered essential to meeting the project's *analytical data requirements* to allow the laboratory the flexibility of selecting the protocols or methods that meet the analytical requirements.

anion **(13.2.2):** An *ion* with a negative charge.

anion exchanger **(14.7.4.2):** An ion-exchange *resin* consisting of chemical groups, bonded to an inert matrix, with a net positive charge. The positive species are electrostatically bonded to negative, labile *ions* bonded to an inert matrix. *Anions* in solution replace the labile *ions* on the exchanger by forming electrostatic bonds with the charged groups. The strength of attraction, which depends on the charge, size, and degree of solvation of the *anion*, provides a means for separating *analyte ions*.

aqueous samples **(10.3.1):** Samples for which the matrix is water, including surface water, groundwater, drinking water, precipitation, or runoff.

arithmetic mean **($\bar{x}$) (1.4.8):** The sum of a series of measured values, divided by the number of values. The *arithmetic mean* is also called the "average." If the measured values are denoted by $x_1, x_2, \ldots, x_N$, the *arithmetic mean* is equal to $(x_1 + x_2 + \cdots + x_N) / N$. (See also *expectation* and *sample mean*.)

assessment team **(9.4):** A team of data assessors (or qualified data assessor) who are technically competent to evaluate the project's activities and the impact of these activities on the quality and usability of data.

audit **(5.3.8):** An assessment to provide assurance that a selected laboratory is capable of or is fulfilling the specifications of the *request for proposals* or *statement of work*. A pre-award *audit* verifies that a laboratory has the ability that it can meet the analytical requirements of the *request for proposals* or *statement of work*. After the award, an *audit* of a laboratory will assess the performance of the laboratory to verify that it is complying with *statement of work* and

( ) Indicates the section in which the term is first used in MARLAP.
Italicized words or phrases have their own definitions in this glossary.

contractual requirements. Thus, the examination of logbooks, charts, or other documentation that are produced as the work progresses.

authoritative sample collection approach **(9.6.2.1):** An approach wherein professional knowledge is used to choose sample locations and times.

auto-oxidation-reduction (disproportionation) **(14.2.3):** An *oxidation-reduction reaction* in which a single chemical species acts simultaneously as an oxidizing and reducing agent.

average **(6.5.1.1):** See *arithmetic mean.*

background, anthropogenic **(3.3.1):** Background radiation levels caused by *radionuclides* in the environment resulting from human activities, such as the atmospheric testing of nuclear weapons.

background, environmental **(3.3.1):** See *background level.* The presence of naturally occurring radiation or *radionuclides* in the environment.

background, instrument **(6.5.5.3):** Radiation detected by an instrument when no *source* is present. The background radiation that is detected may come from *radionuclides* in the materials of construction of the detector, its housing, its electronics and the building as well as the environment and natural radiation.

background level **(2.5):** This term usually refers to the presence of *radioactivity* or radiation in the environment. From an analytical perspective, the presence of background *radioactivity* in samples needs to be considered when clarifying the radioanalytical aspects of the decision or study question. Many *radionuclides* are present in measurable quantities in the environment. Natural background radiation is due to both primordial and *cosmogenic radionuclides.* Anthropogenic background is due to *radionuclides* that are in the environment as a result of human activities, for example, the atmospheric testing of nuclear weapons.

basic ordering agreement (BOA) **(5.1):** A process that serves to pre-certify potential analytical service providers. A list of approved laboratories is assembled and contacted as needed to support specific needs. A task order is used to define a specific scope of work within a BOA.

batch processing **(6.4):** A procedure that involves preparing a group of individual samples together for analysis in such a way that allows the group to be associated with a set of *quality control samples.*

( ) Indicates the section in which the term is first used in MARLAP.
Italicized words or phrases have their own definitions in this glossary.

becquerel (Bq) (1.4.9): Special name for the SI derived unit of activity (of *radionuclides*), equal to one nuclear transformation per second. The traditional unit is the *curie (Ci)*. The relationship between these units is 3.7×10^{10} Bq = 1 Ci.

bias (of an estimator) (1.4.8): If X is an estimator for the true value of parameter θ, then the *bias* of X is $\mu_X - \theta$, where μ_X denotes the expectation of X.

bias (of measurement) (1.4.8): See *systematic error*.

bias (of a measurement process) (1.4.8): The *bias* of a measurement process is a persistent deviation of the mean measured result from the true or accepted reference value of the quantity being measured, which does not vary if a measurement is repeated. See also *bias (of an estimator)* and *bias (of measurement)*.

bioassay (10.2.11.2): A procedure to monitor internal radiation exposure by performing *in vitro* or *in vivo* measurements, primarily urine analysis, fecal analysis, or whole-body counting.

blind sample (18.4.2): A *sample* whose concentration is not known to the analyst. *Blind samples* are used to assess analytical performance. A double-blind sample is a *sample* whose concentration and identity as a *sample* is known to the submitter but not to the analyst. The double-blind sample should be treated as a routine sample by the analyst, so it is important that the double-blind sample is identical in appearance to routine samples.

blunder (7.4.1.1): A mistake made by a person performing an analytical task that produces an a significant error in the result.

branching ratio (7.2.2.2): See *emission probability per decay event*.

breakthrough (14.7.4.1): Appearance of certain *ions* in the output solution (*eluate*) of an ion-exchange column. These *ions* are not bonded to the exchange groups of the column because the groups are already occupied by these or other *ions*, and the *resin* is essentially saturated.

calibration (1.4.8): The set of operations that establish, under specified conditions, the relationship between values indicated by a measuring instrument or measuring system, or values represented by a material measure, and the corresponding known value of a *measurand*.

calibration source (15.1): A prepared *source*, made from a *certified reference material* (standard), that is used for calibrating instruments.

( ) Indicates the section in which the term is first used in MARLAP.
Italicized words or phrases have their own definitions in this glossary.

carrier (14.1): (1) A stable isotopic form of a tracer element or nonisotopic material added to effectively increase the quantity of a tracer element during radiochemical procedures, ensuring conventional behavior of the element in solution. (2) A substance in appreciable amount that, when associated with a tracer of a specified substance, will carry the tracer with it through a chemical or physical process, or prevent the tracer from undergoing nonspecific processes due to its low concentration (IUPAC, 1995). A stable isotope of a *radionuclide* (usually the *analyte*) added to increase the total amount of that element so that a measurable mass of the element is present.

carrier-free tracer (14.2.6): (1) A radioactive isotope tracer that is essentially free from stable (nonradioactive) isotopes of the element in question. (2) Addition of a specific, nonradioactive isotope of an element to change the measured isotopic abundance of the element in the *sample*. Such materials are usually designated as nonisotopic material or marked with the symbol "c.f." (see *radiotracer*).

carrier gas (14.5.1): An inert gas, such as nitrogen or helium, serving as the mobile phase in a gas-liquid chromatographic system. The *carrier gas* sweeps the *sample* in through the system.

cation (13.2.2): An *ion* with a positive charge.

cation exchanger (14.3.4.2): An ion-exchange *resin* consisting of chemical groups, bonded to an inert matrix, with a net negative charge. The negative species are electrostatically bonded to positive, labile *ions*. Cations, in solution, replace the labile *ions* on the exchanger by forming electrostatic bonds with the charged groups. The strength of attraction, which depends on the charge, size, and degree of solvation of the cation, provides a means for separating *analyte ions*.

Cerenkov radiation (14.10.9.10): Cerenkov radiation is emitted in the ultraviolet spectrum when a fast charged particle traverses a dielectric medium (like water) at a velocity exceeding the velocity of light in that medium. It is analogous to the "sonic boom" generated by a craft exceeding the speed of sound.

certified reference material (*CRM*) (1.6, Figure 1.3): A reference material, accompanied by a certificate, one or more of whose property values are certified by a procedure which establishes its traceability to an accurate realization of the unit in which the property values are expressed, and for which each certified value is accompanied by an uncertainty at a stated level of confidence (ISO, 1992).

chain-of-custody (1.4.10): Procedures that provide the means to trace the possession and handling of a *sample* from collection to data reporting.

> ( ) *Indicates the section in which the term is first used in MARLAP.*
> *Italicized words or phrases have their own definitions in this glossary.*

check source **(15.2):** A material used to validate the operability of a radiation measurement device, sometimes used for instrument *quality control*. See *calibration source*, *test source*, and *source, radioactive*.

chelate **(14.3.2):** A *complex ion* or compound that consists of a *ligand* bonded (coordinated) to a metal atom or *ion* through two or more nonmetal atoms forming a ring structure with the metal atom or *ion*. *Ligands* may be inorganic *ions*, such as Cl, F, or carbonate, or organic compounds of two, four, or six functional groups containing atoms of S, N, O, or P.

chelating agent **(14.3.2):** The compound containing the *ligand* that forms a chelate with metal atoms or *ions*.

chemical separations **(1.1):** The removal of all undesirable materials (elements, compounds, etc.) from a *sample* through chemical means so that only the intended *analyte* is isolated and measured.

chemical speciation **(2.5):** The chemical state or form of an *analyte* in a *sample*. When the chemical species of the *analyte* in a *sample* from a new project varies from the chemical species for which an *analytical method* was validated, then the method should be altered and revalidated.

chromatography **(6.6.3.4):** A group of separation techniques based on the unequal distribution (partition) of substances between two immiscible phases, one moving past the other. The mobile phase passes over the surfaces of the *stationary phase*.

coagulation **(14.8.5):** (1) The process in which colloidal particles or macromolecules come together to form larger masses (see *colloid* and *colloidal solution*). (2) Addition of an excess quantity of electrolyte to a *colloidal solution* neutralizing the electrical bilayer of the colloidal particles and permitting their agglomeration to form larger particles that easily settle (precipitate). Also called "flocculation."

coefficient of variation (*CV*) **(19.5.2.2):** The *coefficient of variation* of a nonnegative random variable is the ratio of its *standard deviation* to its *mean*.

coefficient of thermal (volume) expansion **(19E.3):** ratio of the change in volume (of a material) per unit volume to the change in temperature, at constant pressure. If V denotes volume, ρ denotes density, and T denotes temperature, then the coefficient of thermal expansion, β, is given by

( ) Indicates the section in which the term is first used in MARLAP.
Italicized words or phrases have their own definitions in this glossary.

$$\beta = \frac{1}{V}\frac{dV}{dT} = -\frac{1}{\rho}\frac{d\rho}{dT}.$$

collectors (14.8.5): Substances used for the unspecific concentration of trace substances. Colloidal precipitates are excellent collectors because of their great adsorption capacity. Unspecific *carriers* such as manganese dioxide, sulfides, and hydrated oxides are frequently used as *collectors* (also called "scavengers").

colloid (13.2.5): Any form of matter with at least one dimension that is less than one micron but more than one nanometer. This dimension is larger in size than that of a true solution but smaller than particles of an ordinary suspension. They are too small to be observed by a light microscope but larger than molecular size. Colloidal particles are usually aggregates of hundreds or thousands of smaller molecules or macromolecules.

colloidal solution (13.4.1): Sometimes called a "colloidal dispersion." (1) A mixture formed from the dispersion of one phase (dispersed phase) within a second phase (continuous phase) in which one phase has colloidal dimensions. A *colloidal solution* contains dispersed particles with a very high surface-area-to-mass ratio and, thus, a great adsorption capacity. The solution will not usually settle by gravity since the colloidal particles are very small and charged by attraction of *ions* to their surfaces, but they will pass through ordinary filter paper. (2) In radiochemistry, a *colloidal solution* refers to the dispersion of solid particles in the solution phase. (*The mixture is not a true solution because particles of the dispersed phase are larger than typical ions and molecules.*)

column chromatography (14.3.4.2): A chromatographic procedure employing a solid phase packed in a glass or metal column. A liquid phase is passed through the column under pressure supplied by gravity or pumping action. Column chromatography can accommodate larger quantities of materials than other methods of chromatography and, thus, can separate larger loads. It can also provide more separating power with an increased ratio of solid phase to *analyte*.

combined standard uncertainty (1.4.7): *Standard uncertainty* of an *output estimate* calculated by combining the standard uncertainties of the *input estimates*. See also *expanded uncertainty and uncertainty (of measurement)*. The *combined standard uncertainty* of y is denoted by $u_c(y)$.

combined variance (19.3.3): The square of the *combined standard uncertainty*. The *combined variance* of y is denoted by $u^2_c(y)$.

( ) Indicates the section in which the term is first used in MARLAP.
Italicized words or phrases have their own definitions in this glossary.

committed effective dose equivalent (CEDE) **(2.5.2.1):** The sum of the committed dose equivalent to various tissues in the body, each multiplied by the appropriate weighting factor (MARSSIM, 2000). CEDE is expressed in units of sievert (Sv) or rem. See *action level, dose equivalent,* and *total effective dose equivalent.*

common ion **(14.8.3.1):** *Ions* that appear in the equilibrium expressions of reactions. The term is often used to refer to an additional source of the reacting *ions.*

common-ion effect **(14.8.3.1):** An increase in concentration of *ions* participating in a reaction because of the addition of one of the *ions* from another source causing a shift in the equilibrium of the reaction.

comparability **(1.4.11):** A measure of the confidence with which one data set can be compared to another. *Comparability* is one of the five principal *data quality indicators,* which are qualitative and quantitative descriptors used in interpreting the degree of acceptability or utility of data.

completeness **(1.4.11):** A measure of the amount of valid data obtained from a measurement system compared to the amount that was expected to be obtained under correct, normal conditions. *Completeness* is one of the five principal *data quality indicators.* See also *comparability.*

complex **(13.2.4):** Another name for a *coordination compound.*

complex ion **(13.2.4):** An *ion* formed when a metal atom or *ion* forms *coordination bonds* with one or more nonmetal atoms in molecules or *anions.* Examples are $Th(NO_3)_2^{+2}$, $Ra(EDTA)^{-2}$, $U(CO_3)_5^{-6}$, and $Fe(H_2O)_6^{+2}$.

compliance **(8.2.2.2):** In terms of data, *compliance* means that the data passes numerical *quality control tests* based on parameters or limits derived from the *measurement quality objectives* specified in the *statement of work.*

component (of combined standard uncertainty) **(19.2):** The *component* of the combined *standard uncertainty* of an output estimate, $u_c(y)$, generated by the *standard uncertainty* of an input estimate, $u(x_i)$, is the product of the *standard uncertainty,* $u(x_i)$, and the absolute value of the *sensitivity coefficient,* $\partial y / \partial x_i$. The uncertainty component generated by $u(x_i)$ may be denoted by $u_i(y)$.

concentration range **(2.5, Table 2.1):** The minimum and maximum concentration of an *analyte* expected to be present in a *sample* for a given project. While most analytical protocols are

( ) *Indicates the section in which the term is first used in MARLAP.*
Italicized words or phrases have their own definitions in this glossary.

applicable over a fairly large range of concentration for the *radionuclide of interest*, performance over a required concentration range can serve as a measurement quality objective for the protocol selection process and some analytical protocols may be eliminated if they cannot accommodate the expected range of concentration.

conceptual site model (2.3.2): A general approach to planning field investigations that is useful for any type of environmental reconnaissance or investigation plan with a primary focus on the surface and subsurface environment.

consistency (8.4.2): Values that are the same when reported redundantly on different reports or transcribed from one report to another.

control chart (18.1): A graphical representation of data taken from a repetitive measurement or process. *Control charts* may be developed for various characteristics (e.g., *mean, standard deviation*, range, etc.) of the data. A *control chart* has two basic uses: 1) as a tool to judge if a process was *in control*, and 2) as an aid in achieving and maintaining *statistical control*. For applications related to radiation detection instrumentation or radiochemical processes, the *mean* (center line) value of a historical characteristic (e.g., *mean* detector response), subsequent data values and *control limits* placed symmetrically above and below the center line are displayed on a *control chart*. See *statistical control*.

control limit (3.3.7.3): Predetermined values, usually plotted on a *control chart*, which define the acceptable range of the monitored variable. There can be both upper and lower limits; however, when changes in only one direction are of concern, only one limit is necessary. When a measured value exceeds the control limits, one must stop the measurement process, investigate the problem, and take corrective action." See *warning limit*.

coordination bond (14.3.1): (1) The chemical bond between the nonmetal atoms of a *ligand* and a metal atom or *ion*, which forms a coordination compound or *complex ion*. The bond is formed when the *ligand* donates one or more electron pairs to the metal atom or *ion*. (2) In more general terms, a covalent bond formed in which one atom donates both of the shared electrons; often called a coordination-covalent bond.

coordination compound (14.3.1): A compound containing *coordination bonds* in a molecule or *ion*; also called a "*complex*."

coordination number (14.3.1): (1) The number of nonmetal atoms donating electrons to a metal atom or *ion* in the formation of a *complex ion* or coordination compound. For example, the

---

( ) Indicates the section in which the term is first used in MARLAP.
Italicized words or phrases have their own definitions in this glossary.

coordination number is five in $U(CO_3)_5^{-6}$ (2) The number of atoms, *ions*, molecules, or groups surrounding an atom or *ion* in a coordination compound, *complex ion*, or crystal structure.

coprecipitation (14.1): A process used to precipitate a *radionuclide* that is not present in sufficient concentration to exceed the solubility product of the *radionuclide* and precipitate. A stable *ion*, chemically similar to the *radionuclide*, is added in a quantity sufficient to precipitate and carry with it the *radionuclide*.

core group (core team) (2.4.1): A subgroup of the *project planning team*, which includes the project manager and other key members of the *project planning team*, who meet at agreed upon intervals to review the project's progress, respond to unexpected events, clarify questions raised, revisit and revise project requirements as necessary, and communicate the basis for previous assumptions.

correction (8.2.1): A value algebraically added to the uncorrected result of a measurement to compensate for a systematic effect.

correction factor (8.5.1.12): A numerical factor by which the result of an uncorrected result of a measurement is multiplied to compensate for a systematic effect.

corrective action reports (8.2.2.2): Documentation of required steps taken to correct an out of control situation.

correctness (8.4.2): The reported results are based on properly documented and correctly applied algorithms.

correlate (18.4.5): Two *random variables* are *correlated* if their *covariance* is nonzero.

correlation coefficient (19.3.3): The *correlation coefficient* of two *random variables* is equal to their *covariance* divided by the product of their *standard deviations*.

cosmogenic radionuclide (3.3.1): *Radionuclides* that result from the collision of cosmic-ray particles with stable elements in the atmosphere, primarily atmospheric gases. See *background, environmental*.

counting efficiency (15.2.2): The ratio of the events detected (and registered) by a radiation detection system to the number of particle or photons emitted from a radioactive *source*. The counting efficiency may be a function of many variables, such as radiation energy, *source* composition, and *source* or detector geometry.

( ) Indicates the section in which the term is first used in MARLAP.
Italicized words or phrases have their own definitions in this glossary.

counting error **(19.3.5):** See *counting uncertainty; error (of measurement)*. MARLAP uses the term *counting uncertainty* to maintain a clear distinction between the concepts of measurement *error* and *uncertainty*.

counting uncertainty **(18.3.4):** Component of *measurement uncertainty* caused by the random nature of radioactive decay and radiation counting.

count rate **(14A.2.2):** The number of decay particles detected per unit time of a *source*. Generally the *count rate* is uncorrected for detector efficiency. The *count rate* divided by the detector efficiency for a specific particle and energy will yield the *source activity*.

count time **(2.5):** The time interval for the counting of a *sample* or *source* by a radiation detector. Depending upon the context used, this can be either the "clock" time (the entire period required to count the *sample*), or "live" time (the period during which the detector is actually counting). Live time is always less than or equal to clock time.

covariance **(19.3.3):** The *covariance* of two *random variables* X and Y, denoted by $\text{Cov}(X,Y)$ or $\sigma_{X,Y}$, is a measure of the association between them, and is defined as $E([X - \mu_X][Y - \mu_Y])$.

coverage factor **(1.4.7):** The value k multiplied by the *combined standard uncertainty* $u_c(y)$ to give the *expanded uncertainty*, U.

coverage probability **(19.3.6):** Approximate probability that the reported uncertainty interval will contain the value of the *measurand*.

critical level **(20B.1):** See *critical value*.

critical value (S_C) **(3B.2):** In the context of *analyte* detection, the minimum measured value (e.g., of the instrument signal or the *analyte* concentration) required to give confidence that a positive (nonzero) amount of *analyte* is present in the material analyzed. The critical value is sometimes called the *critical level* or *decision level*.

cross-contamination **(3.4, Table 3.1):** Cross-contamination occurs when radioactive material in one *sample* is inadvertently transferred to an uncontaminated sample, which can result from using contaminated sampling equipment and chemicals, and improperly cleaned glassware, crucibles, grinders, etc. *Cross-contamination* may also occur from spills, as well as airborne dusts of contaminated materials.

( ) Indicates the section in which the term is first used in MARLAP.
Italicized words or phrases have their own definitions in this glossary.

crosstalk **(7.4.2.2):** A phenomenon in gas-proportional counting or liquid-scintillation counting when an emission of an alpha particle is recorded as a beta particle count or vice versa. This is due to the ionization effects of the particles at different energies.

cumulative distribution function **(19A.1):** See *distribution function.*

curie (Ci) **(1.4.9):** Traditional non-SI unit of activity (of *radionuclides*), equal to 3.7×10^{10} Bq. Because the curie is such a large value, the more common unit is the *picocurie* (pCi), equal to 10^{-12} Ci.

data assessment **(2.1):** Assessment of environmental data consists of three separate and identifiable phases: data verification, data validation, and *data quality assessment.*

data collection activities **(1.3):** Examples of *data collection activities* include site-characterization activities, site cleanup and compliance-demonstration activities, decommissioning of nuclear facilities, remedial and removal actions, effluent monitoring of licensed facilities, license termination activities, environmental site monitoring, background studies, routine ambient monitoring, and waste management activities.

data life cycle **(1.4.1):** A useful and structured means of considering the major phases of projects that involve data collection activities. The three phases of the *data life cycle* are the planning phase, the implementation phase, and the assessment phase.

data package **(1.4.11):** The information the laboratory should produce after processing samples so that data verification, validation, and quality assessment can be done (see Chapter 16, Section 16.7).

data qualifier **(8.1):** *Data validation* begins with a review of project objectives and requirements, the *data verification* report, and the identified exceptions. If the system being validated is found to be *in control* and applicable to the *analyte* and matrix, then the individual data points can be evaluated in terms of detection, *imprecision*, and *bias*. The data are then assigned *data qualifiers*. Validated data are rejected only when the impact of an exception is so significant that a datum is unreliable.

data quality assessment **(1.1):** The scientific and statistical evaluation of data to determine if data are the right type, quality, and quantity to support their intended use.

( ) Indicates the section in which the term is first used in MARLAP.
Italicized words or phrases have their own definitions in this glossary.

data quality assessment plan **(1.4.1, Figure 1.1)**: A *project plan document* that describes the *data quality assessment* process including *data quality assessment* specifications, requirements, instructions, and procedures.

data quality indicator (DQI) **(3.3.7)**: Qualitative and quantitative descriptor used in interpreting the degree of acceptability or utility of data. The principal DQIs are *precision, bias, representativeness, comparability,* and *completeness.* These five DQIs are also referred to by the acronym PARCC—the "A" refers to *accuracy* rather than *bias.*

data quality objective (DQO) **(1.4.9)**: *DQOs* are qualitative and quantitative statements derived from the *DQO process* that clarify the study objectives, define the most appropriate type of data to collect, determine the most appropriate conditions from which to collect the data, and specify tolerable limits on *decision error rates.* Because DQOs will be used to establish the quality and quantity of data needed to support decisions, they should encompass the total uncertainty resulting from all data collection activities, including analytical and sampling activities.

data quality objective process **(1.6.3)**: A systematic strategic planning tool based on the scientific method that identifies and defines the type, quality, and quantity of data needed to satisfy a specified use. *DQOs* are the qualitative and quantitative outputs from the *DQO process.*

data quality requirement **(2.1)**: See *measurement quality objective.*

data reduction **(1.1)**: The processing of data after generation to produce a *radionuclide* concentration with the required units.

data transcription **(8.5)**: The component of the analytical process involving copying or recording data from measurement logs or instrumentation.

data usability **(1.4.11)**: The scientific and statistical evaluation of data sets to determine if data are of the right type, quality, and quantity to support their intended use (*data quality objectives*). The data quality assessor integrates the *data validation* report, field information, assessment reports, and historical project data to determine *data usability* for the intended decisions.

data validation **(1.1)**: The evaluation of data to determine the presence or absence of an *analyte* and to establish the uncertainty of the measurement process for contaminants of concern. *Data validation* qualifies the usability of each datum (after interpreting the impacts of exceptions identified during data verification) by comparing the data produced with the *measurement quality objectives* and any other *analytical process* requirements contained in the *analytical protocol specifications* developed in the planning process.

( ) Indicates the section in which the term is first used in MARLAP.
Italicized words or phrases have their own definitions in this glossary.

data validation plan (**1.4.1, Figure 1.1**): A *project plan document* that ensures that proper laboratory procedures are followed and data are reported in a format useful for validation and assessment, and will improve the cost-effectiveness of the data collection process.

data verification (**1.2**): Assures that laboratory conditions and operations were compliant with the *statement of work*, *sampling and analysis plan*, and *quality assurance project plan*, and identifies problems, if present, that should be investigated during data validation. *Data verification* compares the material delivered by the laboratory to these requirements (compliance), and checks for consistency and *comparability* of the data throughout the data package and *completeness* of the results to ensure all necessary documentation is available.

decay chain (**3.3.8**): A *decay chain* or "decay series" begins with a parent radionuclide (also called a "parent nuclide"). As a result of the radioactive decay process, one element is transformed into another. The newly formed element, the decay product or progeny, may itself be radioactive and eventually decay to form another nuclide. Moreover, this third decay product may be unstable and in turn decay to form a fourth, fifth or more generations of other radioactive decay products. The final decay product in the series will be a stable element. Elements with extremely long half-lives may be treated as if stable in the majority of cases. Examples of important naturally occurring *decay chains* include the uranium series, the thorium series, and the actinium series. See *radioactive equilibrium*.

decay emissions (**6.2**): The emissions of alpha or beta particles (β^+ or β^-) or gamma rays from an atomic nucleus, which accompany a nuclear transformation from one atom to another or from a higher nuclear energy state to lower one.

decay factor (**14A.2.2**): Also referred to as the "decay-correction factor." The factor that is used to compensate for radioactive decay of a specific *radionuclide* between two points in time.

decay series (**3.3.4**): See *decay chain*.

decision error rate (**1.4.9**): The probability of making a wrong decision under specified conditions. In the context of the *DQO process*, one considers two types of decision errors (*Type I* and *Type II*). The *project planning team* determines the *tolerable decision error rates*.

decision level (**20.2.2**): See *critical value*.

decision performance criteria (**2.1**): Another way to express the concept of using directed project planning as a tool for project management to identify and document the qualitative and quantitative statements that define the project objectives and the acceptable rate of making

( ) Indicates the section in which the term is first used in MARLAP.
Italicized words or phrases have their own definitions in this glossary.

decision errors that will be used as the basis for establishing the quality and quantity of data needed to support the decision. See *data quality objective*.

decision rule (2.3.3): The rule developed during directed planning to get from the problem or concern to the desired decision and define the limits on the *decision error rates* that will be acceptable to the *stakeholder* or customer. Sometimes called a "decision statement." The *decision rule* can take the form of "if ...then..." statements for choosing among decisions or alternative actions. For a complex problem, it may be helpful to develop a *decision tree*, arraying each element of the issue in its proper sequence along with the possible actions. The *decision rule* identifies (1) the *action level* that will be a basis for decision and (2) the statistical parameter that is to be compared to the *action level*.

decision tree (2.5.3): See *decision rule*. Also referred to as a "logic flow diagram" or "decision framework."

decision uncertainty (1.4.7): Refers to uncertainty in the decisionmaking process due to the probability of making a wrong decision because of measurement uncertainties and sampling statistics. *Decision uncertainty* is usually expressed as by the estimated probability of a decision error under specified assumptions.

decommissioning (1.3): The process of removing a facility or site from operation, followed by decontamination, and license termination (or termination of authorization for operation) if appropriate. The process of *decommissioning* is to reduce the residual *radioactivity* in structures, materials, soils, groundwater, and other media at the site to acceptable levels based on acceptable risk, so that the site may be used without restrictions.

deconvolution (8.5.1.11): The process of resolving multiple gamma-spectral peaks into individual components.

deflocculation (14.8.5): The process whereby coagulated particles pass back into the colloidal state. Deflocculation may be accomplished by adding a small amount of electrolyte to produce the electrical double-layer characteristic of colloidal particles. Also called "peptization." Also see *coagulation* and *colloidal solution*.

degrees of freedom (6A.2): In a statistical estimation based on a series of observations, the number of observations minus the number of parameters estimated. See *effective degrees of freedom*.

( ) *Indicates the section in which the term is first used in MARLAP.*
Italicized words or phrases have their own definitions in this glossary.

dentate **(14.3.1):** Term used to categorize *ligands* that describes the number of nonmetal atoms with electron pairs used by a *ligand* for coordinate bond formation (unidentate, bidentate, etc.).

derived concentration guideline level (DCGL) **(2.5.2.1):** A derived radionuclide-specific activity concentration within a *survey unit* corresponding to the release criterion. *DCGLs* are derived from activity/dose relationships through various exposure pathway scenarios.

descriptive statistics **(9.6.4.1):** Statistical methods that are used to determine and use the *mean*, mode, *median, variance*, and correlations among variables, tables, and graphs to describe a set of data.

detection capability **(1.4.7):** The capability of a *measurement process* to distinguish small amounts of *analyte* from zero. It may be expressed in terms of the *minimum detectable concentration*.

detection limit **(2.5, Table 2.1):** The smallest value of the amount or concentration of *analyte* that ensures a specified high probability of detection. Also called *"minimum detectable value."*

deviation reports **(9.2.2.2):** Documentation of any changes from the analysis plan that may affect data utility.

digestion **(6.6):** (1) Heating a precipitate over time; used to form larger crystals after initial precipitation. (2) The dissolution of a *sample* by chemical means, typically through the addition of a strong acid or base.

directed planning process **(1.2):** A systematic framework focused on defining the data needed to support an informed decision for a specific project. Directed planning provides a logic for setting well-defined, achievable objectives and developing a cost-effective, technically sound sampling and analysis design that balances the data user's tolerance for uncertainty in the decision process and the available resources for obtaining data to support a decision. Directed planning helps to eliminate poor or inadequate sampling and analysis designs.

disproportionation (autoxidation-reduction) **(14.2.3):** An oxidation-reduction reaction in which a chemical species is simultaneously oxidized and reduced.

dissolve **(6.5.1.1):** To form a solution by mixing a *solute* with a *solvent*. The particles of the *solute solvent* mix intimately at the atomic, molecular, and ionic levels with the particles of the *solvent*, and the *solute* particles are surrounded by particles of the *solvent*.

( ) Indicates the section in which the term is first used in MARLAP.
Italicized words or phrases have their own definitions in this glossary.

distillation **(12.2.1.2):** Separation of a volatile component(s) of a liquid mixture or solution by boiling the mixture to vaporize the component and subsequent condensation and collection of the components as a liquid.

distribution **(3B.2):** The *distribution* of a *random variable* is a mathematical description of its possible values and their probabilities. The *distribution* is uniquely determined by its *distribution function*.

distribution (partition) coefficient **(15.4.5.5):** An equilibrium constant that represents the ratio of the concentration of a *solute* distributed between two immiscible *solvents*.

distribution function **(19A.1):** The *distribution function*, or *cumulative distribution function*, of a *random variable X* is the function F defined by $F(x) = \Pr[X \le x]$.

dose-based regulation **(2.3.2):** A regulation whose allowable *radionuclide* concentration limits are based on the dose received by an individual or population.

dose equivalent **(2.5.2.1):** A quantity that expresses all radiations on a common scale for calculating the effective absorbed dose. This quantity is the product of absorbed dose (grays or rads) multiplied by a quality factor and any other modifying factors (MARSSIM, 2000). The "quality factor" adjusts the absorbed dose because not all types of ionizing radiation create the same effect on human tissue. For example, a *dose equivalent* of one sievert (Sv) requires 1 gray (Gy) of beta or gamma radiation, but only 0.05 Gy of alpha radiation or 0.1 Gy of neutron radiation. Because the sievert is a large unit, radiation doses often are expressed in millisieverts (mSv). See *committed effective dose equivalent* and *total effective dose equivalent*.

duplicates **(1.4.8):** Two equal-sized samples of the material being analyzed, prepared, and analyzed separately as part of the same batch, used in the laboratory to measure the overall *precision* of the sample measurement process beginning with laboratory subsampling of the field *sample*.

dynamic work plan **(4.4.2):** A type of work plan that specifies the decisionmaking logic to be used in the field to determine where the samples will be collected, when the sampling will stop, and what analyses will be performed. This is in contrast to a work plan that specifies the number of samples to be collected and the location of each *sample*.

effective degrees of freedom (v_{eff}) **(6A.2):** A parameter associated with a combined *standard uncertainty*, $u_c(y)$, analogous to the number of degrees of freedom for a Type A evaluation of *standard uncertainty*, which describes the reliability of the uncertainty estimate and which may

( ) Indicates the section in which the term is first used in MARLAP.
Italicized words or phrases have their own definitions in this glossary.

be used to select the coverage factor for a desired coverage probability. The number of effective degrees of freedom is determined using the *Welch-Satterthwaite formula*.

efficiency **(2.5.4.2):** See *counting efficiency*.

electrodeposition **(14.1):** Depositing (plating or coating) a metal onto the surface of an electrode by electrochemical reduction of its cations in solution.

electronegativity **(14.2.2):** The ability of an atom to attract electrons in a covalent bond.

electron density **(13.2.3):** A term representing the relative electron concentration in part of a molecule. The term indicates the unequal distribution of valence electrons in a molecule. Unequal distribution is the result of *electronegativity* differences of atoms in the bonds of the molecule and the geometry of the bonds; the results is a polar molecule.

eluant **(14.7.4.1):** A liquid or solution acting as the moving phase in a chromatographic system.

eluate **(14.7.4.1):** The liquid or solution that has passed over or through the *stationary phase* in a chromatographic system. The *eluate* may contain components of the analyzed solution, analytes, or impurities. In column chromatography, it is the liquid coming out of the column. The process is referred to as "eluting."

emission probability per decay event (E_e) **(16.2.2):** The fraction of total decay events for which a particular particle or photon is emitted. Also called the "branching fraction" or "branching ratio."

emulsion **(14.4.3):** (1) A *colloidal solution* in which both the dispersed phase and continuous phase are immiscible liquids (2) A permanent *colloidal solution* in which either the dispersed phase or continuous phase is water, usually oil in water or water in oil. See *gel*.

environmental compliance **(4.2):** Agreement with environmental laws and regulations.

environmental data collection process **(2.1):** Consists of a series of elements (e.g., planning, developing *project plan documents*, contracting for services, sampling, analysis, data verification, data validation, and *data quality assessment*), which are directed at the use of the data in decisionmaking.

error (of measurement) **(1.4.7):** The difference between a measured result and the value of the *measurand*. The error of a measurement is primarily a theoretical concept, since its value is never known. See also *random error*, *systematic error*, and *uncertainty (of measurement)*.

( ) Indicates the section in which the term is first used in MARLAP.
Italicized words or phrases have their own definitions in this glossary.

estimator (18B.2): A *random variable* whose value is used to estimate an unknown parameter, θ, is called an *estimator* for θ. Generally, an estimator is a function of experimental data.

exception (8.2.3): A concept in *data verification* meaning a failure to meet a requirement.

excluded particles (14.7.6): Chemical components in a *gel-filtration chromatographic* system that do not enter the solid-phase matrix during separation; these components spend less time in the system and are the first to be eluted in a single fraction during chromatography.

exclusion chromatography (14.7.6): See *gel-filtration chromatography*.

excursion (1.6.2): Departure from the expected condition during laboratory analysis.

expanded uncertainty (1.4.7): "The product, U, of the *combined standard uncertainty* of a measured value y and a *coverage factor* k chosen so that the interval from $y - U$ to $y + U$ has a desired high probability of containing the value of the *measurand*" (ISO, 1995).

expectation (19.2.2): The *expectation* of a *random variable* X, denoted by $E(X)$ or μ_X, is a measure of the center of its *distribution* (a measure of central tendency) and is defined as a probability-weighted average of the possible numerical values. Other terms for the expectation value of X are the *expected value* and the *mean*.

expected value (18.3.2): See *expectation*.

expedited site characterization (2.3.2): A process used to identify all relevant contaminant migration pathways and determine the distribution, concentration, and fate of the contaminants for the purpose of evaluating risk, determining regulatory compliance, and designing remediation systems.

experimental standard deviation (6A.2): A measure of the dispersion of the results of repeated measurements of the same quantity, given explicitly by

$$s(q_k) = \sqrt{\frac{1}{n-1} \sum_{k=1}^{n} (q_k - \bar{q})^2}$$

where $q_1, q_2, ..., q_n$ are the results of the measurements, and $\bar{q}$ is their arithmetic mean (ISO, 1993a).

( ) Indicates the section in which the term is first used in MARLAP.
Italicized words or phrases have their own definitions in this glossary.

external assessment **(4.2):** Part of the evaluation process used to measure the performance or effectiveness of a system and its elements. As an example, this could be information (audit, performance evaluation, inspection, etc.) related to a method's development, validation, and control that is done by personnel outside of the laboratory and is part of the laboratory *quality assurance* program.

extraction chromatography **(14.4.4):** A solid-phase extraction method performed in a chromatographic column that uses a *resin* material consisting of an extractant absorbed onto an inert polymeric support matrix.

false acceptance **(20.2.2):** See *Type II decision error.*

false negative **(20.2.1):** See *Type II decision error.* MARLAP avoids the terms "*false negative*" and "*false positive*" because they may be confusing in some contexts.

false positive **(14.10.9.9):** See *Type I decision error.* MARLAP avoids the terms "*false negative*" and "*false positive*" because they may be confusing in some contexts.

false rejection **(20.2.1):** See *Type I decision error.*

femtogram (fg) **(6.5.5.5):** Unit of mass equal to 10^{-15} grams.

flocculation **(14.8.5):** See *coagulation* and *deflocculation.*

formation constant **(14.3.2):** The equilibrium constant for the formation of a *complex ion* or coordination molecule. The magnitude of the constant represents the stability of the *complex.* Also called "stability constant."

fractional distillation **(14.5.2):** Separation of liquid components of a mixture by repeated *volatilization* of the liquid components and condensation of their vapors within a *fractionation column.* Repeated *volatilization* and condensation produces a decreasing temperature gradient up the column that promotes the collection of the more volatile components (lower boiling point components) at the upper end of the column and return of the less volatile components at the lower end of the column. The process initially enriches the vapors in the more volatile components, and they separate first as lower boiling point fractions.

fractionation column **(14.5.3):** A *distillation* column that allows repeated *volatilization* and condensation steps within the length of the column, accomplishing *fractional distillation* of components of a mixture in one *distillation* process by producing a temperature gradient that

( ) Indicates the section in which the term is first used in MARLAP.
Italicized words or phrases have their own definitions in this glossary.

decreases up the length of the column (see *fractional distillation*). The column is designed with plates or packing material inside the column to increase the surface area for condensation.

frequency plots (9.6.3): Statisticians employ *frequency plots* to display the *imprecision* of a sampling and analytical event and to identify the type of distribution.

fusion (1.4.10): See *sample dissolution*.

full width of a peak at half maximum (FWHM) (8.5.11): A measure of the resolution of a spectral peak used in alpha or gamma spectrometry: the full peak-width energy (FW) at one-half maximum peak height (HM).

full width of a peak at tenth maximum (FWTM) (15.1): A measure of the resolution of a spectral peak used in alpha or gamma spectrometry: the full peak-width energy (FW) at one-tenth maximum peak height (TM).

gas chromatography (GC) (14.5.2): See *gas-liquid phase chromatography*.

gas-liquid phase chromatography (GLPC) (14.7.1): A chromatographic separation process using a mobile gas phase (*carrier gas*) in conjunction with a low-volatility liquid phase that is absorbed onto an inert, solid-phase matrix to produce a *stationary phase*. The components of the analytical mixture are vaporized and swept through the column by the *carrier gas*.

gel (14.7.4.2, **Table 14.9**): (1) A *colloidal solution* that is highly viscous, usually coagulated into a semirigid or jellylike solid. (2) Gelatinous masses formed from the *flocculation* of *emulsions*.

gel-exclusion chromatography (14.7.6): See *gel-filtration chromatography*.

gel-filtration chromatography (14.7.6): A column chromatographic separation process using a solid, inert polymeric matrix with pores that admit molecules less than a certain hydrodynamic size (molecular weight) but exclude larger molecules. The excluded molecules are separated from the included molecules by traveling only outside the matrix and are first eluted in bulk from the column. The included molecules, depending on size, spend different amounts of time in the pores of matrix and are separated by size.

general analytical planning issues (3.3): Activities to be identified and resolved during a *directed planning process*. Typically, the resolution of *general analytical planning issues* normally results, at a minimum, in an *analyte* list, identified matrices of concern, *measurement*

( ) *Indicates the section in which the term is first used in MARLAP.*
Italicized words or phrases have their own definitions in this glossary.

quality objectives, and established frequencies and acceptance criteria for *quality control samples*.

graded approach (2.3): A process of basing the level of management controls applied to an item or work on the intended use of the results and the degree of confidence needed in the quality of the results. MARLAP recommends a *graded approach* to project planning because of the diversity of environmental data collection activities. This diversity in the type of project and the data to be collected impacts the content and extent of the detail to be presented in the *project plan documents*.

gray region (1.6.3): The range of possible values in which the consequences of decision errors are relatively minor. Specifying a *gray region* is necessary because variability in the *target analyte* in a population and *imprecision* in the measurement system combine to produce variability in the data such that the decision may be "too close to call" when the true value is very near the *action level*. The *gray region* establishes the minimum distance from the *action level* where it is most important that the *project planning team* control *Type II errors*.

GUM (1.4.7): *Guide to the Expression of Uncertainty in Measurement* (ISO, 1995).

half-life ($T_{1/2}$ or $t_{1/2}$) (1.4.8): The time required for one-half of the atoms of a particular *radionuclide* in a *sample* to disintegrate or undergo nuclear transformation.

heterogeneity (2.5): (1) "Spatial heterogeneity," a type of distributional heterogeneity, refers to the nonuniformity of the distribution of an *analyte* of concern within a matrix. Spatial heterogeneity affects sampling, sample processing, and sample preparation. See *homogenization*. (2) The "distributional heterogeneity" of a lot depends not only on the variations among particles but also on their spatial distribution. Thus, the distributional heterogeneity may change, for example, when the material is shaken or mixed. (3) The "constitutional" (or "compositional") heterogeneity of a lot is determined by variations among the particles without regard to their locations in the lot. It is an intrinsic property of the lot itself, which cannot be changed without altering individual particles.

high-level waste (HLW) (1.3): (1) irradiated reactor fuel; (2) liquid wastes resulting from the operation of the first-cycle *solvent* extraction system, or equivalent, and the concentrated wastes from subsequent extraction cycles, or equivalent, in a facility for reprocessing irradiated reactor fuel; (3) solids into which such liquid wastes have been converted.

high-pressure liquid chromatography (HPLC) (14.7.7): A column chromatography process using various solid-liquid phase systems in which the liquid phase is pumped through the system at high pressures. The process permits rapid, highly efficient separation when compared to many

( ) Indicates the section in which the term is first used in MARLAP.
Italicized words or phrases have their own definitions in this glossary.

other chromatographic systems and is, therefore, also referred to as "high-performance liquid chromatography."

holdback carrier **(14.8.4.4):** A nonradioactive *carrier* of a *radionuclide* used to prevent that particular radioactive species from contaminating other radioactive species in a chemical operation (IUPAC, 2001).

homogeneous distribution coefficient (D) **(14.8.4.1):** The equality constant in the equation representing the *homogeneous distribution law*. Values of D greater than one represent removal of a foreign *ion* by inclusion during *coprecipitation* (see *homogeneous distribution law*).

homogeneous distribution law **(14.8.4.1):** A description of one mechanism in which *coprecipitation* by inclusion occurs (the less common mechanism). The amount of *ion* coprecipitating is linearly proportional to the ratio of the concentration of the *ion* in solution to the concentration of the coprecipitating agent in solution. Equilibrium between the precipitate and the solution is obtained (during digestion) and the crystals become completely homogeneous with respect to the foreign *ions* (impurities) (see *homogeneous distribution coefficient* and *digestion*).

homogenization **(3.4, Table 3.1):** Producing a uniform distribution of analytes and particles throughout a *sample*.

hydration **(14.3.1):** Association of water molecules with *ions* or molecules in solution.

hydration sphere **(14.3.1):** Water molecules that are associated with *ions* or molecules in solution. The inner-hydration sphere (primary hydration sphere) consists of several water molecules directly bonded to *ions* through ion-dipole interactions and to molecules through dipole-dipole interactions including hydrogen bonding. The outer hydration sphere (secondary hydration sphere) is water molecules less tightly bound through hydrogen bonding to the molecules of the inner-hydration sphere.

hydrolysis: (1) A chemical reaction of water with another compound in which either the compound or water is divided. (2) A reaction of water with *ions* that divides (lyses) water molecules to produce an excess of hydrogen *ions* or excess of hydroxyl *ions* in solution (an acidic or basic solution). Cations form *complex ions* with hydroxyl *ions* as *ligands* producing an acidic solution: $Fe^{+3} + H_2O \rightarrow Fe(OH)^{+2} + H^{+1}$. Anions form covalent bonds with the hydrogen *ion* producing weak acids and a basic solution: $F^{-1} + H_2O \rightarrow HF + OH^{-1}$.

( ) Indicates the section in which the term is first used in MARLAP.
Italicized words or phrases have their own definitions in this glossary.

hypothesis testing **(2.5, Table 2.1):** The use of statistical procedures to decide whether a *null hypothesis* should be rejected in favor of an *alternative hypothesis* or not rejected (see also *statistical test*).

immobile phase **(14.7.1):** See *stationary phase*.

imprecision **(1.4.8):** Variation of the results in a set of *replicate* measurements. This can be expressed as the *standard deviation* or coefficient of variation (*relative standard deviation*) (IUPAC, 1997). See *precision*.

included particle **(14.7.6):** The chemical forms that are separated by *gel-filtration chromatography*. They enter the solid-phase matrix of the chromatographic system and are separated by hydrodynamic size (molecular weight), eluting in inverse order by size.

inclusion **(14.7.1):** Replacement of an *ion* in a crystal lattice by a foreign *ion* similar in size and charge to form a mixed crystal or solid solution. Inclusion is one mechanism by which *ions* are *coprecipitated* with another substance precipitating from solution.

in control **(1.6.2):** The analytical process has met the *quality control* acceptance criteria and project requirements. If the analytical process is *in control*, the assumption is that the analysis was performed within established limits and indicates a reasonable match among matrix, *analyte*, and *method*.

independent **(19.2.2):** A collection of *random variables* $X_1, X_2, ..., X_n$ is *independent* if $\Pr[X_1 \leq x_1, X_2 \leq x_2, ..., X_n \leq x_n] = \Pr[X_1 \leq x_1] \cdot \Pr[X_2 \leq x_2] \cdots \Pr[X_n \leq x_n]$ for all real numbers $x_1, x_2, ..., x_n$. Intuitively, the collection is said to be *independent* if knowledge of the values of any subset of the variables provides no information about the likely values of the other variables.

inferential statistics **(9.6.4.1):** Using data obtained from samples to make estimates about a population (inferential estimations) and to make decisions (*hypothesis testing*). Sampling and *inferential statistics* have identical goals: to use samples to make inferences about a population of interest and to use *sample* data to make defensible decisions.

inner (primary) hydration sphere **(14.3.1):** See *hydration sphere*.

input estimate **(3A.5):** Measured value of an input quantity. See *output estimate*.

( ) Indicates the section in which the term is first used in MARLAP.
Italicized words or phrases have their own definitions in this glossary.

input quantity **(6.5.5.1):** Any of the quantities in a mathematical measurement model whose values are measured and used to calculate the value of another quantity, called the *output quantity*. See *input estimate*.

interferences **(1.4.9):** The presence of other chemicals or *radionuclides* in a *sample* that hinder the ability to analyze for the *radionuclide of interest*. See *method specificity*.

ion-exchange chromatography **(6.6.2.3):** A separation method based on the reversible exchange of *ions* in a mobile phase with *ions* bonded to a solid ionic phase. *Ions* that are bonded less strongly to the solid phase (of opposite charge) are displaced by *ions* that are more strongly bonded. Separation of *analyte ions* depends on the relative strength of bonding to the solid phase. Those less strongly bonded *ions* are released from the solid phase earlier and eluted sooner.

ion-product **(14.8.3.1):** The number calculated by substituting the molar concentration of *ions* that could form a precipitate into the solubility-product expression of the precipitating compound. The *ion-product* is used to determine if a precipitate will form from the concentration of *ions* in solution. If the *ion-product* is larger than the *solubility-product constant*, precipitation will occur; if it is smaller, precipitation will not occur.

isomeric transition **(14.10.9.12):** The transition, via gamma-ray emission (or internal conversion), of a nucleus from a high-energy state to a lower-energy state without accompanying particle emission, e.g., $^{99m}Tc \rightarrow {}^{99}Tc + \gamma$.

isotope **(3.3.4):** Any of two or more nuclides having the same number of protons in their nuclei (same atomic number), but differing in the number of neutrons (different mass numbers, for example ^{58}Co, ^{59}Co, and ^{60}Co). See *radionuclide*.

isotope dilution analysis **(14.10.7):** A method of quantitative analysis based on the measurement of the isotopic abundance of an element after isotopic dilution of the test portion.

key analytical planning issue **(1.6.1):** An issue that has a significant effect on the selection and development of analytical protocols or an issue that has the potential to be a significant contributor of uncertainty to the analytical process and ultimately the resulting data.

laboratory control sample **(2.5.4.2):** A standard material of known composition or an artificial *sample* (created by fortification of a clean material similar in nature to the sample), which is prepared and analyzed in the same manner as the sample. In an ideal situation, the result of an analysis of the *laboratory control sample* should be equivalent to (give 100 percent of) the *target*

( ) Indicates the section in which the term is first used in MARLAP.
Italicized words or phrases have their own definitions in this glossary.

analyte concentration or *activity* known to be present in the fortified sample or standard material. The result normally is expressed as percent *recovery*. See also *quality control sample*.

Laboratory Information Management System (LIMS) (11.2.1): An automated information system used at a laboratory to collect and track data regarding sample analysis, laboratory *quality control* operability information, final result calculation, report generation, etc.

laboratory method **(6.2):** Includes all physical, chemical, and radiometric processes conducted at a laboratory in order to provide an analytical result. These processes may include sample preparation, dissolution, chemical separation, mounting for counting, nuclear instrumentation counting, and analytical calculations. Also called *analytical method*.

law of propagation of uncertainty **(19.1):** See *uncertainty propagation formula*.

level of confidence **(1.4.11):** See *coverage probability*.

ligand **(14.3.1):** A molecule, atom, or *ion* that donates at least one electron pair to a metal atom or *ion* to form a coordination molecule or *complex ion*. See *dentate*.

linearity **(7.2.2.5):** The degree to which the response curve for a measuring device, such as an analytical balance, follows a straight line between the calibration points. The *linearity* is usually specified by the maximum deviation of the response curve from such a straight line.

liquid chromatography (LC) **(14.7.1):** A chromatographic process using a mobile liquid-phase.

liquid-phase chromatography (LPC) **(14.7.1):** A chromatographic process in which the mobile and *stationary phases* are both liquids. Separation is based on relative solubility between two liquid phases. The *stationary phase* is a nonvolatile liquid coated onto an inert solid matrix or a liquid trapped in or bound to a solid matrix. Also called "liquid-partition chromatography."

logarithmic distribution coefficient (λ) **(14.8.4.1):** The equality constant in the equation representing the *Logarithmic Distribution Law*. Values of λ greater than one represent removal of a foreign *ion* by inclusion during *coprecipitation*, and the larger the value, the more effective and selective the process is for a specific *ion*. Generally, the *logarithmic distribution coefficient* decreases with temperature, so *coprecipitation* by inclusion is favored by lower temperature.

Logarithmic Distribution Law **(14.8.4.1):** A description of one mechanism by which *coprecipitation* by inclusion occurs (the more common mechanism). The amount of *ion* coprecipitated is logarithmically proportional to the amount of primary *ion* in the solution during

( ) Indicates the section in which the term is first used in MARLAP.
Italicized words or phrases have their own definitions in this glossary.

crystallization. Crystal are grown in a slow and orderly process, such as precipitation from homogeneous solution, and each crystal surface, as it is formed, is in equilibrium with the solution. As a result, the concentration of a foreign *ion* (impurity) varies continuously from the center to the periphery of the crystal (see *logarithmic distribution coefficient*).

logic statement (2.6): The output from the *directed planning process* about what must be done to obtain the desired answer.

lower limit of detection (LLD) (14.10.9.5): (1) "The smallest concentration of radioactive material in a *sample* that will yield a net count, above the measurement process (MP) blank, that will be detected with at least 95 percent probability with no greater than a 5 percent probability of falsely concluding that a blank observation represents a 'real' signal" (NRC, 1984). (2) "An estimated detection limit that is related to the characteristics of the counting instrument" (EPA, 1980).

low-pressure chromatography (14.7.1): Column chromatography in which a liquid phase is passed through a column under pressure supplied by gravity or a low-pressure pump.

Lucas cell (10.5.4.4): A specially designed, high-efficiency cell for the analysis of radon gas with its progeny. The cell is coated with a zinc sulfide phosphor material that releases ultraviolet light when the alpha particles from radon and its progeny interact with the phosphor.

Marinelli beaker (6.5.3): A counting container that allows the *source* to surround the detector, thus maximizing the geometrical efficiency. It consists of a cylindrical sample container with an inverted well in the bottom that fits over the detector. Also called a "reentrant beaker."

MARLAP Process (1.4): A performance-based approach that develops *Analytical Protocol Specifications*, and uses these requirements as criteria for the analytical protocol selection, development, and evaluation processes, and as criteria for the evaluation of the resulting laboratory data. This process, which spans the three phases of the *data life cycle* for a project, is the basis for achieving MARLAP's basic goal of ensuring that radioanalytical data will meet a project's or program's data requirements or needs.

masking (14.4.3): The prevention of reactions that are normally expected to occur through the presence or addition of a masking agent (reagent).

masking agent (14.4.3): A substance that is responsible for converting a chemical form, which would have otherwise participated in some usual chemical reaction, into a derivative that will not participate in the reaction.

( ) *Indicates the section in which the term is first used in MARLAP.*
Italicized words or phrases have their own definitions in this glossary.

matrix of concern (1.4.10): Those matrices identified during the directed project planning process from which samples may be taken. Typical matrices include: surface soil, subsurface soil, sediment, surface water, ground water, drinking water, process effluents or wastes, air particulates, biota, structural materials, and metals.

matrix-specific analytical planning issue (3.1): Key analytical planning issue specific to that matrix, such as filtration and preservation issues for water samples.

matrix spike (3.3.10): An *aliquant* of a *sample* prepared by adding a known quantity of *target analytes* to specified amount of matrix and subjected to the entire analytical procedure to establish if the method or procedure is appropriate for the analysis of the particular matrix.

matrix spike duplicate (MSD) (9.6.3): A second *replicate* matrix spike prepared in the laboratory and analyzed to evaluate the *precision* of the measurement process.

Maximum Contaminant Level (MCL) (2.5.2.1): The highest level of a contaminant that is allowed in drinking water. MCLs are set as close as feasible to the level believed to cause no human health impact, while using the best available treatment technology and taking cost into consideration. MCLs are enforceable standards.

mean (1.4.8): See *expectation* (compare with *arithmetic mean* and *sample mean*).

mean concentration (2.5.2.3): A weighted average of all the possible values of an *analyte* concentration, where the weight of a value is determined by its probability.

measurand (1.4.7): "Particular quantity subject to measurement"(ISO, 1993a).

measurement performance criteria (1.2): See *measurement quality objectives*.

measurement process (1.3): *Analytical method* of defined structure that has been brought into a state of statistical control, such that its imprecision and bias are fixed, given the measurement conditions (IUPAC, 1995).

measurement quality objective (MQO) (1.4.9): The analytical data requirements of the *data quality objectives* are project- or program-specific and can be quantitative or qualitative. These analytical data requirements serve as *measurement performance criteria* or objectives of the analytical process. MARLAP refers to these performance objectives as *measurement quality objectives (MQOs)*. Examples of quantitative *MQOs* include statements of required *analyte* detectability and the uncertainty of the analytical protocol at a specified *radionuclide* concentra-

( ) Indicates the section in which the term is first used in MARLAP.
Italicized words or phrases have their own definitions in this glossary.

tion, such as the *action level*. Examples of qualitative *MQOs* include statements of the required specificity of the analytical protocol, e.g., the ability to analyze for the *radionuclide of interest* given the presence of interferences.

measurement uncertainty **(1.4.7):** See *uncertainty (of measurement)*.

measurement variability **(2.5.2.2):** The variability in the measurement data for a *survey unit* is a combination of the *imprecision* of the measurement process and the real spatial variability of the *analyte* concentration.

median **(9.6.4.1):** A *median* of a distribution is any number that splits the range of possible values into two equally likely portions, or, to be more rigorous, a *0.5-quantile*. See *arithmetic mean*.

method **(1.4.5):** See *analytical method*.

method blank **(Figure 3.3):** A *sample* assumed to be essentially *target analyte*-free that is carried through the radiochemical preparation, analysis, mounting and measurement process in the same manner as a routine sample of a given matrix.

method control **(6.1):** Those functions and steps taken to ensure that the validated method as routinely used produces data values within the limits of the *measurement quality objectives*. *Method control* is synonymous with process control in most *quality assurance* programs.

method detection limit (MDL) **(3B.4):** "The minimum concentration of a substance that can be measured and reported with 99 percent confidence that the *analyte* concentration is greater than zero ... determined from analysis of a *sample* in a given matrix containing the *analyte*" (40 CFR 136, Appendix B).

method performance characteristics **(3.3.7):** The characteristics of a specific *analytical method* such as *method uncertainty*, *method range*, *method specificity*, and *method ruggedness*. MARLAP recommends developing *measurement quality objectives* for select *method performance characteristics*, particularly for the *uncertainty (of measurement)* at a specified concentration (typically the *action level*).

method range **(1.4.9):** The lowest and highest concentration of an *analyte* that a method can accurately detect.

( ) Indicates the section in which the term is first used in MARLAP.
Italicized words or phrases have their own definitions in this glossary.

method ruggedness **(1.4.9):** The relative stability of method performance for small variations in method parameter values.

method specificity **(1.4.9):** The ability of the method to measure the *analyte* of concern in the presence of interferences.

method uncertainty **(3.3.7):** Method uncertainty refers to the predicted uncertainty of the result that would be measured if the method were applied to a hypothetical laboratory *sample* with a specified *analyte* concentration. Although individual measurement uncertainties will vary from one measured result to another, the required *method uncertainty* is a target value for the individual measurement uncertainties, and is an estimate of *uncertainty (of measurement)* before the sample is actually measured. See also *uncertainty (of measurement)*.

method validation **(5.3):** The demonstration that the radioanalytical method selected for the analysis of a particular *radionuclide* in a given matrix is capable of providing analytical results to meet the project's *measurement quality objectives* and any other requirements in the *analytical protocol specifications*. See *project method validation*.

method validation reference material (MVRM) **(5.5.2):** Reference materials that have the same or similar chemical and physical properties as the proposed project samples, which can be used to validate the laboratory's methods.

metrology **(1.4.7):** The science of measurement.

minimum detectable amount (MDA) **(3B.3):** The minimum detectable value of the amount of analyte in a sample. Same definition as the *minimum detectable concentration* but related to the quantity (activity) of a *radionuclide* rather than the concentration of a *radionuclide*. May be called the "minimum detectable activity" when used to mean the *activity* of a radionuclide (see ANSI N13.30 and N42.23).

minimum detectable concentration (MDC) **(2.5.3):** The *minimum detectable value* of the analyte concentration in a sample. ISO refers to the MDC as the *minimum detectable value of the net state variable*. They define this as the smallest (true) value of the net state variable that gives a specified probability that the value of the response variable will exceed its critical value—i.e., that the material analyzed is not blank.

minimum detectable value **(20.2.1):** An estimate of the smallest true value of the *measurand* that ensures a specified high probability, $1 - \beta$, of detection. The definition of the *minimum*

( ) Indicates the section in which the term is first used in MARLAP.
Italicized words or phrases have their own definitions in this glossary.

detectable value presupposes that an appropriate detection criterion has been specified (see *critical value*).

minimum quantifiable concentration (MQC) **(3.3.7):** The *minimum quantifiable concentration*, or the *minimum quantifiable value* of the *analyte* concentration, is defined as the smallest concentration of *analyte* whose presence in a laboratory *sample* ensures the relative *standard deviation* of the measurement does not exceed a specified value, usually 10 percent.

minimum quantifiable value **(20.2.7):** The smallest value of the *measurand* that ensures the *relative standard deviation* of the measurement does not exceed a specified value, usually 10 percent (see also *minimum quantifiable concentration*).

mixed waste **(1.3):** Waste that contains both radioactive and hazardous chemicals.

mobile phase **(14.7.1):** The phase in a chromatographic system that is moving with respect to the *stationary phase*; either a liquid or a gas phase.

moving phase **(14.7.1):** See *mobile phase*.

net count rate: **(16.3.2):** The *net count rate* is the value resulting form the subtraction of the background count rate (instrument background or appropriate blank) from the total (gross) count rate of a *source* or sample.

nonaqueous samples **(10.3.5):** Liquid-sample matrices consisting of a wide range of organic/ *solvents,* organic compounds dissolved in water, oils, lubricants, etc.

nonconformance **(5.3.7):** An instance in which the contractor does not meet the performance criteria of the contract or departs from contract requirements or acceptable practice.

nuclear decay **(15.3):** A spontaneous nuclear transformation.

nuclear counting **(1.6):** The measurement of alpha, beta or photon emissions from *radionuclides.*

nuclide **(1.1):** A species of atom, characterized by its mass number, atomic number, and nuclear energy state, providing that the mean *half-life* in that state is long enough to be observable (IUPAC, 1995).

( ) Indicates the section in which the term is first used in MARLAP.
Italicized words or phrases have their own definitions in this glossary.

nuclide-specific analysis **(3.3.8.3):** Radiochemical analysis performed to isolate and measure a specific *radionuclide*.

null hypothesis (H_0) **(2.5, Table 2.1):** One of two mutually exclusive statements tested in a statistical *hypothesis test* (compare with *alternative hypothesis*). The *null hypothesis* is presumed to be true unless the test provides sufficient evidence to the contrary, in which case the *null hypothesis* is rejected and the *alternative hypothesis* is accepted.

occlusion **(14.8.3.1):** The mechanical entrapment of a foreign *ion* between subsequent layers during crystal formation. A mechanism of *coprecipitation*.

Ostwald ripening **(14.8.3.2):** Growth of larger crystals during precipitation by first dissolving smaller crystals and allowing the larger crystals to form.

outer (secondary) hydration sphere **(14.3.1):** See *hydration sphere*.

outlier **(9.6.4.1):** A value in a group of observations, so far separated from the remainder of the values as to suggest that they may be from a different population, or the result of an error in measurement (ISO, 1993b).

output estimate **(3A.5):** The calculated value of an output quantity (see *input estimate*).

output quantity **(19.3.2):** The quantity in a mathematical measurement model whose value is calculated from the measured values of other quantities in the model (see *input quantity* and *output estimate*).

oxidation **(6.4):** The increase in oxidation number of an atom in a chemical form during a chemical reaction. Increase in oxidation number is a result of the loss of electron(s) by the atom or the decrease in electron density when the atom bonds to a more electronegative element or breaks a bond to a less electronegative element.

oxidation-reduction (redox) reaction **(10.3.3):** A chemical reaction in which electrons are redistributed among the atoms, molecules, or *ions* in the reaction.

oxidation number **(6.4):** An arbitrary number indicating the relative electron density of an atom or *ion* of an element in the combined state, relative to the electron density of the element in the pure state. The oxidation number increases as the electron density decreases and decreases as the electron density increases.

( ) Indicates the section in which the term is first used in MARLAP.
Italicized words or phrases have their own definitions in this glossary.

oxidation state **(6.4)**: See *oxidation number*.

oxidizing agent **(10.5.2)**: The chemical species in an oxidation-reduction reaction that causes oxidation of another chemical species by accepting or attracting electrons. The oxidizing agent is reduced during the reaction.

paper chromatography **(14.7.1)**: A chromatographic process in which the *stationary phase* is some type of absorbent paper. The *mobile phase* is a pure liquid or solution.

parameter of interest **(2.5, Table 2.1)**: A descriptive measure (e.g., *mean*, median, or proportion) that specifies the characteristic or attribute that the decisionmaker would like to know and that the data will estimate.

PARCC **(3.3.7)**: "Precision, accuracy, representativeness, comparability, and completeness." See *data quality indicators*.

parent radionuclide **(3.3.4)**: The initial *radionuclide* in a *decay chain* that decays to form one or more *progeny*.

partition (distribution) coefficient: See *distribution coefficient*.

peptization: See *deflocculation*.

percentile **(19A.1)**: If X is a random variable and p is a number between 0 and 1, then a $100p^{th}$ percentile of X is any number x_p such that the probability that $X < x_p$ is at most p and the probability that $X \le x_p$ is at least p. For example, if $x_{0.95}$ is a 95th percentile of X then $\Pr[X < x_{0.95}] \le 0.95$ and $\Pr[X \le x_{0.95}] \ge 0.95$. See *quantile*.

performance-based approach **(1.2)**: Defining the analytical data needs and requirements of a project in terms of measurable goals during the planning phase of a project. In a *performance-based approach*, the project-specific analytical data requirements that are determined during a *directed planning process* serve as measurement performance criteria for selections and decisions on how the laboratory analyses will be conducted. The project-specific analytical data requirements are also used for the initial, ongoing, and final evaluation of the laboratory's performance and the laboratory data.

performance-based approach to method selection **(6.1)**: The process wherein a validated method is selected based on a demonstrated capability to meet defined quality and laboratory performance criteria.

( ) Indicates the section in which the term is first used in MARLAP.
Italicized words or phrases have their own definitions in this glossary.

performance evaluation program (5.3.5): A laboratory's participation in an internal or external program of analyzing performance testing samples appropriate for the analytes and matrices under consideration (i.e., *performance evaluation (PE) program* traceable to a national standards body, such as the National Institute of Standards and Technology in the United States).

performance evaluation sample (3.3.10): Reference material samples used to evaluate the performance of the laboratory. Also called *performance testing (PT)* samples or materials.

performance indicator (1.6.2): Instrument- or protocol-related parameter routinely monitored to assess the laboratory's estimate of such controls as chemical yield, instrument background, *uncertainty (of measurement)*, *precision*, and *bias*.

performance testing (PT): See *performance evaluation program*.

picocurie (pCi) (1.4.9): 10^{-12} *curie*.

planchet (10.3.2): A metallic disk (with or without a raised edge) that is used for the analysis of a radioactive material after the material has been filtered, evaporated, electroplated, or dried. Evaporation of water samples for gross alpha and beta analysis often will take place directly in the planchet.

Poisson distribution (18.3.2): A random variable X has the *Poisson distribution* with parameter λ if for any nonnegative integer k,

$$\Pr[X = k] = \frac{\lambda^k e^{-\lambda}}{k!}$$

In this case both the *mean* and *variance* of X are numerically equal to λ. The *Poisson distribution* is often used as a model for the result of a nuclear counting measurement.

polymorphism (14.8.3.1): The existence of a chemical substance in two or more physical forms, such as different crystalline forms.

postprecipitation (14.8.4.3): The subsequent precipitation of a chemically different species upon the surface of an initial precipitate; usually, but not necessarily, including a common *ion* (IUPAC, 1997).

precision (1.4.8): The closeness of agreement between independent test results obtained by applying the experimental procedure under stipulated conditions. *Precision* may be expressed as the *standard deviation* (IUPAC, 1997). See *imprecision*.

( ) *Indicates the section in which the term is first used in MARLAP.*
Italicized words or phrases have their own definitions in this glossary.

prescribed methods **(6.1):** Methods that have been selected by the industry for internal use or by a regulatory agency for specific programs. Methods that have been validated for a specific application by national standard setting organizations, such as ASTM, ANSI, AOAC, etc., may also be used as prescribed methods by industry and government agencies.

primary (inner) hydration sphere **(14.3.1):** See *hydration sphere*.

primordial radionuclide **(3.3.1):** A naturally occurring *radionuclide* found in the earth that has existed since the formation (~4.5 billion years) of the Earth, e.g., ^{232}Th and ^{238}U.

principal decision **(2.7.3):** The *principal decision* or study question for a project is identified during Step 2 of the *data quality objectives* process. The *principal decision* could be simple, like whether a particular discharge is or is not in compliance, or it could be complex, such as determining if an observed adverse health effect is being caused by a nonpoint source discharge.

principal study question **(2.7.3):** See *principal decision*.

probabilistic sampling plan **(9.6.2.3):** Using assumptions regarding average concentrations and variances of samples and matrix by the planning team during the development of the sampling plan.

probability **(1.4.7):** "A real number in the scale 0 to 1 attached to a random event" (ISO, 1993b). The probability of an event may be interpreted in more than one way. When the event in question is a particular outcome of an experiment (or measurement), the probability of the event may describe the relative frequency of the event in many trials of the experiment, or it may describe one's degree of belief that the event occurs (or will occur) in a single trial.

probability density function (pdf) **(19A.1):** A *probability density function* for a *random variable* X is a function $f(x)$ such that the probability of any event $a \leq X \leq b$ is equal to the value of the integral $\int_a^b f(x)\,dx$. The *pdf*, when it exists, equals the derivative of the distribution function.

process knowledge **(1.4.10):** Information about the *radionuclide*(s) of concern derived from historical knowledge about the production of the sampled matrix or waste stream.

progeny **(3.3.4):** The product resulting from the radioactive disintegration or nuclear transformation of its parent *radionuclide*. See *decay chain*.

project method validation **(6.1):** The demonstrated method applicability for a particular project. See *method validation*.

( ) Indicates the section in which the term is first used in MARLAP.
Italicized words or phrases have their own definitions in this glossary.

project narrative statement **(4.3):** Description of environmental data collection activities, such as basic studies or small projects, which only require a discussion of the experimental process and its objectives. Other titles used for project narrative statements are *quality assurance* narrative statement and proposal *quality assurance* plan. Basic studies and small projects generally are of short duration or of limited scope and could include proof of concept studies, exploratory projects, small data collection tasks, feasibility studies, qualitative screens, or initial work to explore assumptions or correlations.

project plan documents **(1.1):** Gives the data user's expectations and requirements, which are developed during the planning process, where the *Analytical Protocol Specifications* (which include the *measurement quality objectives*) are documented, along with the *standard operating procedures*, health and safety protocols and *quality assurance/quality control* procedures for the field and laboratory analytical teams. Project plan, work plan, *quality assurance project plan*, field sampling plan, *sampling and analysis plan*, and *dynamic work plan* are some of the names commonly used for *project plan documents*.

project planning team **(2.1):** Consists of all the parties who have a vested interest or can influence the outcome (*stakeholders*), such as program and project managers, regulators, the public, project engineers, health and safety advisors, and specialists in statistics, health physics, chemical analysis, radiochemical analysis, field sampling, *quality assurance*, *quality control*, data assessment, hydrology and geology, contract management, and field operation. The *project planning team* will define the decision(s) to be made (or the question the project will attempt to resolve) and the inputs and boundaries to the decision using a *directed planning process*.

project quality objectives **(2.1):** See *decision performance criteria* and *data quality objective*.

project specific plan **(4.3):** Addresses design, work processes, and inspection, and incorporates, by citation, site-wide plans that address records management, quality improvement, procurement, and assessment.

propagation of uncertainty **(15.2.5):** See *uncertainty propagation*.

protocol **(1.4.3):** See *analytical protocol*.

protocol performance demonstration **(3.1):** See *method validation*.

qualifiers **(8.1):** Code applied to the data by a data validator to indicate a verifiable or potential data deficiency or *bias* (EPA, 2002).

( ) *Indicates the section in which the term is first used in MARLAP.*
Italicized words or phrases have their own definitions in this glossary.

quality assurance (QA) **(1.3):** An integrated system of management activities involving planning, implementation, assessment, reporting, and quality improvement to ensure that a process, item, or service is of the type and quality needed and expected.

quality assurance project plan (QAPP) **(1.4.11):** A formal document describing in detail the necessary *quality assurance*, *quality control*, and other technical activities that must be implemented to ensure that the results of the work performed will satisfy the stated performance criteria. The QAPP describes policy, organization, and functional activities and the *data quality objectives* and measures necessary to achieve adequate data for use in selecting the appropriate remedy.

quality control (QC) **(1.4.3):** The overall system of technical activities whose purpose is to measure and control the quality of a process or service so that it meets the needs of the users or performance objectives.

quality control sample **(1.4.3):** Sample analyzed for the purpose of assessing *imprecision* and *bias*. See also *blanks, matrix spikes, replicates,* and *laboratory control sample.*

quality control test **(8.5.1):** Comparison of *quality control* results with stipulated acceptance criteria.

quality indicator **(2.5.4.2):** Measurable attribute of the attainment of the necessary quality for a particular environmental decision. *Precision, bias, completeness,* and *sensitivity* are common *data quality indicators* for which quantitative *measurement quality objectives* could be developed during the planning process.

quality system **(9.2.2.3):** The *quality system* oversees the implementation of *quality control samples*, documentation of *quality control sample* compliance or noncompliance with *measurement quality objectives*, audits, surveillances, performance evaluation sample analyses, corrective actions, quality improvement, and reports to management.

quantification capability **(1.4.9):** The ability of a measurement process to quantify the *measurand* precisely, usually expressed in terms of the *minimum quantifiable value.*

quantification limit **(20.2.1):** See *minimum quantifiable value.*

quantile **(6.6.2, Table 6.1):** A *p-quantile* of a *random variable X* is any value x_p such that the probability that $X < x_p$ is at most p and the probability that $X \leq x_p$ is at least p. (See *percentile.*)

( ) Indicates the section in which the term is first used in MARLAP.
Italicized words or phrases have their own definitions in this glossary.

quench **(7.2)**: A term used to describe the process in liquid-scintillation counting when the production of light is inhibited or the light signal is partially absorbed during the transfer of light to the photocathode.

radioactive **(1.1)**: Exhibiting *radioactivity*, or containing *radionuclides*.

radioactive decay **(3A.4)**: "Nuclear decay in which particles or electromagnetic radiation are emitted or the nucleus undergoes spontaneous fission or electron capture." (IUPAC, 1994)

radioactive equilibrium **(3.3.4)**: One of three distinct relationships that arise when a radionuclide decays and creates progeny that are also radioactive: (1) secular equilibrium occurs when *half-life* of the progeny is much less than the *half-life* of the parent (for a single progeny, the total activity reaches a maximum of about twice the initial activity, and then displays the characteristic *half-life* of the parent—usually no change over normal measurement intervals); (2) transient equilibrium occurs when the *half-life* of the progeny is less than the *half-life* of the parent (for a single progeny, total activity passes through a maximum, and then decreases with the characteristic *half-life* of the parent); and (3) no equilibrium occurs when the *half-life* of the progeny is greater than the *half-life* of the parent (total activity decreases continually after time zero).

radioactivity **(2.5.4.1)**: The property of certain nuclides of undergoing *radioactive decay*.

radioanalytical specialist **(2.1)**: Key technical experts who participate on the *project planning team*. *Radioanalytical specialists* may provide expertise in radiochemistry and radiation/nuclide measurement systems, and have knowledge of the characteristics of the analytes of concern to evaluate their fate and transport. They may also provide knowledge about sample transportation issues, preparation, preservation, sample size, subsampling, available analytical protocols, and achievable analytical data quality.

radiochemical analysis **(5.3.5)**: The analysis of a sample matrix for its *radionuclide* content, both qualitatively and quantitatively.

radiocolloid **(14.4.6.2)**: A colloidal form of a *radionuclide* tracer produced by sorption of the *radionuclide* onto a preexisting colloidal impurity, such as dust, cellulose fibers, glass fragments, organic material, and polymeric metal hydrolysis products, or by polycondensation of a monomeric species consisting of aggregates of a thousand to ten million radioactive atoms.

radiological holding time **(6.5)**: The time required to process the *sample*. Also refers to the time differential between the sample collection date and the final sample counting (analysis) date.

( ) Indicates the section in which the term is first used in MARLAP.
Italicized words or phrases have their own definitions in this glossary.

radiolysis **(14.1):** Decomposition of any material as a result of exposure to radiation.

radionuclide **(1.1):** A nuclide that is *radioactive* (capable of undergoing *radioactive decay*).

radionuclide of interest **(1.4.10):** The *radionuclide* or *target analyte* that the planning team has determined important for a project. Also called *radionuclide of concern* or *target radionuclide*.

radiotracer **(6.5.2):** (1) A radioactive isotope of the *analyte* that is added to the *sample* to measure any losses of the *analyte* during the *chemical separations* or other processes employed in the analysis (the chemical yield). (2) A radioactive element that is present in only extremely minute quantities, on the order of 10^{-15} to 10^{-11} Molar.

random effect **(3A.4):** Any effect in a measurement process that causes the measured result to vary randomly when the measurement is repeated.

random error **(3A.4):** A result of a measurement minus the mean that would result from an infinite number of measurements of the same *measurand* carried out under repeatability conditions (ISO, 1993a).

random variable **(19.3.1):** The numerical outcome of an experiment, such as a laboratory measurement, that produces varying results when repeated.

reagent blank **(12.6.5):** Consists of the analytical reagent(s) in the procedure without the *target analyte* or sample matrix, introduced into the analytical procedure at the appropriate point and carried through all subsequent steps to determine the contribution of the reagents and of the involved analytical steps.

recovery **(2.5.4.2):** The ratio of the amount of *analyte* measured in a spiked or *laboratory control sample*, to the amount of *analyte* added, and is usually expressed as a percentage. For a matrix spike, the measured amount of *analyte* is first decreased by the measured amount of *analyte* in the sample that was present before spiking. Compare with *yield*.

redox **(13.2.3):** An acronym for *oxidation-reduction*.

reducing agent **(13.4.1, Table 13.2):** The chemical in an oxidation-reduction reaction that reduces another chemical by providing electrons. The *reducing agent* is oxidized during the reaction.

( ) *Indicates the section in which the term is first used in MARLAP.*
Italicized words or phrases have their own definitions in this glossary.

reducing; reduction **(13.4.1, Table 13.2):** The decrease in oxidation number of an atom in a chemical form during a chemical reaction. The decrease is a result of the gain of electron(s) by an atom or the increase in electron density by an atom when it bonds to a less electronegative element or breaks a bond to a more electronegative element.

regulatory decision limit **(2.5.2.1):** The numerical value that will cause the decisionmaker to choose one of the alternative actions. An example of such a limit for drinking water is the *maximum contaminant level (MCL)*. See *action level*.

rejected result **(8.3.3):** A result that is unusable for the intended purpose. A result should only be rejected when the risks of using it are significant relative to the benefits of using whatever information it carries. *Rejected data* should be qualified as such and not used in the *data quality assessment* phase of the *data life cycle*.

relative standard deviation *(RSD)* **(6.5.5.2):** See *coefficient of variation*.

relative standard uncertainty **(3.3.7.1.2):** The ratio of the *standard uncertainty* of a measured result to the result itself. The relative *standard uncertainty* of x may be denoted by $u_r(x)$.

relative variance **(19A.1):** The *relative variance* of a *random variable* is the square of the coefficient of variation.

release criterion **(1.3):** A regulatory limit expressed in terms of dose or risk. The release criterion is typically based on the *total effective dose equivalent (TEDE)*, the *committed effective dose equivalent (CEDE)*, risk of cancer incidence (morbidity), or risk of cancer death (mortality), and generally can not be measured directly.

repeatability (of results of measurement) **(6.6):** The closeness of the agreement between the results of successive measurements of the same *measurand* carried out under the same "repeatability conditions" of measurement. "Repeatability conditions" include the same measurement procedure, the same observer (or analyst), the same measuring instrument used under the same conditions, the same location, and repetition over a short period of time. *Repeatability* may be expressed quantitatively in terms of the dispersion characteristics of the results (Adapted from ISO, 1993a.).

replicates **(3.3.10):** Two or more *aliquants* of a homogeneous *sample* whose independent measurements are used to determine the *precision* of laboratory preparation and analytical procedures.

( ) Indicates the section in which the term is first used in MARLAP.
Italicized words or phrases have their own definitions in this glossary.

representativeness **(2.5.4):** (1) The degree to which samples properly reflect their parent populations. (2) A representative *sample* is a sample collected in such a manner that it reflects one or more characteristics of interest (as defined by the project objectives) of a population from which it was collected. (3) One of the five principal *data quality indicators* (*precision, bias, representativeness, comparability,* and *completeness*).

reproducibility (of results of measurement) **(6.4):** The closeness of the agreement between the results of measurements of the same *measurand* carried out under changed conditions of measurement. A valid statement of *reproducibility* requires specification of the conditions changed. The changed conditions may include principle of measurement, method of measurement, observer (or analyst), measuring instrument, reference standard, location, conditions of use, and time. *Reproducibility* may be expressed quantitatively in terms of the dispersion characteristics of the results. Results are usually understood to be corrected results. (Adapted from ISO, 1993a.).

request for proposals (RFP) **(5.1):** An advertisement from a contracting agency to solicit proposals from outside providers during a negotiated procurement. See *statement of work*.

required minimum detectable concentration (RMDC) **(8.5.3.2):** An upper limit for the *minimum detectable concentration* required by some projects.

resin **(14.4.5.1):** A synthetic or naturally occurring polymer used in *ion-exchange chromatography* as the solid *stationary phase*.

resolution **(8.5.1.11):** The peak definition of alpha, gamma-ray, and liquid-scintillation spectrometers, in terms of the *full width of a peak at half maximum (FWHM)*, which can be used to assess the adequacy of instrument setup, detector *sensitivity*, and chemical separation techniques that may affect the identification, specification, and quantification of the *analyte*.

response variable **(20.2.1):** The variable that gives the observable result of a measurement—in radiochemistry, typically a gross count or count rate.

robustness **(5.3.9):** The ability of a method to deal with large fluctuations in interference levels and variations in matrix. (See *method ruggedness*.)

ruggedness **(1.4.9):** See *method ruggedness*.

( ) Indicates the section in which the term is first used in MARLAP.
Italicized words or phrases have their own definitions in this glossary.

sample **(1.1):** (1) A portion of material selected from a larger quantity of material. (2) A set of individual samples or measurements drawn from a population whose properties are studied to gain information about the entire population.

sample descriptors **(8.5.1.1):** Information that should be supplied to the laboratory including sample ID, *analytical method* to be used, *analyte*, and matrix.

sample digestion **(1.4.6):** Solubilizing an *analyte* or analytes and its host matrix. Acid digestion, fusion, and microwave digestion are some common *sample digestion* techniques.

sample dissolution **(1.1):** See *sample digestion*.

sample management **(2.7.2):** Includes administrative and *quality assurance* aspects covering sample receipt, control, storage, and disposition.

sample mean **(9.6.4.2):** An estimate of the mean of the *distribution* calculated form a statistical sample of observations. The *sample mean* equals the sum of the observed values divided by the number of values, N. If the observed values are $x_1, x_2, x_3, ..., x_N$, then the *sample mean* is given by

$$sample\ mean = \frac{\sum_{i=1}^{N} x_i}{N}$$

sample population **(3.3.7.1.2):** A set of individual samples or measurements drawn from a population whose properties are studied to gain information about the entire population.

sample processing turnaround time **(5.3.6):** The time differential from the receipt of the sample at the laboratory to the reporting of the analytical results.

sample tracking **(1.4.5):** Identifying and following a *sample* through the steps of the analytical process including: field sample preparation and preservation; sample receipt and inspection; laboratory sample preparation; *sample dissolution*; chemical separation of *radionuclides of interest*; preparation of sample for instrument measurement; instrument measurement; and data reduction and reporting.

sample variance **(9.6.4.2):** An estimate of the *variance* of a distribution calculated from a statistical sample of observations. If the observed values are $x_1, x_2, x_3, ..., x_N$, and the sample mean is $\bar{x}$, then the *sample variance* is given by:

( ) Indicates the section in which the term is first used in MARLAP.
Italicized words or phrases have their own definitions in this glossary.

$$s^2 = \frac{1}{N-1} \sum_{i=1}^{N} (x_i - \bar{x})^2$$

sampling and analysis plan (SAP) **(1.5):** See *project plan documents.*

saturated solution **(14.8.2):** A solution that contains the maximum amount of substance that can dissolve in a prescribed amount of *solvent* at a given temperature. The dissolved substance is in equilibrium with any undissolved substance.

scale of decision **(2.5, Table 2.1):** The spatial and temporal bounds to which the decision will apply. The *scale of decision* selected should be the smallest, most appropriate subset of the population for which decisions will be made based on the spatial or temporal boundaries.

scavengers **(14.8.5):** See *collectors.*

screening method **(6.5.5.3):** An economical gross measurement (alpha, beta, gamma) used in a tiered approach to method selection that can be applied to *analyte* concentrations below an *analyte* level in the *analytical protocol specifications* or below a fraction of the specified *action level.*

secondary (outer) hydration sphere **(14.3.1):** See *hydration sphere.*

self absorption **(6.4):** The absorption of nuclear particle or photon emissions within a matrix during the counting of a *sample* by a detector.

sensitivity **(2.5.4.2):** (1) The ratio of the change in an output to the change in an input. (2) The term "sensitivity" is also frequently used as a synonym for "*detection capability.*" See *minimum detectable concentration.*

sensitivity analysis **(2.5.4):** Identifies the portions of the analytical protocols that potentially have the most impact on the decision.

sensitivity coefficient **(19.4.3):** The *sensitivity coefficient* for an input estimate, x_i, used to calculate an output estimate, $y = f(x_1, x_2, \ldots, x_N)$, is the value of the partial derivative, $\partial f / \partial x_i$, evaluated at $x_1, x_2, \ldots, x_N$. The *sensitivity coefficient* represents the ratio of the change in y to a small change in x_i.

( ) Indicates the section in which the term is first used in MARLAP.
Italicized words or phrases have their own definitions in this glossary.

separation factor **(14.4.3):** In *ion-exchange chromatography*, the ratio of the distribution coefficients for two *ions* determined under identical experimental conditions. Separation factor (α) = $K_{d,1}/K_{d,2}$. The ratio determines the separability of the two *ions* by an ion-exchange system; separation occurs when $\alpha \neq 1$.

serial correlation **(9.6.4.1):** When the characteristic of interest in a *sample* is more similar to that of samples adjacent to it than to samples that are further removed, the samples are deemed to be correlated and are not independent of each other (i.e., there is a *serial correlation* such that samples collected close in time or space have more similar concentrations than those samples further removed.).

sigma (σ) **(3A.3):** The symbol σ and the term "sigma" are properly used to denote a true *standard deviation*. The term "sigma" is sometimes used informally to mean "*standard uncertainty*," and "*k*-sigma" is used to mean an *expanded uncertainty* calculated using the coverage factor *k*.

significance level (*α*) **(6A.2):** In a *hypothesis test*, a specified upper limit for the probability of a *Type I decision error*.

smears **(10.6.1):** See *swipes*.

solid-phase extraction (*SPE*) **(14.4.5):** A *solvent* extraction system in which one of the liquid phases is made stationary by *adsorption* onto a solid support. The other phase is mobile (see *extraction chromatography*).

solid-phase extraction membrane **(14.4.5):** A solid-phase extraction system in which the adsorbent material is embedded into a membrane producing an evenly distributed phase, which reduces the channeling problems associated with columns.

solubility **(14.2.1):** The maximum amount of a particular *solute* that can be dissolved in a particular *solvent* under specified conditions (a *saturated solution*) without precipitating. *Solubility* may be expressed in terms of concentration, molality, mole fraction, etc.

solubility equilibrium **(14.8.3.1):** The equilibrium that describes a solid dissolving in a *solvent* to produce a saturated solution.

solubility-product constant **(14.8.3.1):** The equilibrium constant (K_{sp}) for a solid dissolving in a *solvent* to produce a saturated solution.

( ) Indicates the section in which the term is first used in MARLAP.
Italicized words or phrases have their own definitions in this glossary.

solute (10.3.3.2): The substance that dissolves in a *solvent* to form a solution. A *solute* can be a solid, liquid, or gas. In radiochemistry, it is commonly a solid or liquid.

solution (10.2.9): A homogeneous mixture of one substance with another, usually a liquid with a gas or solid. The particles of the *solute* (molecules, atoms, or *ions*) are discrete and mix with particles of the *solvent* at the atomic, ionic, or molecular level.

solvent (10.2.9): The substance that dissolves the *solute* to form a solution. The *solvent* can be a solid, liquid, or gas; but in radiochemistry, it is commonly a liquid.

solvent extraction (10.5.4.1): A separation process that selectively removes soluble components from a mixture with a solvent. The process is based on the solubility of the components of the mixture in the *solvent* when compared to their solubility in the mixture. In liquid-liquid extraction, the process is based on an unequal distribution (partition) of the *solute* between the two immiscible liquids.

source, radioactive (3.3.4): A quantity of material configured for radiation measurement. See also *calibration source*, *check source*, and *test source*.

spatial variability (2.5.2.2): The nonuniformity of an *analyte* concentration over the total area of a site.

specificity (1.4.9): See *method specificity*.

spike (1.4.8): See *matrix spike*.

spillover (15.4.2.1): See *crosstalk*.

spurious error (18.3.3): A measurement error caused by a human blunder, instrument malfunction, or other unexpected or abnormal event.

stability constant (14.3.2): See *formation constant*.

stakeholder (2.2): Anyone with an interest in the outcome of a project. For a cleanup project, some of the *stakeholders* could be federal, regional, state, and tribal environmental agencies with regulatory interests (e.g., Nuclear Regulatory Commission or Environmental Protection Agency); states with have direct interest in transportation, storage and disposition of wastes, and a range of other issues; city and county governments with interest in the operations and safety at sites as well as economic development and site transition; and site advisory boards, citizens groups,

( ) Indicates the section in which the term is first used in MARLAP.
Italicized words or phrases have their own definitions in this glossary.

licensees, special interest groups, and other members of the public with interest in cleanup activities at the site.

standard deviation (3A.3): The *standard deviation* of a *random variable X*, denoted by σ_X, is a measure of the width of its *distribution*, and is defined as the positive square root of the *variance* of *X*.

standard operating procedure (SOP) (4.1): Routine laboratory procedures documented for laboratory personnel to follow.

standard reference material (SRM) (6A.1): A *certified reference material* issued by the National Institute of Standards and Technology (NIST) in the United States. A SRM is certified by NIST for specific chemical or physical properties and is issued with a certificate that reports the results of the characterization and indicates the intended use of the material.

standard uncertainty (1.4.7): The uncertainty of a measured value expressed as an estimated *standard deviation*, often call a "1-sigma" (1-σ) uncertainty. The *standard uncertainty* of a value *x* is denoted by $u(x)$.

statement of work (SOW) (1.4.11): The part of a *request for proposals*, contract, or other agreement that describes the project's scope, schedule, technical specifications, and performance requirements for all radioanalytical laboratory services.

stationary phase (14.7.4.1): The phase in a chromatographic system that is not moving with respect to the mobile phase. The *stationary phase* can be a solid, a nonvolatile liquid coated onto an inert matrix, or a substance trapped in an inert matrix.

statistical control (1.4.8): The condition describing a process from which all special causes have been removed, evidenced on a *control chart* by the absence of points beyond the *control limits* and by the absence of nonrandom patterns or trends within the *control limits*. A special cause is a source of variation that is intermittent, unpredictable, or unstable. See *control chart, in control*, and *control limits*.

statistical parameter (2.5, Table 2.1): A quantity used in describing the probability distribution of a *random variable*" (ISO, 1993b).

statistical test (4.6.2.3): A statistical procedure to decide whether a *null hypothesis* should be rejected in favor of the *alternative hypothesis* or not rejected." This also can be called a *hypothesis test*.

( ) Indicates the section in which the term is first used in MARLAP.
Italicized words or phrases have their own definitions in this glossary.

subsample **(12.3.1.4):** (1) A portion of a *sample* removed for testing. (2) To remove a portion of a *sample* for testing.

subsampling factor **(19.5.12):** As used in MARLAP, a variable, F_S, inserted into the mathematical model for an analytical measurement to represent the ratio of the *analyte* concentration of the subsample to the *analyte* concentration of the original *sample*. The *subsampling factor* is always estimated to be 1 but has an uncertainty that contributes to the combined *standard uncertainty* of the measured result.

surface adsorption **(14.8.3.3, Table 14.12):** (1) *Adsorption* of particles of a substance onto the surface of another substance. (2) A mechanism of *coprecipitation* in which *ions* are adsorbed from solution onto the surfaces of precipitated particles.

survey **(2.3.2):** "An evaluation of the radiological conditions and potential hazards incident to the production, use, transfer, release, disposal, or presence of radioactive materials or other sources of radiation. When appropriate, such an evaluation includes the a physical survey of the location of radioactive material and measurements or calculations of levels of radiation, or concentrations of quantities of radioactive material present" (Shleien, 1992). A *survey* is a semiquantitative measure of the gross radiological conditions of a material or area (for dose and contamination). A *screen* is a qualitative assessment to determine the type of *radionuclides* (alpha, beta, gamma) and the relative amount (high, medium, low) of each that might be present.

survey unit **(2.5.2.4):** A geographical area consisting of structures or land areas of specified size and shape at a remediated site for which a separate decision will be made whether the unit attains the site-specific reference-based cleanup standard for the designated pollution parameter. *Survey units* are generally formed by grouping contiguous site areas with a similar use history and the same classification of contamination potential. *Survey units* are established to facilitate the survey process and the statistical analysis of survey data. (MARSSIM, 2000)

suspension **(10.3.3.2):** A mixture in which small particles of a solid, liquid, or gas are dispersed in a liquid or gas. The dispersed particles are larger than colloidal particles and produce an opaque or turbid mixture that will settle on standing by gravity and be retained by paper filters. See *colloids* and *colloidal solution*.

swipes **(10.6.1):** A filter pad used to determine the level of general radioactive contamination when it is wiped over a specific area, about 100 cm² in area. Also called *smears* or wipes.

systematic effect **(3A.4):** Any effect in a measurement process that does not vary randomly when the measurement is repeated.

( ) *Indicates the section in which the term is first used in MARLAP.*
Italicized words or phrases have their own definitions in this glossary.

***systematic error* (3A.4):** The mean value that would result from an infinite number of measurements of the same *measurand* carried out under repeatability conditions minus a true value of the *measurand* (ISO, 1993a).

***systematic planning process* (1.4.2):** See *directed planning process*.

***target analyte* (3.3.1):** A *radionuclide* on the *target analyte list*. Also called *radionuclide of interest* or "radionuclide of concern." See *analyte*.

***target analyte list* (3.3.1):** A list of the *radionuclides* of concern for the project.

***target radionuclide* (18.4.1):** See *radionuclide of interest*.

***technical evaluation committee (TEC)* (5.3.9):** A team of technical staff members that assists in the selection of a contract laboratory by reviewing proposals and by auditing laboratory facilities.

***technical proposal* (5.5.1.):** A document, submitted by a laboratory bidding on a contract, which addresses all of the technical and general laboratory requirements within a *request for proposals* and *statement of work*.

***temporal trend* (2.5, Table 2.1):** Effects that time have on the *analyte* concentration in the matrix or *sample*. The *temporal boundaries* describe the time frame the study data will represent (e.g., possible exposure to local residents over a 30-year period) and when samples should be taken (e.g., instantaneous samples, hourly samples, annual average based on monthly samples, samples after rain events).

***tests of detection* (8.3.1):** *Tests of detection* determine the presence or absence of *analytes*. Normally, only numerous *quality control* exceptions and failures in one or more of the *tests of detection* and uncertainty are sufficient reason to reject data.

***tests of unusual uncertainty* (8.3.1):** Part of the validation plan that specifies the level of *measurement uncertainty* considered unusually high and unacceptable.

***test source* (14.10.9.7):** The final radioanalytical processing product or matrix (e.g., precipitate, solution, filter) that is introduced into a measurement instrument. A *test source* is prepared from laboratory sample material for the purpose of determining its radioactive constituents. See *calibration source, check source,* and *source, radioactive*.

( ) Indicates the section in which the term is first used in MARLAP.
Italicized words or phrases have their own definitions in this glossary.

***thin-layer chromatography* (14.7.3):** A chromatographic process in which a thin layer of a *stationary phase* in coated onto a solid support such as a plastic or glass plate. The stationary material is an absorbing solid and the mobile phase is a liquid.

***tolerable decision error rates* (2.3.3):** The limits on *decision error rates* that will be acceptable to the *stakeholder*/customer.

***tolerance limit* (18.3.3):** A value, that may or may not have a statistical basis, which is used as the measure of acceptable or unacceptable values. A *tolerance limit* is sometimes referred to as a "Go/No Go" limit. See *warning limit, control chart.*

***total effective dose equivalent* (*TEDE*) (2.5.2.1):** The sum of the effective dose equivalent (for external exposure) and the committed effective dose equivalent (for internal exposure). TEDE is expressed in units of sievert (Sv) or rem (MARSSIM, 2000). See *action level, dose equivalent,* and *total effective dose equivalent.*

***total propagated uncertainty* (*TPU*) (19.2):** See *combined standard uncertainty,* which is the preferred term.

***traceability* (8.5.1.5):** "Property of the result of a measurement or the value of a standard whereby it can be related to stated references, usually national or international standards, through an unbroken chain of comparisons all having stated uncertainties" (ISO, 1993a).

***tracer* (1.4.8):** See *radiotracer.*

***Type A evaluation (of uncertainty)* (19.3.3):** "Method of evaluation of uncertainty by the statistical analysis of series of observations" (ISO, 1995).

***Type B evaluation (of uncertainty)* (19.3.3):** "Method of evaluation of uncertainty by means other than the statistical analysis of series of observations" (ISO, 1995); any method of uncertainty evaluation that is not a Type A evaluation.

***Type I decision error* (2.5.3):** In a hypothesis test, the error made by rejecting the null hypothesis when it is true. A *Type I decision error* is sometimes called a "*false rejection*" or a "*false positive.*"

***Type II decision error* (2.5.3):** In a hypothesis test, the error made by failing to reject the null hypothesis when it is false. A *Type II decision error* is sometimes called a "*false acceptance*" or a "*false negative.*"

( ) Indicates the section in which the term is first used in MARLAP.
Italicized words or phrases have their own definitions in this glossary.

uncertainty **(1.4.7):** The term "uncertainty" is used with several shades of meaning in MARLAP. In general it refers to a lack of complete knowledge about something of interest; however, in Chapter 19 it usually refers to "*uncertainty (of measurement).*"

uncertainty (of measurement) **(3.3.4):** "Parameter, associated with the result of a measurement, that characterizes the dispersion of the values that could reasonably be attributed to the *measurand*" (ISO, 1993a).

uncertainty interval **(19.3.6):** The interval from $y - U$ to $y + U$, where y is the measured result and U is its *expanded uncertainty*.

uncertainty propagation **(19.1):** Mathematical technique for combining the *standard uncertainties* of the input estimates for a mathematical model to obtain the combined *standard uncertainty* of the output estimate.

uncertainty propagation formula (first-order) **(19.4.3):** the generalized mathematical equation that describes how standard uncertainties and *covariances* of input estimates combine to produce the combined *standard uncertainty* of an output estimate. When the output estimate is calculated as $y = f(x_1, x_2, ..., x_N)$, where f is a differentiable function of the input estimates $x_1, x_2, ..., x_N$, the uncertainty propagation formula may be written as follows:

$$u_c^2(y) = \sum_{i=1}^{N} \left(\frac{\partial f}{\partial x_i} \right)^2 u^2(x_i) + 2 \sum_{i=1}^{N-1} \sum_{j=i+1}^{N} \frac{\partial f}{\partial x_i} \frac{\partial f}{\partial x_j} u(x_i, x_j).$$

This formula is derived by approximating the function $f(x_1, x_2, ..., x_N)$ by a first-order Taylor polynomial. In the *Guide to the Expression of Uncertainty of Measurement*, the uncertainty propagation formula is called the "law of propagation of uncertainty" (ISO, 1995).

unsaturated solution **(14.8.2):** A solution whose concentration of *solute* is less than that of a saturated solution. The solution contains less *solute* than the amount of *solute* will dissolve at the temperature of the solution, and no solid form of the *solute* is present.

validation **(1.1):** See *data validation*.

validation criterion **(2.5.4.2):** Specification, derived from the *measurement quality objectives* and other analytical requirements, deemed appropriate for evaluating data relative to the project's analytical requirements. Addressed in the *validation plan*.

( ) Indicates the section in which the term is first used in MARLAP.
Italicized words or phrases have their own definitions in this glossary.

validation flags **(1.4.11):** Qualifiers that are applied to data that do not meet the acceptance criteria established to assure data meets the needs of the project. See also *data qualifier*.

validation plan **(2.7.4.2):** An integral part of the initial planning process that specifies the data deliverables and *data qualifiers* to be assigned that will facilitate the *data quality assessment*.

variance **(9.6.2.3):** The *variance* of a *random variable* X, denoted by $\text{Var}(X)$, σ_X^2, or $V(X)$, is defined as $E[(X - \mu_X)^2]$, where μ_X denotes the mean of X. The *variance* also equals $E(X^2) - \mu_X^2$.

verification **(1.2):** See *data verification*.

volatility **(10.3.4.1):** The tendency of a liquid or solid to readily become a vapor (evaporates or sublimes) at a given temperature. More volatile substances have higher vapor pressures than less volatile substances.

volatilization **(10.3.3.2, Table 10.1):** A separation method using the volatility of liquids or solids to isolate them from nonvolatile substances, or to isolate a gas from a liquid.

warning limit **(3.3.7.3):** Predetermined values plotted on a *control chart* between the central line and the *control limits*, which may be used to give an early indication of possible problems with the monitored process before they become more significant. The monitored variable will occasionally fall outside the warning limits even when the process is *in control*; so, the fact that a single measurement has exceeded the warning limits is generally not a sufficient reason to take immediate corrective action. See *tolerance limit*.

weight distribution coefficient **(14.7.4.1):** In *ion-exchange chromatography*, the ratio of the weight of an *ion* absorbed on one gram of dry ion-exchange *resin* to the weight of the *ion* that remains in one milliliter of solution after equilibrium has been established. The ratio is a measure of attraction of an *ion* for a *resin*. Comparison of the weight distribution coefficient for *ions* in an analytical mixture is a reflection of the ability of the ion-exchange process to separate the *ions* (see *separation factor*).

Welch-Satterthwaite formula **(19C.2):** An equation used to calculate the *effective degrees of freedom* for the combined *standard uncertainty* of an output estimate when the number of degrees of freedom for the *standard uncertainty* of each input estimate is provided (ISO, 1995).

work plan **(1.6.1):** The primary and integrating plan document when the data collection activity is a smaller supportive component of a more comprehensive project. The *work plan* for a site investigation will specify the number of samples to be collected, the location of each *sample*, and

( ) *Indicates the section in which the term is first used in MARLAP.*
Italicized words or phrases have their own definitions in this glossary.

the analyses to be performed. A newer concept is to develop a *dynamic work plan* that specifies the decisionmaking logic used to determine where the samples will be collected, when the sampling will stop, and what analyses will be performed, rather than specify the number of samples to be collected and the location of each sample.

year: (1) Mean solar or tropical year is 365.2422 days (31,556,296 seconds) and is used for calculations involving *activity* and *half-life* corrections. (2) Calendar year, i.e., 12 months, is usually used in the regulatory sense when determining compliance.

yield **(1.6.2)**: The ratio of the amount of *radiotracer* or *carrier* determined in a sample analysis to the amount of *radiotracer* or *carrier* originally added to a *sample*. The yield is an estimate of the *analyte* during analytical processing. It is used as a correction factor to determine the amount of *radionuclide* (*analyte*) originally present in the sample. *Yield* is typically measured gravimetrically (via a *carrier*) or radiometrically (via a *radiotracer*). Compare with *recovery*.

Sources

American National Standards Institute (ANSI) N13.30. *Performance Criteria for Radiobioassay.* 1996.

American National Standards Institute (ANSI) N42.23. *Measurement and Associated Instrumentation Quality Assurance for Radioassay Laboratories.* 2003.

U.S. Environmental Protection Agency (EPA). 1980. *Upgrading Environmental Radiation Data, Health Physics Society Committee Report HPSR-1*, EPA, 520/1-80-012, EPA, Office of Radiation Programs, Washington, DC.

U.S. Environmental Protection Agency (EPA). 2002. *Guidance on Environmental Data Verification and Data Validation* (EPA QA/G-8). EPA/240/R-02/004. Office of Environmental Information, Washington, DC. Available at www.epa.gov/quality/qa_docs.html.

International Organization for Standardization (ISO). 1992. *Guide 30: Terms and Definitions Used in Connection with Reference Materials.* ISO, Geneva, Switzerland.

International Organization for Standardization (ISO). 1993a. *International Vocabulary of Basic and General Terms in Metrology.* 2nd Edition. ISO, Geneva, Switzerland.

( ) Indicates the section in which the term is first used in MARLAP.
Italicized words or phrases have their own definitions in this glossary.

International Organization for Standardization (ISO). 1993b. *Statistics — Vocabulary and Symbols — Part 1: Probability and General Statistical Terms.* ISO 3534-1. ISO, Geneva, Switzerland.

International Organization for Standardization (ISO). 1995. *Guide to the Expression of Uncertainty in Measurement.* ISO, Geneva, Switzerland.

International Organization for Standardization (ISO). 1997. *Capability of Detection — Part 1: Terms and Definitions.* ISO 11843-1. ISO, Geneva, Switzerland.

International Union of Pure and Applied Chemistry (IUPAC). 1994. "Nomenclature for Radioanalytical Chemistry." *Pure and Applied Chemistry,* 66, p. 2513-2526. Available at www.iupac.org/publications/compendium/R.html.

International Union of Pure and Applied Chemistry (IUPAC). 1995. Nomenclature in Evaluation of Analytical Methods Including Detection and Quantification Capabilities. *Pure and Applied Chemistry* 67:10, pp. 1699–1723. Available at www.iupac.org/reports/1993/6511uden/index.html.

International Union of Pure and Applied Chemistry (IUPAC). 1997. *Compendium of Chemical Terminology: The Gold Book, Second Edition.* A. D. McNaught and A. Wilkinson, eds. Blackwell Science. Available at www.iupac.org/publications/compendium/index.html.

International Union of Pure and Applied Chemistry (IUPAC). 2001. *Nomenclature for Isotope, Nuclear and Radioanalytical Techniques* (Provisional Draft). Research Triangle Park, NC. Available at www.iupac.org/reports/provisional/abstract01/karol_310801.html.

MARSSIM. 2000. *Multi-Agency Radiation Survey and Site Investigation Manual, Revision 1.* NUREG-1575 Rev 1, EPA 402-R-97-016 Rev1, DOE/EH-0624 Rev1. August. Available from www.epa.gov/radiation/marssim/.

U.S. Nuclear Regulatory Commission (NRC). 1984. *Lower Limit of Detection: Definition and Elaboration of a Proposed Position for Radiological Effluent and Environmental Measurements.* NUREG/CR-4007. NRC, Washington, DC.

Shleien, Bernard, ed. 1992. *The Health Physics and Radiological Health Handbook.* Silver Spring, MD: Scinta Inc.

( ) Indicates the section in which the term is first used in MARLAP.
Italicized words or phrases have their own definitions in this glossary.

RC FORM 335
-89)
RCM 1102,
:01, 3202

U.S. NUCLEAR REGULATORY COMMISSION

BIBLIOGRAPHIC DATA SHEET

(See instructions on the reverse)

1. REPORT NUMBER
(Assigned by NRC, Add Vol., Supp., Rev., and Addendum Numbers, If any.)

NUREG-1576, Vol. 1
EPA 402-B-04-001A
NTIS PB2004-105421

. TITLE AND SUBTITLE

Multi-Agency Radiological Laboratory Analytical Protocols Manual (MARLAP)
Volume I: Chapters 1-9 and Appendices A-E

3. DATE REPORT PUBLISHED

MONTH	YEAR
July	2004

4. FIN OR GRANT NUMBER

. AUTHOR(S)

6. TYPE OF REPORT

7. PERIOD COVERED *(Inclusive Dates)*

. PERFORMING ORGANIZATION - NAME AND ADDRESS *(If NRC, provide Division, Office or Region, U.S. Nuclear Regulatory Commission, and mailing address; if contractor, provide name and mailing address.)*

Department of Defense, Washington, DC 20301-3400   Food and Drug Administration, Rockville, MD 20857
Department of Energy, Washington, DC 20585-0119   National Institute of Standards and Technology, Gaithersburg, MD 20899
Department of Homeland Security, Washington, DC 20528   Nuclear Regulatory Commission, Washington, DC 20555-0001
Environmental Protection Agency, Washington, DC 20460-0001   U.S. Geological Survey, Reston, VA 20192

. SPONSORING ORGANIZATION - NAME AND ADDRESS *(If NRC, type "Same as above"; if contractor, provide NRC Division, Office or Region, U.S. Nuclear Regulatory Commission, and mailing address.)*

Same as above

0. SUPPLEMENTARY NOTES

Rateb (Boby) Abu-Eid, NRC Project Manager

1. ABSTRACT *(200 words or less)*

The Multi-Agency Radiological Laboratory Analytical Protocols (MARLAP) manual provides guidance for the planning, implementation, and assessment of projects that require the laboratory analysis of radionuclides. MARLAP's goal is to provide guidance for project planners, managers, and laboratory personnel to ensure that radioanalytical laboratory data will meet a project's or program's data requirements. The manual offers a framework for national consistency in the form of a performance-based approach for meeting data requirements that is scientifically rigorous and flexible enough to be applied to a diversity of projects and programs. The guidance in MARLAP is designed to help ensure the generation of radioanalytical data of known quality, appropriate for its intended use. Examples of data collection activities that MARLAP supports include site characterization, site cleanup and compliance demonstration, decommissioning of nuclear facilities, emergency response, remedial and removal actions, effluent monitoring of licensed facilities, environmental site monitoring, background studies, and waste management activities.

MARLAP is organized into two parts. Part I, Volume 1, is intended for project planners and managers, provides the basic framework of the directed planning process as it applies to projects requiring radioanalytical data for decision making. Part II, Volumes 2 and 3, is intended for laboratory personnel.

2. KEY WORDS/DESCRIPTORS *(List words or phrases that will assist researchers in locating the report.)*

Multi-Agency Radiological Laboratory Analytical Protocols, MARLAP, radiological, laboratory, laboratory sample, analytical protocols, performance-based approach, planning, measurement, quality assurance, survey(s), decommissioning, statistics, waste management, radioanalytical laboratory services, data validation, data quality assessment, data collection

13. AVAILABILITY STATEMENT

unlimited

14. SECURITY CLASSIFICATION

(This Page)

unclassified

(This Report)

unclassified

15. NUMBER OF PAGES

16. PRICE

RC FORM 335 (2-89)

This form was electronically produced by Elite Federal Forms, Inc.

Federal Recycling Program

UNITED STATES
NUCLEAR REGULATORY COMMISSION
WASHINGTON, DC 20555-0001

OFFICIAL BUSINESS

www.ingramcontent.com/pod-product-compliance
Lightning Source LLC
Chambersburg PA
CBHW080227180526
45167CB00006B/2239